ELECTRIC MACHINERY

McGraw-Hill Series in Electrical Engineering

Consulting Editor
Stephen W. Director, Carnegie-Mellon University

Networks and Systems
Communications and Information Theory
Control Theory
Electronics and Electronic Circuits
Power and Energy
Electromagnetics
Computer Engineering and Switching Theory
Introductory and Survey
Radio, Television, Radar, and Antennas

Previous Consulting Editors

Ronald M. Bracewell, Colin Cherry, James F. Gibbons, Willis W. Harman,
Hubert Heffner, Edward W. Herold, John G. Linvill, Simon Ramo,
Ronald A. Rohrer, Anthony E. Siegman, Charles Susskind, Frederick E. Terman,
John G. Truxal, Ernst Weber, and John R. Whinnery

Power and Energy

Consulting Editor
Stephen W. Director, Carnegie-Mellon University

Elgerd: ELECTRIC ENERGY SYSTEMS THEORY: An Introduction
Fitzgerald, Kingsley, and Umans: ELECTRIC MACHINERY
Odum and Odum: ENERGY BASIS FOR MAN AND NATURE
Stevenson: ELEMENTS OF POWER SYSTEM ANALYSIS

fourth edition

ELECTRIC MACHINERY

A. E. Fitzgerald

Late Vice President for Academic Affairs
and Dean of the Faculty
Northeastern University

Charles Kingsley, Jr.

Associate Professor of Electrical Engineering, Emeritus
Massachusetts Institute of Technology

Stephen D. Umans

Principal Research Engineer
Department of Electrical Engineering and Computer Science
Electric Power System Engineering Laboratory
Massachusetts Institute of Technology

McGraw-Hill Book Company

New York St. Louis San Francisco Auckland Bogotá Hamburg
Johannesburg London Madrid Mexico Montreal New Delhi
Panama Paris São Paulo Singapore Sydney Tokyo Toronto

ELECTRIC MACHINERY

1 2 3 4 5 6 7 8 9 0 DOCDOC 8 9 8 7 6 5 4 3 2

ISBN 0-07-021145-0

This book was set in Century Schoolbook by York Graphic Services, Inc.
The editors were T. Michael Slaughter and David A. Damstra;
the designer was Charles A. Carson;
the production supervisor was Leroy A. Young.
New drawings were done by J & R Services, Inc.
R. R. Donnelley & Sons Company was printer and binder.

Library of Congress Cataloging in Publication Data

Fitzgerald, A. E. (Arthur Eugene), date
    Electric machinery.

    (McGraw-Hill series in electrical engineering.
Power and energy)
    Includes index.
    1. Electric machinery. I. Kingsley, Charles,
date   . II. Umans, Stephen D. III. Title.
IV. Series.
TK2181.F5 1983   621.31'042   82-7763
ISBN 0-07-021145-0   AACR2

CONTENTS

9/Polyphase Induction Machines 409

10/Polyphase-Induction-Machine Dynamics and Control 443

11/Fractional-Horsepower AC Motors 479

PREFACE

Like the three preceding editions of this book, the fourth edition presents a unified treatment of the basic theory of electromechanical devices with especial emphasis on the applications, control, and limitations of rotating electric machinery. Much of the text is taken from the best parts of the second and third editions, rearranged in a more logical sequence, and rewritten.

The topic of power electronics has received considerable attention in the past decade. A number of textbooks have been written on the subject, and it is available as a separate course in many electrical engineering curricula. It certainly requires more than a single chapter to do it justice. As a result, the chapter on solid-state motor control has been eliminated from the fourth edition. However, the concepts of solid-state control as applied to electric machinery have been retained in the form of an article discussing dc motor drives in Chapter 6 and an article discussing variable-frequency ac motor drives in Chapter 10.

This bit of reorganization gave us room to add some new material and to reinstate some material which had been included in the second edition. For example, the technique of the transformation to dq0 variables, so useful in the transient analysis of ac machinery, is again included. In addition, the book has been rearranged with the hope that it will be easier to use.

The first four chapters provide a basic introduction to the topic of energy conversion and to rotating electric machinery. The material of the second chapter on electromechanical energy conversion is often felt to be too mathematical for an introductory undergraduate course on electric machinery. In this edition, it has been written in such a fashion that the first two sections can serve as an overview of the topic and the rest of the chapter can be omitted if desired without a loss of understanding in the remainder of the book.

The chapters on dc, synchronous, and induction machinery have been grouped in pairs. In each of these groupings, the first chapter serves as an introduction to the machine type, develops the basic models, and applies them to the steady-state performance of the machine. The second chapter then deals with the issues associated with the control and transient behavior of the machine as well as some of the more subtle issues not required for the development of the basic models.

This book does not require a level of mathematical sophistication beyond that given in undergraduate courses in basic physics and circuit theory. The emphasis is on physical understanding as the basis for mathematical models.

Obviously more material is presented than can be treated in the time available in the typical curriculum. Moreover, the detailed content and order of subject matter are naturally governed by local circumstances and the desires and enthusiasms of individual instructors. For these reasons, particular attention is given to flexibility of use without loss of continuity. Browsing in the book for an hour or two and reading the introductory paragraphs of each chapter should enable an instructor to outline a variety of courses with differing content and sequence. Furthermore the book should serve the student as a reference text during his subsequent professional career.

The untimely death of Dr. A. E. Fitzgerald in July 1978 was a great loss of a coauthor and personal friend of many years. His sound intuitive judgment and lucid writing style are contributions that are hard to replace. However, much of his writing taken from the earlier editions has been retained in this fourth edition.

During the preliminary planning for the fourth edition, Dr. Alexander Kusko, a coauthor of the third edition, decided that he could no longer spare the time from his active consulting business to take part in the revision.

With the loss of these two coauthors I feel very fortunate to have been able to interest Dr. Stephen D. Umans in joining me in this new edition. Without his energy, enthusiasm, and conscientious adherence to a time schedule, the production of the fourth edition would have been difficult. From my point of view I have found working with him to be a very satisfying experience.

Charles Kingsley, Jr.

ELECTRIC MACHINERY

Magnetic Circuits and Transformers

The object of this book is to study the devices used in the interconversion of electric and mechanical energy. Emphasis is placed on electromagnetic rotating machinery, by means of which the bulk of this energy conversion takes place. Attention is also paid to the transformer, which, although not an electromechanical energy-conversion device, is an important component in the overall problem of energy conversion. Moreover, in many respects its analysis uses techniques closely related to those required for rotating machinery.

Practically all transformers and electric machinery use magnetic material for shaping and directing the magnetic fields which act as the medium for transferring and converting energy. Thus the ability to analyze and describe magnetic field quantities is an essential tool for understanding these devices. Magnetic materials play a large role in determining the properties of a piece of electromagnetic equipment and affect its size and efficiency.

This chapter will develop some basic tools for the analysis of magnetic

1

field systems and will provide a brief introduction to the properties of practical magnetic materials. These results will then be applied to the analysis of transformers. In later chapters they will be used in the analysis of rotating machinery.

In this book it is assumed that the reader has basic knowledge of magnetic and electric field theory such as given in a basic physics course for engineering students. Other readers may have had a course on electromagnetic field theory based on Maxwell's equations, but an understanding of Maxwell's equations is not a prerequisite for study of this book. The pertinent basic equations will be introduced in simplified form when required.

1-1 INTRODUCTION TO MAGNETIC CIRCUITS

The complete, detailed solution for magnetic fields in most situations of practical engineering interest involves the solution of Maxwell's equations along with various constitutive relationships which describe material properties. Although in practice exact solutions are often unattainable, various simplifying assumptions permit the attainment of useful engineering solutions.

The first assumption is that for the types of electric machines and transformers treated in this book the frequencies and sizes involved are such that the displacement-current term in Maxwell's equations can be neglected. This term accounts for magnetic fields being produced in space by time-varying electric fields and is associated with electromagnetic radiation. Neglecting this term results in the *magneto-quasi-static* form of Maxwell's equations. By this we mean that the magnetic field quantities are determined solely by the instantaneous values of the source currents and that the time variation of the magnetic fields follow directly from the time variations of the sources.

A second simplifying assumption involves the concept of the *magnetic circuit*. The general solution for the *magnetic field intensity* **H** and the *magnetic flux density* **B** in a structure of complex geometry is extremely difficult. However, a three-dimensional field problem can often be reduced to what is essentially a one-dimensional circuit equivalent, yielding solutions of acceptable engineering accuracy.

A magnetic circuit consists of a structure composed for the most part of high-permeability magnetic material. The presence of high-permeability material causes the magnetic flux to be confined to the paths defined by the structure, much as currents are confined to the conductors of an electric circuit. Use of this concept of the magnetic circuit is illustrated in this article and will be seen to apply quite well to many situations in this book.†

A simple example of a magnetic circuit is shown in Fig. 1-1. The core is

---

†For a more extensive treatment of magnetic circuits see A. E. Fitzgerald, D. E. Higgenbotham, and A. Grabel, "Basic Electrical Engineering." 5th ed., McGraw-Hill, New York, 1981, chap. 13; also E. E. Staff, MIT, "Magnetic Circuits and Transformers," MIT Press, Cambridge, Mass., 1965, chaps. 1 to 3.

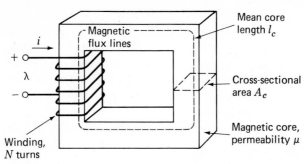

Fig. 1-1. Simple magnetic circuit.

assumed to be composed of magnetic material whose permeability is much greater than that of the surrounding air. The core is of uniform cross section and is excited by a winding of N turns carrying current i amperes (A). This winding produces a magnetic field in the core, as shown in the figure. The magnetic field can be visualized in terms of flux lines, which form closed loops interlinked with the winding. The basic relation between current i and the magnetic field intensity $\mathbf{H}$ states that the line integral of $\mathbf{H}$ around a closed path is equal to the net current enclosed by that path.

As applied to the magnetic circuit of Fig. 1-1, the source of the magnetic field in the core is the ampere-turn product Ni. In magnetic-circuit terminology Ni is the *magnetomotive force* (mmf) $\mathfrak{F}$. Although Fig. 1-1 shows only a single coil, transformers and most rotating machines have at least two windings and Ni is the algebraic sum of the ampere-turns of all the windings. Under the assumption of uniform magnetic flux density across the core cross section, the line integral of $\mathbf{H}$ becomes simply the scalar product $H_c l_c$, of the magnitude of $\mathbf{H}$ along the mean flux path whose length is l_c. Thus, the relationship between the mmf and the magnetic field intensity can be written in magnetic-circuit terminology as

$$\mathfrak{F} = Ni = H_c l_c \qquad (1\text{-}1)$$

The direction of H_c in the core can be found from the *right-hand rule*, which can be stated in two equivalent ways. (1) Imagine a current-carrying conductor held in the right hand with the thumb pointing in the direction of current flow; the fingers then point in the direction of the magnetic field created by that current. (2) Equivalently, if the coil in Fig. 1-1 is grasped in the right hand (figuratively speaking) with the fingers pointing in the direction of the current, the thumb will point in the direction of the magnetic fields.

The relationship between the magnetic field intensity $\mathbf{H}$ and the magnetic flux density $\mathbf{B}$ is a property of the region in which the field exists; thus

$$\mathbf{B} = \mu\mathbf{H} \qquad (1\text{-}2)$$

where μ is the *permeability*. In SI units **B** is in *webers per square meter,* known as *teslas* (T), and μ is in *webers per ampere-turn-meter,* or equivalently *henrys per meter.* In SI units the permeability of free space is $\mu_0 = 4\pi \times 10^{-7}$. The permeability of ferromagnetic material can be expressed in terms of μ_r, its value relative to that of free space, or $\mu = \mu_r\mu_0$. Typical values of μ_r range from 2000 to 80,000 for materials used in transformers and rotating machines. The characteristics of ferromagnetic materials are described in Arts. 1-3 and 1-4. For the present we shall assume that μ_r is a known constant, although it actually varies appreciably with magnetic flux density.

Because of the high permeability of the magnetic core, the magnetic flux is confined almost entirely to the core, the field lines follow that path defined by the core, and the flux density is essentially uniform over a cross section because the cross-sectional area is uniform.

The *magnetic flux* ϕ crossing an area is the surface integral of the normal component of **B**; thus

$$\phi = \int_S \mathbf{B} \cdot d\mathbf{a} \tag{1-3}$$

In SI units ϕ is in *webers.* In terms of field theory the continuity-of-flux equation

$$\oint_S \mathbf{B} \cdot d\mathbf{a} = 0 \tag{1-4}$$

states that the net magnetic flux crossing all surfaces of a three-dimensional closed surface (equal to the surface integral of B over that closed surface) is zero. This is equivalent to saying that all the flux which enters the surface enclosing a volume must leave that volume over some other portion of that surface because magnetic flux lines form closed loops. When flux outside the core is neglected, Eq. 1-3 reduces to the simple scalar equation

$$\phi_c = B_c A_c \tag{1-5}$$

where ϕ_c = the flux in core
$\quad B_c$ = flux density in the core
$\quad A_c$ = cross-sectional area of the core

The area is assumed to be constant throughout the length of the magnetic path. Because the field lines form closed loops, the flux is continuous throughout the length of the core.

Transformers are wound on closed cores like that of Fig. 1-1. Energy-conversion devices which incorporate a moving element must have air gaps in their magnetic circuits. A magnetic circuit with an air gap is shown in Fig. 1-2. When the air-gap length g is much smaller than the dimensions of the adja-

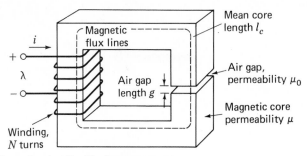

Fig. 1-2. Magnetic circuit with air gap.

cent core faces, the magnetic flux ϕ is constrained essentially to reside in the core and the air gap and is continuous throughout the magnetic circuit.

Thus, the configuration of Fig. 1-2 can be analyzed as a magnetic circuit with two series components, a magnetic core of permeability μ and mean length l_c, and an air gap of permeability μ_0 and length g. In the core the flux density is uniform, and the cross-sectional area is A_c; thus in the core

$$B_c = \frac{\phi}{A_c} \qquad (1\text{-}6)$$

and in the air gap

$$B_g = \frac{\phi}{A_g} \qquad (1\text{-}7)$$

The magnetic field lines bulge outward somewhat as they cross the air gap, as illustrated in Fig. 1-3. The effect of the fringing fields is to increase the effective cross-sectional area A_g of the air gap. Various empirical methods have been developed to account for this effect. A correction for such fringing fields in short air gaps can be made by adding the gap length to each of the two dimensions making up its cross-sectional area. In this book the effect of fring-

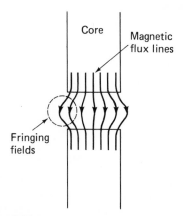

Fig. 1-3. Air-gap fringing fields.

ing fields will usually be ignored. If fringing is neglected, $A_g = A_c$ and

$$B_g = B_c = \frac{\phi}{A_c} \tag{1-8}$$

Application of Eqs. 1-1 and 1-2 to this magnetic circuit yields

$$\mathcal{F} = Ni = H_c l_c + H_g g \tag{1-9}$$

$$\mathcal{F} = \frac{B_c}{\mu} l_c + \frac{B_g}{\mu_0} g \tag{1-10}$$

Here Ni is the total ampere-turns applied to the magnetic circuit. Thus we see that a portion of the mmf is required to excite the magnetic field in the core while the remainder excites the magnetic field in the air gap.

For practical magnetic materials (as discussed in Art. 1-3), B_c and H_c are not simply related by a known permeability μ. In fact, B_c is often a nonlinear, multivalued function of H_c. Thus, although Eq. 1-9 continues to hold, it does not lead directly to a simple expression relating the mmf and the flux densities, such as that of Eq. 1-10. Instead the specifics of the nonlinear B_c-H_c relation must be used, either graphically or analytically. However, in many cases, the concept of core permeability gives results of acceptable engineering accuracy and is frequently used.

From Eq. 1-8, Eq. 1-10 can be rewritten in terms of the total flux ϕ

$$\mathcal{F} = \phi \frac{l_c}{\mu A_c} + \phi \frac{g}{\mu_0 A_c} \tag{1-11}$$

in which fringing at the air gap is neglected and the flux is assumed to go straight across the gap. The terms which multiply the flux in this equation are known as the *reluctance* $\mathcal{R}$, of the core and air gap, respectively,

$$\mathcal{R}_c = \frac{l_c}{\mu A_c} \tag{1-12}$$

$$\mathcal{R}_g = \frac{g}{\mu_0 A_c} \tag{1-13}$$

and thus $$\mathcal{F} = \phi(\mathcal{R}_c + \mathcal{R}_g) \tag{1-14}$$

The fraction of the total mmf required for each portion of the magnetic circuit varies inversely as its reluctance. From Eq. 1–14 we see that the core reluctance becomes small as its permeability increases and can often be made much smaller than that of the air gap; i.e., for $\mu \gg \mu_0$, $\mathcal{R}_c \ll \mathcal{R}_g$. In this case, the flux and hence B can be found from Eq. 1-14 in terms of $\mathcal{F}$ and the air-gap properties alone

$$\phi \approx \frac{\mathscr{F}}{\mathscr{R}_g} = \frac{\mathscr{F}\mu_0 A_c}{g} = Ni\frac{\mu_0 A_c}{g} \tag{1-15}$$

The term which multiplies the mmf is known as the *permeance* $\mathscr{P}$; thus the permeance of the air gap is

$$\mathscr{P}_g = \frac{1}{\mathscr{R}_g} = \frac{\mu_0 A_c}{g} \tag{1-16}$$

As will be seen in Art. 1-3, practical magnetic materials have permeabilities which are not constant but vary with flux level. From Eqs. 1-12 to 1-14 we see that as long as this permeability remains sufficiently large, its variation will not significantly affect the performance of the magnetic circuit.

We have now described the basic principles for reducing a magneto-quasi-static field with simple geometry to a magnetic-circuit model. Our limited purpose in this article is to introduce some of the concepts and terminology used by engineers in solving practical design problems. It should be emphasized that this type of thinking depends quite heavily on engineering judgment and intuition. For example, we have tacitly assumed that the permeability of the "iron" parts of the magnetic circuit is a constant known quantity, although this is not true in general (see Art. 1-3), and that the magnetic field is confined to the core and its air gaps. As we shall see later in this book, when two or more windings are placed on a magnetic circuit, as in a transformer or rotating machine, the fields outside the core, called *leakage fields,* are extremely important in determining the coupling between the windings.

EXAMPLE 1-1

The magnetic circuit as shown in Fig. 1-2 has dimensions $A_c = 9 \text{ cm}^2$; $A_g = 9 \text{ cm}^2$; $g = 0.050 \text{ cm}$; $l_c = 30 \text{ cm}$; $N = 500$ turns. Assume the value $\mu_r = 70,000$ for the iron. Find (a) current i for $B_c = 1 \text{ T}$; (b) flux ϕ and flux linkage $\lambda = N\phi$.

Solution

(a) From Eq. 1-10 the ampere-turns for the circuit are

$$Ni = \frac{B_c l_c}{\mu_r \mu_0} + \frac{B_g g}{\mu_0}$$

Since $\phi = B_c A_c = B_g A_g$, the current is

$$i = \frac{B_c}{\mu_0 N}\left(\frac{l_c}{\mu_r} + g\right) = \frac{1}{(4\pi \times 10^{-7})(500)}(0.04 + 5.0) \times 10^{-4} = 0.80\ \text{A}$$

Note that due to its high permeability, the reluctance of the iron path of 30 cm is only $0.04/5.0 = 0.008$ times the reluctance of the 0.050-cm air gap.

(b) Equation 1-8 gives

$$\phi = B_c A_c = 1(9 \times 10^{-4}) = 9 \times 10^{-4}\ \text{Wb}$$
$$\lambda = N\phi = 500(9 \times 10^{-4}) = 0.45\ \text{Wb-turn}$$

EXAMPLE 1-2

The magnetic structure of a synchronous machine is shown schematically in Fig. 1-4. Assuming that rotor and stator iron have infinite permeability ($\mu \to \infty$), find the air-gap flux ϕ and flux density B_g. For this example $I = 10$ A, $N = 1000$ turns, $g = 1$ cm, and $A_g = 2000$ cm^2.

Solution

Notice that there are two air gaps in series, of total length $2g$, and that by symmetry the flux density in each is equal. Since the iron permeability is here assumed to be infinite, its reluctance is negligible and Eq. 1-15 can be used to

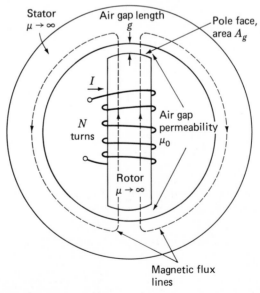

Fig. 1-4. Simple synchronous machine.

find the flux

$$\phi = \frac{NI\mu_0 A_g}{2g} = \frac{1000(10)(4\pi \times 10^{-7})(0.2)}{0.02} = 0.13 \text{ Wb}$$

and

$$B_g = \frac{\phi}{A_g} = \frac{0.13}{0.2} = 0.65 \text{ T}$$

1–2 FLUX LINKAGE, INDUCTANCE, AND ENERGY

When a magnetic field varies with time, an electric field is produced in space. In magnetic structures with windings, such as Fig. 1-2, the varying magnetic field in the core produces an induced voltage† e at the terminals, of value determined by Faraday's law

$$e = N\frac{d\varphi}{dt} = \frac{d\lambda}{dt} \tag{1-17}$$

In general the *flux linkage* of a coil is equal to the surface integral of the normal component of the magnetic flux density integrated over any surface spanned by that coil. Note that the direction of the induced voltage e is defined by Eq. 1-17 so that if the winding terminals were short-circuited, a current would flow in such a direction as to oppose the change of flux linkage.

For a magnetic circuit which has a linear relationship between B and H, because of material of constant permeability or a dominating air gap, we can define the λ-i relationship by the inductance L as

$$L = \frac{\lambda}{i} \tag{1-18}$$

where $\lambda = N\varphi$, the flux linkage, is in weber-turns. The symbol φ is used to indicate the instantaneous value of a time-varying flux.

$$L = \frac{NB_c A_c}{i} = \frac{N^2\mu_0 A_c}{g} \tag{1-19}$$

Inductance is measured in henrys or weber-turns per ampere. Equation 1-19 shows the dimensional form of expressions for inductance. Thus inductance is proportional to the square of the number of turns, the permeability of the magnetic circuit, and its cross-sectional area and inversely proportional to its length.

---

†The term *electromotive force* (emf) is often used instead of induced voltage to represent that component of voltage due to a time-varying flux linkage.

The difficulty in applying the inductance concept in numerical calculations arises from the nonlinear dependence of the permeability μ on magnetic conditions in the core, as will be explained in Arts. 1-3 and 1-4. It must be emphasized that the usefulness of inductance as a parameter depends upon the assumption of a linear relation between flux and mmf. This implies that the effects of the nonlinear magnetic characteristics of the core material can be approximated by some sort of empirical linear relation or that the effects of the core are of secondary importance compared with the effect of an air gap, as shown in Example 1-3.

EXAMPLE 1-3

Find the inductance of the winding on the magnetic circuit of Fig. 1-2. Neglect fringing at the air gap.

Solution

The flux can be found from Eqs. 1-12 to 1-14 as

$$\phi = \frac{Ni}{\mathcal{R}_c + \mathcal{R}_g} = \frac{NA_c i}{l_c/\mu + g/\mu_0}$$

and hence the inductance is

$$L = \frac{N\phi}{i} = \frac{N^2 A_c}{l_c/\mu + g/\mu_0}$$

This can be rewritten as

$$L = \frac{\mu_0 N^2 A_c}{g + (\mu_0/\mu) l_c}$$

which has the characteristic form of Eq. 1-19. Notice that when the air-gap reluctance is much larger than that of the core $[g \gg (\mu_0/\mu) l_c]$, the inductance is determined by the air-gap dimensions alone

$$L = \frac{\mu_0 N^2 A_c}{g}$$

Figure 1-5 shows a magnetic circuit with an air gap and two windings. Note that the reference directions for the currents have been chosen to produce flux in the same direction. The total mmf is

$$\mathcal{F} = Ni = N_1 i_1 + N_2 i_2 \tag{1-20}$$

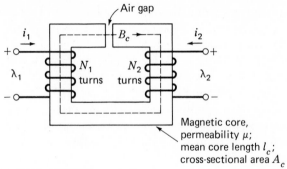

Fig. 1-5. Magnetic circuit with two windings.

and from Eq. 1-15 with the reluctance of the core neglected the core flux ϕ is

$$\phi = (N_1 i_1 + N_2 i_2)\frac{\mu_0 A_c}{g} \tag{1-21}$$

In Eq. 1-21, ϕ is the *resultant core flux* produced by the *simultaneous action* of both mmfs. It is this resultant ϕ which determines the operating point of the core material.

If Eq. 1-21 is broken up into terms attributable to the individual currents, the resultant flux linkages with coil 1 can be expressed as

$$\lambda_1 = N_1 \phi = N_1^2 \frac{\mu_0 A_c}{g} i_1 + N_1 N_2 \frac{\mu_0 A_c}{g} i_2 \tag{1-22}$$

which can be written

$$\lambda_1 = L_{11} i_1 + L_{12} i_2 \tag{1-23}$$

where

$$L_{11} = N_1^2 \frac{\mu_0 A_c}{g} \tag{1-24}$$

is the *self-inductance* of coil 1 and $L_{11} i_1$ is the flux linkage with coil 1 due to its own current i_1. The *mutual inductance* between coils 1 and 2 is

$$L_{12} = N_1 N_2 \frac{\mu_0 A_c}{g} \tag{1-25}$$

and $L_{12} i_2$ is the flux linkage with coil 1 due to the current i_2 in the other coil. Similarly, the flux linkage with coil 2 is

$$\lambda_2 = N_2 \phi = N_1 N_2 \frac{\mu_0 A_c}{g} i_1 + N_2^2 \frac{\mu_0 A_c}{g} i_2 \tag{1-26}$$

or

$$\lambda_2 = L_{21} i_1 + L_{22} i_2 \tag{1-27}$$

where $L_{21} = L_{12}$ is the mutual inductance and

$$L_{22} = N_2^2 \frac{\mu_0 A_c}{g} \tag{1-28}$$

is the self-inductance of coil 2.

It is important to note that the resolution of the resultant flux linkages into the components produced by i_1 and i_2 is based on superposition of the individual effects and therefore implies a linear flux-mmf characteristic (constant permeability).

Substitution of Eq. 1-18 in 1-17 yields

$$e = \frac{d}{dt}(Li) \tag{1-29}$$

for a magnetic circuit with a single winding. For a static magnetic circuit, the inductance is fixed (assuming that material nonlinearities do not cause the inductance to vary), and this equation reduces to the familiar circuit form

$$e = L\frac{di}{dt} \tag{1-30}$$

However, in electromechanical energy-conversion devices inductances are often time-varying, and Eq. 1-29 must be written as

$$e = L\frac{di}{dt} + i\frac{dL}{dt} \tag{1-31}$$

In situations with multiple windings, the total flux linkage of each winding must be used in Eq. 1-17 to find the winding-terminal voltage.

The power at the terminals of a winding on a magnetic circuit is a measure of the rate of energy flow into the circuit through that particular winding. The power is determined from the product of the voltage and the current

$$p = ie = i\frac{d\lambda}{dt} \tag{1-32}$$

and its unit is *watts*, or *joules per second*. Thus the change in magnetic stored energy ΔW in the magnetic circuit in the time interval t_1 to t_2 is

$$\Delta W = \int_{t_1}^{t_2} p \, dt = \int_{\lambda_1}^{\lambda_2} i \, d\lambda \tag{1-33}$$

In SI units, W is measured in joules.

For a single-winding system of constant inductance, the change in magnetic stored energy can be written as

$$\Delta W = \int_{\lambda_1}^{\lambda_2} i \, d\lambda = \int_{\lambda_1}^{\lambda_2} \frac{\lambda}{L} \, d\lambda = \frac{1}{2L}(\lambda_2^2 - \lambda_1^2) \tag{1-34}$$

The total magnetic stored energy at any given value of λ can be found from setting λ_1 equal to zero

$$W = \frac{1}{2L}\lambda^2 = \frac{L}{2}i^2 \tag{1-35}$$

EXAMPLE 1-4

For the magnetic circuit of Example 1-1 and Fig. 1-2 find (*a*) the emf *e* for $B_c = 1 \sin 377t$ T, (*b*) reluctances $\mathcal{R}_c$ and $\mathcal{R}_g$, (*c*) inductance L, and (*d*) energy at $B_c = 1$ T.

Solution

(*a*) The value of λ was found in Example 1-1 as 0.45 Wb · turn for $B_c = 1$ T. Hence, for the sinusoidal variation of B_c the flux linkage is

$$\lambda = 0.45 \sin 377t \qquad \text{Wb} \cdot \text{turn}$$

The emf is given by Eq. 1-17 as

$$e = \frac{d\lambda}{dt} = 170 \cos 377t \qquad \text{V}$$

(*b*) The reluctances can be determined from Eqs. 1-12 and 1-13

$$\mathcal{R}_c = \frac{l_c}{\mu_r \mu_0 A_c} = \frac{0.3}{70{,}000(4\pi \times 10^{-7})(9 \times 10^{-4})} = 3.8 \times 10^3 \text{ A} \cdot \text{turns/Wb}$$

$$\mathcal{R}_g = \frac{g}{\mu_0 A_g} = \frac{5 \times 10^{-4}}{(4\pi \times 10^{-7})(9 \times 10^{-4})} = 44.2 \times 10^4 \text{ A} \cdot \text{turns/Wb}$$

(*c*) The inductance from Eq. 1-18 is

$$L = \frac{\lambda}{i} = \frac{0.45 \sin 377t}{0.80 \sin 377t} = 0.56 \text{ H}$$

(*d*) The energy can be obtained from Eq. 1-35

$$W = \frac{1}{2L}\lambda^2 = \frac{(0.45)^2}{2(0.56)} = 0.18 \text{ J}$$

1-3 PROPERTIES OF MAGNETIC MATERIALS

In the context of electromechanical energy-conversion devices, the importance of magnetic materials is twofold. Through their use it is possible to obtain large magnetic flux densities with relatively low levels of magnetizing force. Since magnetic forces and energy density are increased with increasing flux density, this effect plays a large role in the performance of energy-conversion devices.

In addition, magnetic materials can be used to constrain and direct magnetic fields in well-defined paths. In a transformer they are used to maximize the coupling between the windings as well as to lower the excitation current required for transformer operation. In electric machinery they are used to shape the fields to maximize the desired torque-producing characteristics. Thus a knowledgeable designer can use magnetic materials to achieve specific desirable device characteristics.

Ferromagnetic materials, composed of iron and alloys of iron with cobalt, tungsten, nickel, aluminum, and other metals, are by far the most common magnetic materials. Although these materials are characterized by a wide range of properties, the basic phenomena responsible for their properties are common to them all.

Ferromagnetic materials are composed of a large number of domains, i.e., regions in which the magnetic moments of all the atoms are parallel, giving rise to a net magnetic moment for that domain. In an unmagnetized sample of material the domain magnetic moments are randomly oriented and the net magnetic flux in the material is zero.

When an external magnetizing force is applied to this material, the domain magnetic moments tend to align with the applied magnetic field. As a result, the dipole magnetic moments add to the applied field, resulting in a much larger value of flux density than would exist due to the magnetizing force alone. Thus the effective permeability μ, equal to the ratio of the total magnetic flux density to the applied magnetizing force, is large compared with the permeability of free space μ_0. This behavior continues until all the magnetic moments are aligned with the applied field; at this point they can no longer contribute to increasing the magnetic flux density, and the material is said to be *fully saturated*.

In the absence of an externally applied magnetizing force, the domain magnetic moments naturally align along certain directions associated with the crystal structure of the domain, known as *axes of easy magnetization*.

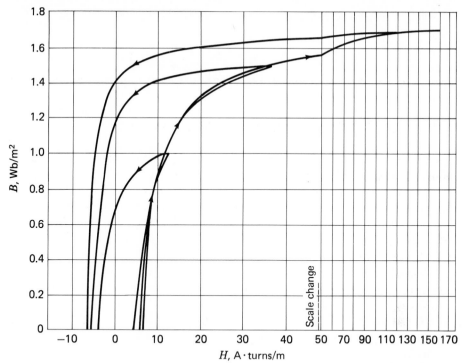

Fig. 1-6. *B-H* loops for M-5 grain-oriented electrical steel 0.012 in thick. Only the top halves of the loops are shown here. (*Armco Inc.*)

Thus if the applied magnetizing force is now reduced, the domain magnetic moments relax to the direction of easy magnetism nearest to that of the applied field. As a result, when the applied field is reduced to zero, the magnetic dipole moments will no longer be totally random in their orientation; they will retain a net magnetization component along the applied-field direction. It is this effect which is responsible for the phenomenon known as *magnetic hysteresis.*

The relationship between *B* and *H* for a ferromagnetic material is both nonlinear and multivalued. In general, the characteristics of the material cannot be described analytically. They are commonly presented in graphical form as a set of empirically determined curves based upon test samples of the material using methods prescribed by the American Society for Testing and Materials (ASTM).†

The most common curve used to describe a magnetic material is the *B-H* curve or *hysteresis loop.* A set of hysteresis loops is shown in Fig. 1-6 for M-5

†Numerical data on a wide variety of magnetic materials are available from the various material manufacturers. One problem in using the references arises from the various systems of units employed. For example, magnetization may be given in oersteds or in ampere-turns per meter and magnetic flux density in gauss, kilogauss, or teslas. A few useful conversion factors are given in Appendix C. The reader is reminded that the equations in this book are based upon SI units.

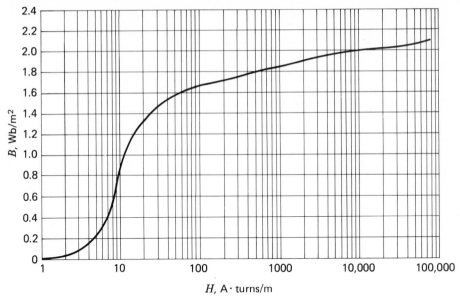

Fig. 1-7. Dc magnetization curve for M-5 grain-oriented electrical steel 0.012 in thick. (*Armco Inc.*)

steel, a typical grain-oriented electrical steel used in electric equipment. These loops show the relationship between the magnetic flux density B and the magnetizing force H. Each curve is obtained while cyclically varying the applied magnetizing force between equal positive and negative values of fixed magnitude. Hysteresis causes these curves to be multivalued. After several cycles the B-H curves form closed loops as shown. The arrows show the paths followed by B with increasing and decreasing H. Notice that with increasing magnitude of H the curves begin to flatten out as the material tends toward saturation. It can be seen that at a maximum flux density of about 1.7 T, this material is heavily saturated.

For many engineering applications it is sufficient to describe the material by the curve drawn through the maximum values of B and H at the tips of the hysteresis loops; this is known as a *dc* or *normal magnetization curve*. A dc magnetization curve for M-5 grain-oriented electrical steel is shown in Fig. 1-7. The dc magnetization curve neglects the hysteretic nature of the material but clearly displays its nonlinear characteristics.

EXAMPLE 1-5

Assume that the core material in Example 1-1 has the dc magnetization curve of Fig. 1-7. Find the current i for $B_c = 1$ T.

Solution

The value of H_c for $B_c = 1\,\mathrm{T}$ is read from Fig. 1-7 as

$$H_c = 12\,\mathrm{A \cdot turns/m}$$

The mmf for the core path is

$$\mathcal{F}_c = H_c l_c = 12(0.3) = 3.6\,\mathrm{A \cdot turns}$$

The mmf for the air gap is

$$\mathcal{F}_g = H_g g = \frac{B_g g}{\mu_0} = \frac{5 \times 10^{-4}}{4\pi \times 10^{-7}} = 396\,\mathrm{A \cdot turns}$$

The current is

$$i = \frac{\mathcal{F}_c + \mathcal{F}_g}{N} = \frac{400}{500} = 0.80\,\mathrm{A}$$

Note that the relative permeability agrees with the value assumed in Example 1-1,

$$\mu_r = \frac{B_c}{\mu_0 H_c} = \frac{1}{(4\pi \times 10^{-7})(12)} = 66{,}000$$

1–4 AC EXCITATION

In ac power systems the waveforms of voltage and flux closely approximate sinusoidal functions of time. This article describes the excitation characteristics and losses associated with steady-state ac operation of magnetic materials. We shall use as our model a closed-core magnetic circuit, i.e., with no air gap, such as that shown in Fig. 1-1 or the transformer of Fig. 1-16. The magnetic path length is l_c, and the cross-sectional area is A_c throughout the length of the core.

We shall assume a sinusoidal variation of the core flux $\varphi(t)$; thus

$$\varphi(t) = \phi_{\max} \sin \omega t = A_c B_{\max} \sin \omega t \qquad (1\text{-}36)$$

where $\phi_{\max} =$ amplitude of core flux φ
$B_{\max} =$ amplitude of flux density B_c
$\omega =$ angular frequency $= 2\pi f$
$f =$ frequency, Hz

From Faraday's law, Eq. 1-17, the voltage induced in the N-turn winding is

$$e(t) = \omega N\phi_{\max} \cos \omega t = E_{\max} \cos \omega t \qquad (1\text{-}37)$$

where
$$E_{\max} = \omega N\phi_{\max} = 2\pi f N A_c B_{\max} \qquad (1\text{-}38)$$

In steady-state ac operation we are usually more interested in the rms values of voltages and currents than in instantaneous or maximum values. The rms value of a sine wave is $1/\sqrt{2}$ times its peak value. Thus the rms value of the induced voltage is

$$E_{\mathrm{rms}} = \frac{2\pi}{\sqrt{2}} f N A_c B_{\max} = 4.44 f N A_c B_{\max} \qquad (1\text{-}39)$$

Because of its importance in the theory of ac machines we shall return to this equation frequently.

To produce the magnetic field in the core requires current in the exciting winding known as the *exciting current* i_φ.† The nonlinear magnetic properties of the core mean that the waveform of the exciting current differs from the sinusoidal waveform of the flux. A curve of the exciting current as a function of time can be found graphically from the magnetic characteristics, as illustrated in Fig. 1-8a. Since B and H are related to φ and i_φ by known geometric constants, the ac hysteresis loop of Fig. 1-8b has been drawn in terms of

†In general, the exciting current is the net ampere-turns acting on the magnetic circuit.

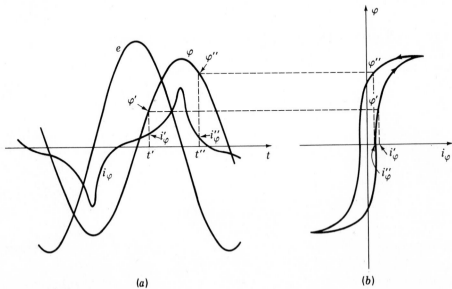

(a) (b)

Fig. 1-8. Excitation phenomena: (a) voltage, flux, and exciting current; (b) corresponding hysteresis loop.

$\varphi = B_c A_c$ and $i_\varphi = H_c l_c / N$. Sine waves of induced voltage e and flux φ in accordance with Eqs. 1-36 and 1-37 are shown in Fig. 1-8a.

At any given time the value of i_φ corresponding to the given value of flux can be found directly from the hysteresis loop. For example, at time t' the flux is φ' and the current is i'_φ; at time t'' the corresponding values are φ'' and i''_φ. Notice that since the hysteresis loop is multivalued, it is necessary to be careful to pick the rising-flux values (φ' in the figure) from the rising-flux portion of the hysteresis loop; similarly for the falling-flux values (φ'' in the figure). Also notice that the waveform of the exciting current is sharply peaked. Its rms value $I_{\varphi,\text{rms}}$ is defined in the standard way as $\sqrt{i_\varphi^2}$ averaged over a cycle. The corresponding rms value H_{rms} of H_c is related to the rms value $I_{\varphi,\text{rms}}$ of the exciting current; thus

$$I_{\varphi,\text{rms}} = \frac{l_c H_{\text{rms}}}{N} \tag{1-40}$$

The ac excitation characteristics of core materials are usually expressed in terms of rms voltamperes rather than a magnetization curve relating B and H. The theory behind this presentation can be explained by combining Eqs. 1-39 and 1-40. Thus the rms voltamperes required to excite the core to a specified flux density is

$$E_{\text{rms}} I_{\varphi,\text{rms}} = 4.44 f N A_c B_{\text{max}} \frac{l_c H_{\text{rms}}}{N} \tag{1-41}$$

$$= 4.44 f A_c l_c B_{\text{max}} H_{\text{rms}} \tag{1-42}$$

For a magnetic material of density ρ_c the weight of the core is $A_c l_c \rho_c$, and the rms voltamperes P_a per unit weight is

$$P_a = \frac{4.44 f}{\rho_c} B_{\text{max}} H_{\text{rms}} \tag{1-43}$$

The excitation voltamperes P_a at a given frequency f is dependent only on B_{max} because H_{rms} is a unique function of B_{max} and is independent of turns and geometry. As a result, the ac excitation requirements for a magnetic material are often given in terms of rms voltamperes per unit weight determined by laboratory tests on closed-core samples of the material. These results are illustrated in Fig. 1-9 for M-5 grain-oriented electrical steel.

The exciting current supplies the mmf required to produce the core flux and the power input associated with the energy in the magnetic field in the core. Part of this energy is dissipated as losses and appears as heat in the core. The rest appears as reactive power associated with the cyclically varying energy stored in the magnetic field. The reactive power is not dissipated in the core; it is cyclically supplied and absorbed by the excitation source and thus

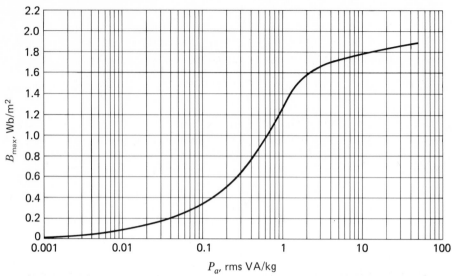

Fig. 1-9. Exciting rms voltamperes per kilogram at 60 Hz for M-5 grain-oriented electrical steel 0.012 in thick. (*Armco Inc.*)

contributes to the current in the source, thereby affecting I^2R losses and voltage drops in the supply system.

Two loss mechanisms are associated with time-varying fluxes in magnetic materials. The first is ohmic I^2R heating, associated with eddy currents. From Faraday's law we see that time-varying magnetic fields give rise to electric fields. In magnetic materials these electric fields result in eddy currents, which circulate in the core material and oppose the change of flux density. To counteract this demagnetizing effect the current in the exciting winding must increase. Thus the dynamic *B-H* loop with ac operation is somewhat "fatter" than the hysteresis loop for slowly varying conditions. To reduce the effects of eddy currents, magnetic structures are usually built up of thin sheets of laminations of the magnetic material. The laminations, which are aligned in the direction of the field lines, are insulated from each other by an oxide layer on their surfaces or by a thin coat of insulating enamel or varnish. This greatly reduces the magnitude of the eddy currents since the layers of insulation interrupt the current paths; the thinner the laminations, the lower the losses. The power loss caused by eddy currents is dissipated as heat in the core. The eddy-current loss increases as the square of the frequency of the flux variation and also as the square of the peak flux density.

The second loss mechanism is due to the hysteretic nature of magnetic material. In a magnetic circuit like that of Fig. 1-1 or the transformer of Fig. 1-16 a time-varying excitation will cause the magnetic material to undergo a cyclic variation such as the hysteresis loop shown in Fig. 1-10.

Equation 1-33 can be used to calculate the energy input to the magnetic

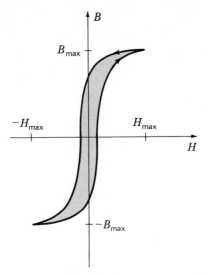

Fig. 1-10. Hysteresis loop; hysteresis loss is proportional to the loop area (*shaded*).

core of Fig. 1-1 as the material undergoes a single cycle. For a single cycle

$$W = \oint i_\varphi \, d\lambda = \oint \frac{H_c l_c}{N} A_c N \, dB_c = A_c l_c \oint H_c \, dB_c \qquad (1\text{-}44)$$

Recognizing that $A_c l_c$ is the volume of the core and that the integral is the area of the ac hysteresis loop, we see that each time the magnetic material undergoes a cycle, there is a net energy input into the material. This energy is required to move around the magnetic dipoles in the material and is dissipated as heat. Thus hysteresis losses are proportional to the area of the hysteresis loop for a given flux level and to the total volume of material. Since there is an energy loss per cycle, hysteresis power loss is proportional to the frequency of the applied excitation.

In general, the losses depend on the metallurgy of the material as well as the flux density and frequency. Information on core loss is typically presented in graphical form. It is plotted in terms of watts per unit weight as a function of flux density; often a family of curves for different frequencies is given. Figure 1-11 shows the core loss P_c for M-5 grain-oriented electrical steel at 60 Hz.

Nearly all transformers and certain sections of electric machines use sheet-steel material that has highly favorable directions of magnetization along which the core loss is low and the permeability is high. The material is termed *grain-oriented steel*. The reason for this property lies in the atomic structure of the simple crystal of the silicon-iron alloy, which is a body-centered cube; each cube has an atom at each corner as well as one in the

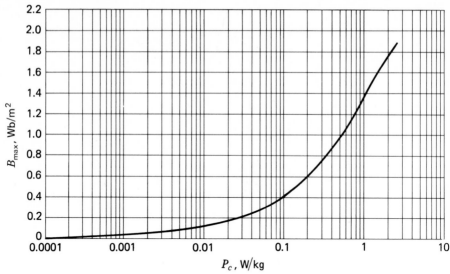

Fig. 1-11. Core loss at 60 Hz in watts per kilogram for M-5 grain-oriented electrical steel 0.012 in thick. (*Armco Inc.*)

center of the cube. In the cube, the easiest axis of magnetization is the cube edge; the diagonal across the cube face is more difficult, and the diagonal through the cube is the most difficult. By suitable manufacturing techniques most of the cube edges are aligned in the rolling direction to make it the favorable direction of magnetization. The behavior in this direction is superior in core loss and required H to nonoriented steels, so that the oriented steels can be operated at higher flux densities than the nonoriented grades. Non-oriented electrical steels are used in applications where the flux does not follow a path which can be oriented with the rolling direction or where low cost is of importance. In these steels the losses are somewhat higher and the permeability very much lower than in grain-oriented steels.

EXAMPLE 1-6

The magnetic core in Fig. 1-12 is made from laminations of M-5 grain-oriented electrical steel. The winding is excited with a voltage to produce a flux density in the steel of $B = 1.5 \sin 377t$ T. The steel occupies 0.94 times the gross core volume. The density of the steel is 7.65 g/cm³. Find (*a*) the applied voltage, (*b*) the peak current, (*c*) the rms exciting current, and (*d*) the core loss.

Solution

(*a*) From Eq. 1-17 the voltage is

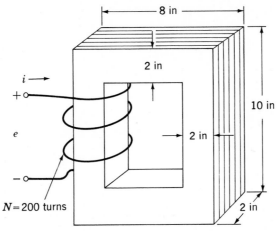

Fig. 1-12. Reactor with laminated steel core.

$$e = N\frac{d\varphi}{dt} = NA_c\frac{dB}{dt}$$

$$= 200(4 \text{ in}^2)(0.94)\frac{1 \text{ m}^2}{(39.4)^2 \text{ in}^2}(1.5)(377\cos 377t)$$

$$= 275\cos 377t \quad \text{V}$$

(b) The magnetic intensity corresponding to $B_{\text{max}} = 1.5$ T is given in Fig. 1-7 as $H = 36$ A · turns/m. Notice that the relative permeability $\mu_r = B/(\mu_0 H) = 33{,}000$ at the flux level of 1.5 T is significantly lower than the value of $\mu_r = 66{,}000$ found in Example 1-5 corresponding to a flux level of 1.0 T.

$$l_c = (6 + 6 + 8 + 8) \text{ in}\frac{1 \text{ m}}{39.4 \text{ in}} = 0.71 \text{ m}$$

The peak current is

$$I = \frac{36(0.71)}{200} = 0.13 \text{ A}$$

(c) The rms current is obtained from the value of P_a of Fig. 1-9 for $B_{\text{max}} = 1.5$ T.

$$P_a = 1.5 \text{ VA/kg}$$

The core volume and weight are

$$V_c = (4\text{ in}^2)(0.94)(28\text{ in}) = 105.5\text{ in}^3$$

$$W_c = (105.5\text{ in}^3)\frac{7.65\text{ g}}{1\text{ cm}^3}\frac{(2.54)^3\text{ cm}^3}{1\text{ in}^3} = 13.2\text{ kg}$$

The total rms voltamperes and current are

$$P_a = (1.5\text{ VA/kg})(13.2\text{ kg}) = 20\text{ VA}$$

$$I_\varphi = \frac{P_a}{E_{\text{rms}}} = \frac{20}{275(0.707)} = 0.10\text{ A}$$

(d) The core-loss density is obtained from Fig. 1-11 as $P_c = 1.2\text{ W/kg}$. The total core loss is

$$P_c = (1.2\text{ W/kg})(13.2\text{ kg}) = 16\text{ W}$$

1-5 INTRODUCTION TO TRANSFORMERS

Before proceeding with a study of electric machinery, it is desirable to discuss certain aspects of the theory of magnetically coupled circuits, with emphasis on transformer action. Although the static transformer is not an energy-conversion device, it is an indispensable component in many energy-conversion systems. As one of the principal reasons for the widespread use of ac power systems, it makes possible electric generation at the most economical generator voltage, power transfer at the most economical transmission voltage, and power utilization at the most suitable voltage for the particular utilization device. The transformer is also widely used in low-power low-current electronic and control circuits for performing such functions as matching the impedances of a source and its load for maximum power transfer, insulating one circuit from another, or isolating direct current while maintaining ac continuity between two circuits.

Moreover, the transformer is one of the simpler devices comprising two or more electric circuits coupled by a common magnetic circuit, and its analysis involves many of the principles essential to the study of electric machinery.

Essentially, a transformer consists of two or more windings interlinked by a mutual magnetic field. If one of these windings, the *primary*, is connected to an alternating-voltage source, an alternating flux will be produced whose amplitude will depend on the primary voltage and number of turns. The mutual flux will link the other winding, the *secondary*, and will induce a voltage in it whose value will depend on the number of secondary turns. By properly

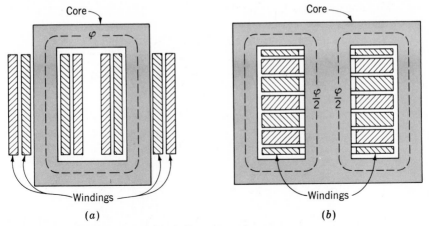

Fig. 1-13. (a) Core-type and (b) shell-type transformer.

proportioning the numbers of primary and secondary turns, almost any desired voltage ratio, or *ratio of transformation,* can be obtained.

Transformer action evidently demands only the existence of alternating mutual flux linking the two windings and is simply utilization of the mutual-inductance concept. Such action will be obtained if an air core is used, but it will be obtained much more effectively with a core of iron or other ferromagnetic material, because most of the flux is then confined to a definite path linking both windings and having a much higher permeability than that of air. Such a transformer is commonly called an *iron-core transformer.* Most transformers are of this type. The following discussion will be concerned almost wholly with iron-core transformers.

In order to reduce the losses caused by eddy currents in the core, the magnetic circuit usually consists of a stack of thin laminations, two common types of construction being shown in Fig. 1-13. In the *core type* (Fig. 1-13a) the windings are wound around two legs of a rectangular magnetic core; in the *shell type* (Fig. 1-13b) the windings are wound around the center leg of a three-legged core. Silicon-steel laminations 0.014 in thick are generally used for transformers operating at frequencies below a few hundred hertz. Silicon steel has the desirable properties of low cost, low core loss, and high permeability at high flux densities (1.0 to 1.5 T). The cores of small transformers used in communication circuits at high frequencies and low energy levels are sometimes made of compressed powdered ferromagnetic alloys such as permalloy.

Most of the flux is confined to the core and therefore links both windings. Although leakage flux which links one winding without linking the other is a small fraction of the total flux, it has an important effect on the behavior of the transformer. Leakage is reduced by subdividing the windings into sections placed as close together as possible. In the core-type construction, each winding consists of two sections, one section on each of the two legs of the core, the primary and secondary windings being concentric coils. In the shell-type con-

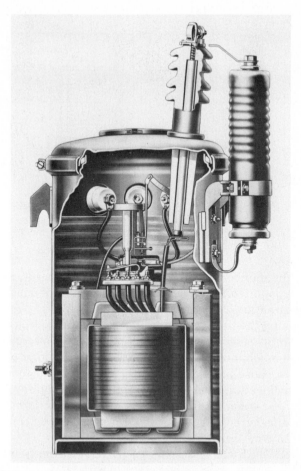

Fig. 1-14. Cutaway view of self-protected transformer typical of sizes 2 to 25 kVA, 7200:240/120 V. Only one high-voltage insulator and lightning arrester is needed because one side of the 7200-V line and one side of the primary are grounded. (*General Electric Company.*)

struction, variations of the concentric-winding arrangement may be used, or the windings may consist of a number of thin "pancake" coils assembled in a stack with primary and secondary coils interleaved.

Figure 1-14 illustrates the internal construction of a *distribution transformer* such as is used in public-utility systems to provide the appropriate voltage at the consumer's premises. A large power transformer is shown in Fig. 1-15.

1–6 NO–LOAD CONDITIONS

Figure 1-16 shows a transformer with its secondary circuit open and an alternating voltage v_1 applied to its primary terminals. In order to simplify the

Fig. 1-15. A 660-MVA 3-phase 50-Hz transformer used to step up generator voltage of 20 kV to transmission voltage of 405 kV. (*CEM Le Havre, French Member of the Brown Boveri Corporation.*)

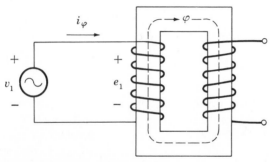

Fig. 1-16. Transformer with open secondary.

drawings it is common practice on schematic diagrams of transformers to show the primary and secondary windings as if they were on separate legs of the core, as in Fig. 1-16, even though the windings are actually interleaved in practice. A small steady-state current i_φ, called the *exciting current,* exists in the primary and establishes an alternating flux in the magnetic circuit. This flux induces an emf in the primary equal to

$$e_1 = \frac{d\lambda_1}{dt} = N_1 \frac{d\varphi}{dt} \qquad (1\text{-}45)$$

where λ_1 = flux linkage with primary
  φ = flux (here assumed all confined to core)
  N_1 = number of turns in primary winding

The voltage e_1 is in volts when φ is in webers. This counter emf together with the voltage drop in the primary resistance R_1 must balance the applied voltage v_1; thus

$$v_1 = R_1 i_\varphi + e_1 \qquad (1\text{-}46)$$

In most power apparatus the no-load resistance drop is very small indeed, and the induced emf e_1 very nearly equals the applied voltage v_1. Furthermore, the waveforms of voltage and flux are very nearly sinusoidal. The analysis can then be greatly simplified, as we have shown in Art. 1-4. Thus, if the instantaneous flux is

$$\varphi = \phi_{max} \sin \omega t \qquad (1\text{-}47)$$

the induced voltage is

$$e_1 = N_1 \frac{d\varphi}{dt} = \omega N_1 \phi_{max} \cos \omega t \qquad (1\text{-}48)$$

where ϕ_{max} is the maximum value of the flux and $\omega = 2\pi f$, the frequency being f Hz. For the positive directions shown in Fig. 1-16, the induced emf leads the flux by 90°. The rms value of the induced emf is

$$E_1 = \frac{2\pi}{\sqrt{2}} f N_1 \phi_{max} = 4.44 f N_1 \phi_{max} \qquad (1\text{-}49)$$

If the resistive voltage drop is negligible, the counter emf equals the applied voltage. Under these conditions, if a sinusoidal voltage is applied to a winding, a sinusoidally varying core flux must be established whose maximum value ϕ_{max} satisfies the requirement that E_1 in Eq. 1-49 equal the rms value V_1 of the applied voltage; thus

$$\phi_{\max} = \frac{V_1}{4.44 f N_1} \qquad (1\text{-}50)$$

The flux is determined solely by the applied voltage, its frequency, and the number of turns in the winding. This important relation applies not only to transformers but also to any device operated with sinusoidal alternating impressed voltage, so long as the resistance drop is negligible. The magnetic properties of the core determine the exciting current. It must adjust itself so as to produce the mmf required to create the flux demanded by Eq. 1-50.

Because of the nonlinear magnetic properties of iron, the waveform of the exciting current differs from the waveform of the flux. A curve of the exciting current as a function of time can be found graphically from the ac hysteresis loop, as shown in Fig. 1-8.

If the exciting current is analyzed by Fourier-series methods, it will be found to comprise a fundamental and a family of odd harmonics. The fundamental can, in turn, be resolved into two components, one in phase with the counter emf and the other lagging the counter emf by 90°. The fundamental in-phase component accounts for the power absorbed by hysteresis and eddy-current losses in the core. It is called the *core-loss component* of the exciting current. When the core-loss component is subtracted from the total exciting current, the remainder is called the *magnetizing current.* It comprises a fundamental component lagging the counter emf by 90°, together with all the harmonics. The principal harmonic is the third. For typical power transformers, the third harmonic usually is about 40 percent of the exciting current.

Except in problems concerned directly with the effects of harmonics, the peculiarities of the exciting-current waveform usually need not be taken into account, because the exciting current itself is small. For example, the exciting current of a typical power transformer is about 5 percent of full-load current. Consequently the effects of the harmonics usually are swamped out by the sinusoidal-current requirements of other linear elements in the circuit. The exciting current can then be represented by its *equivalent sine wave,* which has the same rms value and frequency and produces the same average power as the actual wave. Such representation is essential to the construction of a phasor diagram. In Fig. 1-17, the phasors $\widehat{E}_1$ and $\widehat{\Phi}$, respectively, represent the induced emf and the flux. The phasor $\widehat{I}_\varphi$ represents the equivalent sinusoidal exciting current. It lags the induced emf $\widehat{E}_1$ by a phase angle θ_c such that

$$P_c = E_1 I_\varphi \cos \theta_c \qquad (1\text{-}51)$$

where P_c is the core loss. The component $\widehat{I}_c$ in phase with $\widehat{E}_1$ represents the core-loss current. The component $\widehat{I}_m$ in phase with the flux represents an equivalent sine wave having the same rms value as the magnetizing current. Typical core-loss and exciting voltampere characteristics of high-quality sili-

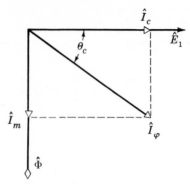

Fig. 1-17. No-load phasor diagram.

con steel used for power- and distribution-transformer laminations are shown in Figs. 1-9 and 1-11.

EXAMPLE 1-7

In Example 1-6 the core loss and exciting voltamperes for the core of Fig. 1-12 at $B_{\text{max}} = 1.5$ T and 60 Hz were found to be

$$P_c = 16 \text{ W} \qquad (VI)_{\text{rms}} = 20 \text{ VA}$$

and the induced voltage was $275/\sqrt{2} = 194$ V rms when the winding had 200 turns.

Find the power factor, the core-loss current I_c, and the magnetizing current I_m.

Solution

$$\text{Power factor } \cos \theta_c = \frac{16}{20} = 0.80 \qquad \theta_c = 36.9 \qquad \sin \theta_c = 0.60$$

$$\text{Exciting current } I_\varphi = \frac{20}{194} = 0.10 \text{ A rms}$$

$$\text{Core-loss component } I_c = \frac{16}{194} = 0.082 \text{ A rms}$$

$$\text{Magnetizing component } I_m = I_\varphi \sin \theta_c = 0.060 \text{ A rms}$$

1–7 EFFECT OF SECONDARY CURRENT; IDEAL TRANSFORMER

As a first approximation to a quantitative theory, consider a transformer with a primary winding of N_1 turns and a secondary winding of N_2 turns, as shown

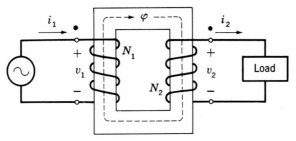

Fig. 1-18. Ideal transformer and load.

schematically in Fig. 1-18. Notice that the secondary current is defined as positive out of the winding; thus positive secondary current creates an mmf in the opposite direction from that created by positive primary current. Let the properties of this transformer be idealized in that the winding resistances are negligible, and assume that all the flux is confined to the core and links both windings, core losses are negligible, and the permeability of the core is so high that only a negligible exciting current is required to establish the flux. These properties are closely approached but never actually attained in practical transformers. A hypothetical transformer having these properties is often called an *ideal transformer*.

When a time-varying voltage v_1 is impressed on the primary terminals, a core flux φ must be established such that the counter emf e_1 equals the impressed voltage when winding resistance is negligible. Thus

$$v_1 = e_1 = N_1 \frac{d\varphi}{dt} \tag{1-52}$$

The core flux also links the secondary and produces an induced emf e_2 and an equal secondary terminal voltage v_2, given by

$$v_2 = e_2 = N_2 \frac{d\varphi}{dt} \tag{1-53}$$

From the ratio of Eqs. 1-52 and 1-53

$$\frac{v_1}{v_2} = \frac{N_1}{N_2} = \frac{I_2}{I_1} \tag{1-54}$$

Thus an ideal transformer changes voltages in the direct ratio of the turns in its windings.

Now let a load be connected to the secondary. A current i_2 and an mmf $N_2 i_2$ are then present in the secondary. Unless this secondary mmf is counteracted in the primary, the core flux will be radically changed and the balance between impressed voltage and counter emf in the primary will be disturbed.

Hence, a compensating primary mmf and current i_1 must be called into being such that

$$N_1 i_1 = N_2 i_2 \qquad (1\text{-}55)$$

This is the means by which the primary knows of the presence of current in the secondary. Note that for the reference directions shown in Fig. 1-18 the mmfs of i_1 and i_2 are in opposite directions and therefore compensate. The net mmf acting on the core therefore is zero, in accordance with the assumption that the exciting current of an ideal transformer is zero. From Eq. 1-55

$$\frac{i_1}{i_2} = \frac{N_2}{N_1} \qquad (1\text{-}56)$$

Thus an ideal transformer changes currents in the inverse ratio of the turns in its windings. Also notice from Eqs. 1-54 and 1-56 that

$$v_1 i_1 = v_2 i_2 \qquad (1\text{-}57)$$

i.e., instantaneous power input equals instantaneous power output, a necessary condition because all causes of active- and reactive-power losses in the transformer have been neglected.

For further study, consider the case of a sinusoidal applied voltage and an impedance load. Phasor symbolism can then be used. The circuit is shown in simplified form in Fig. 1-19a, in which the dot-marked terminals of the transformer correspond to the similarly marked terminals in Fig. 1-18. The dot

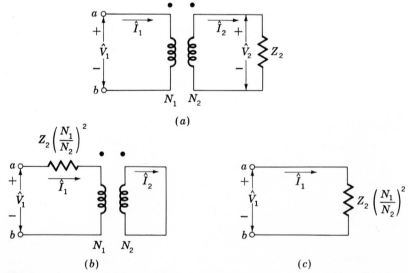

Fig. 1-19. Three circuits which are identical at terminals ab when the transformer is ideal.

markings indicate terminals of corresponding polarity; i.e., if one follows through the primary and secondary windings of Fig. 1-18 beginning at their dot-marked terminals, one will find that both windings encircle the core in the same direction with respect to the flux. Therefore, if one compares the voltages of the two windings, the voltages from a dot-marked to an unmarked terminal will be of the same instantaneous polarity for primary and secondary. In other words, the voltages $\widehat{V}_1$ and $\widehat{V}_2$ in Fig. 1-19a are in phase. Also, the currents $\widehat{I}_1$ and $\widehat{I}_2$ are in phase. The fact that their mmfs must balance is accounted for by their being in opposite directions through the windings.

In phasor form, Eqs. 1-54 and 1-56 can be expressed as

$$\widehat{V}_1 = \frac{N_1}{N_2}\widehat{V}_2 \quad \text{and} \quad \widehat{V}_2 = \frac{N_2}{N_1}\widehat{V}_1 \tag{1-58}$$

$$\widehat{I}_1 = \frac{N_2}{N_1}\widehat{I}_2 \quad \text{and} \quad \widehat{I}_2 = \frac{N_1}{N_2}\widehat{I}_1 \tag{1-59}$$

From these equations

$$\frac{\widehat{V}_1}{\widehat{I}_1} = \left(\frac{N_1}{N_2}\right)^2 \frac{\widehat{V}_2}{\widehat{I}_2} = \left(\frac{N_1}{N_2}\right)^2 Z_2 \tag{1-60}$$

where Z_2 is the complex impedance of the load. Consequently, as far as its effect is concerned, an impedance Z_2 in the secondary circuit can be replaced by an equivalent impedance Z_1 in the primary circuit, provided that

$$Z_1 = \left(\frac{N_1}{N_2}\right)^2 Z_2 \tag{1-61}$$

Thus, the three circuits of Fig. 1-19 are indistinguishable as far as their performance viewed from terminals ab is concerned. Transferring an impedance from one side of a transformer to the other in this fashion is called *referring the impedance* to the other side. In a similar manner, voltages and currents can be *referred* to one side or the other by using Eqs. 1-58 and 1-59 to evaluate the equivalent voltage and current on that side.

To sum up, *in an ideal transformer voltages are transformed in the direct ratio of turns, currents in the inverse ratio, and impedances in the direct ratio squared; power and voltamperes are unchanged.*

1-8 TRANSFORMER REACTANCES AND EQUIVALENT CIRCUITS

The departures in an actual transformer from the ideal properties assumed in Art. 1-7 must be included to a greater or lesser degree in most analyses of transformer performance. A more complete theory must take into account the effects of winding resistances, magnetic leakage, and exciting current. Sometimes the capacitances of the windings also have important effects, notably in problems involving transformer behavior at frequencies above the audio range or during rapidly changing transient conditions such as those encountered in pulse transformers and in power-system transformers as a result of voltage surges caused by lightning or switching transients. The analysis of these high-frequency problems is beyond the scope of the present treatment, however, and accordingly the capacitances of the windings are neglected in the following analyses.

Two methods of analysis by which the departures from the ideal can be taken into account are (1) an equivalent-circuit technique based on physical reasoning and (2) a mathematical attack based on the classical theory of magnetically coupled circuits. Both methods are in everyday use, and both have very close parallels in the theories of rotating machines. Because it offers an excellent example of the thought process involved in translating physical concepts into a quantitative theory, the equivalent-circuit technique is presented here.

The total flux linking the primary winding can be divided into two components: the resultant mutual flux, confined essentially to the iron core and produced by the combined effect of the primary and secondary currents, and the primary leakage flux, which links only the primary. These components are identified in the elementary transformer shown in Fig. 1-20, where for simplicity the primary and secondary windings are shown on opposite legs of the core. In an actual transformer with interleaved windings the details of the flux map are more complicated, but the essential features remain the same. Because the leakage path is largely in air, the leakage flux and the voltage induced by it vary linearly with primary current $\hat{I}_1$. The effect on the primary

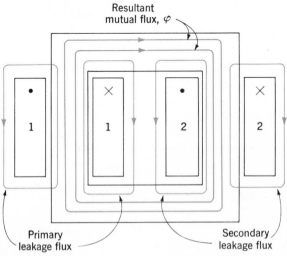

Fig. 1-20. Component fluxes.

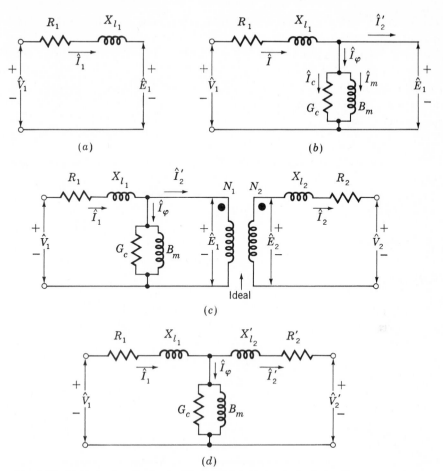

Fig. 1-21. Steps in development of the transformer equivalent circuit.

circuit is the same as that of flux linkages anywhere in the circuit leading up to the transformer primary and can be simulated by assigning to the primary a *leakage inductance* (equal to the leakage-flux linkages with the primary per unit of primary current) or *leakage reactance* X_{l1} (equal to $2\pi f$ times leakage inductance). In addition there will be a voltage drop in the primary effective resistance R_1.

The impressed voltage $\widehat{V}_1$ is then opposed by three phasor voltages: the $\widehat{I}_1R_1$ drop in the primary resistance, the $\widehat{I}_1X_{l1}$ drop arising from primary leakage flux, and the counter emf $\widehat{E}_1$ induced in the primary by the resultant mutual flux. All these voltages are appropriately included in the equivalent circuit in Fig. 1-21a.

The resultant mutual flux links both the primary and secondary windings and is created by their combined mmfs. It is convenient to treat these mmfs by considering that the primary current must meet two requirements of the magnetic circuit: it must not only (1) counteract the demagnetizing effect. of the secondary current but also (2) produce sufficient mmf to create the resultant mutual flux. According to this physical picture, it is convenient to resolve the primary current into two components, a load component and an exciting component. The *load component* $\widehat{I}_2'$ is defined as the component current in the primary which would exactly counteract the mmf of the secondary current $\widehat{I}_2$. Thus, for opposing currents

$$\widehat{I}_2' = \frac{N_2}{N_1}\widehat{I}_2 \tag{1-62}$$

It equals the secondary current referred to the primary as in an ideal transformer. The *exciting component* i_φ is defined as the additional primary current required to produce the resultant mutual flux. It is a nonsinusoidal current of the nature described in Art. 1-6.

The exciting current can be treated as an equivalent sinusoidal current $\widehat{I}_\varphi$, in the manner described in Art. 1-6 and can be resolved into a core-loss component $\widehat{I}_c$ in phase with the counter emf $\widehat{E}_1$ and a magnetizing component $\widehat{I}_m$ lagging $\widehat{E}_1$ by 90°. In the equivalent circuit (Fig. 1-21b) the equivalent sinusoidal exciting current is accounted for by means of a shunt branch connected across $\widehat{E}_1$, comprising a noninductive resistance whose conductance is G_c in parallel with a lossless inductance whose susceptance is B_m. Alternatively, a series combination of resistance and reactance can be connected across $\widehat{E}_1$. In the parallel combination (Fig. 1-21b) the power $E_1^2 G_c$ accounts for the core loss due to the resultant mutual flux. When G_c is assumed constant, the core loss is thereby assumed to vary as E_1^2 or (for sine waves) as $\phi_{\max}^2 f^2$, where $\phi_{\max}$ is the maximum value of the resultant mutual flux. The magnetizing susceptance B_m varies with the saturation of the iron. When the inductance corresponding to B_m is assumed constant, the magnetizing current is thereby assumed to be independent of frequency and directly proportional to the resultant mutual flux. Both G_c and B_m are usually determined at rated voltage and frequency; they are then assumed to remain constant for the small departures from rated values associated with normal operation.

The resultant mutual flux $\widehat{\Phi}$ induces an emf $\widehat{E}_2$ in the secondary, and since this flux links both windings, the induced-voltage ratio is

$$\frac{\widehat{E}_1}{\widehat{E}_2} = \frac{N_1}{N_2} \tag{1-63}$$

just as in an ideal transformer. This voltage transformation and the current transformation of Eq. 1-62 can be accounted for by introducing an ideal transformer in the equivalent circuit, as in Fig. 1-21c. The emf $\widehat{E}_2$ is not the secondary terminal voltage, however, because of the secondary resistance and because the secondary current $\widehat{I}_2$ creates *secondary leakage flux* (see Fig.

1-20). The secondary terminal voltage $\widehat{V}_2$ differs from the induced voltage $\widehat{E}_2$ by the voltage drops due to secondary resistance R_2 and *secondary leakage reactance* X_{l2}, as in the portion of the equivalent circuit (Fig. 1-21c) to the right of $\widehat{E}_2$.

The actual transformer therefore is equivalent to an ideal transformer plus external impedances. By referring all quantities to the primary or secondary the ideal transformer in Fig. 1-21c can be moved out to the right or left, respectively, of the equivalent circuit. This is almost invariably done, and the equivalent circuit is usually drawn as in Fig. 1-21d, with the ideal transformer not shown and all voltages, currents, and impedances referred to the same side.

In Fig. 1-21d the referred values are indicated with primes, for example, X'_{l2} and R'_2, to distinguish them from the actual values of Fig. 1-21c. In what follows we shall almost always deal with referred values, and the primes will be omitted. One must simply keep in mind the side of the transformers to which all quantities have been referred. The circuit of Fig. 1-21d is called the equivalent *T circuit* for a transformer.

EXAMPLE 1-8

A 50-kVA 2400:240-V 60-Hz distribution transformer has a leakage impedance of $0.72 + j0.92\ \Omega$ in the high-voltage winding and $0.0070 + j0.0090\ \Omega$ in the low-voltage winding. At rated voltage and frequency the admittance Y_φ of the shunt branch accounting for the exciting current is $(0.324 - j2.24) \times 10^{-2}$ mho when viewed from the low-voltage side. Draw the equivalent circuit referred to (a) the high-voltage side and (b) the low-voltage side and label the impedances numerically.

Solution

The circuits are given in Fig. 1-22a and b, respectively, with the high-voltage side numbered 1 and the low-voltage side numbered 2. The voltages given on the nameplate of a power-system transformer are based on the turns ratio and neglect the small leakage-impedance voltage drops under load. Since this is a 10-to-1 transformer, impedances are referred by multiplying or dividing by 100. The value of an impedance referred to the high-voltage side is greater than its value referred to the low-voltage side. Since admittance is the reciprocal of impedance, an admittance is referred from one side to the other by use of the reciprocal of the referring factor for impedance. The value of an admittance referred to the high-voltage side is smaller than its value referred to the low-voltage side.

The ideal transformer may be explicitly drawn, as shown dotted in Fig. 1-22, or it may be omitted in the diagram and remembered mentally, making the unprimed letters the terminals.

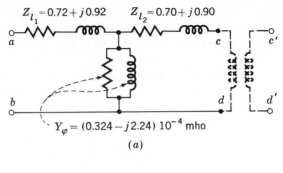

$$Z_{l_1} = 0.72 + j0.92 \qquad Z_{l_2} = 0.70 + j0.90$$

$$Y_\varphi = (0.324 - j2.24)\,10^{-4}\ \text{mho}$$

(a)

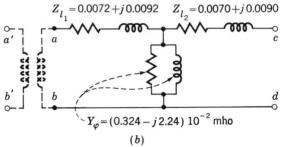

$$Z_{l_1} = 0.0072 + j0.0092 \qquad Z_{l_2} = 0.0070 + j0.0090$$

$$Y_\varphi = (0.324 - j2.24)\,10^{-2}\ \text{mho}$$

(b)

Fig. 1-22. Equivalent circuits for transformer of Example 1-8.

1-9 ENGINEERING ASPECTS OF TRANSFORMER ANALYSIS

In engineering analyses involving the transformer as a circuit element, it is customary to adopt one of several approximate forms of the equivalent circuit of Fig. 1-21 rather than the full circuit. The approximations chosen in a particular case depend largely on physical reasoning based on orders of magnitude of the neglected quantities. The more common approximations are presented in this article. In addition, test methods are given for determining the transformer constants.

a. Approximate Equivalent Circuits; Power Transformers

The approximate equivalent circuits commonly used for constant-frequency power-transformer analyses are summarized for comparison in Fig. 1-23. All quantities in these circuits are referred to either the primary or the secondary, and the ideal transformer is not shown.

The computational labor involved often can be appreciably reduced by moving the shunt branch representing the exciting current out from the middle of the T circuit to either the primary or the secondary terminals, as in Fig. 1-23a and b. These are *cantilever circuits*. The series branch is the combined resistance and leakage reactance referred to the same side. This impedance is sometimes called the *equivalent impedance* and its components the *equiva-*

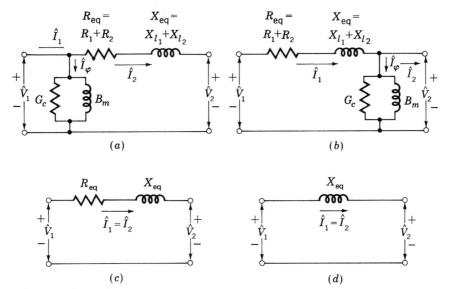

Fig. 1-23. Approximate equivalent circuits.

lent resistance R_{eq} and *equivalent reactance* X_{eq}, as shown in Fig. 1-23a and b. Error is introduced by neglect of the voltage drop in the primary or secondary leakage impedance caused by the exciting current, but this error is insignificant in most problems involving power-system transformers.

Further simplification results from neglecting the exciting current entirely, as in Fig. 1-23c, in which the transformer is represented as an equivalent series impedance. If the transformer is large (several hundred kilovoltamperes or over), the equivalent resistance R_{eq} is small compared with the equivalent reactance X_{eq} and can frequently be neglected, giving Fig. 1-23d. The circuits of Fig. 1-23c and d are sufficiently accurate for most ordinary power-system problems. Finally, in situations where the currents and voltages are determined almost wholly by the circuits external to the transformer or when a high degree of accuracy is not required, the entire transformer impedance can be neglected and the transformer considered to be ideal, as in Art. 1-7.

The circuits of Fig. 1-23 have the additional advantage that the total equivalent resistance R_{eq} and equivalent reactance X_{eq} can be found from a very simple test, whereas measurement of the values of the component leakage reactances X_{l1} and X_{l2} is a difficult experimental task.

EXAMPLE 1-9

The 50-kVA 2400:240-V transformer whose constants are given in Example 1-8 is used to step down the voltage at the load end of a feeder whose imped-

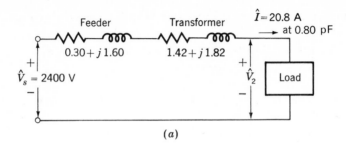

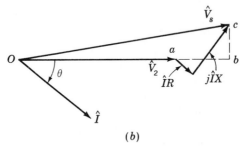

(b)

Fig. 1-24. Equivalent circuit and phasor diagram, Example 1-9.

ance is $0.30 + j1.60 \, \Omega$. The voltage V_s at the sending end of the feeder is 2400 V.

Find the voltage at the secondary terminals of the transformer when the load connected to its secondary draws rated current from the transformer and the power factor of the load is 0.80 lagging. Neglect the voltage drops in the transformer and feeder caused by the exciting current.

Solution

The circuit with all quantities referred to the high-voltage (primary) side of the transformer is shown in Fig. 1-24a, where the transformer is represented by its equivalent impedance, as in Fig. 1-23c. From Fig. 1-22a, the value of the equivalent impedance is $Z_{eq} = 1.42 + j1.82 \, \Omega$, and the combined impedance of the feeder and transformer in series is $Z = 1.72 + j3.42 \, \Omega$. From the transformer rating, the load current referred to the high-voltage side is $I = 50,000/2400 = 20.8 \, \text{A}$.

The phasor diagram referred to the high-voltage side is shown in Fig. 1-24b, from which

$$Ob = \sqrt{V_s^2 - (bc)^2} \quad \text{and} \quad V_2 = Ob - ab$$

Note that

$$bc = IX \cos \theta - IR \sin \theta \qquad ab = IR \cos \theta + IX \sin \theta$$

where R and X are the combined resistance and reactance, respectively. Thus

$$bc = 20.8(3.42)(0.80) - 20.8(1.72)(0.60) = 35.5 \text{ V}$$
$$ab = 20.8(1.72)(0.80) + 20.8(3.42)(0.60) = 71.4 \text{ V}$$

Substitution of numerical values shows that Ob very nearly equals V_s, or 2400 V. Then $V_2 = 2329$ V referred to the high-voltage side. The actual voltage at the secondary terminals is 2329/10, or

$$V_2 = 233 \text{ V}$$

---

b. Approximate Equivalent Circuits; Variable-Frequency Transformers

Small iron-core transformers operating in the audio-frequency range (hence called *audio-frequency transformers*) are often used as coupling devices in electronic circuits for communications, measurements, and control. Their principal functions are either to step up voltage, thereby contributing to the overall voltage gain in amplifiers, or to act as impedance-transforming devices bringing about the optimum relation between the apparent impedance of a load and the impedance of a source. They may also serve other auxiliary functions, such as providing a path for direct current through the primary while keeping it out of the secondary circuit.

Application of transformers for impedance matching makes direct use of the impedance-transforming property shown in Eq. 1-61. Oscillators and amplifiers, for example, give optimum performance when working into a definite order of magnitude of load impedance, and transformer coupling can be used to change the apparent impedance of the actual load to this optimum. A tranformer so used is called an *output transformer.*

When the frequency varies over a wide range, it is important that the output voltage be as closely instantaneously proportional to the input voltage as possible. Ideally, this means that voltages should be amplified equally and phase shift should be zero for all frequencies. The *amplitude-frequency characteristic* (often abbreviated to *frequency characteristic*) is a curve of the ratio of the load voltage on the secondary side to the internal source voltage on the primary side plotted as a function of frequency, a flat characteristic being the most desirable. The *phase characteristic* is a curve of the phase angle of the load voltage relative to the source voltage plotted as a function of frequency, a small phase angle being desirable. These characteristics depend not only on the transformer but also on the constants of the entire primary and secondary circuits.

As an example of the use of engineering approximations, consider an amplifier coupled to its load through an output transformer. The amplifier is considered as being equivalent to a source of voltage E_G in series with an internal resistance R_G and the load is considered as a resistance R_L. This is

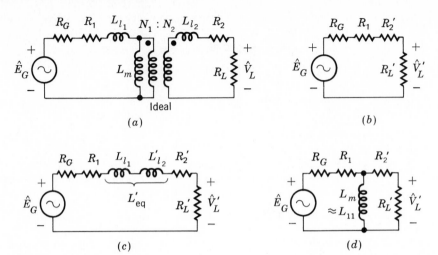

Fig. 1-25. Equivalent circuits of an output transformer: (a) complete equivalent circuit, (b) approximate equivalent in the middle range of audio frequencies, (c) high-frequency equivalent, (d) low-frequency equivalent.

shown in Fig. 1-25a, where the transformer is represented by the equivalent circuit of Fig. 1-21c, with core loss neglected. Sometimes the stray capacitances of the windings must be taken into account at high audio frequencies, especially when the source impedance is higher than a few thousand ohms.

The analysis of a properly designed circuit breaks down into three frequency ranges.

Intermediate. At intermediate frequencies (around 500 Hz) none of the inductances is important, and the equivalent circuit reduces to a network of resistances, as shown in Fig. 1-25b, where all quantities have been referred to the primary, as indicated by the prime superscripts. Analysis of this circuit shows that the ratio of load voltage V_L to source voltage E_G is

$$\frac{V_L}{E_G} = \frac{N_2}{N_1} \frac{R'_L}{R'_{se}} \tag{1-64}$$

where

$$R'_{se} = R_G + R_1 + R'_2 + R'_L \tag{1-65}$$

In this middle range (which usually extends over several octaves) the voltage ratio is very nearly constant; i.e., the amplitude characteristic is flat, and the phase shift is zero.

High. As the frequency is increased, however, the leakage reactances of the transformer become increasingly important. The equivalent circuit in the high audio range is shown in Fig. 1-25c. Analysis of this circuit shows that at high freqencies

$$\frac{V_L}{E_G} = \frac{N_2}{N_1} \frac{R'_L}{R'_{se}} \frac{1}{\sqrt{1 + (\omega L'_{eq}/R'_{se})^2}} \tag{1-66}$$

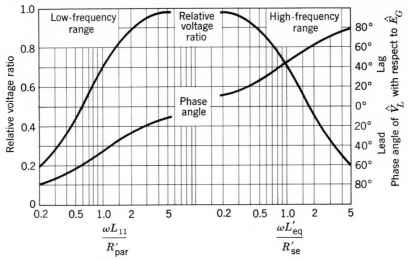

Fig. 1-26. Normalized frequency characteristics of output transformers.

where L'_{eq} is the equivalent leakage inductance. The voltage ratio relative to its midrange value is

$$\text{Relative voltage ratio} = \frac{1}{\sqrt{1 + (\omega L'_{eq}/R'_{se})^2}} \qquad (1\text{-}67)$$

The phase angle by which the load voltage lags the source voltage is

$$\theta = \tan^{-1}\frac{\omega L'_{eq}}{R'_{se}} \qquad (1\text{-}68)$$

Curves of the relative voltage ratio and phase angle as functions of the reactance-to-resistance ratio $\omega L'_{eq}/R'_{se}$ are shown in the right-hand half of Fig. 1-26.

Low. At low frequencies the leakage reactances are negligible, but the shunting effect of the magnetizing branch becomes increasingly important as its reactance decreases. The inductance of the magnetizing branch very nearly equals the self-inductance L_{11} of the primary. The equivalent circuit at low frequencies is shown in Fig. 1-25d, from which

$$\frac{V_L}{E_G} = \frac{N_2}{N_1}\frac{R'_L}{R'_{se}}\frac{1}{\sqrt{1 + (R'_{par}/\omega L_{11})^2}} \qquad (1\text{-}69)$$

where

$$R'_{par} = \frac{(R_G + R_1)(R'_2 + R'_L)}{R_G + R_1 + R'_2 + R'_L} \qquad (1\text{-}70)$$

The voltage ratio relative to its midrange value is

$$\text{Relative voltage ratio} = \frac{1}{\sqrt{1 + (R'_{\text{par}}/\omega L_{11})^2}} \tag{1-71}$$

and the phase angle by which the load voltage leads the source voltage is

$$\theta = \tan^{-1}\frac{R'_{\text{par}}}{\omega L_{11}} \tag{1-72}$$

Curves of the relative voltage ratio and phase angle as functions of the reactance-to-resistance ratio $\omega L_{11}/R'_{\text{par}}$ are shown in the left-hand half of Fig. 1-26.

The points at which the relative voltage ratio is 0.707 are called the *half-power points*. From Eq. 1-67 the upper half-power point occurs at a frequency f_h for which the equivalent leakage reactance $\omega_h L'_{\text{eq}}$ equals the series resistance R'_{se}, or

$$f_h = \frac{R'_{\text{se}}}{2\pi L'_{\text{eq}}} \tag{1-73}$$

and from Eq. 1-71 the lower half-power point occurs at a frequency f_l for which the self-reactance ωL_{11} equals the parallel resistance R'_{par}, or

$$f_l = \frac{R'_{\text{par}}}{2\pi L_{11}} \tag{1-74}$$

The bandwidth is describable in terms of the ratio

$$\frac{f_h}{f_l} = \frac{R'_{\text{se}}}{R'_{\text{par}}}\frac{L_{11}}{L'_{\text{eq}}} \tag{1-75}$$

A broad bandwidth requires a high ratio of self-inductance to leakage inductance or a coefficient of coupling as close as possible to unity.

Stray capacitances of the windings may have a significant effect on the frequency characteristics at high audio frequencies. This effect tends to produce a peak in the voltage ratio in the high-frequency range at the frequency where the capacitances are in resonance with the leakage inductances. The height of the peak depends on the circuit impedances as well as the transformer inductances and stray capacitances. These capacitances are difficult to calculate accurately. The frequency-response characteristics of a specific circuit can usually best be determined by tests. The foregoing analysis should be regarded more as an explanation of the effects of the transformer inductances than as an accurate analysis.

c. Short-Circuit and Open-Circuit Tests

Two very simple tests serve to determine the constants of the equivalent circuits of Fig. 1-23 and the power losses in a transformer. These consist in measuring the input voltage, current, and power to the primary, first with the secondary short-circuited and then with the secondary open-circuited.

With the secondary short-circuited, typically a primary voltage of only 2 to 12 percent of the rated value need be impressed to obtain full-load current. For convenience, the high-voltage side is usually taken as the primary in this test. If $\hat{V}_{sc}$, $\hat{I}_{sc}$, and P_{sc} are the impressed voltage, primary current, and power input, the short-circuit impedance Z_{sc} and its resistance and reactance components R_{sc} and X_{sc} referred to the primary are

$$Z_{sc} = \frac{\hat{V}_{sc}}{\hat{I}_{sc}} \tag{1-76}$$

$$R_{sc} = \frac{P_{sc}}{I_{sc}^2} \tag{1-77}$$

$$X_{sc} = \sqrt{|Z_{sc}|^2 - R_{sc}^2} \tag{1-78}$$

The equivalent circuit with the secondary terminals short-circuited is shown in Fig. 1-27. The voltage induced in the secondary by the resultant core flux equals the secondary leakage-impedance voltage drop, and at rated current this voltage is only about 1 to 6 percent of rated voltage. At the correspondingly low value of core flux, the exciting current and core losses are entirely negligible. The exciting admittance, shown dashed in Fig. 1-27, can then be omitted, and the primary and secondary currents are very nearly equal when referred to the same side. The power input very nearly equals the total I^2R loss in the primary and secondary windings, and the impressed voltage equals the drop in the combined primary and secondary leakage impedance Z_{eq}. The equivalent resistance and reactance referred to the primary very nearly equal the short-circuit resistance and reactance of Eqs. 1-77 and 1-78, respectively. The equivalent impedance can, of course, be referred from one side to the other in the usual manner. On the rare occasions when the equivalent T circuit in Fig. 1-21d must be resorted to, approximate values of the individual primary and secondary resistances and leakage reactances can be

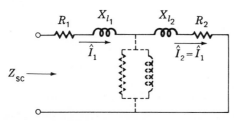

Fig. 1-27. Equivalent circuit with short-circuited secondary.

obtained by assuming that $R_1 = R_2 = 0.5\, R_{eq}$ and $X_{l1} = X_{l2} = 0.5\, X_{eq}$ when all impedances are referred to the same side.

With the secondary open-circuited and rated voltage impressed on the primary, an exciting current of only 2 to 6 percent of full-load current is obtained. If the transformer is to be used at other than its rated voltage, the test should be taken at that voltage. For convenience, the low-voltage side is usually taken as the primary in this test. The voltage drop in the primary leakage impedance caused by the small exciting current is entirely negligible, and the primary impressed voltage $\widehat{V}_1$ very nearly equals the emf $\widehat{E}_1$ induced by the resultant core flux. Also, the primary I^2R loss caused by the exciting current is entirely negligible, so that the power input P_1 very nearly equals the core loss P_c. Thus the exciting admittance $Y_\varphi = G_c - jB_m$ in Fig. 1-21d very nearly equals the open-circuit admittance $Y_{oc} = G_{oc} - jB_{oc}$ determined from the impressed voltage $\widehat{V}_1$, exciting current $\widehat{I}_\varphi$, and power input P_1 measured in the primary with the secondary open-circuited; thus the exciting admittance and its conductance and susceptance components are very nearly

$$Y_\varphi = Y_{oc} = \frac{\widehat{I}_\varphi}{\widehat{V}_1} \tag{1-79}$$

$$G_c = G_{oc} = \frac{P_1}{V_1^2} \tag{1-80}$$

$$B_m = G_{oc} = \sqrt{|Y_{oc}|^2 - G_{oc}^2} \tag{1-81}$$

The values so obtained are, of course, referred to the side which was used as the primary in this test. With the approximate equivalent circuits of Fig. 1-23c and d the open-circuit test is used only to obtain core loss for efficiency computations and to check the magnitude of the exciting current. Sometimes the voltage at the terminals of the open-circuited secondary is measured as a check on the turns ratio.

EXAMPLE 1-10

With the instruments located in the high-voltage side and the low-voltage side short-circuited, the short-circuit test readings for the 50-kVA 2400:240-V transformer of Example 1-8 are 48 V, 20.8 A, and 617 W. An open-circuit test with the low-voltage side energized gives instrument readings on that side of 240 V, 5.41 A, and 186 W. Determine the efficiency and the voltage regulation at full load, 0.80 power factor lagging.

Solution

From the short-circuit test, the equivalent impedance, resistance, and reactance of the transformer (referred to the high-voltage side as denoted by the subscript H) are

$$|Z_{eq,H}| = \frac{48}{20.8} = 2.31\ \Omega \qquad R_{eq,H} = \frac{617}{(20.8)^2} = 1.42\ \Omega$$

$$X_{eq,H} = \sqrt{(2.31)^2 - (1.42)^2} = 1.82\ \Omega$$

Full-load high-tension current is

$$I_H = \frac{50{,}000}{2400} = 20.8\ \text{A}$$

$$
\begin{aligned}
\text{Loss} = I_H^2 R_{eq,H} = (20.8)^2(1.42) = \quad &617\ \text{W}\\
\text{Core loss} = \quad &\underline{186}\\
\text{Total losses at full load} = \quad &803\\
\text{Output} = 50{,}000(0.80) = \quad &\underline{40{,}000}\\
\text{Input} = \quad &40{,}803\ \text{W}
\end{aligned}
$$

$$\frac{\text{Losses}}{\text{Input}} = \frac{803}{40{,}803} = 0.0197$$

The *efficiency* of any power-transmitting device is

$$\text{Efficiency} = \frac{\text{power output}}{\text{power input}}$$

which can also be expressed as

$$\text{Efficiency} = \frac{\text{input} - \text{losses}}{\text{input}} = 1 - \frac{\text{losses}}{\text{input}}$$

For the specific load, the efficiency of this transformer is

$$\text{Efficiency} = 1 - 0.0197 = 0.980$$

The *voltage regulation* of a transformer is the change in secondary terminal voltage from no load to full load and is usually expressed as a percentage of the full-load value. The equivalent circuit of Fig. 1-23c will be used with everything still referred to the high-voltage side. The primary voltage is assumed to be adjusted so that the secondary terminal voltage has its rated value at full load, or $V_{2H} = 2400$ V referred to the high-voltage side. The required value of the primary voltage $\widehat{V}_{1H}$ can be computed from the phasor

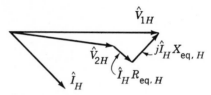

Fig. 1-28. Phasor diagram for Example 1-10.

diagram shown in Fig. 1-28.

$$\widehat{V}_{1H} = \widehat{V}_{2H} + \widehat{I}_H(R_{eq,H} + jX_{eq,H})$$
$$= 2400 + 20.8(0.80 - j0.60)(1.42 + j1.82)$$
$$= 2446 + j13$$

The magnitude of $\widehat{V}_{1H}$ is 2446 V. If this voltage were held constant and the load removed, the secondary voltage on open circuit would rise to 2446 V referred to the high-voltage side. Then

$$\text{Regulation} = \frac{2446 - 2400}{2400}(100) = 1.92\%$$

1-10 TRANSFORMERS IN THREE-PHASE CIRCUITS†

Three single-phase transformers can be connected to form a 3-phase bank in any of the four ways shown in Fig. 1-29. In all four parts of this figure the windings at the left are the primaries, those at the right are the secondaries, and any primary winding is mated in one transformer with the secondary winding drawn parallel to it. Also shown are the voltages and currents resulting from balanced impressed primary line-to-line voltages V and line currents I when the ratio of primary to secondary turns N_1/N_2 is a and ideal transformers are assumed. It will be noted that, for fixed line-to-line voltages and total kVA, the kVA rating of each transformer is one-third of the kVA rating of the bank, regardless of the connections used, but the voltage and current ratings of the individual transformers depend on the connections.

The Y-Δ connection is commonly used in stepping down from a high voltage to a medium or low voltage. One of the reasons is that a neutral is thereby provided for grounding on the high-voltage side, a procedure which can be shown to be desirable in most cases. Conversely, the Δ-Y connection is commonly used for stepping up to a high voltage. The Δ-Δ connection has the advantage that one transformer can be removed for repair or maintenance

†The relationship between 3-phase and single-phase quantities is discussed in Appendix A.

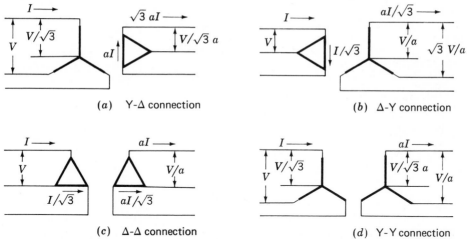

(a) Y-Δ connection

(b) Δ-Y connection

(c) Δ-Δ connection

(d) Y-Y connection

Fig. 1-29. Common 3-phase transformer connections; the transformer windings are indicated by the heavy lines.

while the remaining two continue to function as a 3-phase bank with the rating reduced to 58 percent of that of the original bank; this is known as the *open-delta,* or V, *connection.* The Y-Y connection is seldom used because of difficulties with exciting-current phenomena.

Instead of three single-phase transformers, a 3-phase bank may consist of one 3-phase transformer having all six windings on a common multilegged core and contained in a single tank. Advantages of 3-phase transformers are that they cost less, weigh less, require less floor space, and have somewhat higher efficiency. A photograph of the internal parts of a large 3-phase transformer is shown in Fig. 1-30.

Circuit computations involving 3-phase transformer banks under balanced conditions can be made by dealing with only one of the transformers or phases and recognizing that conditions are the same in the other two phases except for the phase displacements associated with a 3-phase system. It is usually convenient to carry out the computations on a per phase Y line-to-neutral basis, since transformer impedances can then be added directly in series with transmission-line impedances. The impedances of transmission lines can be referred from one side of the transformer bank to the other by use of the square of the ideal line-to-line voltage ratio of the bank. In dealing with Y-Δ or Δ-Y banks all quantities can be referred to the Y-connected side. In dealing with Δ-Δ banks in series with transmission lines, it is convenient to replace the Δ-connected impedances of the transformers by equivalent Y-connected impedances. It is well known that a balanced Δ-connected circuit of Z_Δ Ω/phase is equivalent to a balanced Y-connected circuit of Z_Y Ω/phase if

$$Z_Y = \tfrac{1}{3}Z_\Delta \tag{1-82}$$

Fig. 1-30. A 200-MVA 3-phase 50-Hz 3-winding 210/80/10.2-kV transformer removed from its tank. The 210-kV winding has an on-load tap changer for adjustment of the voltage. (*Brown Boveri Corporation.*)

EXAMPLE 1-11

Three single-phase 50-kVA 2400:240-V transformers identical with that of Example 1-10 are connected Y-Δ in a 3-phase 150-kVA bank to step down the voltage at the load end of a feeder whose impedance is $0.15 + j1.00 \, \Omega$/phase. The voltage at the sending end of the feeder is 4160 V line to line. On their secondary sides the transformers supply a balanced 3-phase load through a feeder whose impedance is $0.0005 + j0.0020 \, \Omega$/phase. Find the line-to-line voltage at the load when the load draws rated current from the transformers at a power factor of 0.80 lagging.

Solution

The computations can be made on a per phase Y basis by referring everything to the high-voltage Y-connected side of the transformer bank. The voltage at the sending end of the feeder is equivalent to a source voltage V_s of

$$V_s = \frac{4160}{\sqrt{3}} = 2400 \text{ V to neutral}$$

From the transformer rating, the rated current on the high-voltage side is 20.8 A/phase Y. The low-voltage feeder impedance referred to the high-voltage side by means of the square of the ideal line-to-line voltage ratio of the bank is

$$\left(\frac{4160}{240}\right)^2 (0.0005 + j0.0020) = 0.15 + j0.60 \ \Omega$$

and the combined series impedance of the high- and low-voltage feeders referred to the high-voltage side is

$$Z_{\text{feeder}} = 0.30 + j1.60 \ \Omega/\text{phase Y}$$

From Example 1-10 the equivalent impedance of the transformer bank referred to its high-voltage Y-connected side is

$$Z_{\text{eq},H} = 1.42 + j1.82 \ \Omega/\text{phase Y}$$

The equivalent circuit for 1 phase referred to the Y-connected primary side then is exactly the same as Fig. 1-24a, and the solution on a per phase basis is exactly the same as the solution of Example 1-9, whence the load voltage referred to the high-voltage side is 2329 V to neutral. The actual load voltage is

$$V_{\text{load}} = 233 \text{ V line to line}$$

This is the line-to-line voltage because the secondaries are Δ-connected.

EXAMPLE 1-12

The three transformers of Example 1-11 are connected Δ-Δ and supplied with power through a 2400-V (line-to-line) 3-phase feeder whose reactance is 0.80 Ω/phase. The equivalent reactance of each transformer referred to its high-voltage side is 1.82 Ω. All circuit resistances will be neglected. At its sending end the feeder is connected to the secondary terminals of a 3-phase Y-Δ connected transformer whose 3-phase rating is 500 kVA, 24,000:2400 V (line-to-line voltages). The equivalent reactance of the sending-end transformers is 2.76 Ω/phase Δ referred to the 2400-V side. The voltage applied to the primary terminals is 24,000 V line to line.

A 3-phase short circuit occurs at the 240-V terminals of the receiving-end transformers. Compute the steady-state short-circuit current in the 2400-V feeder wires, in the primary and secondary windings of the receiving-end transformers, and at the 240-V terminals.

Solution

The computations will be made on an equivalent line-to-neutral basis with everything referred to the 2400-V feeder. The source voltage then is

$$\frac{2400}{\sqrt{3}} = 1385 \text{ V to neutral}$$

Sending-end transformers (from Eq. 1-82):

$$X_{eq} = \frac{2.76}{3} = 0.92 \text{ equivalent } \Omega/\text{phase Y}$$

Receiving-end transformers:

$$X_{eq} = \frac{1.82}{3} = 0.61 \text{ equivalent } \Omega/\text{phase Y}$$

$$\text{Feeder } X = 0.80 \text{ }\Omega/\text{phase Y}$$

$$\text{Total reactance} = 0.92 + 0.61 + 0.80 = 2.33 \text{ }\Omega$$

$$\text{Current in 2400-V feeder} = \frac{1385}{2.33} = 594 \text{ A}$$

$$\text{Current in 2400-V windings} = \frac{594}{\sqrt{3}} = 342 \text{ A}$$

$$\text{Current in 240-V windings} = 3420 \text{ A}$$

$$\text{Current at the 240-V terminals} = 5940 \text{ A}$$

1-11 THE PER UNIT SYSTEM

Computations relating to machines, transformers, and systems of machines are often carried out in per unit form, i.e., with all pertinent quantities expressed as decimal fractions of appropriately chosen base values. All the usual computations are then carried out in these per unit values instead of the familiar volts, amperes, ohms, etc.

There are two advantages to the system. One is that the constants of machines and transformers lie in a reasonably narrow numerical range when expressed in a per unit system related to their rating. The correctness of their values is thus subject to a rapid approximate check. The other is that the analyst is relieved of the worry of referring circuit quantities to one side or the other of transformers. For complicated systems involving many transformers of different turns ratios, this advantage is a significant one in that a possible cause of serious mistakes is removed. The per unit system is also very useful in simulating machine systems on analog and digital computers for transient and dynamic analyses.

Quantities such as voltage V, current I, power P, reactive power Q, volt-amperes VA, resistance R, reactance X, impedance Z, conductance G, susceptance B, and admittance Y can be translated to and from per unit form as follows:

$$\text{Quantity in per unit} = \frac{\text{actual quantity}}{\text{base value of quantity}} \qquad (1\text{-}83)$$

where *actual quantity* refers to the value in volts, amperes, ohms, etc. To a certain extent, base values can be chosen arbitrarily, but certain relations between them must be observed for the normal electrical laws to hold in the per unit system. Thus, for a single-phase system,

$$P_{\text{base}}, \ Q_{\text{base}}, \ VA_{\text{base}} = V_{\text{base}}I_{\text{base}} \qquad (1\text{-}84)$$

$$R_{\text{base}}, \ X_{\text{base}}, \ Z_{\text{base}} = \frac{V_{\text{base}}}{I_{\text{base}}} \qquad (1\text{-}85)$$

In normal usage, values of VA_{base} and V_{base} are chosen first, values of I_{base} and all other quantities in Eqs. 1-84 and 1-85 being thereby established.

The value of VA_{base} must be the same over the entire system concerned. When a transformer is encountered, the values of V_{base} are different on each side and must be in the same ratio as the turns on the transformer. Usually, the rated or nominal voltages of the respective sides are chosen. The process of referring quantities to one side of the transformer is then taken care of automatically by using Eqs. 1-84 and 1-85 in finding and interpreting per unit values.

This can be seen with reference to the equivalent circuit of Fig. 1-21c. If the base voltages of the primary and secondary are chosen to be in the ratio of the turns of the ideal transformer, the per unit ideal transformer will have a unity turns ratio and hence can be eliminated. The procedure thus becomes one of translating all quantities to per unit values, using these values in all the customary circuit-analysis techniques, and translating the end results back to the more usual forms.

When only one electric device, such as a transformer, is involved, the device's own rating is generally used for the voltampere base. When expressed in per unit on the rating as a base, the characteristics of power and distribution transformers do not vary much over a wide range of ratings. For example, the exciting current usually is between 0.02 and 0.06 per unit, the equivalent resistance usually is between 0.005 and 0.02 per unit (the smaller values applying to large transformers), and the equivalent reactance usually is between 0.015 and 0.10 per unit (the larger values applying to large high-voltage transformers). Similarly, the per unit values of synchronous- and induction-machine constants fall within a relatively narrow range. When several devices are involved, however, an arbitrary choice of voltampere base must usually be made so that the same base will be used for the overall system. Per unit values can be changed from one base to another by the following relations:

$$(P, Q, VA)_{\text{pu on base 2}} = (P, Q, VA)_{\text{pu on base 1}} \frac{VA_{\text{base 1}}}{VA_{\text{base 2}}} \tag{1-86}$$

$$(R, X, Z)_{\text{pu on base 2}} = (R, X, Z)_{\text{pu on base 1}} \frac{(V_{\text{base 1}})^2 \, VA_{\text{base 2}}}{(V_{\text{base 2}})^2 \, VA_{\text{base 1}}} \tag{1-87}$$

$$V_{\text{pu on base 2}} = V_{\text{pu on base 1}} \frac{V_{\text{base 1}}}{V_{\text{base 2}}} \tag{1-88}$$

$$I_{\text{pu on base 2}} = I_{\text{pu on base 1}} \frac{V_{\text{base 2}}}{V_{\text{base 1}}} \frac{VA_{\text{base 1}}}{VA_{\text{base 2}}} \tag{1-89}$$

EXAMPLE 1-13

The exciting current measured on the low-voltage side of a 50-kVA 2400: 240-V transformer is 5.41 A. Its equivalent impedance referred to the high-voltage side is $1.42 + j1.82 \, \Omega$. Taking the transformer rating as a base, express in per unit on the low- and high-voltage side (a) the exciting current and (b) the equivalent impedance.

Solution

The base values of voltages and currents are

$$V_{\text{base},H} = 2400 \text{ V} \qquad V_{\text{base},X} = 240 \text{ V} \qquad I_{\text{base},H} = 20.8 \text{ A} \qquad I_{\text{base},X} = 208 \text{ A}$$

where subscripts H and X indicate the high- and low-voltage sides, respectively.

From Eq. 1-85

$$Z_{\text{base},H} = \frac{2400}{20.8} = 115.2 \ \Omega \qquad Z_{\text{base},X} = \frac{240}{208} = 1.152 \ \Omega$$

(a) From Eq. 1-83

$$I_{\varphi X} = \frac{5.41}{208} = 0.0260 \text{ per unit}$$

The exciting current referred to the high-voltage side is 0.541 A. Its per unit value is

$$I_{\varphi H} = \frac{0.541}{20.8} = 0.0260 \text{ per unit}$$

The per unit values are the same referred to either side. The turns ratios required to refer currents in amperes from one side of the transformer to the other are taken care of in the per unit system by the base values for currents on the two sides when the voltampere base is the same on both sides and the voltage bases are in the ratio of the turns.

(b) From Eq. 1-83 and the value for $Z_{\text{base},H}$

$$Z_{\text{eq},H} = \frac{1.42 + j1.82}{115.2} = 0.0123 + j0.0158 \text{ per unit}$$

The equivalent impedance referred to the low-voltage side is $0.0142 + j0.0182 \ \Omega$. Its per unit value is

$$Z_{\text{eq},X} = \frac{0.0142 + j0.0182}{1.152} = 0.0123 + j0.0158 \text{ per unit}$$

The per unit values referred to the high- and low-voltage sides are the same, the referring factors being taken care of in per unit by the base values.

When applied to 3-phase problems, the base values for the per unit system are chosen so that the relations for a balanced 3-phase system hold between them:

$$(P_{\text{base}}, Q_{\text{base}}, VA_{\text{base}}) \text{ 3-phase} = 3VA_{\text{base per phase}} \qquad (1\text{-}90)$$

$$V_{\text{base, line to line}} = \sqrt{3}\, V_{\text{base, line to neutral}} \qquad (1\text{-}91)$$

$$I_{\text{base, per phase }\Delta} = \frac{1}{\sqrt{3}} I_{\text{base, per phase Y}} \qquad (1\text{-}92)$$

In dealing with 3-phase systems the 3-phase VA base and the line-to-line voltage base are usually chosen first. The base values for phase voltages and currents then follow from Eqs. 1-90 to 1-92. Equations 1-86 to 1-89 still apply to the base values per phase. For example, the base value for Y-connected impedances is given by Eq. 1-87 with V_{base} taken as the base voltage to neutral and I_{base} taken as the base current per phase Y; the base value for Δ-connected impedances is also given by Eq. 1-87 but with V_{base} taken as the base line-to-line voltage and I_{base} taken as the base current per phase Δ. Division of Eq. 1-91 by Eq. 1-92 shows that

$$Z_{\text{base, per phase }\Delta} = 3 Z_{\text{base, per phase Y}} \qquad (1\text{-}93)$$

The factors of $\sqrt{3}$ and 3 relating Δ and Y quantities in volts, amperes, and ohms in a balanced 3-phase system are thus automatically taken care of in per unit by the base values. Such 3-phase problems can be solved in per unit as if they were single-phase problems without paying any attention to the details of the transformer connections except in translating volt-ampere-ohm values into and out of the per unit system.

EXAMPLE 1-14

Solve Example 1-12 in per unit on a 3-phase 150-kVA rated-voltage base. The reactance of the 3-phase 500-kVA sending-end transformer is 0.08 per unit on its rating as a base. The reactance of the 2400-V feeder is 0.80 Ω/phase. The reactance of the receiving-end transformers is 0.0158 per unit, as computed in Example 1-13.

Solution

Convert reactance of sending-end transformer to 150-kVA base

$$X_{\text{sending end}} = 0.08\frac{150}{500} = 0.024 \text{ per unit}$$

For the 2400-V feeder

$$V_{\text{base}} = \frac{2400}{\sqrt{3}} = 1385 \text{ V line to neutral}$$

$$I_{\text{base}} = \frac{50{,}000}{1385} = 36.1 \text{ A/phase Y}$$

$$Z_{\text{base}} = \frac{1385}{36.1} = 38.3 \ \Omega/\text{phase Y}$$

Therefore $$X_{\text{feeder}} = \frac{0.80}{38.3} = 0.021 \text{ per unit}$$

For the receiving-end transformers

$$X_{\text{receiving end}} = 0.0158 \text{ per unit}$$
$$\text{Total reactance} = \text{sum} = 0.0608 \text{ per unit}$$
$$\text{Short-circuit current} = \frac{1.00}{0.0608} = 16.4 \text{ per unit}$$

The currents in various parts of the circuit can now be computed. For example, the current in the 2400-V feeder wires is 16.4(36.1) = 594 A. Compare with the result in Example 1-12.

Note: It can readily be shown that the reactance of 0.08 per unit for the sending-end transformers is consistent with the value of 2.76 Ω/phase given in Example 1-12.

1–12 AUTOTRANSFORMERS; MULTICIRCUIT TRANSFORMERS

The principles discussed in the foregoing articles have been developed with specific reference to 2-winding transformers. They are also generally applicable to transformers with other than two separate windings. Aspects relating to autotransformers and multiwinding transformers are considered in this article.

a. Autotransformers

Viewed from the terminals, substantially the same transformation effect on voltages, currents, and impedances can be obtained with the connections of Fig. 1-31a as in the normal transformer with two separate windings shown in Fig. 1-31b. In Fig. 1-31a the winding bc is common to both the primary and secondary circuits. This type of transformer is called an *autotransformer*. It is really nothing but a normal transformer connected in a special way. The only difference structurally is that winding ab must be provided with extra insulation. The performance of an autotransformer is governed by the fundamental considerations already discussed for transformers having two separate windings. Autotransformers have lower leakage reactances, lower losses, and smaller exciting current and cost less than 2-winding transformers when the voltage ratio does not differ too greatly from 1 to 1. A disadvantage is the direct electric connection between the high- and low-voltage sides.

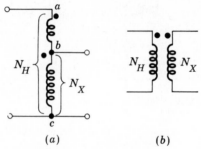

(a)   (b)

Fig. 1-31. (a) Autotransformer; (b) 2-winding transformer.

EXAMPLE 1-15

The 2400:240-V 50-kVA transformer of Examples 1-10 and 1-11 is connected as an autotransformer, as shown in Fig. 1-32, in which ab is the 240-V winding and bc is the 2400-V winding. (It is assumed that the 240-V winding has enough insulation to withstand a voltage of 2640 V to ground.)

(a) Compute the voltage ratings V_H and V_X of the high- and low-tension sides, respectively, when the transformer is connected as an autotransformer.

(b) Compute the kVA rating as an autotransformer.

(c) Data with respect to the losses are given in Example 1-10. Compute the full-load efficiency as an autotransformer at 0.80 power factor.

Solution

(a) Since the 2400-V winding bc is connected to the low-tension circuit, $V_X = 2400$ V. When $V_{bc} = 2400$ V, a voltage $V_{ab} = 240$ V in phase with V_{bc} will be induced in winding ab (leakage-impedance voltage drops being neglected). The voltage of the high-tension side therefore is

$$V_H = V_{ab} + V_{bc} = 2640 \text{ V}$$

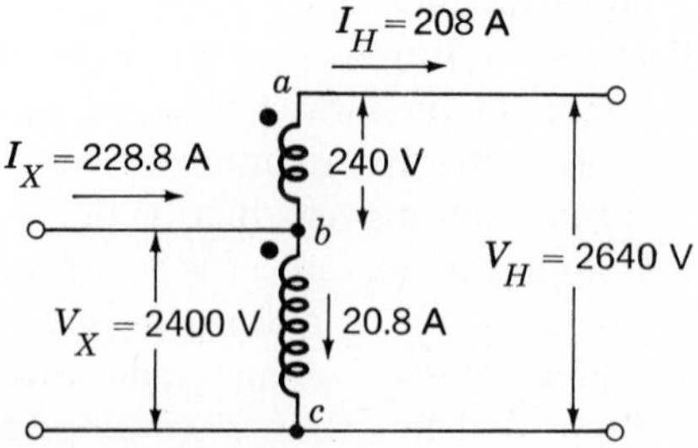

Fig. 1-32. Autotransformer for Example 1-15.

(b) From the rating of 50 kVA as a normal 2-winding transformer, the rated current of the 240-V winding is $50,000/240 = 208$ A. Since the 240-V winding is in series with the high-tension circuit, the rated current of this winding is the rated current I_H on the high-tension side as an autotransformer. The kVA rating as an autotransformer therefore is

$$\frac{V_H I_H}{1000} = \frac{2640(208)}{1000} = 550 \text{ kVA}$$

The rating can also be computed on the low-tension side in a manner which highlights the current-transforming properties. Thus, if the current in the 240-V winding has its rated value of 208 A, the current in the 2400-V winding must produce an equal and opposite mmf (exciting current being neglected) and therefore must be 20.8 A in the arrow direction (Fig. 1-32). The current I_X on the low-tension side as an autotransformer therefore is

$$I_X = 208 + 20.8 = 228.8 \text{ A}$$

and the kVA rating is

$$\frac{V_X I_X}{1000} = \frac{2400(228.8)}{1000} = 550 \text{ kVA}$$

Note that this transformer, whose rating as a normal 2-winding transformer is 50 kVA, is capable of handling 550 kVA as an autotransformer. The higher rating as an autotransformer is a consequence of the fact that all of the 550 kVA does not have to be transformed by electromagnetic induction. In fact, all that the transformer has to do is to boost a current of 208 A through a potential rise of 240 V, corresponding to a rating of 50 kVA.

(c) When connected as an autotransformer with the currents and voltages shown in Fig. 1-32, the losses are the same as in Example 1-10, namely, 803 W. But the output as an autotransformer at 0.80 power factor is $0.80(550,000) = 440,000$ W. The efficiency therefore is

$$1 - \frac{803}{440,803} = 0.9982$$

The efficiency is so high because the losses are those incident to transforming only 50 kVA.

b. Multicircuit Transformers

Transformers having three or more windings, known as *multicircuit,* or *multiwinding, transformers,* are often used to interconnect three or more circuits which may have different voltages. For these purposes a multicircuit transformer costs less and is more efficient than an equivalent number of two-circuit transformers. A transformer having a primary and two secondaries is generally used to supply power to electronic units. The distribution transformers used to supply power for domestic purposes usually have two 120-V secondaries connected in series. Lighting circuits are connected across each of the 120-V windings, while electric ranges, domestic hot-water heaters, and other similar loads are supplied with 240-V power from the series-connected secondaries. A large distribution system may be supplied through a 3-phase bank of multicircuit transformers from two or more transmission systems having different voltages. The 3-phase transformer banks used to interconnect two transmission systems of different voltages often have a third, or *tertiary,* set of windings to provide voltage for auxiliary power purposes in the substation or to supply a local distribution system. Static capacitors or synchronous condensers may be connected to the tertiary windings for power-factor correction or voltage regulation. Sometimes Δ-connected tertiary windings are put on 3-phase banks to provide a circuit for the third harmonics of the exciting current.

Some of the problems arising in the use of multicircuit transformers concern the effects of leakage impedances on voltage regulation, short-circuit currents, and division of load among circuits. These problems can be solved by an equivalent-circuit technique similar to that used in dealing with two-circuit transformers.

The equivalent circuits of multiwinding transformers are more complicated than in the 2-winding case because they must take into account the leakage impedances associated with each pair of windings. In these equivalent circuits all quantities are referred to a common base, either by use of the appropriate turns ratios as referring factors or by expressing all quantities in per unit. The exciting current usually is neglected.

1–13 SUMMARY

Electromechanical devices of the magnetic field type use ferromagnetic materials for guiding and concentrating the fields. Because the permeability of ferromagnetic material is several thousand times that of the surrounding space, most of the magnetic flux is confined to fairly definite paths determined by the magnetic material. The frequencies are low enough to permit the magnetic fields to be considered quasi-static.

As a result, the solution for the magnetic fields in these devices can be obtained in a straightforward fashion by using the techniques of magnetic-circuit analysis. These techniques reduce the complex three-dimensional field solution to what is essentially a one-dimensional problem. As in all engineering solutions, a certain amount of experience and judgment is required, but the technique gives useful results in many situations of practical engineering interest.

Ferromagnetic materials are available with a wide variety of characteristics. In general, their behavior is nonlinear and their *B-H* characteristics are often represented in the form of a family of hysteresis loops. Losses, both hysteresis and eddy current, are a function of the flux level and frequency as well as the composition and manufacturing process used. A basic understanding of the nature of these phenomena is extremely useful in the application of these materials to practical devices.

Although no electromechanical energy conversion is involved in a static transformer, its performance has many points of similarity with the electrical behavior of ac rotating machines. The following discussion of these points will therefore constitute a brief preview of the thought processes involved in ac machine theory presented in subsequent chapters.

In both transformers and rotating machines, a magnetic field is created by the combined action of the currents in the windings. In an iron-core transformer most of this flux is confined to the core and links all the windings. This resultant mutual flux induces voltages in the windings proportional to their numbers of turns and provides the voltage-changing property. In rotating machines most of the flux crosses the air gap, like the core flux in a transformer, and links all the windings on both stator and rotor. The voltages induced in the windings by this resultant mutual air-gap flux are similar to those induced by the resultant core flux in a transformer. The difference is that mechanical motion together with electromechanical energy conversion is involved in rotating machines. The torque associated with this energy-conversion process is created by the interaction of the air-gap flux with the magnetic field of the rotor currents.

In addition to the useful mutual fluxes, in both transformers and rotating machines there are leakage fluxes which link one winding without linking the other. Although the detailed picture of the leakage fluxes in rotating machines is more complicated than in transformers, their effects are essentially the same. In both, the leakage fluxes induce voltages in ac windings which are accounted for as leakage-reactance voltage drops. In both, the leakage-flux paths are mostly in air, and the leakage fluxes are nearly linearly proportional to the currents producing them. The leakage reactances therefore are often assumed to be constant, independent of the degree of saturation of the main magnetic circuit.

From the viewpoint of the winding, the induced-voltage phenomena in transformers and rotating machines are essentially the same, although the internal phenomena causing the time variations in flux linkages are different. In a rotating machine the time variation in flux linkages is caused by relative motion of the field and the winding, and the induced voltage is sometimes referred to as a speed voltage. Speed voltages accompanied by mechanical motion are a necessary counterpart of the electromechanical energy conversion. In a static transformer, however, the time variation of flux linkages is caused by the growth and decay of a stationary magnetic field, no mechanical motion is involved, and no electromechanical energy conversion takes place.

The resultant core flux in a transformer induces a counter emf in the primary, which, together with the primary resistance and leakage-reactance voltage drops, must balance the applied voltage. Since the resistance and leakage-reactance voltage drops usually are small, the counter emf must approximately equal the applied voltage and the core flux must adjust itself accordingly. Exactly similar phenomena must take place in the armature windings of an ac motor; the resultant air-gap flux wave must adjust itself to generate a counter emf approximately equal to the applied voltage. In both transfomers and rotating machines the net mmf of all the currents must accordingly adjust itself to create the resultant flux required by this voltage balance. In any ac electromagnetic device in which the resistance and leakage-reactance voltage drops are small the resultant flux is very nearly determined by the applied voltage and frequency, and the currents must adjust themselves accordingly as to produce the mmf required to create this flux.

In a transformer the secondary current is determined by the voltage induced in the secondary, the secondary leakage impedance, and the electric load. In an induction motor, the secondary (rotor) current is determined by the voltage induced in the secondary, the secondary leakage impedance, and the mechanical load on its shaft. Essentially the same phenomena take place in the primary winding of the transformer and in the armature (stator) windings of induction and synchronous motors. In all three, the primary, or armature, current must adjust itself so that the combined mmf of all currents creates the flux required by the applied voltage.

Further examples of these basic similarities can be cited. Except for friction and windage, the losses in transformers and rotating machines are essentially the same. Tests for determining the losses and equivalent-circuit constants are essentially the same: an open-circuit, or no-load, test gives information regarding the excitation requirements and core losses (and friction and windage in rotating machines), while a short-circuit test together with dc resistance measurements gives information regarding leakage reactances and winding resistances. The handling of the effects of magnetic saturation is another example: in both transformers and ac rotating machines the leakage reactances are usually assumed to be unaffected by saturation, and the saturation of the main magnetic circuit is assumed to be determined by the resultant mutual or air-gap flux.

PROBLEMS

1-1 A simple magnetic circuit is shown in Fig. 1-33. Neglect magnetic leakage and fringing. Assuming that the iron permeability is infinite and that the

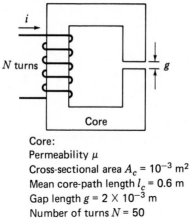

Core:
Permeability μ
Cross-sectional area $A_c = 10^{-3}$ m^2
Mean core-path length $l_c = 0.6$ m
Gap length $g = 2 \times 10^{-3}$ m
Number of turns $N = 50$

Fig. 1-33. Magnetic core for Prob. 1-1.

current $i = 1$ A, calculate (a) the total flux ϕ, (b) the flux linkages λ of the coil, and (c) the coil inductance L.

1-2 Repeat Prob. 1-1 for a finite iron permeability of $\mu = 2000\ \mu_0$.

1-3 A square voltage wave having a fundamental frequency of 60 Hz and equal positive and negative half cycles of amplitude E volts is impressed on a resistanceless winding of 1000 turns surrounding a closed iron core of 10^{-3} m^2 cross section.

    (a) Sketch curves of voltage and flux as functions of time.
    (b) Find the maximum permissible value of E if the maximum flux density is not to exceed 1.00 T.

1-4 A 250-turn coil is wound around a core of circular cross section 1.75 in^2 and mean path length of 6 in. This magnetic circuit also includes an air gap 0.05 in long.

    (a) Calculate the reluctance of the magnetic circuit if the core permeability is assumed to be infinite. What is the corresponding air-gap flux density for a coil current of 5 A?
    (b) Upon measurement it is found that for a current of 5 A the air-gap flux density is 0.95 T. What is the actual reluctance of the magnetic circuit and the core permeability?

1-5 A proposed energy-storage mechanism consists of a coil wound around the large toroidal form shown in Fig. 1-34. There are N turns, each of circular cross section of radius a. The radius of the toroid is r, as measured to the center of each circular turn. The geometry of this device is such that the magnetic field can be considered to be zero everywhere outside the toroid. Under the assumption that $a \ll r$, the H field inside the torus can be consid-

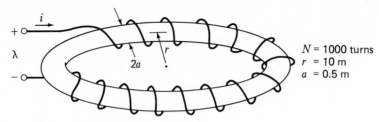

Fig. 1-34. Toroidal winding for Prob. 1-5.

ered to be directed around the torus and of uniform magnitude

$$H = \frac{Ni}{2\pi r}$$

(a) Calculate the coil inductance L.
(b) The coil is to be charged to a magnetic flux density of 1.5 T. Calculate the total stored magnetic energy in the torus when this flux density is achieved.
(c) If the coil is to be charged at a uniform rate, that is, $di/dt =$ constant, calculate the terminal voltage required to achieve the required flux density in 20 s. Assume the coil resistance to be negligible.

1-6 Figure 1-35 shows an inductor wound on a high-permeability laminated iron core of rectangular cross section. Assume that the permeability of the iron is infinite. Neglect magnetic leakage and fringing in the air gap g. The winding is insulated copper wire whose resistivity is ρ $\Omega \cdot$ m. Assume that the fraction f_w of the winding space is available for copper, the rest of the space being used for insulation.

(a) Estimate the mean length l of a turn of the winding.

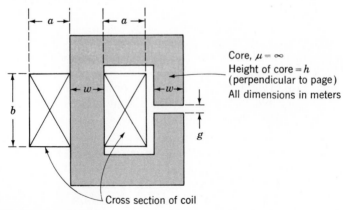

Fig. 1-35. Iron-core inductor.

(b) Derive an expression for the electric power input to the coil for a specified steady flux density B. This expression should be in terms of B, ρ, μ_0, l, f_w, and the given dimensions. Note that the expression is independent of the number of turns if the winding factor f_w is assumed to be independent of the turns.

(c) Derive an expression for the magnetic stored energy in terms of B and the given dimensions.

(d) From parts (b) and (c) derive an expression for the time constant L/R of the coil.

1-7 The inductor of Fig. 1-35 has the following dimensions:

$$a = h = w = 1 \text{ cm} \qquad b = 2 \text{ cm} \qquad g = 0.2 \text{ cm}$$

The winding factor is $f_w = 0.7$. The resistivity of copper is $1.73 \times 10^{-6} \, \Omega \cdot$ cm. The coil is to be operated with a constant applied voltage of 50 V and the air-gap flux density is to be 1.0 T. Find the power input to the coil, the coil current, the number of turns, the coil resistance, the inductance, the time constant, and the wire size to the nearest standard size.

1-8 From the dc magnetization curve of Fig. 1-7 it is possible to calculate the relative permeability $\mu_r = B_c/(\mu_0 H_c)$ for the core material as a function of flux level B_c. For the core of Fig. 1-2 of dimensions given in Example 1-1, calculate the maximum flux density such that the core-material reluctance never exceeds 5 percent of the reluctance of the total magnetic circuit.

1-9 The coils of the magnetic-circuit device shown in Fig. 1-36 are connected in series so that the mmf's of paths A and B both tend to set up flux in the

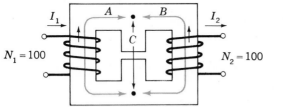

Cross section area of A and B legs = 2 in^2

Cross section area of C legs = 4 in^2

Length A path = 6 in
Length B path = 6 in
Length C path = 2 in
Air gap = 0.15 in

Fig. 1-36. Magnetic circuit for Prob. 1-9.

center leg C in the same direction. The material is M-5 grade, 0.012-in steel, stacking factor 0.94. Neglect fringing and leakage.

(a) How many amperes are required to set up a flux density of 0.6 T?

(b) How many joules of energy are stored in the magnetic field in the air gap?

1-10 Data for the top half of a symmetrical hysteresis loop for the Prob. 1-3 are:

B, T	0	0.2	0.4	0.6	0.7	0.8	0.9	1.0	0.95	0.9	0.8	0.7	0.6
H, A·turns/m	48	52	58	73	85	103	135	193	80	42	2	−18	−29

The mean length of the flux paths in the core is 0.30 m. Find graphically the hysteresis loss in watts for a maximum flux density of 1.00 T at a frequency of 60 Hz.

1-11 A potential transformer is made up of a 1000-turn primary coil and an open-circuited 80-turn secondary coil wound around a closed core of cross-sectional area 4.9 in^2. The core material can be said to saturate when the rms applied flux density reaches 1.2 T. What maximum 60-Hz rms primary voltage is possible without reaching this saturation level? What is the corresponding 50-Hz voltage?

1-12 A source which can be represented by a voltage source of 5 V rms in series with an internal resistance of 2000 Ω is connected to a 50-Ω load resistance through an ideal transformer. Plot the power in milliwatts supplied to the load as a function of the transformer ratio, covering ratios ranging from 0.1 to 10.0.

1-13 A single-phase transformer has a nameplate voltage rating of 7.97 kV/280 V, which is based upon its winding turns ratio. It has been determined that its primary (7.97-kV) leakage inductance is 168 mH and its primary magnetizing inductance is 110 H. For an applied primary voltage of 7970 V, calculate the resultant open-circuit secondary voltage.

1-14 The resistances and leakage reactances in ohms of a 10-kVA 60-Hz 2400:240-V distribution transformer are

$$R_1 = 4.20 \qquad R_2 = 0.0420 \qquad X_{l1} = 5.50 \qquad X_{l2} = 0.0550$$

where subscript 1 denotes the 2400-V winding and subscript 2 the 240-V winding. Each quantity is referred to its own side of the transformer.

   (a) Find the equivalent impedance referred to (1) the high- and (2) the low-voltage side.
   (b) Consider the transformer to deliver its rated kilovoltamperes at 0.80 power factor lagging to a load on the low-tension side with 240 V across the load. Find the high-tension terminal voltage.

1-15 A single-phase load is supplied through a 33,000-V feeder whose impedance is $105 + j360 \, \Omega$ and a 33,000:2400-V transformer whose equivalent impedance is $0.26 + j1.08 \, \Omega$ referred to its low-voltage side. The load is 180 kW at 0.85 leading power factor and 2250 V.

(a) Compute the voltage at the sending end of the feeder.
(b) Compute the voltage at the primary terminals of the transformer.
(c) Compute the power and reactive-power input at the sending end of the feeder.

1-16 An audio-frequency output transformer has a primary-to-secondary turns ratio of 31.6. Its primary inductance measured with the secondary open is 19.6 H and measured with the secondary short-circuited is 0.207 H. The winding resistances are negligible. This transformer is used to connect an 8-Ω resistance load to a source which can be represented by a variable-frequency internal emf in series with an internal impedance of 5000-Ω resistance. Compute the following that relate to the frequency characteristics of the circuit: (a) the upper half-power frequency, (b) the lower half-power frequency, (c) the geometric mean of these frequencies, (d) the ratio of load voltage to source voltage at the frequency of part (c).

1-17 The following data were obtained for a 20-kVA 60-Hz 2400:240-V distribution transformer tested at 60 Hz:

	Voltage, V	Current, A	Power, W
With high-voltage winding open-circuited	240	1.066	126.6
With low-voltage terminals short-circuited	57.5	8.34	284

(a) Compute the efficiency at full-load current and rated terminal voltage at 0.8 power factor.
(b) Assume that the load power factor is varied while the load current and secondary terminal voltage are held constant. Use a phasor diagram to determine the load power factor for which the regulation is greatest. What is this regulation?

1-18 A 500-kVA 60-Hz transformer with an 11,000-V primary winding takes 3.35 A and 2960 W at no load, rated voltage and frequency. Another transformer has a core with all its linear dimensions $\sqrt{2}$ times as large as the corresponding dimensions of the first transformer. Core material and lamination thickness are the same in both transformers. If the primary windings of both transformers have the same number of turns, what no-load current and power will the second transformer take with 22,000 V at 60 Hz impressed on its primary?

1-19 The high-voltage terminals of a 3-phase bank of three single-phase transformers are connected to a 3-wire 3-phase 13,800-V (line-to-line) system. The low-voltage terminals are connected to a 3-wire 3-phase substation load rated at 1500 kVA and 2300 V line to line. Specify the voltage, current, and kVA ratings of each transformer (both high- and low-voltage windings) for the following connections:

	High-voltage windings	Low-voltage windings
(a)	Y	Δ
(b)	Δ	Y
(c)	Y	Y
(d)	Δ	Δ

1-20 Three 750-kVA 7970/440-V single-phase transformers are to be connected in a 3-phase bank. Each transformer has a 60-Hz series reactance of 0.12 per unit divided equally between primary and secondary. The primaries are to be Y-connected directly to a 13.8-kV 60-Hz 3-phase source.

(a) Calculate the magnitude of the primary and secondary current in rms amperes if the secondaries are Y-connected and the secondary terminals are all shorted together.

(b) Repeat the calculation for a Δ-connected secondary. What is the magnitude of the terminal current to the short circuit?

1-21 A 15 kV:200-kV 150-MVA transformer has primary and secondary impedances of $0.01 + j0.05$ per unit each. The magnetizing impedance is $j120$ per unit. All quantities are in per unit on the transformer base. Calculate the primary and secondary resistances and inductances and the magnetizing inductance in ohms and henrys.

1-22 A Δ-Y-connected bank of three identical 100-kVA 2400:120-volt 60-Hz transformers is supplied with power through a feeder whose impedance is $0.80 + j0.30 \ \Omega$ per phase. The voltage at the sending end of the feeder is held constant at 2400 V line to line. The results of a single-phase short-circuit test on one of the transformers with its low-voltage terminals short-circuited are

$$V_H = 52.0 \text{ V} \qquad f = 60 \text{ Hz} \qquad I_H = 41.6 \text{ A} \qquad P = 950 \text{ W}$$

(a) Determine the secondary line-to-line voltage when the bank delivers rated current to a balanced 3-phase 1.00-power-factor load.

(b) Compute the currents in the transformer primary and secondary windings and in the feeder wires if a solid 3-phase short circuit occurs at the secondary line terminals.

1-23 A 480:120-V 5-kVA 2-winding transformer is to be used as an autotransformer to supply a 480-V circuit from a 600-V source. When tested as a 2-winding transformer at rated load, 0.80 power factor lagging, its efficiency is 0.965.

(a) Show a diagram of connections as an autotransformer.
(b) Determine its kVA rating as an autotransformer.
(c) Find its efficiency as an autotransformer at full load, 0.80 power factor lagging.

1-24 A 3-phase generator step-up transformer is rated 26 kV:345 kV, 1000 MVA and has a series impedance of $0.005 + j0.09$ per unit on this base. It is connected to a 26-kV 800-MVA generator which can be represented as a voltage source in series with a reactance of $j1.70$ per unit on the generator base.

(a) Convert the per unit generator reactance to the step-up transformer base.
(b) The unit is supplying 500 MW, unity power factor, 345 kV, to the system at the transformer terminals. Draw a phasor diagram for this condition, using the transformer high-side voltage as the reference phasor.
(c) Calculate the transformer low-side voltage and the generator internal voltage behind its reactance in kilovolts for the conditions of part (b). Find the generator output power in megawatts and the power factor.

Electromechanical
Energy-Conversion
Principles

We are concerned here with the electromechanical energy-conversion process, which takes place through the medium of the electric or magnetic field of the conversion device. Although the various conversion devices operate on similar principles, their structures depend upon their function. Devices for measurement and control are frequently referred to as *transducers*; they generally operate under linear input-output conditions and with relatively small signals. The many examples include torque motors, microphones, pickups, and loudspeakers. A second category of devices encompasses force-producing devices and includes solenoids, relays, and electromagnets. The third category includes continuous-energy-conversion equipment such as motors and generators.

This chapter is devoted to the principles of electromechanical energy conversion and the analysis of the devices which accomplish this function. Emphasis will be placed upon the analysis of systems which use magnetic fields as the conversion medium since the remaining chapters of the book deal with such devices. However, the analytical techniques for electric field systems are quite similar.

The purpose of this analysis is threefold: (1) to assist us in understanding how energy conversion takes place, (2) to provide us with the techniques for designing and optimizing the devices for specific requirements, and (3) to show how to develop models of electromechanical energy-conversion devices that can be used in analyzing their performance as components in engineering systems. Transducers and force-producing devices are treated in this chapter; continuous-energy-conversion devices are treated in the rest of the book.

The concepts and techniques presented in this chapter are quite powerful and can be applied to a wide range of engineering situations involving electro-mechanical energy conversion. However, they are somewhat mathematical in nature, perhaps more so than the reader may find comfortable in a first introduction to electric machinery. With this in mind, Arts. 2-1 and 2-2 present a quantitative discussion of the forces in electromechanical systems; the analytical details are developed in the remaining articles. Based upon the techniques developed in this chapter, torque expressions are developed as needed throughout the book. Thus the reader may elect to omit the later sections of this chapter at the expense of accepting these expressions on faith when they are presented.

2–1 FORCES AND TORQUES IN MAGNETIC FIELD SYSTEMS

The Lorentz force law

$$\mathbf{F} = q(\mathbf{E} + \mathbf{v} \times \mathbf{B}) \qquad (2\text{-}1)$$

gives the force $\mathbf{F}$ on a particle of charge q in the presence of electric and magnetic fields. In SI units, $\mathbf{F}$ is given in newtons, q in coulombs, $\mathbf{E}$ in volts per meter, $\mathbf{B}$ in teslas, and $\mathbf{v}$, which is the velocity of the particle relative to the magnetic field, in meters per second.

Thus in a pure electric field system, the force is determined simply by the charge on the particle and the electric field

$$\mathbf{F} = q\mathbf{E} \qquad (2\text{-}2)$$

The force acts in the direction of the electric field and is independent of the particle motion.

In magnetic field systems the situation is somewhat more complex. Here the force

$$\mathbf{F} = q(\mathbf{v} \times \mathbf{B}) \qquad (2\text{-}3)$$

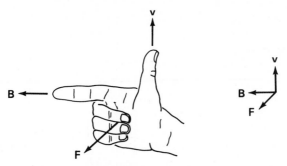

Fig. 2-1. Right-hand rule for determining the direction of the Lorentz force $\mathbf{F} = q\,\mathbf{v} \times \mathbf{B}$.

is determined by the magnitude of the charge on the particle and the magnitude of the $\mathbf{B}$ field as well as the velocity of the particle. In fact the direction of the force is always perpendicular to both the direction of particle motion and the magnetic field direction. Mathematically, this is given by the vector cross product $\mathbf{v} \times \mathbf{B}$ in Eq. 2-3. The magnitude of this cross product is equal to the product of the magnitudes of $\mathbf{v}$ and $\mathbf{B}$ and the sine of the angle between them; its direction can be found from the right-hand rule, which states that when the thumb of the right hand points in the direction of $\mathbf{v}$ and the index finger points in the direction of $\mathbf{B}$, the force points in the direction normal to the palm of the hand, as shown in Fig. 2-1.

For situations where large numbers of charged particles are in motion, it is convenient to rewrite Eq. 2-3 in terms of the current density $\mathbf{J}$, in which case the force is a force density

$$\mathbf{F} = \mathbf{J} \times \mathbf{B} \qquad \text{N/m}^3 \qquad\qquad (2\text{-}4)$$

For currents flowing in conducting media, Eq. 2-4 can be used to find the force density acting on the material itself. A considerable amount of physics is hidden in this result since the mechanism by which the force is transferred from the moving charges to the conducting medium is a complex one.

EXAMPLE 2-1

A nonmagnetic rotor containing a single-turn coil is placed in a uniform magnetic field of magnitude B_0, as shown in Fig. 2-2. The coil sides are at radius R, and the wire carries a current I as indicated. Find the θ-directed torque as a function of rotor position α when $I = 10\,\text{A}$, $B_0 = 0.5\,\text{T}$, and $R = 0.1\,\text{m}$. Assume that the rotor is 0.6 m long.

Solution

The force per unit length on a wire carrying a current I can be found by multiplying Eq. 2-4 by the cross-sectional area of the wire. When we recognize

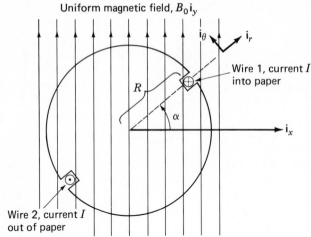

Fig. 2-2. Single-coil rotor, Example 2-1.

that the product of the cross-sectional area and the current density is simply the current **I**, the force per unit length acting on the wire is given by

$$\mathbf{F} = \mathbf{I} \times \mathbf{B}$$

Thus, for wire 1 carrying current I into the paper, the θ-directed force is given by

$$F_{1\theta} = IB_0 l \sin \alpha$$

and for wire 2 (which carries current in the opposite direction and is located 180° away from wire 1)

$$F_{2\theta} = IB_0 l \sin \alpha$$

where l is the length of the rotor. The torque T acting on the rotor is given by the sum of the force–moment-arm products for each wire

$$T = 2IB_0 Rl \sin \alpha = 2(10)(0.5)(0.1)(0.6)(\sin \alpha) = 0.6 \sin \alpha \qquad \text{N} \cdot \text{m}$$

For situations in which the forces act only upon current-carrying elements and which are of simple geometry, Eq. 2-4 is generally the simplest and easiest way to calculate the forces acting on the system. Very few practical situations fall into this class, however. In fact, as discussed in Chap. 1, most electromechanical energy-conversion devices contain magnetic material; in these systems forces act directly on the magnetic material and clearly cannot be calculated from Eq. 2-4.

Techniques for calculating the detailed, localized forces acting on magnetic materials are extremely complex and require detailed knowledge of the field distribution throughout the structure. Fortunately, most electromechanical energy-conversion devices are constructed of rigid, nondeforming structures. In these devices, it is the net force or torque that is of importance, and the details of the localized force distribution are of secondary interest. For example, in a properly designed motor the net accelerating torque acting on the rotor determines the motor's characteristics; accompanying forces, which act to squash or make the solid rotor oval, play no significant role and generally are not even calculated.

To understand the behavior of rotating machinery, a simple physical picture is quite useful. Associated with the rotor structure is a magnetic field, and similarly with the stator; one can picture them as a set of north and south poles associated with each structure. Just as a compass needle tries to align with the earth's magnetic field, these two sets of fields attempt to align, and torque is associated with their displacement from alignment. Thus in a motor, the stator magnetic field rotates ahead of that of the rotor, pulling on it and performing work. The opposite is true for a generator; here the rotor does the work on the stator.

Various techniques have evolved to calculate the net forces of concern in electromechanical energy conversion. The technique developed in this chapter and used throughout the book is known as the *energy method* and is based upon the principle of *conservation of energy*. The basis for this method can be understood with reference to Fig. 2-3a, where a magnetic field–based electromechanical energy-conversion device is indicated schematically as a lossless

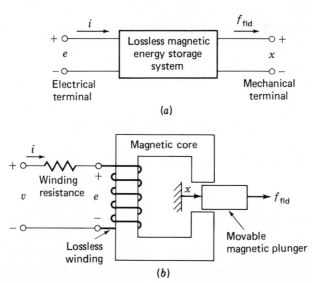

Fig. 2-3. (a) Schematic magnetic field electromechanical energy-conversion device; (b) simple force-producing device.

magnetic-energy storage system with two terminals. The electric terminal has the terminal variables voltage e and current i, and the mechanical terminal has the terminal variables force f_{fld} and position x. This sort of representation is valid in situations where the loss mechanism can be separated from the energy-storage mechanism; in these cases the electrical losses such as ohmic losses can be included as external elements connected to the electric terminals and the mechanical losses such as friction can be included external to the mechanical terminals. Figure 2-3b shows an example of such a system; a simple force-producing device with a single coil forms the electric terminal, and a movable plunger serves as the mechanical terminal.

The interaction between the electric and mechanical terminals, i.e., the electromechanical energy conversion, occurs through the medium of the magnetic stored energy. Since the energy-storage system is lossless, it is a simple matter to write that the time rate of change of W_{fld}, the stored energy in the magnetic field, is equal to the electric power input less the mechanical power output of the energy-storage system

$$\frac{dW_{fld}}{dt} = ei - f_{fld}\frac{dx}{dt} \tag{2-5}$$

Using Eq. 1-17 and multiplying Eq. 2-5 by dt, we get

$$dW_{fld} = i\,d\lambda - f_{fld}\,dx \tag{2-6}$$

As shown in Art. 2-4, Eq. 2-6 permits us to solve for the force simply as a function of the flux λ and the mechanical-terminal position x. This result comes about as a consequence of our ability to separate the losses out of the physical problem, resulting in a lossless energy-storage element, as in Fig. 2-3a.

Equations 2-5 and 2-6 form the basis for the energy method. This technique is quite powerful in its ability to calculate forces and torques in complex situations of electromechanical energy-conversion systems. The reader should recognize that this power comes at the expense of a detailed picture of the force-producing mechanism. The forces themselves are produced by such well-known physical phenomena as the Lorentz force on current-carrying elements, described by Eq. 2-4, or the interaction of the magnetic fields with the dipoles in the magnetic material.

2–2 ENERGY BALANCE

The principle of conservation of energy states that energy is neither created nor destroyed; it is merely changed in form. For example, a golf ball leaves the tee with a certain amount of kinetic energy; this energy is eventually dissipated as heat due to air friction or rolling friction by the time the ball comes to rest on the fairway (or with luck, the green). Similarly, the kinetic energy of a hammer is eventually dissipated as heat as a nail is driven into a piece of wood. For isolated systems with clearly identifiable boundaries, this fact permits us to keep track of energy in a simple fashion; the net flow of energy into the system across its boundary is equal to the time rate of change of energy stored in the system.

Although this result, which is a statement of the first law of thermodynamics, is quite general, we shall apply it in this chapter to electromechanical systems whose predominant electric energy-storage mechanism is in magnetic fields. In such systems, one can account for energy transfer as

$$
\begin{pmatrix} \text{Energy input} \\ \text{from electric} \\ \text{sources} \end{pmatrix} = \begin{pmatrix} \text{mechanical} \\ \text{energy} \\ \text{output} \end{pmatrix} + \begin{pmatrix} \text{increase in energy} \\ \text{stored in} \\ \text{magnetic field} \end{pmatrix}
$$

$$
+ \begin{pmatrix} \text{energy} \\ \text{converted} \\ \text{into heat} \end{pmatrix} \quad (2\text{-}7)
$$

Equation 2-7 is written so that the electric- and mechanical-energy terms have positive values for motor action. The equation applies equally well to generator action: these terms then simply have negative values.

In these systems the conversion of energy into heat occurs by such mechanisms as ohmic heating due to current flow in the windings of the electric terminals and mechanical friction due to motion of the system components forming the mechanical terminals. As discussed in Art. 2-1, it is generally possible mathematically to separate these loss mechanisms from the energy-storage mechanism. In this case, the device can be represented as a lossless magnetic energy-storage system with electric and mechanical terminals, as shown in Fig. 2-4. The loss mechanisms can then be represented by external elements connected to these terminals, resistances to the electric terminals and mechanical dashpots to the mechanical terminals. Figure 2-4 is quite general in the sense that there is no limit to the number of electric or mechanical terminals. For this type of system the magnetic field serves as the coupling medium between the electric and mechanical terminals.

The ability to identify a lossless energy-storage system is the essence of the energy method. It is important to recognize that this is done mathematically as part of the modeling process. It is not possible, of course, to take the

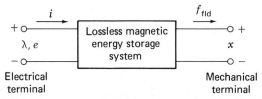

Fig. 2-4. Lossless magnetic energy-storage system with electric and mechanical terminals.

resistance out of windings or the friction out of bearings. Instead we are making use of the fact that a model in which this is done is a valid representation of the physical system.

For the lossless magnetic energy-storage system of Fig. 2-4, Eq. 2-7 becomes

$$dW_{elec} = dW_{mech} + dW_{fld} \qquad (2\text{-}8)$$

where dW_{elec} = differential electric-energy input
dW_{mech} = differential mechanical-energy output
dW_{fld} = differential change in magnetic stored energy

In time dt, from Eq. 1-32,

$$dW_{elec} = ei\, dt \qquad (2\text{-}9)$$

Here e is the voltage induced in the electric terminals by the changing magnetic stored energy. It is through this reaction voltage that the external electric circuit supplies power to the coupling magnetic field and hence to the mechanical output terminals. Thus the basic energy-conversion process is one involving the coupling field and its action and reaction on the electric and mechanical systems.

Combining Eqs. 2-8 and 2-9 results in

$$dW_{elec} = ei\, dt = dW_{mech} + dW_{fld} \qquad (2\text{-}10)$$

Equation 2-10, together with Faraday's law for induced voltage, Eq. 1-17, forms the basis for the energy method; the following sections illustrate its use in the analysis of electromagnetic energy-conversion devices.

2–3 ENERGY AND FORCE
IN SINGLY EXCITED MAGNETIC FIELD SYSTEMS

In Chap. 1 we were concerned primarily with the closed-core magnetic circuits used for transformers. Energy in those devices is stored in the leakage fields and to some extent in the core itself; the stored energy does not enter directly into the transformation process. In this chapter we are dealing with energy conversion; the magnetic circuits have air gaps between the stationary and moving members in which considerable energy is stored in the magnetic field. This field acts as the energy-conversion medium, and its energy is the reservoir between the electric and mechanical systems.

Consider the electromagnetic relay shown schematically in Fig. 2-5. The resistance of the excitation coil is shown as an external resistance R, and the mechanical-terminal variables are shown as a force f_{fld} produced by the magnetic field directed from the relay to the external mechanical system and a

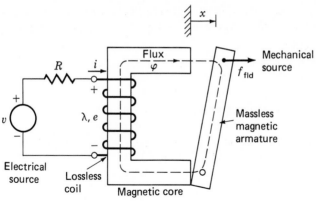

Fig. 2-5. Schematic electromagnetic relay.

displacement x; mechanical losses can be included as external elements connected to the mechanical terminal. Similarly, the moving armature is shown as being massless; its mass represents mechanical-energy storage and can be included as an external mass connected to the mechanical terminal. As a result, the magnetic core and armature constitute a lossless magnetic energy system, as in Fig. 2-4.

This relay structure is essentially the same as the magnetic structures analyzed in Chap. 1. From Faraday's law (Eq. 1-17) we can solve for the induced voltage e in terms of the flux linkages λ

$$e = \frac{d\lambda}{dt} \tag{2-11}$$

and thus from Eq. 2-9

$$dW_{\text{elec}} = i \, d\lambda \tag{2-12}$$

In Chap. 1 we saw that the magnetic circuit of Fig. 2-5 can be described by an inductance L which is a function of the geometry of the magnetic structure and the permeability of the magnetic material. Electromagnetic energy-conversion devices contain air gaps in their magnetic circuits to separate the moving parts. As discussed in Art. 1-1, in most such cases the reluctance of the air gap is much larger than that of the magnetic material. Thus the predominant energy storage occurs in the air gap, and the properties of the magnetic circuit are determined by the dimensions of the air gap.

Because of the simplicity of the resulting relations, magnetic nonlinearity and core losses are often neglected in the analysis of practical devices. The final results of such approximate analyses can, if necessary, be corrected for the effects of these neglected factors by semiempirical methods. Consequently, analyses are carried out under the assumption that the flux and mmf are directly proportional, as in air, for the entire magnetic circuit. Thus the flux linkages λ and current i are considered to be linearly related by a factor which depends solely upon the geometry and hence on the armature position x

$$\lambda = L(x)i \qquad (2\text{-}13)$$

where the explicit dependence of L on x has been indicated.

Since the magnetic force f_{fld} has been defined as acting from the relay upon the external mechanical system and dW_{mech} is defined as the mechanical energy output of the relay, we can write

$$dW_{\text{mech}} = f_{\text{fld}}\, dx \qquad (2\text{-}14)$$

Thus, using Eqs. 2-12 and 2-14, we can write Eq. 2-8 as

$$dW_{\text{fld}} = i\, d\lambda - f_{\text{fld}}\, dx \qquad (2\text{-}15)$$

Since our magnetic energy-storage system is lossless, it is a *conservative system* and the value of W_{fld} is uniquely specified by the values of λ and x. Thus, W_{fld} *is the same independent of how λ and x are brought to their final values.* Consider Fig. 2-6, in which two separate paths are shown over which Eq. 2-15 can be integrated to find W_{fld} at the point (λ_0, x_0). Path 1 is the general case and is difficult to integrate unless both i and f_{fld} are known explicitly as a function of λ and x. However, path 2 gives the same result and is much easier to integrate. From Eq. 2-15

$$W_{\text{fld}}(\lambda_0, x_0) = \int_{\text{path } 2a} dW_{\text{fld}} + \int_{\text{path } 2b} dW_{\text{fld}} \qquad (2\text{-}16)$$

Notice that on path 2a, $d\lambda = 0$ and $f_{\text{fld}} = 0$ since $\lambda = 0$. Thus from Eq. 2-15, $dW_{\text{fld}} = 0$ on path 2a. On path 2b, $dx = 0$, and thus, using Eq. 2-15, Eq. 2-16

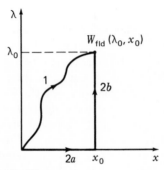

Fig. 2-6. Integration paths for W_{fld}.

reduces to

$$W_{fld}(\lambda_0, x_0) = \int_0^{\lambda_0} i(\lambda, x_0) \, d\lambda \qquad (2\text{-}17)$$

For a linear system in which λ is proportional to i, as in Eq. 2-13, Eq. 2-17 gives

$$W_{fld}(\lambda_0, x_0) = \int_0^{\lambda_0} i(\lambda, x_0) \, d\lambda = \int_0^{\lambda_0} \frac{\lambda}{L(x_0)} \, d\lambda = \frac{1}{2} \frac{\lambda_0^2}{L(x_0)} \qquad (2\text{-}18)$$

The energy can also be expressed in terms of the energy density of the magnetic field integrated over the volume V of the magnetic field. In this case

$$W_{fld} = \int_V \tfrac{1}{2}(\mathbf{H} \cdot \mathbf{B}) \, dV \qquad (2\text{-}19)$$

For a magnetic medium with constant permeability, this reduces to

$$W_{fld} = \int_V \left(\frac{1}{2} \frac{B^2}{\mu} \right) dV \qquad (2\text{-}20)$$

EXAMPLE 2-2

The relay shown in Fig. 2-7a is made from infinitely permeable magnetic material with a movable plunger, also of infinitely permeable material. The

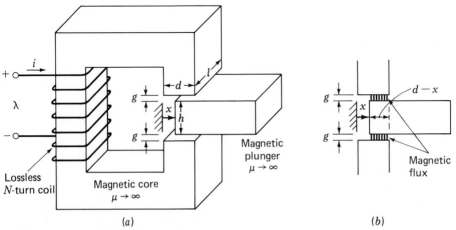

Fig. 2-7. (a) Relay with movable plunger; (b) detail showing air-gap configuration with plunger partially removed.

height of the plunger is much greater than the air-gap length ($h \gg g$). Calculate the magnetic stored energy W_{fld} as a function of plunger position ($0 < x < d$) for $N = 1000$ turns, $g = 0.002$ m, $d = 0.15$ m, $l = 0.1$ m, and $i = 10$ A.

Solution

Equation 2-18 can be used to solve for W_{fld} when λ is known. For this situation, i is held constant, and thus it would be useful to have an expression for W_{fld} as a function of i and x. This can be obtained quite simply by substituting Eq. 2-13 into Eq. 2-18

$$W_{\text{fld}} = \tfrac{1}{2} L(x) i^2$$

The inductance is given by (see Example 1-3)

$$L(x) = \frac{N^2 \mu_0 A_{\text{gap}}}{2g}$$

where A_{gap} is the gap cross-sectional area. From Fig. 2-7b it can be seen to be

$$A_{\text{gap}} = l(d - x) = ld\left(1 - \frac{x}{d}\right)$$

Thus

$$L(x) = \frac{N^2 \mu_0 ld(1 - x/d)}{2g}$$

and

$$W_{\text{fld}} = \frac{1}{2} \frac{N^2 \mu_0 ld(1 - x/d)}{2g} i^2$$

$$= \frac{1}{2} \frac{(1000)^2 (4\pi \times 10^{-7})(0.15)(0.1)}{2(0.002)} \times 10^2 \left(1 - \frac{x}{d}\right)$$

$$= 236\left(1 - \frac{x}{d}\right) \quad \text{J}$$

In this article we have seen the relationship between the magnetic stored energy and the electric- and mechanical-terminal variables for a system which can be represented in terms of a lossless magnetic energy-storage element. If we had chosen for our example a device with a rotating mechanical terminal instead of a linearly displacing one, the results would have been identical except that force is replaced by torque and linear displacement by angular displacement. In the next article we see how knowledge of the magnetic stored energy permits us to solve for the mechanical force.

2–4 DETERMINATION OF MAGNETIC FORCE; COENERGY

As discussed in the previous article, the magnetic stored energy W_{fld} is determined uniquely by λ and x. The lossless magnetic energy-storage system is a conservative system, and W_{fld} is a *state function,* determined uniquely by the values of the independent *state variables* λ and x. This can be shown explicitly by writing Eq. 2-15 in the form

$$dW_{fld}(\lambda, x) = i\, d\lambda - f_{fld}\, dx \qquad (2\text{-}21)$$

For any function $F(x_1, x_2)$ the total differential of F with respect to the two independent state variables x_1 and x_2 can be written

$$dF(x_1, x_2) = \frac{\partial F}{\partial x_1}\, dx_1 + \frac{\partial F}{\partial x_2}\, dx_2 \qquad (2\text{-}22)$$

It is extremely important to recognize that the partial derivatives in Eq. 2-22 are each taken holding the opposite state variable constant.

Since Eq. 2-22 is valid for any state function F, it is certainly valid for W_{fld}; thus we can write

$$dW_{fld}(\lambda, x) = \frac{\partial W_{fld}}{\partial \lambda}\, d\lambda + \frac{\partial W_{fld}}{\partial x}\, dx \qquad (2\text{-}23)$$

Since λ and x are independent variables, Eqs. 2-21 and 2-23 must be equal for all values of $d\lambda$ and dx and thus we see that

$$i = \frac{\partial W_{fld}(\lambda, x)}{\partial \lambda} \qquad (2\text{-}24)$$

where the partial derivative is taken holding x constant and

$$f_{fld} = -\frac{\partial W_{fld}(\lambda, x)}{\partial x} \qquad (2\text{-}25)$$

in this case holding λ constant while taking the partial derivative.

This is the result we have been seeking. Once we know W_{fld} as a function of λ and x, Eq. 2-24 can be used to solve for $i(\lambda, x)$. More importantly, Eq. 2-25 can be used to solve for the mechanical force $f_{fld}(\lambda, x)$. It cannot be overemphasized that the partial derivative of Eq. 2-25 is taken holding the flux linkages λ constant. This is easily done provided W_{fld} is known as a function of λ and x. This is purely a mathematical requirement and has nothing to do with whether or not λ is held fixed in the actual device.

The force f_{fld} is determined from Eq. 2-25 directly in terms of the electrical state variable λ. If we then want to express the force as a function of i, we can do so by substituting the appropriate expression for λ as a function of i in Eq. 2-25. Alternatively, one can use a state function other than energy, namely, the *coenergy,* to obtain the force directly as a function of current. The selection of the state function is purely a matter of convenience; they both give the same result, but one or the other may be simpler analytically, depending upon the desired result and the initial description of the system being analyzed.

The coenergy W'_{fld} is defined as a function of i and x such that

$$W'_{\mathrm{fld}}(i, x) = i\lambda - W_{\mathrm{fld}}(\lambda, x) \tag{2-26}$$

and the desired result can be obtained from the energy expression of Eq. 2-21. The transformation is carried out by using the differential of $i\lambda$

$$d(i\lambda) = i\,d\lambda + \lambda\,di \tag{2-27}$$

and the differential of $dW_{\mathrm{fld}}(\lambda, x)$ from Eq. 2-21. From Eq. 2-26

$$dW'_{\mathrm{fld}}(i, x) = d(i\lambda) - dW_{\mathrm{fld}}(\lambda, x) \tag{2-28}$$

Substitution of Eqs. 2-21 and 2-27 into Eq. 2-28 results in

$$dW'_{\mathrm{fld}}(i, x) = \lambda\,di + f_{\mathrm{fld}}\,dx \tag{2-29}$$

Coenergy $W'_{\mathrm{fld}}(i, x)$ is a state function of the two independent variables i and x. Thus, its differential can be expressed as

$$dW'(i, x) = \frac{\partial W'_{\mathrm{fld}}}{\partial i}\,di + \frac{\partial W'_{\mathrm{fld}}}{\partial x}\,dx \tag{2-30}$$

Equations 2-29 and 2-30 must be equal for all values of di and dx; thus

$$\lambda = \frac{\partial W'_{\mathrm{fld}}(i, x)}{\partial i} \tag{2-31}$$

$$f_{\mathrm{fld}} = \frac{\partial W'_{\mathrm{fld}}(i, x)}{\partial x} \tag{2-32}$$

Equation 2-32 gives the mechanical force directly in terms of i and x. The reader is cautioned that the partial derivative in Eq. 2-32 is taken holding i constant; thus W'_{fld} must be known as a function of i and x. Equations 2-25 and 2-32 are equivalent and can both be used to calculate the force for a given system; user preference and convenience often dictate the choice. As discussed, for a system with a rotating mechanical terminal, similar expressions for the mechanical torque are derived and the partial derivatives are taken with respect to angular displacement.

By arguments directly analogous to those leading to Eq. 2-17 the co-energy for any given values of i and x can be found

$$W'_{fld}(i_0, x_0) = \int_0^{i_0} \lambda(i, x_0) \, di \tag{2-33}$$

For a linear system in which λ and i are proportional and which can be described by a position-dependent inductance, as in Eq. 2-13, the coenergy is

$$W'_{fld}(i, x) = \int_0^i L(x)i' \, di' = \tfrac{1}{2} L(x)i^2 \tag{2-34}$$

By analogy with Eq. 2-20, the coenergy can be expressed in field terms for a medium with constant permeability μ

$$W'_{fld} = \int_V \tfrac{1}{2} \mu H^2 \, dV \tag{2-35}$$

EXAMPLE 2-3

For the relay of Example 2-2, find the force on the plunger as a function of x when the coil current is held constant at 10 A. Use both the energy and coenergy expressions.

Solution

From Example 2-2

$$L(x) = \frac{N^2 \mu_0 l d (1 - x/d)}{2g}$$

The energy, expressed as a function of λ and x is, from Eq. 2-18,

$$W_{fld} = \frac{1}{2} \frac{\lambda^2}{L(x)} = \frac{g\lambda^2}{N^2 \mu_0 l d (1 - x/d)}$$

and from Eq. 2-25

$$f_{fld} = -\frac{\partial W_{fld}}{\partial x} = -\frac{g\lambda^2}{N^2 \mu_0 l d^2 (1 - x/d)^2}$$

This can now be expressed in terms of i using Eq. 2-13

$$f_{\text{fld}} = -\frac{gL^2(x)i^2}{N^2\mu_0 ld^2(1-x/d)^2} = -\frac{N^2\mu_0 l}{4g}i^2$$

and we see that for constant current the force is independent of the plunger position and acts to pull the plunger back into the relay.

Alternately, the force can be found from the coenergy. From Eq. 2-34

$$W'_{\text{fld}} = \tfrac{1}{2}L(x)i^2 = \frac{N^2\mu_0 ld(1-x/d)}{4g}i^2$$

and from Eq. 2-32

$$f_{\text{fld}} = \frac{\partial W'_{\text{fld}}}{\partial x} = -\frac{N^2\mu_0 l}{4g}i^2$$

Notice that this gives the same answer, but because of the form of the inductance expression and specifically because we want a solution in terms of current, the choice of coenergy yielded a much simpler solution. Notice also that if one attempts to substitute $\lambda = L(x)i$ in the expression for W'_{fld} before the derivative is taken, one gets the wrong answer (by a minus sign). This is because it is λ and not i that must be held constant when the partial derivative of Eq. 2-25 is taken.

Finally, for $i = 10$ A

$$f_{\text{fld}} = -\frac{N^2\mu_0 l}{4g}i^2 = -\frac{(1000)^2(4\pi \times 10^{-7})(0.1)(10^2)}{4(0.002)} = -1570 \text{ N}$$

For a linear system the energy and coenergy are numerically equal; for example, $\tfrac{1}{2}Li^2 = \tfrac{1}{2}\lambda^2/L$ or $(\mu/2)H^2 = (1/2\mu)B^2$. For a nonlinear system in which λ and i or B and H are not proportional, the two functions are not even numerically equal. A graphical interpretation of the energy and coenergy for a nonlinear system is shown in Fig. 2-8. The area between the λ-i curve and the vertical axis given by the integral of $i\,d\lambda$ is the energy. The area to the horizontal axis given by the integral of $\lambda\,di$ is the coenergy. The sum for the singly excited system is, by definition

$$W_{\text{fld}} + W'_{\text{fld}} = \lambda i \qquad (2\text{-}36)$$

The force of the field in the device of Fig. 2-5 for some particular value of x and i or λ must be independent of whether it is calculated from the energy or coenergy. A graphical illustration will demonstrate this point. Assume that

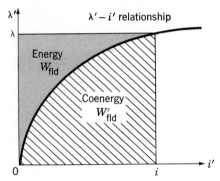

Fig. 2-8. Graphical interpretation of energy and coenergy in a singly excited system.

the armature is at position x so that the device is operating at point a in Fig. 2-9a. The partial derivative of Eq. 2-25 can be interpreted as the limit of $-\Delta W_{\text{fld}}/\Delta x$ with λ constant as $\Delta x \to 0$. If we allow a change Δx, the change $-\Delta W_{\text{fld}}$ is shown by the shaded area in Fig. 2-9a. Hence, the force $f_{\text{fld}} = (\text{shaded area})/\Delta x$ as $\Delta x \to 0$. On the other hand, the partial derivative of Eq. 2-32 can be interpreted as the limit of $\Delta W'_{\text{fld}}/\Delta x$ with i constant as $\Delta x \to 0$. This perturbation of the device is shown in Fig. 2-9b; the force $f_{\text{fld}} = (\text{shaded area})/\Delta x$ as $\Delta x \to 0$. The shaded areas only differ by the small triangle abc of sides Δi and $\Delta \lambda$, so that in the limit the shaded areas resulting from Δx at constant λ or at constant i are equal. Thus the force of the field is independent of whether the determination is made with energy or coenergy.

Equations 2-25 and 2-23 express the mechanical force of electrical origin in terms of partial derivatives of the energy and coenergy functions $W_{\text{fld}}(\lambda, x)$ and $W'_{\text{fld}}(i, x)$. It is important to note two things about them: (1) the variables

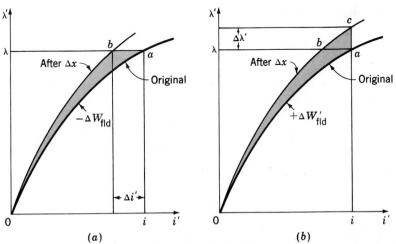

Fig. 2-9. Effect of Δx on energy and coenergy of singly excited device: (a) change of energy, λ held constant; (b) change of coenergy, i held constant.

in terms of which they must be expressed and (2) their algebraic signs. Physically, of course, the force depends on the dimension x and the magnetic field. The field can be specified in terms of flux linkage λ, or current i, or related variables. As seen in Example 2-3, the selection of the energy or coenergy function as a basis for analysis is a matter of convenience; the choice depends upon the initial description of the system and the desired variables in the result.

The algebraic signs in Eqs. 2-25 and 2-32 show that the field force acts in a direction to decrease the magnetic field stored energy at constant flux or to increase the coenergy at constant current. In a singly excited device the force acts to increase the inductance by pulling on members so as to reduce the reluctance of the magnetic path linking the winding.

EXAMPLE 2-4

The magnetic circuit shown in Fig. 2-10 is made of cast steel. The rotor is free to turn about a vertical axis. The dimensions are shown in the figure.

(a) Derive an expression in SI units for the torque acting on the rotor in terms of the dimensions and the magnetic field in the two air gaps. Neglect the effects of fringing.

(b) The maximum flux density in the overlapping portions of the air gaps is limited to approximately 130 kilolines/in^2, because of saturation in the steel. Compute the maximum torque in inch-pounds for $r_1 = 1.00$ in, $h = 1.00$ in, and $g = 0.10$ in.

Solution

The torque T can be obtained from the derivative of field coenergy with respect to angle θ, as in Eq. 2-32.

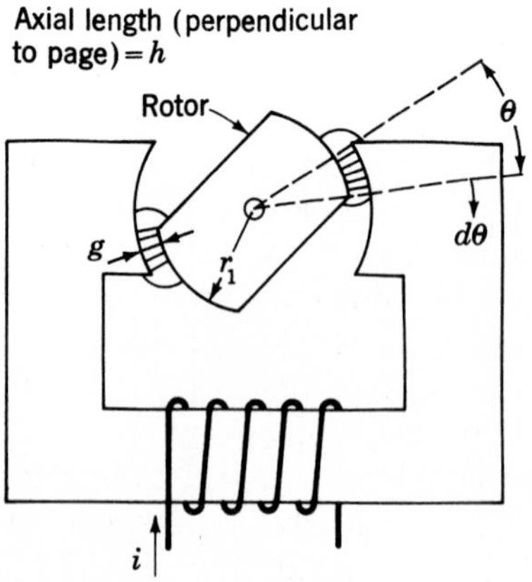

Fig. 2-10. Magnetic system, Example 2-4.

(a) The air-gap field intensity H_{ag} will be used as the independent variable because it is related to the terminal variable i by a constant; that is, $Ni = 2gH_{ag}$. The field coenergy density is $\mu_0 H_{ag}^2/2$ (Eq. 2-35), and the volume of the two overlapping air gaps is $2gh(r_1 + 0.5g)\theta$. Consequently, the field coenergy is

$$W'_{ag} = \mu_0 H_{ag}^2 gh(r_1 + 0.5g)\theta$$

and

$$T_{fld} = \frac{\partial W'_{ag}(H_{ag}, \theta)}{\partial \theta} = \mu_0 H_{ag}^2 gh(r_1 + 0.5g) = \frac{B_{ag}^2 gh(r_1 + 0.5g)}{\mu_0}$$

The torque acts in a direction to align the rotor with the stator pole faces.

(b) Convert the flux density and dimensions to SI units.

$$B_{ag} = \frac{130,000}{6.45} \times 10^4 \times 10^{-8} = 2.02 \text{ T}$$

$$g = 0.1(2.54 \times 10^{-2}) = 0.00254 \text{ m}$$

$$h = r_1 = 1.00(2.54 \times 10^{-2}) = 0.0254 \text{ m}$$

$$\mu_0 = 4\pi \times 10^{-7}$$

Substitution of these numerical values gives

$$T = 5.56 \text{ N} \cdot \text{m}$$
$$= 5.56(0.738)(12) = 49.3 \text{ in} \cdot \text{lb}$$

Moving-iron devices are used in a wide variety of applications for producing mechanical force or torque. Some of them, such as lifting magnets and magnetic chucks, are required merely to hold a piece of ferromagnetic material; others, such as iron-core solenoids, relays, and contactors, are required to exert a force through a specified distance; still others, such as iron-vane instruments, are required to produce a rotational torque against a restraining spring so that the deflection of a pointer is indicative of the steady-state value of the current or voltage of the circuit to which they are connected; with others, such as moving-iron telephone receivers and the electromagnets used for controlling the operation of hydraulic motors in servomechanisms, the force or torque should be very nearly proportional to an electric signal, and the dynamic response should be as rapid as possible. The dynamics of electromagnetically coupled systems are discussed in Art. 2-6.

2–5 MULTIPLY EXCITED MAGNETIC FIELD SYSTEMS

Many electromechanical devices have multiple sets of electric terminals. In measurement systems it is often desirable to obtain torques proportional to two electric signals; power as the product of voltage and current is a common example. Similarly, most electromechanical energy-conversion devices consist of multiply excited magnetic field systems.

Analysis of these systems follows directly from the techniques discussed in the previous articles. This article will illustrate these techniques based upon a system with two electric terminals. The model of a simple system with two sets of electric terminals and one mechanical terminal is shown in Fig. 2-11; in this case there is a rotating mechanical terminal with terminal variables torque T_{fld} and angular displacement θ. Since there are three terminals, the system must be described in terms of three independent variables; these can be the mechanical angle θ along with the flux linkages λ_1 and λ_2, the currents i_1 and i_2, or a hybrid set including one current and one flux.†

When the fluxes are used, the differential energy function $dW_{\text{fld}}(\lambda_1, \lambda_2, \theta)$ corresponding to Eq. 2-21 is

$$dW_{\text{fld}}(\lambda_1, \lambda_2, \theta) = i_1\,d\lambda_1 + i_2\,d\lambda_2 - T_{\text{fld}}\,d\theta \qquad (2\text{-}37)$$

and in direct analog to Eqs. 2-24 and 2-25

$$i_1 = \frac{\partial W_{\text{fld}}(\lambda_1, \lambda_2, \theta)}{\partial \lambda_1} \qquad (2\text{-}38)$$

$$i_2 = \frac{\partial W_{\text{fld}}(\lambda_1, \lambda_2, \theta)}{\partial \lambda_2} \qquad (2\text{-}39)$$

and
$$T_{\text{fld}} = -\frac{\partial W_{\text{fld}}(\lambda_1, \lambda_2, \theta)}{\partial \theta} \qquad (2\text{-}40)$$

The energy W_{fld} can be found by integrating Eq. 2-37. As in the singly excited case, this is most conveniently done by holding λ_1 and λ_2 fixed at

†See, for example, H. H. Woodson and J. R. Melcher, "Electromechanical Dynamics," pt I., chap. 3, Wiley New York, 1968.

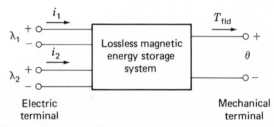

Fig. 2-11. Multiply excited magnetic energy-storage system.

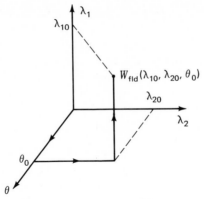

Fig. 2-12. Integration path to obtain $W_{\text{fld}}(\lambda_{10}, \lambda_{20}, \theta_0)$.

zero and integrating first over θ; T_{fld} is zero, and thus this integral is zero. One can then integrate over λ_2 (holding λ_1 zero) and finally over λ_1. Thus $W_{\text{fld}}(\lambda_{1_0}, \lambda_{2_0}, \theta_0)$ can be found as

$$W_{\text{fld}}(\lambda_{1_0}, \lambda_{2_0}, \theta_0) = \int_0^{\lambda_{2_0}} i_2(\lambda_1 = 0, \lambda_2, \theta = \theta_0)\, d\lambda_2 +$$

$$\int_0^{\lambda_{1_0}} i_1(\lambda_1, \lambda_2 = \lambda_{2_0}, \theta = \theta_0)\, d\lambda_1 \quad (2\text{-}41)$$

The path of integration is illustrated in Fig. 2-12 and is directly analogous to that of Fig. 2-6. One could of course interchange the order of integration for λ_2 and λ_1. It is extremely important to recognize, however, that the state variables are integrated over a specific path and that, for example, λ_1 is maintained at zero while integrating over λ_2 in Eq. 2-41. This is explicitly indicated in Eq. 2-41 and can also be seen from Fig. 2-12. Failure to observe this fact is one of the most common errors made in analyzing these systems.

In a linear system the relationships between λ and i are specified in terms of inductances

$$\lambda_1 = L_{11} i_1 + L_{12} i_2 \qquad (2\text{-}42)$$

$$\lambda_2 = L_{21} i_1 + L_{22} i_2 \qquad (2\text{-}43)$$

and
$$L_{12} = L_{21} \qquad (2\text{-}44)$$

Here the inductances are in general functions of angular position θ. These equations can be inverted to obtain expressions for the i's as a function of the θ's

$$i_1 = \frac{L_{22}\lambda_1 - L_{12}\lambda_2}{D} \qquad (2\text{-}45)$$

$$i_2 = \frac{-L_{21}\lambda_1 + L_{11}\lambda_2}{D} \tag{2-46}$$

where

$$D = L_{11}L_{22} - L_{12}L_{21} \tag{2-47}$$

The energy can be found from Eq. 2-41

$$W_{\text{fld}}(\lambda_{1_0}, \lambda_{2_0}, \theta_0) = \int_0^{\lambda_{2_0}} \frac{L_{11}\lambda_2}{D} \, d\lambda_2 + \int_0^{\lambda_{1_0}} \frac{L_{22}\lambda_1 - L_{12}\lambda_{2_0}}{D} \, d\lambda_1$$

$$= \frac{1}{2D} L_{11}\lambda_{2_0}^2 + \frac{1}{2D} L_{22}\lambda_{1_0}^2 - \frac{L_{12}}{D}\lambda_{1_0}\lambda_{2_0} \tag{2-48}$$

Similarly, when the currents are used to describe the system, the differential of the coenergy is

$$d\, W'_{\text{fld}}(i_1, i_2, \theta) = \lambda_1 \, di_1 + \lambda_2 \, di_2 + T_{\text{fld}} \, d\theta \tag{2-49}$$

and

$$\lambda_1 = \frac{\partial W'_{\text{fld}}(i_1, i_2, \theta)}{\partial i_1} \tag{2-50}$$

$$\lambda_2 = \frac{\partial W'_{\text{fld}}(i_1, i_2, \theta)}{\partial i_2} \tag{2-51}$$

$$T_{\text{fld}} = + \frac{\partial W'_{\text{fld}}(i_1, i_2, \theta)}{\partial \theta} \tag{2-52}$$

Analogous to Eq. 2-41, the coenergy can be found

$$W'_{\text{fld}}(i_{1_0}, i_{2_0}, \theta_0) = \int_0^{i_{2_0}} \lambda_2(i_1 = 0, i_2, \theta_0) \, di_2 + \int_0^{i_{1_0}} \lambda_1(i_1, i_2 = i_{2_0}, \theta_0) \, di_1 \tag{2-53}$$

For the linear system of Eqs. 2-42 to 2-44

$$W'_{\text{fld}}(i_1, i_2, \theta) = \tfrac{1}{2}L_{11}i_1^2 + \tfrac{1}{2}L_{22}i_2^2 + L_{12}i_1 i_2 \tag{2-54}$$

Systems with more than two pairs of electric terminals are handled in analogous fashion; one independent variable is chosen for each terminal pair.

EXAMPLE 2-5

In the system shown in Fig. 2-13 the inductances in henrys are given as $L_{11} = (3 + \cos 2\theta) \times 10^{-3}$; $L_{12} = 0.1 \cos \theta$; $L_{22} = 30 + 10 \cos 2\theta$. Find the torque $T_{\text{fld}}(\theta)$ for the current $i_1 = 1$ A; $i_2 = 0.01$ A.

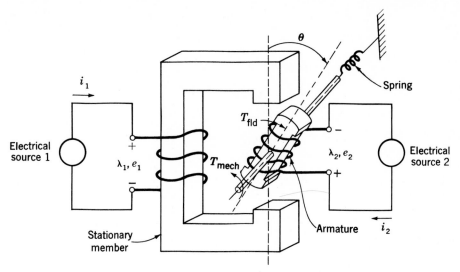

Fig. 2-13. Multiply excited magnetic system, Example 2-5.

Solution

Since the expression for torque is needed as a function of the currents i_1 and i_2 rather than the flux linkages, we shall use the coenergy of the system as the state function. When the inductances are given, the coenergy of the system from Eq. 2-54 is

$$W'_{\text{fld}} = \tfrac{1}{2}L_{11}i_1^2 + L_{12}i_1i_2 + \tfrac{1}{2}L_{22}i_2^2$$

The torque is given by Eq. 2-52 as

$$T_{\text{fld}} = +\frac{\partial W'_{\text{fld}}}{\partial \theta} = -1 \times 10^{-3}i_1^2 \sin 2\theta - 0.1i_1i_2 \sin \theta - 10i_2^2 \sin 2\theta$$

At $i_1 = 1$ A and $i_2 = 0.01$ A, the torque is

$$T_{\text{fld}} = -2 \times 10^{-3} \sin 2\theta - 10^{-3} \sin \theta \qquad \text{N} \cdot \text{m}$$

Notice that the torque expression consists of terms of two types. One term is proportional to the product of the two currents and to the sine of the angular displacement. This torque is due to the mutual interaction between the rotor and stator currents; it acts in a direction to align the rotor and stator so as to maximize the coenergy. Alternately, it can be thought of as being due to the tendency of two magnetic fields (in this case those of the rotor and stator) to align.

The torque expression also has two terms each proportional to the sine of twice the angle and to the square of one of the coil currents. These terms

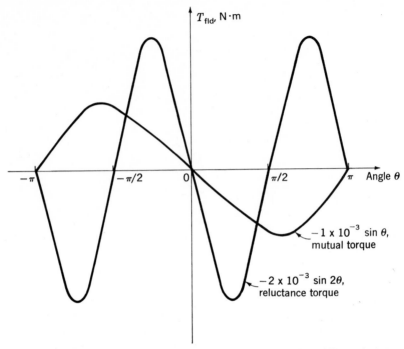

Fig. 2-14. Plot of torque components for multiply excited device of Example 2-5.

correspond to the torques one sees in singly excited systems. Here the torque is due to the fact that the coenergy is a function of rotor position (corresponding to the position of the movable plunger in Art. 2-4) and the torque acts in a direction to align the rotor so as to maximize the coenergy. The 2θ variation is due to the corresponding variation in the self-inductances, which in turn is due to the variation of the air-gap reluctance; notice that rotating the rotor by 180° from any given position gives the same air-gap reluctance and hence twice the angle variation. This torque component is known as the *reluctance torque*. The two torque components (mutual and reluctance) are plotted in Fig. 2-14.

2–6 DYNAMIC EQUATIONS

We have derived expressions for the forces produced in electromechanical energy-conversion devices as functions of the electrical variables and the mechanical displacement. These expressions were derived for a conservative energy-conversion system; any losses were assigned to the electric and mechanical sources. These conversion devices always operate as a coupling means between an electric system and a mechanical system. Hence, we should be interested not only in how the conversion device itself operates but in how

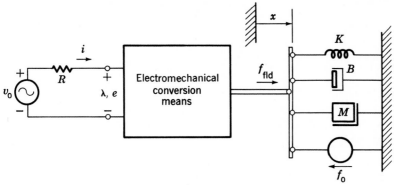

Fig. 2-15. Model of a singly excited electromechanical system.

the complete electromechanical system operates with the conversion device contained within it.

The model of a simple conversion system in Fig. 2-15 shows the basic portions, the details of which may vary from system to system. The system shown consists of three parts: an external electric portion, the energy-conversion means, and the external mechanical portion. The electric portion is merely represented by a voltage source v_0 and resistance R; the source could alternatively be represented by a current source and a parallel conductance G. The electric equation is thus

$$v_0 = iR + \frac{d\lambda}{dt} \qquad (2\text{-}55)$$

If the flux linkage λ can be expressed as $\lambda = L(x)i$, the external equation becomes

$$v_0 = iR + L(x)\frac{di}{dt} + i\frac{dL(x)}{dx}\frac{dx}{dt} \qquad (2\text{-}56)$$

The term $L\,di/dt$ is the self-inductance voltage term, and the term $i(dL/dx)(dx/dt)$ is the velocity-voltage term, which is responsible for the energy transfer between the external electric system and the energy-conversion means. For a multiply excited system, electric equations corresponding to Eq. 2-55 are written for each input pair. If the expressions for the λ's are to be expanded in terms of inductances, as in Eq. 2-56, both self- and mutual-inductance terms will be required.

The mechanical portion of Fig. 2-15 shows symbols for a spring (spring constant K), a damper (damping constant B), a mass M, and an external mechanical excitation force f_0. The forces and displacement x are related as follows:

Spring: $$f_K = K(x - x_0) \qquad (2\text{-}57)$$

Damper: $$f_D = B\frac{dx}{dt} \qquad (2\text{-}58)$$

Mass: $$f_M = M\frac{d^2x}{dt^2} \qquad (2\text{-}59)$$

where x_0 is the value of x with the spring unstretched and the applied mechanical force $f_0 = 0$. Force equilibrium thus requires

$$f_{\text{fld}} = f_K + f_D + f_M + f_0 = K(x - x_0) + B\frac{dx}{dt} + M\frac{d^2x}{dt^2} + f_0 \qquad (2\text{-}60)$$

The differential equations for the overall system of Fig. 2-15 for arbitrary inputs $v_0(t)$ and $f_0(t)$ are

$$v_0(t) = iR + L(x)\frac{di}{dt} + i\frac{dL(x)}{dt}\frac{dx}{dt} \qquad (2\text{-}61)$$

$$f_0(t) = -M\frac{d^2x}{dt^2} - B\frac{dx}{dt} - K(x - x_0) + f_{\text{fld}}(x, i) \qquad (2\text{-}62)$$

The functions $L(x)$ and $f_{\text{fld}}(x, i)$ depend upon the properties of the energy-conversion device and are calculated as previously discussed.

EXAMPLE 2-6

Figure 2-16 shows in cross section a cylindrical solenoid magnet in which the cylindrical plunger of mass M moves vertically in brass guide rings of thickness t and mean diameter d. The permeability of brass is the same as that of free space and is $\mu_0 = 4\pi = 10^{-7}$ H/m in SI units. The plunger is supported by a spring whose elastance is K. Its unstretched length is l_0. A mechanical load force f_t is applied to the plunger from the mechanical system connected to it, as shown in Fig. 2-16. Assume that frictional force is linearly proportional to velocity and that the coefficient of friction is B. The coil has N turns and resistance R. Its terminal voltage is v_t, and its current is i. The effects of magnetic leakage and reluctance of the steel are negligible.

Derive the dynamic equations of motion of the electromechanical system, i.e., the differential equations expressing the dependent variables i and x in terms of v_t, f_t, and the given constants and dimensions.

Solution

We begin by expressing the inductance as a function of x. The coupling terms, i.e., magnetic force f_{fld} and induced emf e, can then be expressed in terms of x

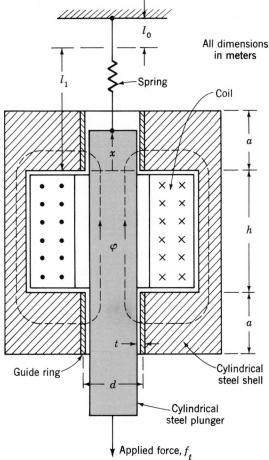

Fig. 2-16. Solenoid magnet, Example 2-6.

and i and these relations substituted in the equations for the mechanical and electric systems.

The reluctance of the magnetic circuit is that of the two guide rings in series, with the flux directed radially through them, as shown by the dashed flux lines φ in Fig. 2-16. Because $t \ll d$, the flux density in the guide rings is very nearly constant with respect to radial distance. In a region where the flux density is constant the reluctance is

$$\frac{\text{Length of flux path in direction of field}}{\mu(\text{area perpendicular to field})}$$

The reluctance of the upper gap is

$$\mathfrak{R}_1 = \frac{t}{\mu_0 \pi x d}$$

in which it is assumed that the field is concentrated in the area between the upper end of the plunger and the lower end of the upper guide ring. Similarly the reluctance of the lower gap is

$$\mathcal{R}_2 = \frac{t}{\mu_0 \pi a d}$$

The total reluctance is

$$\mathcal{R} = \mathcal{R}_1 + \mathcal{R}_2 = \frac{t}{\mu_0 \pi\, d}\left(\frac{1}{a} + \frac{1}{x}\right) = \frac{t}{\mu_0 \pi a d}\frac{a + x}{x}$$

Hence, the inductance is

where $\quad L = \dfrac{N^2}{\mathcal{R}} = \dfrac{\mu_0 \pi a d N^2}{t}\dfrac{x}{a + x} = L'\dfrac{x}{a + x}\quad$ and $\quad L' = \dfrac{\mu_0 \pi a d N^2}{t}$

The magnetic force acting upward on the plunger in the positive direction of x is

$$f_{\text{fld}} = \frac{+\partial W'_{\text{fld}}(i, x)}{\partial x} = +\tfrac{1}{2}i^2\frac{dL}{dx} = +\tfrac{1}{2}L'\frac{ai^2}{(a + x)^2}$$

The counter emf induced in the coil is

$$e = \frac{d}{dt}(Li) = L\frac{di}{dt} + i\frac{dL}{dx}\frac{dx}{dt}$$

$$e = L'\frac{x}{a + x}\frac{di}{dt} + L'\frac{ai}{(a + x)^2}\frac{dx}{dt}$$

Substitution of the magnetic force in the differential equation of motion of the mechanical system gives

$$-f_t + \tfrac{1}{2}L'\frac{ai^2}{(a + x)^2} = M\frac{d^2x}{dt^2} + B\frac{dx}{dt} - K(l_1 - x) + Mg$$

The voltage equation for the electric system is

$$v_t = Ri + e = Ri + L'\frac{x}{a + x}\frac{di}{dt} + L'\frac{ai}{(a + x)^2}\frac{dx}{dt}$$

These last two equations are the desired results. They are valid only so long as the upper end of the plunger is well within the upper guide ring, say, between the limits $0.1a < x < 0.9a$. This is the normal working range of the solenoid.

2–7 ANALYTICAL TECHNIQUES

We have described relatively simple devices in this chapter. The devices have one or two pairs of electric terminals and one mechanical terminal, which is usually constrained to incremental motion. More complicated devices capable of continuous energy conversion will be treated in the following chapters. The analytical techniques to be discussed here apply to the simple devices, but the principles are applicable to the more complicated devices as well.

a. Reasons for Analysis

Some of the devices described in this chapter are used with gross motion, such as in relays and solenoids, where the devices operate under essentially "on" and "off" conditions. Analysis of these devices is carried out to determine force as a function of displacement and reaction on the electric source. Such calculations have already been made in this chapter. If the details of the motion are required, such as the displacement as a function of time after energizing the device, nonlinear differential equations of the form of Eqs. 2-61 and 2-62 must be solved.

Compared with gross-motion devices, other devices are used with small motion, as in loudspeakers, pickups, and transducers of various kinds, for converting electric signals into mechanical signals and vice versa. The relationship between the electrical and mechanical variables is made linear either by the design of the device or by restricting the excursion of the signals to a linear range. In such case the differential equations are linear and can be solved for transient response, frequency response, or equivalent-circuit representation, as required.

b. Gross Motion

The differential equations for a singly excited device as derived in Example 2-6 are of the form

$$\tfrac{1}{2}L'\frac{ai^2}{(a+x)^2} = M\frac{d^2x}{dt^2} + B\frac{dx}{dt} + K(x - l_1) + f_t \qquad (2\text{-}63)$$

$$v_t = iR + L'\frac{x}{a+x}\frac{di}{dt} + \frac{L'ai}{(a+x)^2}\frac{dx}{dt} \qquad (2\text{-}64)$$

A typical problem using these differential equations is to find the excursion $x(t)$ when a prescribed voltage $v_t = V$ is applied at $t = 0$. An even simpler problem is to find the time required for the armature to move from its position $x(0)$ at $t = 0$ to a given displacement $x = X$ when a voltage $v_t = V$ is applied at $t = 0$. There is no general solution for these differential equations; they involve products of variables x and i and their derivatives and are termed nonlinear. They can be solved using numerical step-by-step techniques or a digital computer. Either a program can be prepared for the solution of the equations or a standard library program can be called on and the coefficients for the particular problem inserted into that program.

In many cases the gross-motion problem can be simplified and a solution found by relatively simple methods. For example, when the winding of the device is connected to the voltage source with a relatively large resistance, the iR term dominates on the right-hand side of Eq. 2-64 compared with the di/dt transformer-voltage term and the dx/dt velocity-voltage term. The current i can then be assumed equal to V/R and inserted directly into Eq. 2-63. The same assumption can be made when the winding is driven from a transistor which acts as a current source to the winding. With the assumption that $i = V/R$, two cases can be solved easily.

Case 1. We can handle those devices in which the dynamic motion is governed by damping rather than inertia, e.g., relays having dashpots or dampers to slow down the motion or devices purposely having low inertia. In such case, the differential equation 2-63 reduces to

$$f(x) = \tfrac{1}{2}L'\frac{a}{(a + x)^2}\left(\frac{V}{R}\right)^2 - K(x - l_1) = B\frac{dx}{dt} \qquad (2\text{-}65)$$

where $f(x)$ is the difference between the force of electric origin and the spring force in the device of Fig. 2-16. The velocity at any value of x is merely $dx/dt = f(x)/B$; the time t to reach $x = X$ is given by

$$t = \int_0^X \frac{B}{f(x)}\,dx \qquad (2\text{-}66)$$

The integration of Eq. 2-66 can be carried out analytically or graphically.

Case 2. The dynamic motion is governed by the inertia rather than the damping. Where additional damping is not introduced, the inertia usually governs the motion. In such a case, the differential equation 2-63 reduces to

$$f(x) = M\frac{d^2x}{dt^2} \qquad (2\text{-}67)$$

The acceleration at any value of x is $d^2x/dt^2 = f(x)/M$, and the velocity $v(x)$ at any value of x is obtained from

$$f(x) = \frac{M}{2}\frac{d}{dx}\left(\frac{dx}{dt}\right)^2 \qquad (2\text{-}68)$$

$$v(x) = \frac{dx}{dt} = \left[\frac{2}{M}\int_0^X f(x)\,dx\right]^{1/2} \qquad (2\text{-}69)$$

The integration of Eq. 2-69 can be carried out analytically or graphically to find $v(x)$ and also to find the time t to reach any value of x.

c. Linearization

Devices which are characterized by nonlinear differential equations such as Eqs. 2-63 and 2-64 will yield nonlinear responses to input signals when used as transducers. To obtain linear behavior such devices must be restricted to small excursions of displacement and electrical quantities about an equilibrium position. The equilibrium position is determined either by a bias mmf produced by current or a permanent magnet acting against a spring or by a pair of windings producing mmfs whose forces cancel at the equilibrium point. The equilibrium point must be stable; the armature following a small disturbance should return to the equilibrium position. The equilibrium condition is determined with the time derivatives set to zero in Eqs. 2-63 and 2-64; this occurs for

$$\tfrac{1}{2} L' \frac{aI_0^2}{(a + X_0)^2} = K(X_0 - l_1) + f_{t0} \tag{2-70}$$

$$V_0 = I_0 R \tag{2-71}$$

where $i = I_0$ and $x = X_0$ at equilibrium.

The incremental operation can be described by expressing the variables as $i = I_0 + i_1$, $f_t = f_{t_0} + f_1$, $v_t = V_0 + v_1$ and $x = X_0 + x_1$ and canceling the products of increments as being of second order. Equations 2-63 and 2-64 thus become

$$\frac{1}{2} \frac{L'a(I_0 + i_1)^2}{(a + X_0 + x_1)^2} = M \frac{d^2 x_1}{dt^2} + B \frac{dx_1}{dt} + K(X_0 + x_1 - l_1) + f_{t0} + f_1 \tag{2-72}$$

$$V_0 + v_1 = (I_0 + i_1)R + \frac{L'(X_0 + x_1)}{a + X_0 + x_1} \frac{di_1}{dt} + \frac{L'a(I_0 + i_1)}{(a + X_0 + x_1)^2} \frac{dx_1}{dt} \tag{2-73}$$

The equilibrium terms cancel out, and retaining only first-order incremental terms results in a set of linear differential equations in terms of just the incremental variables of first order as

$$\frac{L'aI_0 i_1}{(a + X_0)^2} = M \frac{d^2 x_1}{dt^2} + B \frac{dx_1}{dt} + K'x_1 + f_1 \tag{2-74}$$

$$v_1 = i_1 R + \frac{L'(X_0)}{a + X_0} \frac{di_1}{dt} + \frac{L'aI_0}{(a + X_0)^2} \frac{dx_1}{dt} \tag{2-75}$$

The constant K' represents the effect of the spring force and the component of magnetic field force proportional to x_1. The equations can be written in more compact form in terms of the self-inductance L_0 at the equilibrium point and a coefficient K_0 of energy conversion as

$$K_0 i_1 = M\frac{d^2 x_1}{dt^2} + B\frac{dx_1}{dt} + K' x_1 + f_1 \tag{2-76}$$

$$v_1 = i_1 R + L_0 \frac{di_1}{dt} + K_0 \frac{dx_1}{dt} \tag{2-77}$$

d. Transfer Functions and Block Diagrams

In the study of systems involving more than a few relatively simple equations an analytical attack by means of the equations themselves becomes cumbersome and confusing. It is then helpful to represent the system by either a block diagram or a signal flowchart. This technique is useful in analytical studies of system characteristics such as stability, accuracy, and frequency response. The technique has been highly developed in connection with studies of feedback control systems. Closely related to the analytical technique based on block diagrams is the technique of programming a problem for solution on an analog computer.

The *block diagram* is a pictorial representation of the equations of the system. Each block represents a mathematical operation. The blocks are then interconnected in accordance with the dictates of the system. There is no need to solve simultaneous equations in setting up the block diagram. The block diagram itself is a chart of the procedure to be followed in combining the simultaneous equations. Often useful information can be obtained from the block diagram without making a complete analytical solution.

The symbols needed in block diagrams of linear systems are shown in Fig. 2-17. In Fig. 2-17a, X is an *input variable,* and Y is an *output variable.* They may be functions of time or complex-amplitude functions of frequency. The operator A is the *transfer function* representing the mathematical operation performed on X to obtain Y. If X and Y are time functions, A is a differential operator $A(p)$, where p is the derivative operator d/dt. If X and Y are complex amplitudes, A is a complex operator $A(s)$, where s is complex frequency. For steady-state sinusoidal variables, X and Y are phasors and $s = j\omega$. The symbol in Fig. 2-17b represents addition or subtraction.

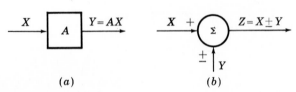

(a)   (b)

Fig. 2-17. Block-diagram symbols.

EXAMPLE 2-7

Draw a block diagram in the complex-frequency domain for the circuit of Fig. 2-18 with input $e_1(t)$ and output $e_2(t)$.

Solution

The differential equations can be written in the form

$$e_1(t) - e_2(t) = Lpi + Ri \qquad e_2(t) = \frac{1}{Cp}i \qquad p \equiv \frac{d}{dt}$$

When transformed to complex quantities and rearranged, the equations become

$$\frac{E_1(s) - E_2(s)}{Ls + R} = I(s) \qquad E_2(s) = \frac{1}{Cs}I(s)$$

The block diagram is shown in Fig. 2-18b.

In a linear system the response of an output variable to any one input can be found by considering all other inputs to be zero. The overall response to

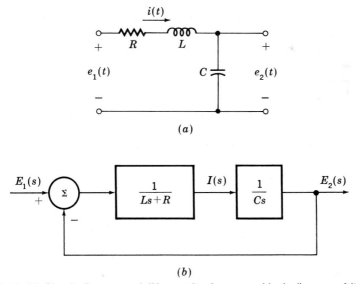

(a)

(b)

Fig. 2-18. (a) Circuit diagram and (b) complex-frequency block diagram of its transfer function, Example 2-7.

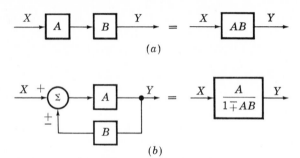

Fig. 2-19. Basic operations of block-diagram algebra.

several inputs can be found by superposition. The blocks can be combined by means of a few simple rules. Two basic ones are shown in Fig. 2-19. For example, in Fig. 2-19b,

$$Y = A(X \pm BY) \qquad \text{or} \qquad Y = \frac{A}{1 \mp AB}X \qquad (2\text{-}78)$$

2–8 SUMMARY

Magnetic and electrostatic fields are seats of energy storage. Whenever the energy in the field is influenced by the configuration of the mechanical parts constituting the boundaries of the field, mechanical forces are created which tend to move the mechanical elements so that energy is transmitted from the field to the mechanical system.

The singly excited magnetic system is considered first in Art. 2-3. When the electric and mechanical loss elements are removed from the electromechanical conversion device and combined with the external electric and mechanical systems, the device can be considered as a conservative system. Its energy is thus a state function and is described by its independent variables of λ and x. The energy can be specified in terms of field quantities, electric-circuit quantities, and magnetic-circuit quantities.

In Art. 2-4 the state function is enlarged in concept to include the coenergy as a function of i and x. The force of field origin acting on the movable members is derived as the partial derivative of the energy or coenergy expressed in the suitable independent variables. The force always acts to increase the self-inductance of the system, and the values of force are independent of the method of analysis. The energy or coenergy of the system can be introduced through both the electric and mechanical terminals, depending upon the constraints placed upon the variables.

Most rotating machinery and signal-handling transducers are built with multiple windings or excitation sources. The analysis is extended in Art. 2-5 for such magnetic field systems in terms of energy and coenergy. The use of self- and mutual inductances for expressing the state functions is introduced.

Energy-conversion devices operate between electric and mechanical systems. The behavior is described in Art. 2-6 by the differential equations which include the coupling terms between the systems. The equations are usually nonlinear and can be solved by computer or numerical methods if necessary. Usually, the solution is required for special conditions which can be handled as discussed in Art. 2-7. The most useful condition is that for small signal amplitudes, where the resulting equations are linear.

This chapter has been concerned with basic principles applying broadly to the electromechanical energy-conversion process, with emphasis on magnetic field coupling. Basically, rotating machines and linear-motion transducers work in the same way. The remainder of this text is devoted almost entirely to the rotating-machine aspects of electromechanical energy conversion, both the dynamic characteristics of the machines as system components and their behavior under steady-state conditions. The rotating machine is a fairly complicated assembly of electric and magnetic circuits with a number of variations of machine types. For all of them, the electromechanical coupling terms (torque and induced voltage) can be found by the principles developed in this chapter.

PROBLEMS

2-1 An inductor is made from a magnetic core with an air gap as shown in Fig. 2-20. Based upon the usual assumptions, the coil inductance is

$$L = \frac{\mu_0 N^2 A}{g_0}$$

and the winding resistance is known to be R. The applied voltage $v(t)$ varies linearly from 0 to V_0 in t_0 seconds, after which it remains constant.

(a) Calculate the magnetic stored energy in the inductor after all electric transients have ended.

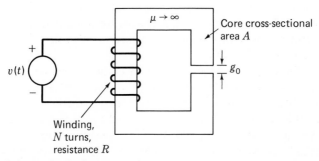

Fig. 2-20. Inductor for Prob. 2-1.

(b) With the voltage held constant at V_0, the air-gap length is varied from g_0 to g_1. Again after all electric transients have ended, calculate the resultant change in magnetic stored energy.

2-2 The voltage source of Fig. 2-20 is made to vary sinusoidally in time

$$v(t) = V_0 \cos \omega t$$

With the air gap held fixed at g_0, calculate the time-averaged magnetic stored energy $\langle W_{\mathrm{fld}} \rangle$ in the inductor and the instantaneous and time-averaged power output of the voltage source as a function of the winding resistance R. Compare with the time-averaged power dissipated in the resistor.

2-3 An RC circuit is connected to a battery as shown in Fig. 2-21. The capacitor C is initially uncharged. At time $t = 0$ switch S_1 is closed.

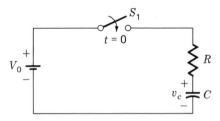

Fig. 2-21. RC circuit for Prob. 2-3.

(a) What is the energy stored in the capacitor as a function of time? [*Hint*: $W_{\mathrm{fld}} = \frac{1}{2}(q^2/C)$ where $q = Cv_c$.] What is the final value of stored energy in the capacitor, $W_{\mathrm{fld}}(t = \infty)$?

(b) What is the power dissipated in the resistor as a function of time? What is the total energy dissipated in the resistor?

(c) What is the battery output power as a function of time? What is the total energy output from the battery?

2-4 The time constant L/R of the field winding of a 10-kW 1150-r/min dc shunt generator is 0.15 s. At normal operating conditions the i^2R loss in its field winding is 350 W. Compute the energy stored in its magnetic field in watt-seconds at normal operating conditions.

2-5 The cylindrical iron-clad solenoid magnet shown in Fig. 2-22 is used for tripping circuit breakers, for operating valves, and in other applications in which a relatively large force is applied to a member which moves a relatively short distance. When the coil current is zero, the plunger drops against a stop such that the gap g is 0.50 in. When the coil is energized by a direct current of sufficient magnitude, the plunger is raised until it hits another stop set so that g is 0.10 in. The plunger is supported so that it can move freely in the vertical

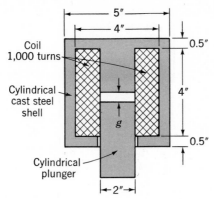

Fig. 2-22. Plunger magnet.

direction. The air gap between the shell and the plunger can be assumed to be uniform and 0.01 in long. For this problem neglect magnetic leakage and fringing in the air gaps. The exciting coil has 1000 turns and carries a constant current of 3.0 A. If the mmf in the iron is neglected, compute (a) the flux densities in teslas between the working faces of the center core and plunger for gaps g of 0.10, 0.20, and 0.50 in; (b) the corresponding values of the energy stored in the magnetic field in watt-seconds; and (c) the corresponding values of the coil inductance in henrys.

2-6 Data for the magnetization curve of the iron portion of the magnetic circuit of the plunger magnet of Prob. 2-5 are given below:

Flux, kilolines	100	150	200	240	250	260	270	275
mmf, A · turns	60	95	150	250	305	425	600	725

Plot magnetization curves for the complete magnetic circuit (flux in webers versus total mmf in ampere-turns) for (a) gap $g = 0.05$ in and (b) gap $g = 0.10$ in. (c) From these curves find graphically the magnetic field energy and coenergy for each of the gaps in parts (a) and (b) with 3.0-A coil current.

2-7 As shown in Fig. 2-23 an electromagnet is to be used to lift a 120-kg slab of iron. The surface roughness of the iron is such that when the iron and electromagnet are in contact, there is a minimum air gap of 0.1 cm in each leg. The coil resistance is 3 Ω. Calculate the minimum coil voltage which must be used in order to lift the slab against the force of gravity. Neglect the reluctance of the iron.

2-8 An iron-clad plunger magnet is shown in Fig. 2-22. The air gap between the shell and the plunger can be assumed to be uniform and 0.01 in long.

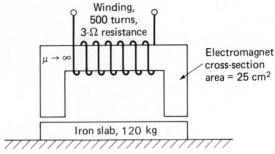

Fig. 2-23. Electromagnet lifting an iron slab, Prob. 2-7.

Magnet leakage and fringing may be neglected. The coil has 1000 turns and carries a constant current of 3.0 A.

> (a) If the plunger is allowed to move slowly, reducing the gap g from 0.50 to 0.10 in, how much mechanical work in joules will be done by the plunger?
>
> (b) For the conditions of part (a) how much energy will be supplied by the electric source (in excess of heat loss in the coil)?

2-9 For the plunger magnet of Prob. 2-8 find a numerical expression for the magnetic force f in newtons acting on the plunger as a function of the gap g in meters with a constant coil current of 3.0 A. Plot a curve of f as a function of g. From the area under this force-displacement curve compute the mechanical work done by the plunger when it moves slowly so as to reduce the gap g from 0.50 to 0.10 in. Compare with the result of part (a) of Prob. 2-8.

2-10 A long, thin solenoid or radius r_0 and height h is shown in Fig. 2-24. The magnetic field inside such a solenoid is axially directed and essentially uni-

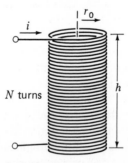

Fig. 2-24. Solenoidal coil, Prob. 2-10.

form and equal to $H = Ni/h$. The magnetic field outside the solenoid is negligible. Calculate the radial pressure in newtons per square meter acting on the sides of the solenoid for constant coil current $i = I_0$.

2-11 An inductor is made up of a 500-turn coil on a core of cross-sectional area 10 cm² and gap length 0.03 cm. The coil is connected directly to a 120-V 60-Hz voltage source. Neglect coil resistance and leakage inductance and assume the core reluctance to be negligible. Calculate the force acting on the core acting to close the air gap. How does this force vary if the air-gap length is doubled?

2-12 Figure 2-25 shows the general nature of the slot-leakage flux produced by current i in a rectangular conductor embedded in a rectangular slot in iron.

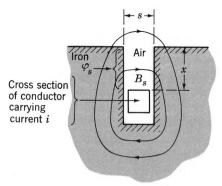

Fig. 2-25. Conductor in a slot, Prob. 2-12.

Assume that the slot-leakage flux φ_s goes straight across the slot in the region between the top of the conductor and the top of the slot.

(a) Derive an expression for the flux density B_s in the region between the top of the conductor and the top of the slot.

(b) Derive an expression for the slot-leakage flux φ_s crossing the slot above the conductor, in terms of the height x of the slot above the conductor, the slot width s, and the embedded length l perpendicular to the paper.

(c) Derive an expression for the force f created by this magnetic field on a conductor of length l. Use SI units. In what direction does this force act on the conductor?

(d) Compute the force in pounds on a conductor 1.0 ft long in a slot 1.0 in wide when the current in the conductor is 1000 A.

2-13 An electromechanical system in which electric energy storage is in electric fields can be analyzed by techniques directly analogous to those derived in this chapter for magnetic field systems. Consider such a system in which it is possible to separate the loss mechanism mathematically from those of energy storage in electric fields. Then the system can be represented as in Fig. 2-26.

Fig. 2-26. Lossless electric energy-storage system.

For a single electric terminal, Eq. 2-8 applies, where

$$dW_{elec} = vi\, dt = v\, dq$$

where v is the electric-terminal voltage and q is the net charge associated with electric energy storage. Thus, by analogy to Eq. 2-15,

$$dW_{fld} = v\, dq - f_{fld}\, dx$$

(a) Derive an expression for electric stored energy $W_{fld}(q, x)$ analogous to that for magnetic stored energy in Eq. 2-17.

(b) Derive an expression for the force of electric origin f_{fld} analogous to that of Eq. 2-25. State clearly which variable must be held constant when the derivative is taken.

(c) By analogy to the derivation of Eqs. 2-26 to 2-33, derive an expression for the coenergy $W'_{fld}(v, x)$ and the corresponding force of electric origin.

2-14 A capacitor (Fig. 2-27) is made of two conducting plates of area A separated in air by a spacing x. The terminal voltage is v, and the charge on the plates is q. The capacitance C, defined as the ratio of charge to voltage, is given by

$$C = \frac{q}{v} = \frac{\epsilon_0 A}{x}$$

where ϵ_0 is the dielectric constant of free space (in SI units $\epsilon_0 = 8.85 \times 10^{-12}$ F/m).

(a) Using the results of Prob. 2-13, calculate the energy $W_{fld}(q, x)$ and the coenergy $W'_{fld}(v, x)$.

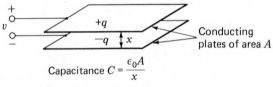

Capacitance $C = \dfrac{\epsilon_0 A}{x}$

Fig. 2-27. Capacitor plates, Prob. 2-14.

(b) The terminals of the capacitor are connected to a source of constant voltage V_0. Calculate the force required to maintain the plates separated by a constant spacing $x = \delta$.

2-15 A rotary motion converter is composed of a stator and rotor, each with a single coil. The rotor is free to turn. Coil 1, the stator coil, has a constant self-inductance $L_{11} = 0.5$ H, as does coil 2, the rotor coil, $L_{22} = 3.0$ H. The mutual inductance between the coils varies with the angular displacement of the rotor as $L_{12} = 1.2 \cos \theta$ H. Calculate the torque as a function of angle θ when the current in coil 1 is 3.5 A and that in coil 2 is 0.75 A.

2-16 The rotor of Prob. 2-15 is replaced by a modified single-coil rotor such that L_{11} now varies with angle as $L_{11} = 0.5 + 0.2 \cos 2\theta$ H while L_{22} and L_{12} remain unchanged. Calculate the torque as a function of θ for $i_1 = 3.5$ A and (a) $i_2 = 0$ A and (b) $i_2 = 0.75$ A.

2-17 The self-inductance L_{11} and L_{22} and the absolute value of the mutual inductance L_{12} of the device shown in Fig. 2-28 are given below for two angular positions θ_0 of the rotor, where θ_0 is measured from a horizontal reference axis to the axis of the rotor:

θ_0	L_{11}	L_{22}	L_{12}
45°	0.60	1.10	0.30
75°	1.00	2.00	1.00

The inductances are given in henrys and may be assumed to vary linearly with θ_0 over the range $45° < \theta_0 < 75°$.

For each of the following cases, compute the electromagnetic torque in newton-meters when the rotor is stationary at an angular position $\theta_0 = 60°$ (approximately the position shown in Fig. 2-28), and state whether this torque tends to turn the rotor clockwise or counterclockwise.

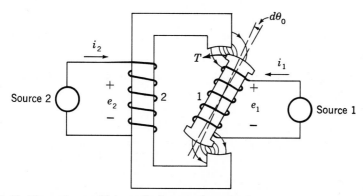

Fig. 2-28. Elementary multiply excited magnetic system.

(a) $i_1 = 10.0$ A, $i_2 = 0$.

(b) $i_1 = 0$, $i_2 = 10.0$ A

(c) $i_1 = 10.0$ A, and $i_2 = 10.0$ A in arrow directions (Fig. 2-28).

(d) $i_1 = 10.0$ A in arrow direction, and $i_2 = 10.0$ A in reverse direction.

(e) $i_1 = 10.0$ A rms sinusoidal alternating current, with coil 2 short-circuited. In this case the resistance of coil 2 may be neglected, and it is the time average of the torque which is wanted.

2-18 Two coils, one mounted on a stator and the other on a rotor, have self- and mutual inductances of

$$L_{11} = 0.20 \text{ mH} \qquad L_{22} = 0.10 \text{ mH} \qquad L_{12} = 0.05 \cos\theta \text{ mH}$$

where θ is the angle between the axes of the coils. The coils are connected in series and carry a current

$$i = \sqrt{2}\,I \sin \omega t$$

(a) Derive an expression for the instantaneous torque T on the rotor as a function of the angular position θ.

(b) Give an expression for the time-average torque T_{av} as a function of θ.

(c) Compute the numerical value of T_{av} for $I = 10$ A and $\theta = 90°$.

(d) Sketch three curves of T_{av} versus θ for currents $I = 5, 7.07$, and 10 A, respectively.

(e) A helical restraining spring which tends to hold the rotor at $\theta = 90°$ is now attached to the rotor. The restraining torque of the spring is proportional to the angular deflection from $\theta = 90°$ and is 0.004 N · m when the rotor is turned to $\theta = 0$. Show on the curves of part (d) how you could find the angular position of the rotor-plus-spring combination for coil currents $I = 5, 7.07$, and 10 A, respectively. Sketch a curve showing θ versus I. A reasonable-looking sketch with estimated numerical values is all that is required.

Note: This problem illustrates the principles of the dynamometer-type ac ammeter.

2-19 Figure 2-29 shows in cross section a cylindrical-plunger magnet. The plunger is constrained to move in the x direction and is restrained by a spring whose spring constant is k_s. The position of the plunger when the coil is unexcited is D. The mass of the plunger is M. Friction is negligible. The cross-sectional area of the working air gap is A. Neglect magnetic leakage and fringing. Also neglect the reluctance of the iron. The coil has N turns and negligible resistance.

The plunger is operating under steady-state sinusoidal conditions. The flux density in the air gap is

$$B(t) = B_m \sin \omega t$$

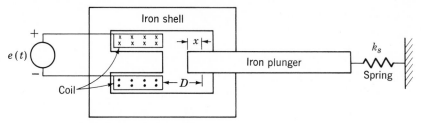

Fig. 2-29. Solenoid magnet, Prob. 2-19.

(a) Find an expression for the magnetic force acting on the plunger in terms of B_m, ω, and t.
(b) Write the equation for the coil voltage $e(t)$ in terms of B_m, ω, and t.
(c) Write the differential equation of motion of the plunger in terms of B_m, ω, t, and x.
(d Assume that the system is in the steady state. Solve for the position of the plunger $x(t)$. Note that the solution must contain a constant part X_0 and a time-varying part $x_1(t)$.

2-20 Two windings, one mounted on a stator and the other on a rotor, having self- and mutual inductances in henrys of

$$L_{11} = 2.20 \text{ H} \qquad L_{22} = 1.00 \text{ H} \qquad L_{12} = 1.414 \cos \theta_0 \text{ H}$$

where θ_0 is the angle between the axes of the windings. The resistances of the windings may be neglected. Winding 2 is short-circuited, and the current in winding 1 as a function of time is $i_1 = 14.14 \sin \omega t$.

(a) If the rotor is stationary, derive an expression for the numerical value in newton-meters of the instantaneous torque on the rotor in terms of the angle θ_0.
(b) Compute the average torque in newton-meters when $\theta_0 = 45°$.
(c) If the rotor is allowed to move, will it rotate continuously or will it tend to come to rest? If the latter, at what value of θ_0?

2-21 Consider a solenoid magnet similar to that shown in Fig. 2-16, except that the length of the cylindrical steel plunger is $a + h$. Derive the dynamic equations of motion of the system.

2-22 The following dimensions and data apply to the solenoid magnet of Prob. 2-21.

$$a = 2 \text{ cm} \qquad h = 6 \text{ cm} \qquad d = 2 \text{ cm} \qquad t = 0.05 \text{ cm}$$
$$\text{Turns } N = 1000 \qquad \text{Coil resistance } R = 100 \text{ }\Omega$$
$$\text{Mass of plunger } M = 0.2 \text{ kg} \qquad \text{Spring constant } K = 6.25 \text{ N/cm}$$

With $i = 0$, the plunger is at rest at $x = 0.25$ cm. The external applied force is

0, and friction is negligible. The voltage applied to the coil has a quiescent value of 10 V.

    (a) Find the quiescent operating point. Is it stable?

    (b) Linearize the dynamic equations of motion for small incremental changes in the applied voltage. Give numerical values of the parameters in SI units.

2-23 A solenoid has a winding resistance of $2\,\Omega$ and an inductance of $L = 3.2/x$ mH, where x is the plunger displacement in centimeters. A dc voltage of 6 V is suddenly applied to the winding terminals. Calculate the force in newtons as a function of time required to hold the plunger stationary at $x = 0.5$ cm.

2-24 The plunger of the solenoid in Prob. 2-23 has a mass of 0.01 kg and is connected to a spring whose force upon the plunger is given by $f = 1.25(x - 0.1)$ N (for x in meters).

    (a) Find the steady-state displacement X_0 for a terminal voltage of 6 V.

    (b) Write the dynamic equations of motion for the system.

    (c) Linearize these dynamic equations for incremental changes in plunger position assuming that the terminal voltage remains constant.

    (d) If the plunger position is made to vary incrementally as $x = X_0 + 0.1X_0 \cos \omega t$, where X_0 is as found in part (a) and $\omega = 2\pi$ rad/s, use the linearized equations to find the time-varying component of current in the coil. Find the time average of the corresponding power dissipation in the coil. How is this power dissipation supplied?

2-25 A loudspeaker is made using a magnetic core of infinite permeability and circular symmetry, as shown in Fig. 2-30a and b. The air-gap length g is

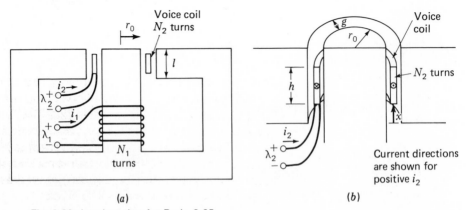

(a)   (b)

Fig. 2-30. Loudspeaker for Prob. 2-25.

much less than the radius r_0 of the central core. The voice coil is constrained to move only in the x direction and is attached to the speaker cone, which is not shown in the figure. A constant radial magnetic field is produced in the air by a direct current in coil 1, $i_1 = I_1$. An audio-frequency signal $i_2 = I_2 \cos \omega t$ is then applied to the voice coil. Assume the voice coil to be of negligible thickness and composed of N_2 turns uniformly distributed over its height h. Also assume that its displacement is such that it remains in the air gap $(0 \leq x \leq l - h)$.

(a) Calculate the force on the voice coil using the Lorentz law as in Example 2-1.
(b) Calculate the self-inductance of each coil.
(c) Calculate the mutual inductance between the coils. *Hint*: Assume that current is applied to the voice coil and calculate the flux linkages of coil 1. Note that these flux linkages vary with displacement x.
(d) Calculate the force on the voice coil from the coenergy W'_{fld}.

3

Rotating Machines: Basic Concepts

The object of this chapter is to discuss the techniques and approximations involved in reducing a physical machine to a simple mathematical model and to give some simple concepts relating to the basic machine types. The models developed in this chapter are sufficient to illustrate the basic principles involved. Later chapters will refine these models into various forms applicable to the analysis of practical engineering situations.

3-1 ELEMENTARY CONCEPTS

Faraday's law, $e = d\lambda/dt$, describes quantitatively the induction of voltages by a time-varying magnetic field. Electromagnetic energy conversion takes place when the change in flux is associated with mechanical motion. In rotating machines voltages are generated in windings or groups of coils by rotating

Fig. 3-1. Armature of a dc motor. (*General Electric Company.*)

these windings mechanically through a magnetic field, by mechanically rotating a magnetic field past the winding, or by designing the magnetic circuit so that the reluctance varies with rotation of the rotor. By any of these methods the flux linking a specific coil is changed cyclically, and a time-varying voltage is generated. A group of such coils interconnected so that their generated voltages all make a positive contribution to the desired result is called an *armature winding*. The armature of a dc machine is shown in Fig. 3-1; the armature is the rotating member, or *rotor*. Figure 3-2 shows the armature of an ac machine, in this case a synchronous generator. Here the armature is the stationary member, or *stator*.

In general, these coils are wound on iron cores. This is done to maximize the coupling between the coils, to increase the magnetic energy density associated with the electromechanical interaction, and to shape and distribute the magnetic fields according to the requirements of each particular machine design. Because the armature iron is subjected to a time-varying magnetic flux, eddy currents will be induced in it. To minimize this eddy-current loss, the armature iron is built up of thin laminations, as illustrated in Fig. 3-3 for the armature of an ac machine. The magnetic circuit is completed through the iron of the other machine member, and excitation coils, or *field windings,* are placed on that member to act as the primary source of flux. Permanent mag-

Fig. 3-2. Stator of a 190-MVA 3-phase 12-kV 375-r/min hydroelectric generator. The conductors have hollow passages through which cooling water is circulated. (*Brown Boveri Corporation.*)

nets may be used in small machines, and developments in permanent-magnet technology are resulting in their use in larger machines.

Rotating electric machines take many forms and are known by many names—dc, synchronous, permanent-magnet, induction, hysteresis, etc. Al-

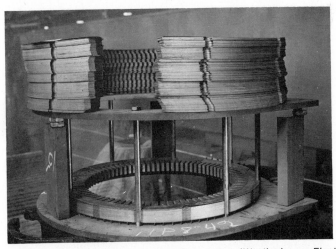

Fig. 3-3. Partially completed stator core for an ac motor. (*Westinghouse Electric Corporation.*)

though these machines appear to be quite dissimilar and require a variety of analytical techniques, the physical principles governing their behavior are quite similar, and in fact these machines can often be explained from the same physical picture. For example, analysis of a dc machine shows that associated with both the rotor and stator are magnetic flux distributions which are fixed in space and that the torque-producing characteristic of the dc machine stems from the tendency of these flux distributions to align. An induction machine, in spite of many fundamental differences, works on exactly the same principle; one can identify flux distributions associated with the rotor and stator, which rotate in synchronism and which are separated by some torque-producing angular displacement.

Although mathematical analytical techniques and models are essential to the analysis of electric machines, physical insight is a very useful engineering tool for the analysis and application of these devices. One object of this and subsequent chapters is to guide the reader in the development of such insight.

3-2 INTRODUCTION TO AC AND DC MACHINES

a. Elementary Synchronous Machines

Preliminary ideas of generator action can be gained by discussing the armature-induced voltage in the very much simplified ac synchronous generator shown in Fig. 3-4. With rare exceptions, the armature winding of a synchronous machine is on the stator, and the field winding is on the rotor, as in Fig. 3-4. The field winding is excited by direct current conducted to it by means of carbon *brushes* bearing on *slip rings* or *collector rings*. Constructional factors usually dictate this orientation of the two windings: it is advantageous to have the low-power field winding on the rotor. The armature winding, consist-

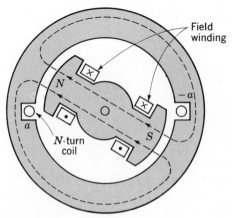

Fig. 3-4. Elementary synchronous generator.

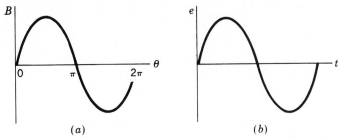

Fig. 3-5. (a) Space distribution of flux density and (b) corresponding waveform of the generated voltage.

ing of a single coil of N turns, is indicated in cross section by the two coil sides a and $-a$ placed in diametrically opposite narrow slots on the inner periphery of the stator of Fig. 3-4. The conductors forming these coil sides are parallel to the shaft of the machine and are connected in series by end connections (not shown in the figure). The rotor is turned at a constant speed by a source of mechanical power connected to its shaft. Flux paths are shown by dotted lines in Fig. 3-4.

The radial distribution of the air-gap flux density B is shown in Fig. 3-5a as a function of the space angle θ around the air-gap periphery. The flux-density wave of practical machines can be made to approximate a sinusoidal distribution by properly shaping the pole faces. As the rotor revolves, the flux waveform sweeps by the coil sides a and $-a$. The resulting coil voltage (Fig. 3-5b) is a time function having the same waveform as the spatial distribution B. The coil voltage passes through a complete cycle of values for each revolution of the 2-pole machine of Fig. 3-4. Its frequency in cycles per second (hertz) is the same as the speed of the rotor in revolutions per second; i.e., the electric frequency is synchronized with the mechanical speed, and this is the reason for the designation *synchronous* machine. Thus a 2-pole synchronous machine must revolve at 3600 r/min to produce a 60-Hz voltage.

A great many synchronous machines have more than two poles. As a specific example, Fig. 3-6 shows the most elementary 4-pole single-phase alternator. The field coils are connected so that the poles are of alternate north and south polarity. There are two complete wavelengths or cycles in the flux distribution around the periphery, as shown in Fig. 3-7. The armature winding now consists of two coils $a_1, -a_1$ and $a_2, -a_2$ connected in series by their end connections. The span of each coil is $\frac{1}{2}$ wavelength of flux. The generated voltage now goes through two complete cycles per revolution of the rotor. The frequency f in hertz then is twice the speed in revolutions per second.

When a machine has more than two poles, it is convenient to concentrate on a single pair of poles and to recognize that the electric, magnetic, and mechanical conditions associated with every other pole pair are repetitions of those for the pair under consideration. For this reason it is convenient to express angles in *electrical degrees* or *electrical radians* rather than in mechanical units. One pair of poles in a P-pole machine or one cycle of flux

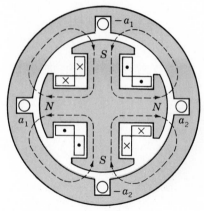

Fig. 3-6. Elementary 4-pole synchronous generator.

distribution equals 360 electrical degrees or 2π electrical radians. Since there are $P/2$ complete wavelengths or cycles in one complete revolution, it follows that

$$\theta = \frac{P}{2}\theta_m \qquad (3\text{-}1)$$

where θ is the angle in electrical units and θ_m is the mechanical angle. The coil voltage of a P-pole machine passes through a complete cycle every time a pair of poles sweeps by, or $P/2$ times each revolution. The frequency of the voltage wave is therefore

$$f = \frac{P}{2}\frac{n}{60} \qquad \text{Hz} \qquad (3\text{-}2)$$

where n is the mechanical speed in revolutions per minute and $n/60$ is the speed in revolutions per second. The radian frequency ω of the voltage wave is

$$\omega = \frac{P}{2}\omega_m \qquad (3\text{-}3)$$

where ω_m is the mechanical speed in radians per second.

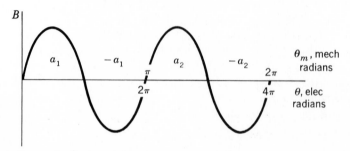

Fig. 3-7. Space distribution of flux density in a 4-pole synchronous generator.

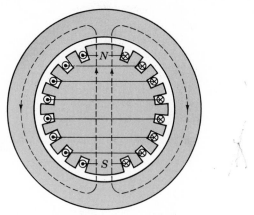

Fig. 3-8. Elementary 2-pole cylindrical-rotor field winding.

The rotors shown in Figs. 3-4 and 3-6 have *salient,* or projecting, poles with *concentrated* windings. Figure 3-8 shows diagrammatically a *non-salient-pole,* or *cylindrical, rotor.* The field winding is a *distributed* winding placed in slots and arranged to produce an approximately sinusoidal 2-pole field. The constructional reasons for some synchronous generators having salient-pole rotor structures and others having cylindrical rotors can be appreciated with the help of Eq. 3-2. Most power systems in the United States operate at a frequency of 60 Hz. A salient-pole construction is characteristic of hydroelectric generators because hydraulic turbines operate at relatively low speeds and a relatively large number of poles are required to produce the desired frequency; the salient-pole construction is better adapted mechanically to this situation. The rotor of a large hydroelectric generator is shown in Fig. 3-9. Steam turbines and gas turbines, on the other hand, operate best at relatively high speeds, and turbine-driven alternators or turbine generators are commonly 2- or 4-pole cylindrical-rotor machines. The rotors are made from a single steel forging or from several forgings, as shown in Figs. 3-10 and 3-11.

With very few exceptions, synchronous generators are 3-phase machines because of the advantages of 3-phase systems for generation, transmission, and heavy-power utilization. For the production of a set of three voltages phase-displaced by 120 electrical degrees in time, it follows that a minimum of three coils phase-displaced 120 electrical degrees in space must be used. An elementary 3-phase 2-pole machine with one coil per phase is shown in Fig. 3-12a. The three phases are designated by the letters a, b, and c. In an elementary 4-pole machine, a minimum of two such sets of coils must be used, as illustrated in Fig. 3-12b; in an elementary P-pole machine, P/2 such sets must be used. The two coils in each phase of Fig. 3-12b are connected in series so that their voltages add, and the three phases may then be either Y- or Δ-connected. Figure 3-12c shows how the coils are interconnected to form a Y connection.

When a synchronous generator supplies electric power to a load, the ar-

Fig. 3-9. Water-cooled rotor of the 190-MVA hydroelectric generator whose stator is shown in Fig. 3-2. (*Brown Boveri Corporation*.)

mature current creates a component flux wave in the air gap which rotates at synchronous speed, as will be shown in Art. 3-5. This flux reacts with the flux created by the field current, and electromagnetic torque results from the tendency of the two magnetic fields to align themselves. In a generator this torque opposes rotation, and mechanical torque must be applied from the

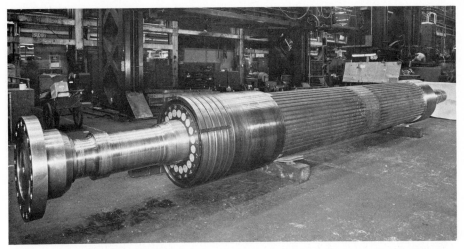

Fig. 3-10. Rotor of a 2-pole 3600-r/min turbine generator. (*Westinghouse Electric Corporation.*)

prime mover in order to sustain rotation. This electromagnetic torque is the mechanism through which the synchronous generator converts mechanical into electric energy.

The counterpart of the synchronous generator is the *synchronous motor*. A cutaway view of a 3-phase 60-Hz synchronous motor is shown in Fig. 3-13.

Fig. 3-11. Parts of multipiece rotor for a 1333-MVA 3-phase 4-pole 1800-r/min turbine generator. The separate forgings will be shrunk on the shaft before final machining and milling slots for the windings. Total weight of the rotor is 435,000 lb. This is the same generator shown in Fig. 4-8. (*Brown Boveri Corporation.*)

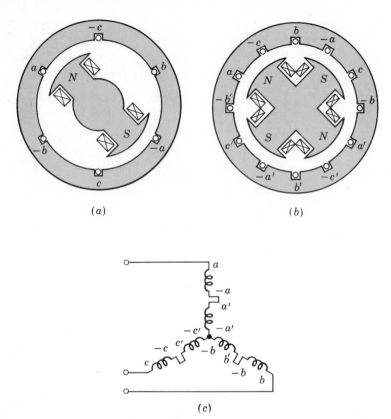

Fig. 3-12. Elementary 3-phase generators: (*a*) 2-pole, (*b*) 4-pole, (*c*) Y connection of the windings.

Alternating current is supplied to the armature winding (usually the stator), and dc excitation is supplied to the field winding (usually the rotor). The magnetic field of the armature currents rotates at synchronous speed. In order to produce a steady electromagnetic torque, the magnetic fields of stator and rotor must be constant in amplitude and stationary with respect to each other. In a synchronous motor the steady-state speed is determined by the number of poles and the frequency of the armature current, exactly as in Eq. 3-2 or 3-3 for a synchronous generator. Thus a synchronous motor operated from a constant-frequency ac source must run at a constant steady-state speed.

In a motor the electromagnetic torque is in the direction of rotation and balances the opposing torque required to drive the mechanical load. The flux produced by currents in the armature of a synchronous motor rotates ahead of that produced by the rotor, thus pulling on the rotor and doing work. This is the opposite of the situation in a synchronous generator, where the rotor does work as its flux pulls on that of the armature, which is lagging behind. In both generators and motors an electromagnetic torque and a rotational volt-

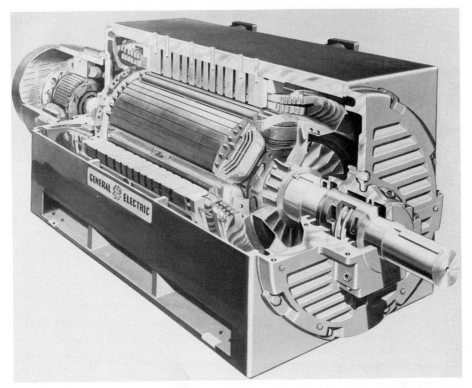

Fig. 3-13. Cutaway view of a high-speed synchronous motor. The exciter shown on the left end of the rotor is a small ac generator with a rotating semiconductor rectifier assembly. (*General Electric Company.*)

age are produced. These are the essential phenomena for electromechanical energy conversion.

b. Elementary Induction Machines

A second type of ac machine is the *induction machine,* in which there are alternating currents in both the stator and rotor windings. The most common example is the induction motor, in which alternating current is supplied directly to the stator and by induction, i.e., transformer action, to the rotor. The induction machine may be regarded as a generalized transformer in which electric power is transformed between rotor and stator together with a change of frequency and a flow of mechanical power. Although the induction motor is the most common of all motors, it is seldom used as a generator; its perform-

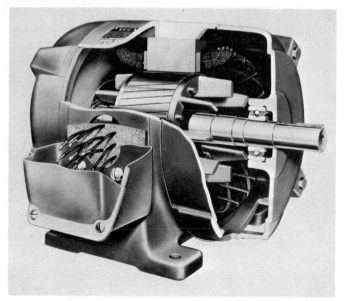

Fig. 3-14. Cutaway view of a squirrel-cage induction motor. (*Westinghouse Electric Corporation.*)

ance characteristics as a generator are unsatisfactory for most applications. The induction machine may also be used as a frequency changer.

In the induction motor the stator winding is essentially the same as that of a synchronous machine. However, the rotor winding is electrically closed on itself and frequently has no external connections; currents are induced in it by transformer action from the stator winding. A cutaway view of a squirrel-cage induction motor is shown in Fig. 3-14. Here the rotor "windings" are actually solid aluminum bars which are cast into the slots in the rotor and which are short-circuited together by cast aluminum rings at each end of the rotor. This type of rotor construction results in induction motors which are relatively inexpensive and highly reliable, factors contributing to their immense popularity and widespread application.

As in a synchronous motor, the armature flux in the induction motor leads that of the rotor and produces an electromagnetic torque. It is essential to recognize that the rotor and stator fluxes rotate in synchronism with each other and that this torque is related to the relative displacement between them. However, unlike a synchronous machine, the rotor does not itself rotate synchronously; it is the slipping of the rotor through the synchronous armature flux that gives rise to the induced rotor currents and hence the torque. An induction motor operates at some speed less than the synchronous mechanical speed of Eq. 3-3. A typical speed-torque characteristic for an induction motor is shown in Fig. 3-15.

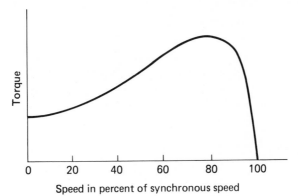

Fig. 3-15. Typical induction-motor speed-torque characteristic.

c. Elementary DC Machines

The armature winding of a dc generator is on the rotor with current conducted from it by means of carbon brushes. The field winding is on the stator and is excited by direct current. A cutaway view of a dc motor is shown in Fig. 3-16.

A very elementary 2-pole dc generator is shown in Fig. 3-17. The armature winding, consisting of a single coil of N turns, is indicated by the two coil sides a and $-a$ placed at diametrically opposite points on the rotor with the

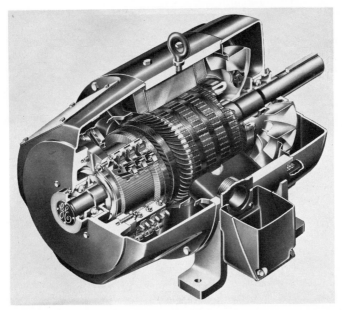

Fig. 3-16. Cutaway view of a typical integral-horsepower dc motor. (*General Electric Company.*)

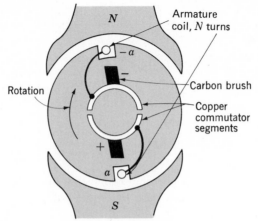

Fig. 3-17. Elementary dc machine with commutator.

conductors parallel to the shaft. The rotor is normally turned at a constant speed by a source of mechanical power connected to the shaft. The air-gap flux distribution usually approximates a flat-topped wave rather than the sine wave found in ac machines and is shown in Fig. 3-18a. Rotation of the coil generates a coil voltage which is a time function having the same waveform as the spatial flux-density distribution.

Although the ultimate purpose is the generation of a direct voltage, the voltage induced in an individual armature coil is an alternating voltage, which

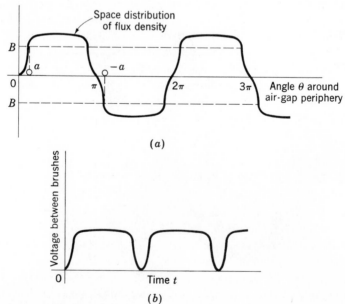

Fig. 3-18. (a) Space distribution of air-gap flux density in an elementary dc machine and (b) waveform of voltage between brushes.

must therefore be rectified. Rectification is sometimes provided externally, e.g., by semiconductor rectifiers. The machine then is an ac generator plus external rectifiers. In the conventional dc machine rectification is provided mechanically by means of a *commutator,* which is a cylinder formed of copper segments insulated from each other by mica and mounted on, but insulated from, the rotor shaft. Stationary carbon brushes held against the commutator surface connect the winding to the external armature terminals. The commutator and brushes can readily be seen in Fig. 3-16. The need for commutation is the reason why the armature windings of dc machines are placed on the rotor.

For the elementary dc generator the commutator takes the form shown in Fig. 3-17. For the direction of rotation shown, the commutator at all times connects the coil side which is under the south pole to the positive brush and that under the north pole to the negative brush. The commutator provides full-wave rectification, transforming the voltage waveform between brushes to that of Fig. 3-18b and making available a unidirectional voltage to the external circuit. The dc machine of Fig. 3-17 is, of course, simplified to the point of being unrealistic in the practical sense, and later it will be essential to examine the action of more realistic commutators.

Figure 3-17 shows that the effect of direct current in the field winding of a dc machine is to create a magnetic flux distribution which is stationary with respect to the stator. Similarly, the effect of the commutator is such that when direct current flows through the brushes, the armature creates a magnetic flux distribution which is also fixed in space and whose axis, determined by the design of the machine and the position of the brushes, is typically perpendicular to the axis of the field flux. Just as in the ac machines discussed previously, it is the interaction of these two flux distributions that creates the torque of the dc machine. If the machine is acting as a generator, this torque opposes rotation. If it is acting as a motor, the electromagnetic torque acts in the direction of the rotation. Remarks similar to those already made concerning the roles played by the generated voltage and electromagnetic torque in the energy-conversion process in synchronous machines apply equally well to dc machines.

3–3 MMF OF DISTRIBUTED WINDINGS

Most armatures have distributed windings, i.e., windings which are spread over a number of slots around the air-gap periphery, as in Figs. 3-1 and 3-2. The individual coils are interconnected so that the result is a magnetic field having the same number of poles as the field winding.

The study of the magnetic fields of distributed windings can be approached by study of the magnetic field of the single N-turn coil which spans 180 electrical degrees and is defined as a *full-pitch coil,* as shown in Fig. 3-19a. The dots and crosses indicate current toward and away from the reader, re-

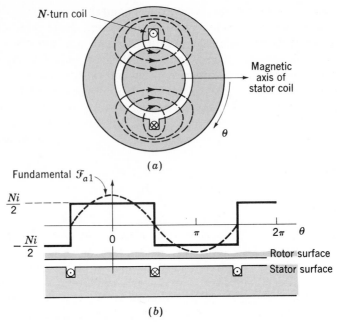

Fig. 3-19. The mmf of a concentrated full-pitch coil.

spectively. For simplicity, a concentric cylindrical rotor is shown. The general nature of the magnetic field produced by the current in the coil is shown by the dashed lines in Fig. 3-19a. Since the permeability of the armature and field iron is much greater than that of air, it is sufficiently accurate for our present purposes to assume that all the reluctance of the magnetic circuit is in the two air gaps. From symmetry of the structure it is evident that the magnetic field intensity H in the air gap at angle θ under one pole is the same in magnitude as that at $\theta + \pi$ under the opposite pole but the fields are in the opposite direction.

As discussed in Art. 1-1, the mmf $\mathcal{F}$ of any closed path in a magnetic circuit is defined as the net ampere-turns Ni enclosed by that path. From Ampère's law the mmf is equal to the line integral of $\mathbf{H}$ around that path

$$\mathcal{F} = \oint \mathbf{H} \cdot d\mathbf{l} \tag{3-4}$$

This is a more general form of the magnetic-circuit expression of Eq. 1-1. The usefulness of the concept of mmf stems from the fact that in well-defined magnetic circuits mmf can be associated with various portions of a path in such a fashion that it is related by some simple geometric coefficient to the magnetic field intensity which exists over that portion of the path.

Around any of the closed paths shown by the flux lines in Fig. 3-19a the mmf is Ni. The assumption that all the reluctance of this magnetic circuit is in the air gap leads to the result that the line integral of $\mathbf{H}$ inside the iron is negligibly small, and thus it is reasonable to neglect the mmf drops associated with portions of the magnetic circuit inside the iron. Symmetry has shown that the air-gap **H** fields on opposite sides of the rotor are equal in magnitude but opposite in direction. It follows that the air-gap mmf should be similarly distributed.

Figure 3-19*b* shows the air gap and winding in developed form, i.e., laid out flat. The air-gap mmf distribution is shown by the steplike distribution of amplitude $\pm Ni/2$. On the assumption of narrow slot openings, the mmf jumps abruptly by Ni in crossing from one side to the other of a coil. This mmf distribution is discussed again in Art. 3-4, where the resultant magnetic fields are evaluated.

a. AC Machines

In the design of ac machines, serious efforts are made to distribute the winding so as to produce a close approximation to a sinusoidal space distribution of mmf. We shall focus attention on the fundamental component.

The rectangular mmf wave of the concentrated full-pitch coil of Fig. 3-19*b* can be resolved into a Fourier series comprising a fundamental component and a series of odd harmonics. The fundamental component $\mathcal{F}_{a1}$ is

$$\mathcal{F}_{a1} = \frac{4}{\pi} \frac{Ni}{2} \cos \theta \qquad (3\text{-}5)$$

where θ is measured from the axis of the stator coil, as shown by the dashed sinusoid in Fig. 3-19*b*. It is a sinusoidal space wave of amplitude

$$F_{1,\text{peak}} = \frac{4}{\pi} \frac{Ni}{2} \qquad (3\text{-}6)$$

with its peak aligned with the magnetic axis of the coil.

Now consider the effect of distributing a winding in several slots. For example, Fig. 3-20*a* shows phase *a* of the armature winding of a somewhat simplified 2-pole 3-phase ac machine. Phases *b* and *c* occupy the empty slots. The windings of the three phases are identical and are located with their magnetic axes 120 electrical degrees apart. We shall direct attention to the mmf of phase *a* alone, postponing the discussion of the effects of all three phases until Art. 3-5. The winding is arranged in two layers, each coil of n_c turns having one side in the top of a slot and the other coil side in the bottom of a slot a pole pitch away. The two-layer arrangement simplifies the geometrical problem of getting the end turns of the individual coils past each other.

Figure 3-20*b* shows one pole of this winding laid out flat. The mmf wave is a series of steps each of height $2n_c i_c$ equal to the ampere-conductors in the slot, where i_c is the coil current. Its space-fundamental component is shown by the sinusoid. It can be seen that the distributed winding produces a closer

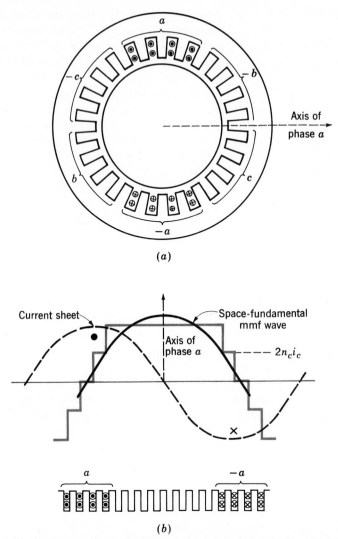

Fig. 3-20. The mmf of one phase of a distributed 2-pole 3-phase winding with full-pitch coils.

approximation to a sinusoidal mmf wave than the concentrated coil of Fig. 3-19.

The resultant fundamental mmf wave of a distributed winding is less than the sum of the fundamental components of the individual coils because the magnetic axes of the individual coils are not aligned with the resultant. The modified form of Eq. 3-5 for a distributed P-pole winding having N_{ph} series turns per phase is

$$\mathcal{F}_{a1} = \frac{4}{\pi} k_w \frac{N_{ph}}{P} i_a \cos \theta \qquad (3\text{-}7)$$

in which the factor $4/\pi$ arises from the Fourier-series analysis of the sawtooth mmf wave of a concentrated full-pitch coil, as in Eq. 3-5, and the winding factor k_w takes into account the distribution of the winding. The factor $k_w N_{ph}$ is the effective series turns per phase for the fundamental mmf. (See Appendix B for details.) Through the use of fractional-pitch coils and other artifices the effects of space harmonics in ac machines can be made small.

Equation 3-7 describes the space-fundamental component of the mmf wave produced by current in phase a. It is equal to the mmf wave produced by a finely divided sinusoidally distributed current sheet placed on the inner periphery of the stator, as shown by the sine wave labeled "current sheet" in Fig. 3-20b. The mmf is a standing wave whose *spatial* distribution around the periphery is described by $\cos \theta$. Its peak is along the magnetic axis of phase a and its peak amplitude is proportional to the instantaneous current i_a. Accordingly, if the current $i_a = I_m \cos \omega t$, the *time* maximum of the peak is

$$F_{max} = \frac{4}{\pi} k_w \frac{N_{ph}}{P} I_m \tag{3-8}$$

In Art. 3-5 we shall study the effect of currents in all three phases.

In a directly analogous fashion, rotor windings are often distributed in slots to reduce the effects of space harmonics. Figure 3-21a shows the rotor of a typical 2-pole round-rotor generator. Although the winding is symmetrical with respect to the rotor axis, the number of turns per slot can be varied to control the various harmonics. In Fig. 3-21b there are fewer turns in the slots nearest the pole face. In addition, the designer can vary the spacing of the slots. As for distributed armature windings, the fundamental mmf wave of a P-pole rotor winding $\mathcal{F}_{r1}$, can be expressed in terms of the total number of turns N_r, the winding current I_r, and a winding factor k_r

$$\mathcal{F}_{r1} = \frac{4}{\pi} k_r \frac{N_r}{P} I_r \cos \theta \tag{3-9}$$

Its peak amplitude is

$$F_{1,peak} = \frac{4}{\pi} k_r \frac{N_r}{P} I_r \tag{3-10}$$

b. DC Machines

Because of the restrictions imposed on the winding arrangement by the commutator, the mmf wave of a dc machine armature approximates a sawtooth waveform more nearly than the sine wave of ac machines. For example, Fig. 3-22 shows diagrammatically in cross section the armature of a 2-pole dc machine. (In practice a larger number of slots would probably be used.) The current directions are shown by dots and crosses. The armature winding is equivalent to a coil wrapped around the armature and producing a magnetic

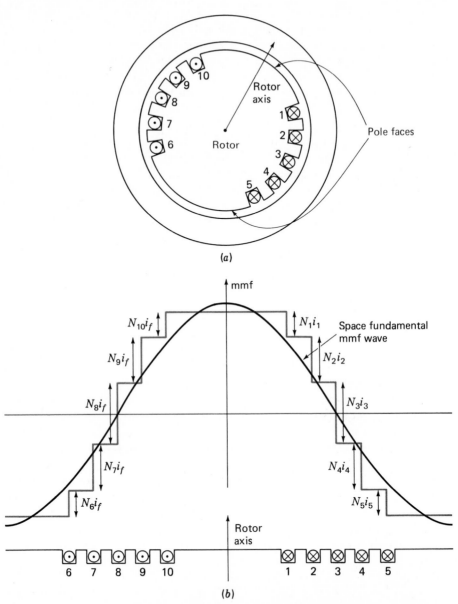

Fig. 3-21. The mmf of a distributed winding on the rotor of a round rotor generator.

field whose axis is vertical. As the armature rotates, the coil connections to the external circuit are changed through commutator action, so that the magnetic field of the armature is always perpendicular to that of the field winding and a continuous unidirectional torque results. Commutator action is described in Art. 5-2.

Figure 3-23a shows this winding laid out flat. The mmf wave is shown in

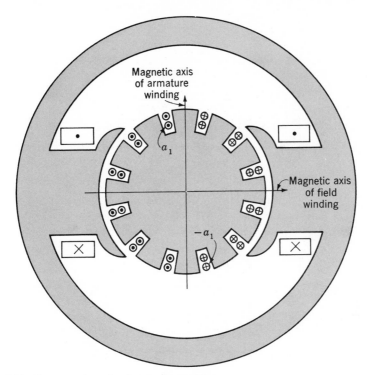

Fig. 3-22. Cross section of a 2-pole dc machine.

Fig. 3-23b. On the assumption of narrow slots, it consists of a series of steps. The height of each step equals the number of ampere-conductors $2n_c i_c$ in a slot, where n_c is the number of turns in each coil and i_c is the coil current, a two-layer winding and full-pitch coils being assumed. The peak value of the mmf wave is along the magnetic axis of the armature, midway between the field poles. This winding is equivalent to a coil of $12n_c i_c$ A · turns distributed around the armature. On the assumption of symmetry at each pole, the peak value of the mmf wave at each armature pole is $6n_c i_c$ A · turns.

This mmf wave can be represented approximately by the sawtooth wave drawn in Fig. 3-23b and repeated in Fig. 3-23c. For a more realistic winding with a larger number of armature slots per pole, the triangular distribution becomes a close approximation. It is the exact equivalent of the mmf wave of a uniformly distributed current sheet wrapped around the armature and carrying current in the dot and cross directions, as shown by the rectangular space distribution of current density in Fig. 3-23c.

For our preliminary study, it is convenient to resolve the mmf waves of distributed windings into a Fourier series. The fundamental component of the sawtooth mmf wave of Fig. 3-23c is shown by the sine wave. Its peak value is $8/\pi^2 = 0.81$ times the height of the sawtooth wave. The fundamental mmf wave is the exact equivalent of the mmf wave of a sinusoidally distributed

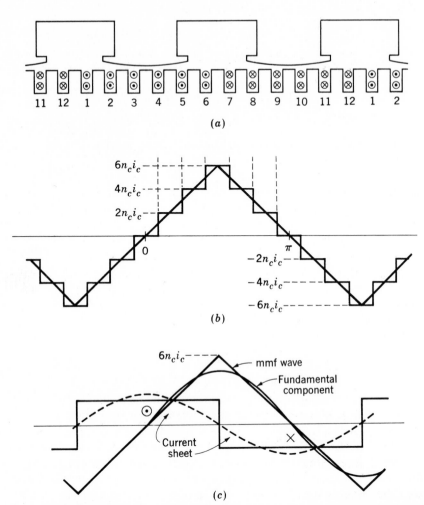

Fig. 3-23. (a) Developed sketch of the dc machine of Fig. 3-22, (b) mmf wave, (c) equivalent sawtooth mmf wave, its fundamental component, and equivalent rectangular current sheet.

current sheet wrapped around the armature whose peak value equals the fundamental component of the rectangular current sheet of Fig. 3-23c. This sinusoidally distributed current sheet is shown dashed in Fig. 3-23c.

It should be noted that the mmf wave depends only on the winding arrangement and symmetry of the magnetic structure at each pole. The flux-density wave, however, depends not only on the mmf but also on the magnetic boundary conditions, primarily the length of the air gap, the effect of the slot openings, and the shape of the pole face. The designer takes these effects into account by means of flux maps, but these details need not concern us here.

Machines often have a magnetic structure with more than two poles. For

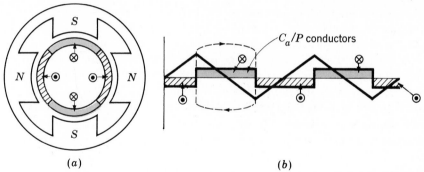

Fig. 3-24. (a) Cross section of a 4-pole dc machine and (b) development of current sheet and mmf wave.

example, Fig. 3-24a shows schematically a 4-pole dc machine. The field winding produces alternate north-south-north-south polarity, and the armature conductors are distributed in four belts of slots carrying currents alternately toward and away from the reader, as symbolized by the crosshatched areas. This machine is shown in laid-out form in Fig. 3-24b. The corresponding sawtooth armature-mmf wave is also shown. On the assumption of symmetry of the winding and field poles each successive pair of poles is like every other pair of poles. Magnetic conditions in the air gap can then be determined by examining any pair of adjacent poles, that is, 360 electrical degrees.

The peak value of the sawtooth armature-mmf wave is

$$F_a = \frac{1}{2} \frac{C_a}{P} \frac{i_a}{m} \qquad \text{A} \cdot \text{turns/pole} \qquad (3\text{-}11)$$

where C_a = total number of conductors in armature winding.
P = number of poles
i_a = armature current
m = number of parallel paths through armature winding

Thus i_a/m is the current in each conductor. This equation comes directly from the line integral around the dotted closed path in Fig. 3-24b which crosses the air gap twice and encloses C_a/P conductors, each carrying current i_a/m in the same direction. In more compact form

$$F_a = \frac{N_a}{P} i_a \qquad (3\text{-}12)$$

where $N_a = C_a/m$ is the number of series armature turns. From the Fourier series for the sawtooth mmf wave of Fig. 3-24b the peak value of the space fundamental is

$$F_{a1,\text{peak}} = \frac{8}{\pi^2} \frac{N_a}{P} i_a \qquad (3\text{-}13)$$

We shall base our preliminary investigations of both ac and dc machines on the assumption of sinusoidal space distributions of mmf. This model will be found to give very satisfactory results for most problems involving ac machines because their windings can be distributed so as to minimize the effects of harmonics. Because of the restrictions placed on the winding arrangement by the commutator, the mmf waves of dc machines inherently approach more nearly a sawtooth waveform. Nevertheless the theory based on a sinusoidal model brings out the essential features of dc machine theory. The results can readily be modified whenever necessary to account for any significant discrepancies.

We shall also use a 2-pole machine as our mathematical model. The results can immediately be extrapolated to a P-pole machine when it is recalled that electrical angles θ and electrical angular velocities ω are related to mechanical angles θ_m and mechanical angular velocities ω_m through Eqs. 3-1 and 3-3.

For a preliminary study we shall further simplify our model by assuming that the stator and rotor air-gap surfaces are smooth, concentric cylinders.

3–4 MAGNETIC FIELDS IN ROTATING MACHINERY

The behavior of electric machinery is determined by the magnetic fields created by currents in the various windings of the machine. This section discusses how these magnetic fields and currents are related.

a. Machines with Uniform Air Gaps

Figure 3-25a shows a single full-pitch N-turn coil in a magnetic structure with a concentric cylindrical rotor. The mmf of this configuration is shown in Fig. 3-25b. For such a structure, with a uniform air gap of length g at radius r_r (very much larger than g) it is quite accurate to assume that the magnetic field $\mathbf{H}$ in the air gap is directed only radially and has constant magnitude across the air gap.

As discussed in Art. 3-3, the mmf distribution of Fig. 3-25b is equal to the line integral of $\mathbf{H}$ across the air gap. For this case of constant radial H this integral is simply equal to the product of the radial magnetic field H times the air-gap length g, and thus H can be found simply by dividing the mmf by the air-gap length

$$H = \frac{\mathcal{F}}{g} \tag{3-14}$$

In Fig. 3-25c the radial H field and mmf are seen to be identical in form, simply related by a factor of $1/g$.

The fundamental space-harmonic component of H can be found directly

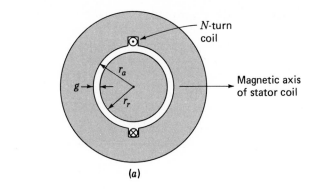

(a)

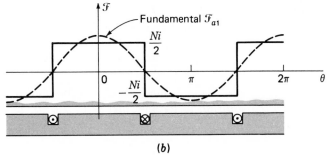

(b)

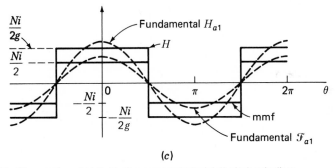

(c)

Fig. 3-25. The mmf and H field of a concentrated full-pitch winding.

from the fundamental component $\mathcal{F}_{a1}$, given by Eq. 3-5

$$H_{a1} = \frac{\mathcal{F}_{a1}}{g} = \frac{4}{\pi} \frac{Ni}{2g} \cos \theta \tag{3-15}$$

It is a sinusoidal space wave of amplitude

$$H_{1,\text{peak}} = \frac{4}{\pi} \frac{Ni}{2g} \tag{3-16}$$

For a distributed winding such as that of Fig. 3-20, the air-gap magnetic field intensity is easily found once the mmf is known. Thus the fundamental component of H can be found from Eq. 3-7.

$$H_{a1} = \frac{4}{\pi} k_w \frac{N_{ph}}{Pg} i_a \cos \theta \qquad (3\text{-}17)$$

where N_{ph} = number of series turns per phase
θ = electrical angle measured with respect to magnetic axis of winding
k_w = winding factor

and the equation has been written for the general case of a P-pole machine. It is useful to notice that the space-fundamental mmf and H produced by a distributed winding of winding factor k_w and N_{ph}/P series turns per pole produces the same fundamental mmf and H as a single full-pitch winding of $k_w N_{ph}/P$ turns per pole. In the analysis of machines with distributed windings this result is useful since in considering space-fundamental quantities it permits the distributed solution to be obtained from the single N-turn full-pitch coil solution simply by replacing N by the effective number of turns of the distributed winding.

b. Machines with Nonuniform Air Gaps

Figure 3-26a shows the structure of a typical dc machine, and Fig. 3-26b shows the structure of a typical salient-pole synchronous machine. Both machines consist of magnetic structures with extremely nonuniform air gaps. In such

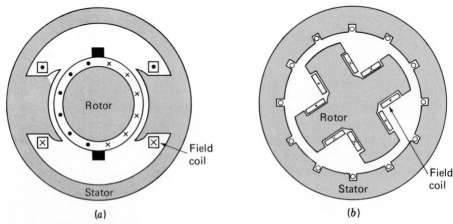

(a)   (b)

Fig. 3-26. Structure of typical salient-pole machines: (a) dc machine, (b) salient-pole synchronous machine.

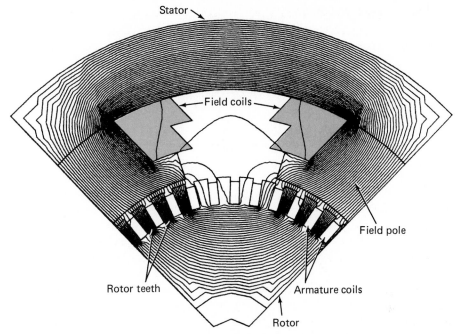

Fig. 3-27. Finite-element solution of the magnetic filed distribution in a salient-pole dc generator. Field coils excited, no current in armature coils. (*General Electric Company*)

cases the magnetic fields are somewhat more complex than for a uniform air gap. In fact, the effect of slots, which clearly violate the assumption of an absolutely uniform air gap, may be significant enough in some cases to require modification of the uniform-air-gap results.

Detailed analysis of the magnetic field distributions in such complex situations requires complete solutions of the field problem. For example, Fig. 3-27 shows the magnetic field distribution in a salient-pole dc generator (obtained by a digital-computer-based finite-element solution). However, experience has shown that through various simplifying assumptions, analytical techniques which yield reasonably accurate results can be developed. These techniques are illustrated in later chapters, where the effects of saliency on both dc and ac machines are discussed.

3–5 ROTATING MMF WAVES IN AC MACHINES

To understand the theory and operation of polyphase ac machines it is necessary to study the nature of the mmf wave produced by a polyphase winding. Attention will be focused on a 2-pole machine or one pair of a P-pole winding. To develop insight into the polyphase situation it is helpful to begin with an analysis of a single-phase winding.

a. Single-Phase Winding

Figure 3-28a shows the space-fundamental mmf distribution of a single-phase winding, where, from Eq. 3-9,

$$\mathcal{F}_{a1} = \frac{4}{\pi} k_w \frac{N_{ph}}{P} i_a \cos\theta \qquad (3\text{-}18)$$

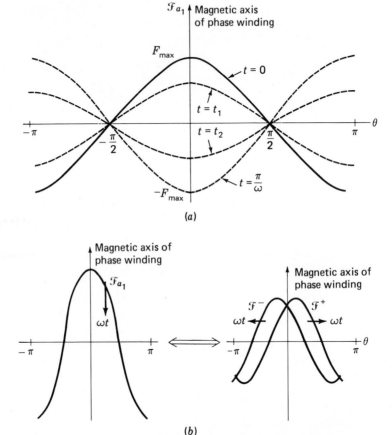

Fig. 3-28. Single-phase-winding fundamental mmf: (a) mmf distribution of a single-phase winding at various times, (b) total mmf $\mathcal{F}_{a1}$ decomposed into two traveling waves $\mathcal{F}^+$ and $\mathcal{F}^-$, (c) phasor decomposition of $\mathcal{F}_{a1}$.

When this winding is excited by a sinusoidally varying current in time

$$i_a = I_a \cos \omega t \qquad (3\text{-}19)$$

the mmf distribution is given by

$$\mathcal{F}_{a1} = (F_{\max} \cos \theta) \cos \omega t \qquad (3\text{-}20)$$

Equation 3-20 has been written in a form to emphasize the fact that the result is an mmf distribution of maximum amplitude

$$F_{\max} = \frac{4}{\pi} k_w \frac{N_{\text{ph}}}{P} I_a \qquad (3\text{-}21)$$

which remains fixed in space but whose amplitude varies sinusoidally in time at frequency ω, as shown in Fig. 3-28a.

Use of a common trigonometric identity† permits Eq. 3-20 to be rewritten in the form

$$\mathcal{F}_{a1} = F_{\max} \left[\tfrac{1}{2} \cos (\theta - \omega t) + \tfrac{1}{2} \cos (\theta + \omega t) \right] \qquad (3\text{-}22)$$

which shows that the mmf of a single-phase winding can be resolved into two rotating mmf waves each of amplitude one-half the maximum amplitude of $\mathcal{F}_{a1}$, with one, $\mathcal{F}^+$, traveling in the $+\theta$ direction and the other, $\mathcal{F}^-$, traveling in the $-\theta$ direction, both with angular velocity ω

$$\mathcal{F}^+ = \tfrac{1}{2} F_{\max} \cos (\theta - \omega t) \qquad (3\text{-}23)$$
$$\mathcal{F}^- = \tfrac{1}{2} F_{\max} \cos (\theta + \omega t) \qquad (3\text{-}24)$$

This decomposition is shown graphically in Fig. 3-28b and in a phasor representation in Fig. 3-28c.

The fact that the mmf of a single-phase winding excited by a source of alternating current can be resolved into rotating traveling waves is an important conceptual step in understanding ac machinery. As shown in the next section, in polyphase ac machinery the windings are equally displaced in space phase and the winding currents are similarly displaced in time phase, with the result that the negative-traveling flux waves of the various windings add to zero while the positive-traveling flux waves reinforce, giving a single positive-traveling flux wave. In single-phase ac machinery the machine is designed so that the effects of the positive-traveling flux wave are maximized and those of the negative-traveling flux wave are minimized.

†$\cos \alpha \cos \beta = \tfrac{1}{2} \cos (\alpha - \beta) + \tfrac{1}{2} \cos (\alpha + \beta)$.

b. Polyphase Winding

In this section we shall study the mmf patterns of 3-phase windings such as those found on the stator of 3-phase induction and synchronous machines. Once again attention will be focused on a 2-pole machine or one pair of poles of a P-pole winding. The analyses presented can readily be extended to a polyphase winding with any number of phases.

In a 3-phase machine the windings of the individual phases are displaced from each other by 120 electrical degrees in space around the air-gap circumference, as shown by the coils a, $-a$, b, $-b$, and c, $-c$ in Fig. 3-29. The concentrated full-pitch coils shown here may be considered to represent distributed windings producing sinusoidal mmf waves centered on the magnetic axes of the respective phases. The three-component sinusoidal mmf waves are accordingly displaced 120 electrical degrees in *space*. But each phase is excited by an alternating current which varies in magnitude sinusoidally with *time*. Under balanced 3-phase conditions the instantaneous currents are

$$i_a = I_m \cos \omega t \tag{3-25}$$

$$i_b = I_m \cos (\omega t - 120°) \tag{3-26}$$

$$i_c = I_m \cos (\omega t - 240°) \tag{3-27}$$

where I_m is the maximum value of the current and the time origin is arbitrarily taken as the instant when the phase a current is a positive maximum. The phase sequence is assumed to be abc. The instantaneous currents are shown in Fig. 3-30. The dots and crosses in the coil sides, Fig. 3-29, indicate the reference directions for positive phase currents.

The mmf of phase a has been shown to be

$$\mathcal{F}_{a1} = \mathcal{F}_{a1}^{+} + \mathcal{F}_{a1}^{-} \tag{3-28}$$

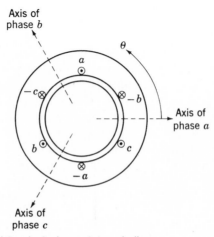

Fig. 3-29. Simplified 2-pole 3-phase stator winding.

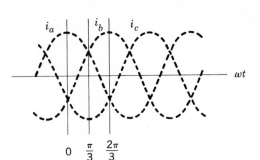

Fig. 3-30. Instantaneous 3-phase currents.

where
$$\mathcal{F}_{a1}^{+} = \tfrac{1}{2} F_{\max} \cos(\theta - \omega t) \qquad (3\text{-}29)$$
$$\mathcal{F}_{a1}^{-} = \tfrac{1}{2} F_{\max} \cos(\theta + \omega t) \qquad (3\text{-}30)$$

and
$$F_{\max} = \frac{4}{\pi} k_w \frac{N_{\mathrm{ph}}}{P} I_m \qquad (3\text{-}31)$$

Similarly, for phases b and c,

$$\mathcal{F}_{b1} = \mathcal{F}_{b1}^{+} + \mathcal{F}_{b1}^{-} \qquad (3\text{-}32)$$
$$\mathcal{F}_{b1}^{+} = \tfrac{1}{2} F_{\max} \cos(\theta - \omega t) \qquad (3\text{-}33)$$
$$\mathcal{F}_{b1}^{-} = \tfrac{1}{2} F_{\max} \cos(\theta + \omega t - 240°) \qquad (3\text{-}34)$$

and
$$\mathcal{F}_{c1} = \mathcal{F}_{c1}^{+} + \mathcal{F}_{c1}^{-} \qquad (3\text{-}35)$$
$$\mathcal{F}_{c1}^{+} = \tfrac{1}{2} F_{\max} \cos(\theta - \omega t) \qquad (3\text{-}36)$$
$$\mathcal{F}_{c1}^{-} = \tfrac{1}{2} F_{\max} \cos(\theta + \omega t - 120°) \qquad (3\text{-}37)$$

The total mmf is the sum of the contributions from each of the three phases

$$\mathcal{F}(\theta, t) = \mathcal{F}_{a1} + \mathcal{F}_{b1} + \mathcal{F}_{c1} \qquad (3\text{-}38)$$

This summation can be performed quite easily in terms of the positive- and negative-traveling waves. The negative-traveling waves sum to zero

$$\begin{aligned}
\mathcal{F}^{-}(\theta, t) &= \mathcal{F}_{a1}^{-} + \mathcal{F}_{b1}^{-} + \mathcal{F}_{c1}^{-} \\
&= \tfrac{1}{2} F_{\max}[\cos(\theta + \omega t) + \cos(\theta + \omega t - 120°) + \cos(\theta + \omega t - 240°)] \\
&= 0 \qquad\qquad (3\text{-}39)
\end{aligned}$$

and the positive-traveling waves reinforce

$$\mathcal{F}^{+}(\theta, t) = \mathcal{F}_{a1}^{+} + \mathcal{F}_{b1}^{+} + \mathcal{F}_{c1}^{+} = \tfrac{3}{2} F_{\max} \cos(\theta - \omega t) \qquad (3\text{-}40)$$

Thus, the result of displacing the three windings by 120° in space phase and displacing the winding currents by 120° in time phase is a single positive traveling wave

$$\mathcal{F}(\theta, t) = \tfrac{3}{2}F_{\text{max}} \cos(\theta - \omega t) \qquad (3\text{-}41)$$

The wave described by Eq. 3-41 is a sinusoidal function of the space angle θ. It has a constant amplitude and a space-phase angle ωt which is a linear function of time. The angle ωt provides rotation of the entire wave around the air gap at the constant angular velocity ω. Thus, at a fixed time t_x the wave is a sinusoid in space with its positive peak displaced ωt_x electrical radians from the fixed point on the winding which is the origin for θ; at a later instant t_y the positive peak of the same wave is displaced ωt_y from the origin, and the wave has moved $\omega(t_y - t_x)$ around the gap. At $t = 0$ the current in phase a is a maximum, and the positive peak of the resultant mmf wave is located at the axis of phase a; one-third of a cycle later the current in phase b is a maximum, and the positive peak is located at the axis of phase b; and so on. The angular velocity of the wave is $\omega = 2\pi f$ electrical radians per second. For a P-pole machine the rotational speed is

$$\omega_m = \frac{2}{P}\omega \qquad \text{rad/s} \qquad (3\text{-}42)$$

or

$$n = \frac{120f}{P} \qquad \text{r/min} \qquad (3\text{-}43)$$

results which are consistent with Eqs. 3-2 and 3-3.

In general it can be shown that a rotating field of constant amplitude will be produced by a q-phase winding excited by balanced q-phase currents when the respective phases are wound $2\pi/q$ electrical radians apart in space. The constant amplitude will be $q/2$ times the maximum contribution of any one phase, and the speed will be $\omega = 2\pi f$ electrical radians per second.

A polyphase winding excited by balanced polyphase currents is thus seen to produce the same general effect as spinning a permanent magnet about an axis perpendicular to the magnet, or as the rotation of dc-excited field poles.

c. Graphical Analysis of Polyphase MMF

For balanced 3-phase currents as given by Eqs. 3-25 to 3-27, the resultant rotating mmf can also be shown graphically. Consider the state of affairs at $t = 0$, Fig. 3-30, the moment when the phase a current is at its maximum value I_m. The mmf of phase a then has its maximum value F_{max}, as shown by the vector $\mathbf{F}_a = \mathbf{F}_{\text{max}}$ drawn along the magnetic axis of phase a in Fig. 3-31a. At this moment the currents i_b and i_c are both $I_m/2$ in the negative direction, as shown by the dots and crosses in Fig. 3-31a indicating the *actual instantaneous* directions. The corresponding mmf's of phases b and c are shown by the vectors $\mathbf{F}_b$ and $\mathbf{F}_c$, both equal to $\mathbf{F}_{\text{max}}/2$ drawn in the negative direction along

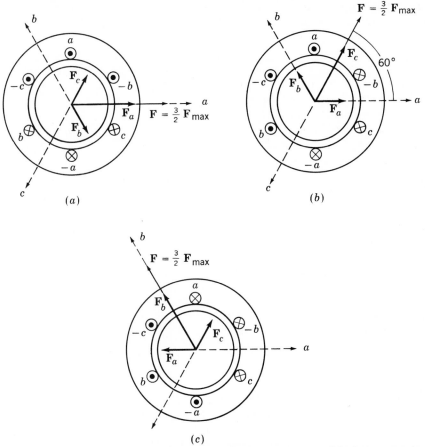

Fig. 3-31. The production of a rotating magnetic field by means of 3-phase currents.

the magnetic axes of phases b and c, respectively. The resultant, obtained by adding the individual contributions of the three phases, is a vector $\mathbf{F} = \frac{3}{2}\mathbf{F}_{max}$ centered on the axis of phase a. It represents a sinusoidal space wave with its positive half wave centered on the axis of phase a and having an amplitude $\frac{3}{2}$ times that of the phase a contribution alone.

At a later time $\omega t_1 = \pi/3$ (Fig. 3-30) the currents in phases a and b are a positive half maximum and that in phase c is a negative maximum. The individual mmf components and their resultant are now shown in Fig. 3-31b. The resultant has the same amplitude as at $t = 0$, but it has now rotated counterclockwise 60 electrical degrees in space. Similarly, at $\omega t_2 = 2\pi/3$ (when the phase b current is a positive maximum and the phase a and c currents are a negative half maximum) the same resultant mmf distribution is again obtained, but it has rotated counterclockwise 60 electrical degrees still farther and is now aligned with the magnetic axis of phase b (see Fig. 3-31c). As time passes, then, the resultant mmf wave retains its sinusoidal form and amplitude but shifts progressively around the air gap. This shift corresponds to a field rotating uniformly around the circumference of the air gap. Results consistent with this conclusion can be obtained by sketching the distribution at any arbitrary instant of time.

In one cycle the resultant mmf must be back in the position of Fig. 3-31a. The mmf wave therefore makes one revolution per cycle in a 2-pole machine. In a P-pole machine the wave travels 1 wavelength in $2/P$ revolutions per cycle.

3–6 GENERATED VOLTAGE

The general nature of the induced voltage has already been discussed in Art. 3-2. Quantitative expressions for the induced voltage will now be determined.

a. AC Machines

An elementary ac machine is shown in cross section in Fig. 3-32. The coils on both the rotor and stator have been shown as single multiple-turn full-pitch coils. The analysis can readily be extended to distributed windings. The field winding on the rotor is assumed to produce a sinusoidal space wave of flux density B at the stator surface. The rotor is spinning at a constant angular velocity of ω electrical radians per second. Although a 2-pole machine is shown, the derivations presented here are for the general case of a P-pole machine.

When the rotor poles are in line with the magnetic axis of the stator coil, the flux linkage with the stator N-turn coil is $N\Phi$, where Φ is the air-gap flux

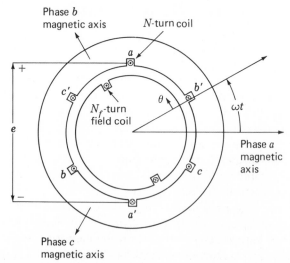

Fig. 3-32. Cross-sectional view of an elementary 3-phase ac machine.

per pole. For the assumed sinusoidal flux-density wave

$$B = B_{\text{peak}} \cos \theta \tag{3-44}$$

where B_{peak} is its peak value at the rotor pole center and θ is measured in electrical radians from the rotor pole axis. The air-gap flux per pole is the integral of the flux density over the pole area; thus, for a 2-pole machine

$$\Phi = \int_{-\pi/2}^{+\pi/2} B_{\text{peak}} \cos \theta \; lr \; d\theta = 2B_{\text{peak}} lr \tag{3-45}$$

where l is the axial length of the stator and r is its radius at the air gap. For a P-pole machine

$$\Phi = \frac{2}{P} 2B_{\text{peak}} lr \tag{3-46}$$

because the pole area is $2/P$ times that of a 2-pole machine of the same length and diameter.

As the rotor turns, the flux linkage varies as the cosine of the angle α between the magnetic axes of the stator coil and rotor. With the rotor spinning at constant angular velocity ω, the flux linkage with the stator coil is

$$\lambda = N\Phi \cos \omega t \tag{3-47}$$

where time t is arbitrarily chosen as zero when the peak of the flux-density wave coincides with the magnetic axis of the stator coil. By Faraday's law, the voltage induced in the stator coil is

$$e = \frac{d\lambda}{dt} = N\frac{d\Phi}{dt} \cos \omega t - \omega N\Phi \sin \omega t \tag{3-48}$$

The polarity of this induced voltage is such that if the stator coil were short-circuited, the induced voltage would cause a current to flow in the direction that would oppose any change in the flux linkage of the stator coil. Although Eq. 3-48 was derived on the assumption that only the field winding was producing air-gap flux, the equation applies equally well to the general situation where Φ is the net air-gap flux per pole produced by currents on both the rotor and the stator.

The first term on the right-hand side of Eq. 3-48 is a transformer voltage and is present only when the amplitude of the flux-density wave changes with time. The second term is a *speed voltage* generated by the relative motion of the air-gap flux wave and the stator coil. In the normal steady-state operation of most rotating machines the amplitude of the air-gap flux wave is constant, and the generated voltage is simply the speed voltage. The term *electromotive force* (emf) is often used for the speed voltage. Thus

$$e = \omega N \Phi \sin \omega t \tag{3-49}$$

Equation 3-49 was derived directly from Faraday's law. Alternatively, the voltage e induced in a conductor of length l moving with linear velocity v in a non-time-varying magnetic field of flux density B is given by the "cutting-of-flux" equation

$$e = Blv \tag{3-50}$$

where B, l, and v are mutually perpendicular. Properly interpreted, this equation also applies to rotating machines and gives a useful alternative basis for visualizing induced voltages. To show that Eq. 3-50 applies to rotating machines we shall now derive it from Eq. 3-49.

Substitution of Eq. 3-46 in Eq. 3-49 gives

$$e = \omega N \frac{2}{P} 2 B_{\text{peak}} lr \sin \omega t \tag{3-51}$$

$$e = \frac{2}{P} \omega r (2lN) B_{\text{peak}} \sin \omega t \tag{3-52}$$

Now $B_{\text{peak}} \sin \omega t$ is the flux density B_{coil} at the stator coil side (Fig. 3-32), $2\omega/P$ is the mechanical angular velocity ω_m, and $2lN$ is the total active length of conductors in the two coil sides; thus, for a concentrated full-pitch coil

$$e = B_{\text{coil}}(2lN)r\omega_m = B_{\text{coil}}(2lN)v \tag{3-53}$$

where $v = r\omega_m$ is the linear velocity of the conductor relative to the field. Even though the conductors are embedded in slots, the voltage can be computed by the cutting-of-flux concept, Eq. 3-50, just as if the conductors were lifted out of the slots and placed directly in the air-gap field.

In the normal steady-state operation of ac machines we are usually interested in the rms values of voltages and currents rather than their instantaneous values. From Eq. 3-49 the maximum value of the induced voltage is

$$E_{\text{max}} = \omega N \Phi = 2\pi f N \Phi \tag{3-54}$$

and its rms value is

$$E_{\text{rms}} = \frac{2\pi}{\sqrt{2}} f N \Phi = 4.44 f N \Phi \tag{3-55}$$

where f is the frequency in hertz. These equations are identical in form to the corresponding emf equations for a transformer. Relative motion of a coil and a constant-amplitude spatial flux-density wave in a rotating machine produces the same voltage effect as a time-varying flux in association with stationary coils does in a transformer. Rotation, in effect, introduces the time element and transforms a *space* distribution of flux density into a *time* variation of voltage.

The voltage induced is a single-phase voltage. For the production of a set of 3-phase voltages it follows that three coils displaced 120 electrical degrees in space must be used, as shown in elementary form in Fig. 3-12. Equation 3-55 then gives the rms voltage per phase when N is the total series turns per phase. All these elementary windings are full-pitch concentrated windings because the two sides of any coil are 180 electrical degrees apart and all the turns of that coil are concentrated in one pair of slots. In actual ac machine windings the armature coils of each phase are distributed in a number of slots, as discussed in Art. 3-3. A distributed winding makes better use of the iron and copper and improves the waveform. For distributed windings a reduction factor k_w must be applied because the emf's induced in the individual coils of any one phase group are not in time phase. Their phasor sum is then less than their numerical sum when they are connected in series. (See Appendix B for details. For most 3-phase windings, k_w is about 0.85 to 0.95.) For distributed windings Eq. 3-55 becomes

$$E_{\text{rms}} = 4.44 f k_w N_{\text{ph}} \Phi \qquad \text{V rms/phase} \qquad (3\text{-}56)$$

where N_{ph} is the number of series turns per phase.

EXAMPLE 3-1

A 2-pole 3-phase Y-connected 60-Hz round-rotor synchronous generator has a field winding on the rotor with N_f distributed turns and winding factor k_f, an armature winding on the stator with N_a turns per phase (line to neutral) and winding factor k_a. The air-gap length is g, and the mean air-gap radius is r. The active length of the armature is l. The dimensions and winding data are

$$N_f = 46 \text{ series turns} \qquad k_f = 0.90 \qquad N_a = 24 \text{ series turns/phase}$$
$$k_a = 0.833 \qquad r = 0.50 \text{ m} \qquad g = 0.075 \text{ m} \qquad l = 4.0 \text{ m}$$

The field current $I_f = 1500$ A dc. The rotor is driven by a steam turbine at a speed of 3600 r/min. Compute (a) the peak fundamental mmf $F_{1,\text{peak}}$ produced by the field winding, (b) the peak fundamental flux density $B_{1,\text{peak}}$ in the air gap, (c) the fundamental flux per pole Φ, and (d) the rms value of the open-circuit voltage generated in the armature.

Solution

(a) From Eq. 3-10

$$F_{1,\text{peak}} = \frac{4}{\pi}\frac{k_f N_f}{P}I_f = \frac{4}{\pi}\frac{0.90(46)}{2}(1500)$$

$$= \frac{4}{\pi}(20.7)(1500) = \frac{4}{\pi}(31{,}000) \quad \text{A} \cdot \text{turns/pole}$$

(b) Using Eq. 3-14, we get

$$B_{1,\text{peak}} = \frac{\mu_0 F_{1,\text{peak}}}{g} = \frac{4\pi \times 10^{-7}}{7.5 \times 10^{-2}}\frac{4}{\pi}(31{,}000) = 0.661 \text{ T}$$

Because of the effect of the slots containing the armature winding, most of the air-gap flux is confined to the stator teeth. The flux density in the teeth at a pole center is higher than the value calculated in part (b), probably by a factor of about 2. In a detailed design this flux density would be calculated to determine whether the teeth would be excessively saturated.

(c) From Eq. 3-45 or 3-46

$$\Phi = 2B_{1,\text{peak}}lr = 2(0.661)(4.0)(0.50) = 2.64 \text{ Wb}$$

(d) From Eq. 3-55 with $f = 60$ Hz

$$E_{\text{rms}} = 4.44\,fk_a N_a\, \Phi = 4.44(60)(0.833)(24)(2.64)$$
$$= 14.1 \text{ kV rms/phase line to neutral}$$

The three phases are connected in Y, as in Fig. 3-12. The voltage between line terminals is

$$\sqrt{3}\,(14.1 \text{ kV}) = 24.4 \text{ kV}$$

b. DC Machines

Even if the ultimate purpose is the generation of a direct voltage, it is evident that the speed voltage generated in an armature coil is an alternating voltage. The alternating waveform must therefore be rectified. Mechanical rectification is provided by the commutator, the device already described in elementary form in Art. 3-2c. For the single coil of Fig. 3-17 the commutator provides full-wave rectification. With the continued assumption of sinusoidal flux distribution, the voltage waveform between brushes is transformed to that of

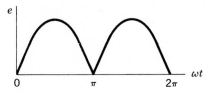

Fig. 3-33. Voltage between brushes in an elementary dc machine.

Fig. 3-33. The average, or dc, value of the voltage between brushes is

$$E_a = \frac{1}{\pi} \int_0^{\pi} \omega N \Phi \sin \omega t \, d(\omega t) = \frac{2}{\pi} \omega N \Phi \qquad (3\text{-}57)$$

For dc machines it is usually more convenient to express the voltage E_a in terms of the mechanical speed ω_m rad/s or n r/min. Substitution of Eq. 3-3 in Eq. 3-57 for a P-pole machine then yields

$$E_a = \frac{PN}{\pi} \Phi \omega_m = 2PN\Phi \frac{n}{60} \qquad (3\text{-}58)$$

The single-coil dc winding implied here is, of course, unrealistic in the practical sense, and it will be essential later to examine the action of commutators more carefully. Actually, Eq. 3-58 gives correct results for the more practical distributed dc armature windings as well, provided N is taken as the total number of turns in series between armature terminals. Usually, the voltage is expressed in terms of the total number of active conductors C_a and the number m of parallel paths through the armature winding. Because it takes two coil sides to make a turn and $1/m$ of these are connected in series, the number of series turns $N = C_a/2m$. Substitution in Eq. 3-58 then gives

$$E_a = \frac{PC_a}{2\pi m} \Phi \omega_m = \frac{PC_a}{m} \Phi \frac{n}{60} \qquad (3\text{-}59)$$

3-7 TORQUE IN NON–SALIENT–POLE MACHINES

The behavior of any electromagnetic device as a component in an electrome-chanical system can be described in terms of its Kirchhoff-law voltage equations and its electromagnetic torque. The purpose of this article is to derive the voltage and torque equations for an idealized elementary machine, results which can readily be extended later to more complex machines. We shall derive these equations from two viewpoints and show that basically they stem from the same ideas.

The first viewpoint is essentially the same as that of Art. 2-5. The machine will be regarded as a circuit element whose inductances depend on the angular position of the rotor. The flux linkages λ and magnetic field coenergy will be expressed in terms of the currents and inductances. The torque can then be found from the partial derivative of the magnetic field energy or coenergy with respect to angle and the terminal voltages from the sum of the resistance drops Ri and the Faraday-law voltages $d\lambda/dt$. The result will be a set of nonlinear differential equations describing the dynamic performance of the machine.

The second viewpoint regards the machine as two groups of windings producing magnetic fields in the air gap, one group on the stator and the other on the rotor. By making suitable assumptions regarding these fields, simple expressions can be derived for the flux linkages and magnetic field energy stored in the air gap in terms of the field quantities. The torque and generated voltage can then be found in terms of these quantities. Thus torque is expressed explicitly as the tendency for two magnetic fields to line up in the same way as permanent magnets tend to align themselves, and generated voltage is expressed as the result of relative motion between a field and a winding. These expressions lead to a simple physical picture of the normal steady-state behavior of rotating machines.

a. The Coupled-Circuit Viewpoint

Consider the elementary machine of Fig. 3-34 with one winding on the stator and one on the rotor. These windings are distributed over a number of slots so that their mmf waves can be approximated by space sinusoids. In Fig. 3-34a the coil sides s, $-s$ and r, $-r$ mark the positions of the centers of the belts of conductors comprising the distributed windings. An alternative way of drawing these windings is shown in Fig. 3-34b, which also shows reference directions for voltages and currents. Here it is assumed that current in the arrow direction produces a magnetic field in the air gap in the arrow direction, so that a single arrow defines reference directions for both current and flux. The

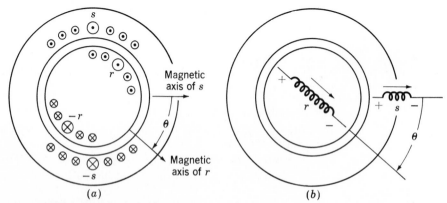

Fig. 3-34. Elementary 2-pole machine with smooth air gap: (a) winding distribution, (b) schematic representation.

stator and rotor are concentric cylinders, and slot openings are neglected. Consequently, our elementary model does not include the effects of salient poles, which will be investigated in later chapters. We shall also assume that the reluctances of the stator and rotor iron are negligible.

On these assumptions the stator and rotor self-inductances L_{ss} and L_{rr} are constant, but the stator-rotor mutual inductance depends on the angle θ between the magnetic axes of the stator and rotor windings. The mutual inductance is a positive maximum when $\theta = 0$ or 2π, is zero when $\theta = \pm\pi/2$, and is a negative maximum when $\theta = \pm\pi$. On the assumption of sinusoidal mmf waves and a uniform air gap, the space distribution of the air-gap flux wave is sinusoidal, and the mutual inductance is

$$\mathcal{L}_{sr}(\theta) = L_{sr} \cos \theta \qquad (3\text{-}60)$$

where the script letter $\mathcal{L}$ is used to denote an inductance which is a function of the electrical angle θ. The italic capital letter L denotes a constant value. Thus L_{sr} is the value of the mutual inductance when the magnetic axes of stator and rotor are aligned. In terms of the inductances, the stator and rotor flux linkages λ_s and λ_r are

$$\lambda_s = L_{ss}i_s + \mathcal{L}_{sr}(\theta)i_r = L_{ss}i_s + L_{sr}i_r \cos \theta \qquad (3\text{-}61)$$
$$\lambda_r = \mathcal{L}_{sr}(\theta)i_s + L_{rr}i_r = L_{sr}i_s \cos \theta + L_{rr}i_r \qquad (3\text{-}62)$$

where the inductances can be calculated as in Appendix B. In matrix notation

$$\begin{bmatrix} \lambda_s \\ \lambda_r \end{bmatrix} = \begin{bmatrix} L_{ss} & \mathcal{L}_{sr}(\theta) \\ \mathcal{L}_{sr}(\theta) & L_{rr} \end{bmatrix} \begin{bmatrix} i_s \\ i_r \end{bmatrix} \qquad (3\text{-}63)$$

The terminal voltages v_s and v_r are

$$v_s = R_s i_s + p\lambda_s \qquad (3\text{-}64)$$
$$v_r = R_r i_r + p\lambda_r \qquad (3\text{-}65)$$

where R_s and R_r are the winding resistances and p is the time-derivative operator d/dt. When the rotor is revolving, θ must be treated as a variable. Differentiation of Eqs. 3-61 and 3-62 and substitution of the results in Eqs. 3-64 and 3-65 then gives

$$v_s = R_s i_s + L_{ss}p i_s + L_{sr}(\cos \theta)p i_r - L_{sr}i_r(\sin \theta)p\theta \qquad (3\text{-}66)$$
$$v_r = R_r i_r + L_{rr}p i_r + L_{sr}(\cos \theta)p i_s - L_{sr}i_s(\sin \theta)p\theta \qquad (3\text{-}67)$$

where $p\theta$ is the instantaneous speed ω in *electrical* radians per second. In a 2-pole machine θ and ω are equal to the instantaneous shaft angle θ_m and speed ω_m, respectively. In a P-pole machine they are related by Eqs. 3-1 and

3-3. The second and third terms on the right-hand sides of 3-66 and 3-67 are $L\,di/dt$ induced voltages like those induced in stationary coupled circuits such as the windings of transformers. The fourth terms are caused by mechanical motion and are proportional to the instantaneous speed. They are the coupling terms relating the interchange of power between the electric and mechanical systems.

The electromagnetic torque can be found from the coenergy in the magnetic field in the air gap. From Eq. 2-50

$$W'_{\text{fld}} = \tfrac{1}{2}L_{ss}i_s^2 + \tfrac{1}{2}L_{rr}i_r^2 + L_{sr}i_s i_r \cos\theta \tag{3-68}$$

and from Eq. 2-52

$$T = +\frac{\partial W'_{\text{fld}}(\theta_m, i_s, i_r)}{\partial\theta_m} = +\frac{\partial W'_{\text{fld}}(\theta, i_s, i_r)}{\partial\theta}\frac{d\theta}{d\theta_m} \tag{3-69}$$

where T is the electromagnetic torque acting in the positive direction of θ_m and the derivative must be taken with respect to actual *mechanical* angle θ_m because we are dealing here with mechanical variables. Differentiation of Eqs. 3-68 and 3-1 for a P-pole machine then gives

$$T = -\frac{P}{2}L_{sr}i_s i_r \sin\theta = -\frac{P}{2}L_{sr}i_s i_r \sin\frac{P}{2}\theta_m \tag{3-70}$$

with T in newton-meters. The negative sign in Eq. 3-70 means that the electromagnetic torque acts in the direction to bring the magnetic fields of stator and rotor into alignment.

Equations 3-66, 3-67, and 3-70 are a set of three differential equations relating the electrical variables v_s, i_s, v_r, i_r and the mechanical variables T and θ_m. These equations, together with the constraints imposed on the electrical variables by the networks connected to the terminals (sources or loads and external impedances) and the constraints imposed on the mechanical variables (applied torques and inertial, frictional, and spring torques), determine the performance of the device as a coupling element. They are nonlinear differential equations and are difficult to solve except under special circumstances. We are not concerned with their solution here as we are using them merely as steps in the development of the theory of rotating machines.

Now consider a uniform-air-gap machine with several stator and rotor windings. The same general principles that apply to the elementary model of Fig. 3-34 also apply to the multiwinding machine. Each winding has its own self-inductance and mutual inductances with other windings. The self-inductances and mutual inductances between pairs of windings on the same side of the air gap are constant on the assumption of a uniform gap and negligible magnetic saturation. But the mutual inductances between pairs of stator and rotor windings vary as the cosine of the angle between the magnetic axes of the windings. The torque results from the tendency of the magnetic field of the rotor windings to line up with that of the stator windings. It can be expressed as the sum of terms like Eq. 3-70.

EXAMPLE 3-2

Consider the elementary rotating machine of Fig. 3-34. Its shaft is coupled to a mechanical device which can be made to absorb or deliver mechanical torque over a wide range of speeds. This machine can be connected and operated in several ways. For example, suppose the rotor winding is excited with direct current I_r and the stator winding is connected to an ac source which can either absorb or deliver electric power. Let the stator current be

$$i_s = I_s \cos \omega_s t$$

where $t = 0$ is arbitrarily chosen as the moment when the stator current has its peak value.

(a) Derive an expression for the magnetic torque developed by the machine as the speed is varied by control of the mechanical device connected to its shaft.
(b) Find the speed at which average torque will be produced if the stator frequency is 60 Hz.
(c) With the assumed current-source excitations what are the voltages induced in the stator and rotor windings at synchronous speed?

Solution

(a) From Eq. 3-70 for a 2-pole machine

$$T = -L_{sr} i_s i_r \sin \theta_m$$

For the conditions of this problem

$$T = -L_{sr} I_s I_r \cos \omega_s t \sin (\omega_m t + \delta)$$

where ω_m is the clockwise angular velocity impressed on the rotor by the mechanical drive and δ is the angular position of the rotor at $t = 0$. Using a trigonometric identity,† we have

$$T = -\tfrac{1}{2} L_{sr} I_s I_r \{ \sin [(\omega_m + \omega_s)t + \delta] + \sin [(\omega_m - \omega_s)t + \delta] \}$$

---

†$\sin \alpha \cos \beta = \tfrac{1}{2} \sin (\alpha + \beta) + \tfrac{1}{2} \sin (\alpha - \beta)$.

The torque consists of two sinusoidally time-varying terms of frequencies $\omega_m + \omega_s$ and $\omega_m - \omega_s$. As shown in Art. 3-5, the single stator winding in the machine of Fig. 3-34 creates two flux waves, one traveling in the positive θ direction with angular velocity ω_s and the second in the negative θ direction also with angular velocity ω_s. It is the interaction of the rotor with these two flux waves which results in the two components of the torque expression.

(b) Except when $\omega_m = \pm\omega_s$, the torque averaged over a sufficiently long time is zero. But if $\omega_m = +\omega_s$, the rotor is traveling in synchronism with the positive-traveling stator flux wave and the torque becomes

$$T = -\tfrac{1}{2}L_{sr}I_sI_r[\sin(2\omega_s t + \delta) + \sin \delta]$$

The first sine term is a double-frequency component whose average value is zero. The second term is the average torque

$$T_{av} = -\tfrac{1}{2}L_{sr}I_sI_r \sin \delta$$

The other possibility is $\omega_m = -\omega_s$, which merely means rotation in the counterclockwise direction, the rotor now traveling in synchronism with the negative-traveling stator flux wave. The negative sign in the expression for T_{av} means that the magnetic torque tends to reduce δ. The machine is an idealized single-phase synchronous machine. (With polyphase synchronous machines the direction of rotation is determined by the phase sequence, as shown in Art. 3-5b.)

With stator frequency of 60 Hz

$$\omega_m = \omega_s = 2\pi(60)\ \text{rad/s} = 60\ \text{r/s} = 3600\ \text{r/min}$$

(c) From the second and fourth terms of Eq. 3-66 with $\omega_m = \omega_s$ the voltage induced in the stator is

$$e_s = -\omega_s L_{ss}I_s \sin \omega_s t - \omega_s L_{sr}I_r \sin(\omega_s t + \delta)$$

From the third and fourth terms of Eq. 3-67 the voltage induced in the rotor is

$$e_r = -\omega_s L_{sr}I_s[\sin \omega_s t \cos(\omega_s t + \delta) + \cos \omega_s t \sin(\omega_s t + \delta)]$$
$$= -\omega_s L_{sr}I_s \sin(2\omega_s t + \delta)$$

The stator current induces a double-frequency voltage in the rotor.

b. The Magnetic Field Viewpoint

In the foregoing discussion the characteristics of the device as viewed from its electric and mechanical terminals have been expressed in terms of its inductances. This viewpoint gives very little insight into the internal phenomena and gives no conception of the effects of physical dimensions. An alternative formulation in terms of the interacting magnetic fields in the air gap should serve to supply some of these missing features.

Currents in the machine windings create magnetic flux in the air gap between the stator and rotor, the flux paths being completed through the stator and rotor iron. This condition corresponds to the appearance of magnetic poles on both stator and rotor, centered on their respective magnetic axes, as shown in Fig. 3-35a for a 2-pole machine with a smooth air gap. Torque is produced by the tendency of the two component magnetic fields to line up their magnetic axes. The simple physical picture is like that of two bar magnets pivoted at their centers on the same shaft. The torque is proportional to the product of the amplitudes of the stator and rotor mmf waves. It is also a function of the angle δ_{sr} between their magnetic axes. We shall show that for a smooth-air-gap machine the torque is proportional to $\sin \delta_{sr}$.

Most of the flux produced by the stator and rotor windings (roughly 90 percent in typical machines) crosses the air gap and links both windings; this flux is termed the *mutual flux*. Small percentages of the flux, however, do not cross the air gap but link only the rotor or stator winding; these are, respectively, the *rotor leakage flux* and the *stator leakage flux*. They comprise slot and toothtip leakage, end-turn leakage, and space harmonics in the air-gap field. It is only the mutual flux which is of direct concern in torque production. The leakage fluxes do affect machine performance however, by virtue of the voltages they induce in their own windings. Their effect on the electrical

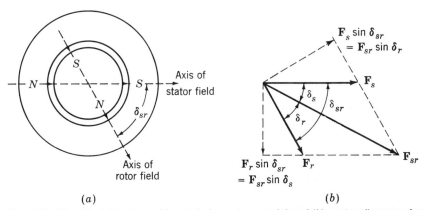

(a) (b)

Fig. 3-35. Simplified 2-pole machine: (a) elementary model and (b) vector diagram of mmf waves. Torque is produced by the tendency of the rotor and stator magnetic fields to align.

characteristics is accounted for by means of *leakage inductances* like those of a transformer. This effect, however, is an auxiliary one rather than a fundamental part of torque production.

Our analysis, then, will be in terms of the resultant mutual flux. We shall derive an expression for the magnetic coenergy stored in the air gap in terms of the stator and rotor mmfs and the angle δ_{sr} between their magnetic axes. The torque can then be found from the partial derivative of the coenergy with respect to angle δ_{sr}.

We shall now assume that the tangential component of the magnetic field in the air gap is negligible compared with the radial component; in other words, the mutual flux goes straight across the gap. We shall also assume that the radial length g of the gap (the clearance between rotor and stator) is small compared with the radius of the rotor or stator. On this assumption there is negligible difference between the flux density at the rotor surface, at the stator surface, or at any intermediate radial distance in the air gap, say, halfway between. The air-gap field then reduces to a radial field H or B whose intensity varies with angle around the periphery. The line integral of H across the gap then is simply Hg and equals the resultant mmf $\mathfrak{F}_{sr}$ of the stator and rotor windings; thus

$$Hg = \mathfrak{F}_{sr} \tag{3-71}$$

where the script $\mathfrak{F}$ denotes the mmf wave as a function of angle around the periphery.

The mmf waves of stator and rotor are spatial sine waves with δ_{sr} the phase angle between their magnetic axes in electrical degrees. They can be represented by the space vectors $\mathbf{F}_s$ and $\mathbf{F}_r$ drawn along the magnetic axes of the stator- and rotor-mmf waves, as in Fig. 3-35b. The resultant mmf $\mathbf{F}_{sr}$ acting across the air gap, also a sine wave, is the vector sum. From the trigonometric formula for the diagonal of a parallelogram, its peak value is found from

$$F_{sr}^2 = F_s^2 + F_r^2 + 2F_s F_r \cos \delta_{sr} \tag{3-72}$$

in which the F's are the peak values of the mmf waves. The resultant radial H field is a sinusoidal space wave whose peak value H_{peak} is, from Eq. 3-71,

$$H_{\text{peak}} = \frac{F_{sr}}{g} \tag{3-73}$$

Now consider the magnetic field coenergy stored in the air gap. The coenergy density at a point where the magnetic field intensity is H is $(\mu_0/2)H^2$ in SI units. The coenergy density averaged over the volume of the air gap is $\mu_0/2$ times the average value of H^2. The average value of the square of a sine wave is half its peak value. Hence,

$$\text{Average coenergy density} = \frac{\mu_0}{2}\frac{H^2_{\text{peak}}}{2} = \frac{\mu_0}{4}\left(\frac{F_{sr}}{g}\right)^2 \tag{3-74}$$

The total coenergy is

$$W'_{\text{fld}} = (\text{average coenergy density})\,(\text{volume of air gap})$$

$$= \frac{\mu_0}{4}\left(\frac{F_{sr}}{g}\right)^2 \pi Dlg = \frac{\mu_0 \pi Dl}{4g}F^2_{sr} \tag{3-75}$$

where D = average diameter of air gap, m
l = axial length of air gap, m
g = air-gap clearance, m
μ_0 = permeability of free space = $4\pi \times 10^{-7}$ H/m

From Eq. 3-72 the coenergy stored in the air gap can now be expressed in terms of the peak amplitudes of the stator- and rotor-mmf waves and the space-phase angle between them; thus

$$W'_{\text{fld}} = \frac{\mu_0 \pi Dl}{4g}(F^2_s + F^2_r + 2F_s F_r \cos \delta_{sr}) \tag{3-76}$$

An expression for the electromagnetic torque T can now be obtained in terms of the interacting magnetic fields by taking the partial derivative of the field coenergy with respect to angle. For a 2-pole machine

$$T = +\frac{\partial W'_{\text{fld}}}{\partial \delta_{sr}} = -\frac{\mu_0 \pi Dl}{2g}F_s F_r \sin \delta_{sr} \tag{3-77}$$

For a P-pole machine Eq. 3-77 gives the torque per pair of poles. The torque for a P-pole machine then is

$$T = -\frac{P}{2}\frac{\mu_0}{2}\frac{\pi Dl}{g}F_s F_r \sin \delta_{sr} \tag{3-78}$$

This important equation states that the torque is proportional to the peak values of the stator- and rotor-mmf waves F_s and F_r and the sine of the electrical space-phase angle δ_{sr} between them. The minus sign means that the fields tend to align themselves. Equal and opposite torques are exerted on stator and rotor. The torque on the stator is simply transmitted through the frame of the machine to the foundation.

One can now compare the results of Eq. 3-78 with that of Eq. 3-70. Recognizing that F_s is proportional to i_s and F_r is proportional to i_r, one sees that they are similar in form. In fact, they must be equal, as can be verified by substitution of the appropriate expressions for F_s, F_r (Art. 3-3), and L_{sr} (Appendix B). It should be pointed out that these results have been derived with the assumption that the iron reluctance is negligible. However, the two techniques are equally valid for finite iron permeability.

On referring to Fig. 3-35b it can be seen that $F_r \sin \delta_{sr}$ is the component of the F_r wave in electrical space quadrature with the F_s wave. Similarly $F_s \sin \delta_{sr}$ is the component of the F_s wave in quadrature with the F_r wave. Thus, the torque is proportional to the product of one magnetic field and the component of the other in quadrature with it, like the cross product of vector analysis. Also note that in Fig. 3-35b

$$F_s \sin \delta_{sr} = F_{sr} \sin \delta_r \tag{3-79}$$

and
$$F_r \sin \delta_{sr} = F_{sr} \sin \delta_s \tag{3-80}$$

The torque can then be expressed in terms of the *resultant* mmf wave F_{sr} by substitution of either Eq. 3-79 or 3-80 in Eq. 3-78; thus

$$T = -\frac{P}{2} \frac{\pi}{2} \frac{\mu_0 Dl}{g} F_s F_{sr} \sin \delta_s \tag{3-81}$$

$$T = -\frac{P}{2} \frac{\pi}{2} \frac{\mu_0 Dl}{g} F_r F_{sr} \sin \delta_r \tag{3-82}$$

Comparison of Eqs. 3-78, 3-81, and 3-82 shows that the torque can be expressed in terms of the component magnetic fields due to *each* current acting alone, as in Eq. 3-78, or in terms of the *resultant* field and *either* of the components, as in Eqs. 3-81 and 3-82, *provided that we use the corresponding angle between the axes of the fields*. Ability to reason in any of these terms is a convenience in machine analysis.

In Eqs. 3-78, 3-81, and 3-82 the fields have been expressed in terms of the peak values of their mmf waves. When magnetic saturation is neglected, the fields can, of course, be expressed in terms of their flux-density waves or in terms of total flux per pole. Thus the peak value B of the field due to a sinusoidally distributed mmf wave in a uniform-air-gap machine is $\mu_0 F/g$, where F is the peak value of the mmf wave. For example, the resultant mmf F_{sr} produces a resultant flux-density wave whose peak value is $\mu_0 F_{sr}/g$. Thus,

$$T = -\frac{P}{2} \frac{\pi Dl}{2} B_{sr} F_r \sin \delta_r \tag{3-83}$$

One of the inherent limitations in the design of electromagnetic apparatus is the saturation flux density of magnetic materials. Because of saturation in the armature teeth the peak value B_{sr} of the resultant flux-density wave in the air gap is limited to about 1 T (64.5 kilolines/in²). The maximum permissible value of the mmf wave is limited by temperature rise of the winding and other design requirements. Because the resultant flux density and mmf appear explicitly in Eq. 3-83, this equation is in a convenient form for design purposes.

Alternative forms arise when it is recognized that the resultant flux per pole is

$$\Phi = \text{(average value of } B \text{ over a pole)(pole area)} \qquad (3\text{-}84)$$

and that the average value of a sinusoid over $\frac{1}{2}$ wavelength is $2/\pi$ times its peak value. Thus

$$\Phi = \frac{2}{\pi} B \frac{\pi D l}{P} = \frac{2Dl}{P} B \qquad (3\text{-}85)$$

where B is the peak value of the corresponding flux-density wave. For example, substitution of Eq. 3-85 in Eq. 3-83 gives

$$T = -\frac{\pi}{2} \left(\frac{P}{2}\right)^2 \Phi_{sr} F_r \sin \delta_r \qquad (3\text{-}86)$$

where Φ_{sr} is the resultant flux produced by the combined effect of the stator and rotor mmfs.

To recapitulate, we now have several forms in which the torque of a uniform-air-gap machine can be expressed in terms of its magnetic fields. *All of them are merely statements that the torque is proportional to the interacting fields and the sine of the electrical space angle between their magnetic axes.* The negative sign indicates that the electromagnetic torque acts in a direction to decrease the displacement angle between the fields. In a preliminary discussion of machine types, Eq. 3-86 will be the preferred form.

One further remark can be made concerning the torque equations and the thought process leading up to them. There is no restriction that the mmf wave or flux-density wave need remain stationary in space. They may remain stationary, or they may be traveling waves, as shown in Art. 3-5. If the magnetic fields of stator and rotor are constant in amplitude and travel around the air gap at the same speed, a steady torque will be produced by the efforts of the stator and rotor fields to align themselves in accordance with the torque equations.

3-8 SUMMARY

In this chapter we have presented a brief and elementary description of three basic types of rotating machines: synchronous, induction, and dc machines. In all of them the basic principles are essentially the same. Voltages are generated by relative motion of a magnetic field with respect to a winding, and torques are produced by the interaction of the magnetic fields of the stator and rotor windings. The characteristics of the various machine types are determined by the methods of connection and excitation of the windings, but the basic principles are still essentially similar.

The basic analytical tools for studying rotating machines are expressions for the generated voltages and for the electromagnetic torque. Taken together, they express the coupling between the electric and mechanical systems. In order to develop a reasonably quantitative theory without the confusion arising from too much detail, we have made several simplifying approximations. In the study of ac machines we have assumed sinusoidal time variations of voltages and currents and sinusoidal space waves of air-gap flux density and mmf. On examination of the mmf of distributed ac windings, we found that the space-fundamental component is the most important, Eqs. 3-7 and 3-8, but that the mmf of dc machine armature windings is more nearly a sawtooth wave, Eq. 3-10. For our preliminary study in this chapter, however, we have assumed sinusoidal mmf distributions for both ac and dc machines. We shall examine this assumption more thoroughly for dc machines in Chap. 5. Faraday's law results in Eq. 3-56 for the rms voltage generated in an ac machine winding or Eq. 3-59 for the average voltage generated between brushes in a dc machine.

On examination of the mmf wave of a 3-phase winding, we found that balanced 3-phase currents produce a constant-amplitude magnetic field rotating in the air gap at synchronous speed, as shown in Fig. 3-31 and Eq. 3-36. The importance of this fact cannot be overestimated, for it means that the double-frequency time-varying torque inherently associated with the double-frequency component of the instantaneous power in a single-phase system is eliminated in balanced 3-phase machines, as shown in Appendix A. Imagine a multimegawatt 60-Hz generator or a multihorsepower motor subjected to a multimegawatt instantaneous power pulsation at 120 Hz! The discovery of rotating fields led to the invention of the simple, rugged reliable self-starting polyphase induction motor, which will be analyzed in Chap. 9. (A single-phase induction motor will not start itself; it needs an auxiliary starting winding, as shown in Chap. 11.) It also made possible the construction of multimegawatt synchronous generators. The most significant reason for the almost universal use of 3-phase generation, transmission, and utilization (for loads above a few kilowatts) is the rotating field in the generators and motors.

Having assumed sinusoidally distributed magnetic fields in the air gap, we then derived expressions for the magnetic torque. The simple physical picture is like that of two magnets, one on the stator and one pivoted on the rotor, as shown schematically in Fig. 3-35a. The torque acts in the direction to align the magnets. To get a reasonably close quantitative analysis without being hindered by details, we assumed a smooth air gap and neglected the reluctance of the magnetic paths in the iron parts, with the mental note to look into these assumptions in later chapters.

In Art. 3-7 we derived expressions for the magnetic torque from two viewpoints, both based on the fundamental principles of Chap. 2. The first viewpoint regards the machine as a group of magnetically coupled circuits with inductances which depend on the angular position of the rotor, as in Art. 3-7a. The second regards the machine from the viewpoint of the magnetic fields in the air gap, as in Art. 3-7b. It is shown that the torque can be expressed as the product of the stator field, the rotor field, and the sine of the angle between their magnetic axes, as in Eq. 3-78 or any of the forms derived from Eq. 3-78. The two viewpoints are supplementary, and ability to reason in terms of both of them is helpful in reaching an understanding of how machines work.

This chapter has been concerned with basic principles underlying rotating-machine theory. By itself it is obviously incomplete. Many questions remain unanswered. How do we apply these principles to the determination of the characteristics of synchronous, induction, and dc machines? What are some of the practical problems that arise from the use of iron, copper, and insulation in physical machines? What are some of the economic and engineering considerations affecting rotating-machine applications? What are the physical factors limiting the conditions under which a machine can operate successfully? Chapter 4 discusses some of these problems. Taken together, Chaps. 3 and 4 serve as an introduction to the more detailed treatments of rotating machines in Chaps. 5 to 11.

PROBLEMS

3-1 A 3-phase motor runs off a 60-Hz supply and is used to drive a pump. It is noted (using a stroboscope) that the motor speed decreases from 1195 r/min at no load to 1145 r/min at full load.

(a) Is this a synchronous or induction motor?
(b) How many poles does the motor have?

3-2 The object of this problem is to illustrate how the armature windings of certain machines, i.e., dc machines, can be approximately represented by uniform current sheets, the degree of correspondence growing better as the winding is distributed in a greater number of slots around the armature periphery. For this purpose, consider an armature with eight slots uniformly distributed over 360 electrical degrees or one pair of poles. The air gap is of uniform length, the slot openings are very small, and the reluctance of the iron is negligible.

Lay out 360 electrical degrees of the armature with its slots in developed form in the manner of Fig. 3-23a, and number the slots 1 to 8 from left to right. The winding consists of eight single-turn coils, each carrying a direct current of 10 A. Coil sides which can be placed in slots 1 to 4 carry current directed into the paper; those which can be placed in slots 5 to 8 carry current out of the paper.

(a) Consider that all eight coils are placed with one side in slot 1 and the other in slot 5. The remaining slots are empty. Draw the rectangular mmf wave produced by these coils.

(b) Next consider that four coils have one side in slot 1 and the other in slot 5, while the remaining four have one side in slot 3 and the other in slot 7. Draw the component rectangular mmf waves produced by each group of coils and superimpose the components to give the resultant mmf wave.

(c) Now consider that two coils are placed in slots 1 and 5, two in 2 and 6, two in 3 and 7, and two in 4 and 8. Again superimpose the component rectangular waves to produce the resultant wave. Note that the task can be systematized and simplified by recognizing that the mmf wave is symmetrical about its axis and takes a step at each slot which is definitely related to the number of ampere-conductors in the slot.

(d) Let the armature now consist of 16 slots per 360 electrical degrees with one coil side in each slot. Draw the resultant mmf wave.

(e) Approximate each of the resultant waves of parts (a) to (d) by isosceles triangles, noting that the representation grows better as the winding is more finely distributed.

3-3 This problem investigates the advantages of short-pitching the stator coils of an ac machine. Figure 3-36a shows a single full-pitch coil in a 2-pole machine. Figure 3-36b shows a fractional-pitch coil; the coil sides are β rad apart rather than π rad (180°), as in the full-pitch case.

For an air-gap radial flux distribution of the form

$$B_r = \sum_{n \text{ odd}} B_n \cos n\theta$$

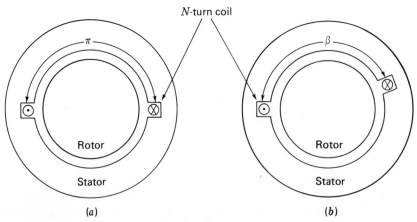

Fig. 3-36. Problem 3-3: (a) full-pitch coil, (b) fractional-pitch coil.

where $n = 1$ corresponds to the fundamental space harmonic, $n = 3$ the third harmonic, etc., the flux linkage of each coil is the integral of B_r over the surface spanned by that coil. Thus for the nth space harmonic, the ratio of the fractional-pitch-coil flux linkage to that of the full-pitch coil is

$$\frac{\int_{-\beta/2}^{\beta/2} B_n \cos n\theta \, d\theta}{\int_{-\pi/2}^{\pi/2} B_n \cos n\theta \, d\theta} = \frac{\int_{-\beta/2}^{\beta/2} \cos n\theta \, d\theta}{\int_{-\pi/2}^{\pi/2} \cos n\theta \, d\theta}$$

It is common to fractional-pitch the coils of an ac machine by $30°$ ($\beta = 5\pi/6 = 150°$). For $n = 1, 3, 5$ calculate the fractional reduction in flux linkage due to short-pitching.

3-4 A 4-pole 60-Hz synchronous machine has a rotor winding with a total of 120 series turns and a winding factor $k_r = 0.92$. The rotor length is 2.1 m, the rotor radius is 45 cm and the air-gap length is 3.0 cm. Calculate the rotor winding current required to achieve a peak fundamental air-gap flux density of 1.3 T. What is the corresponding flux per pole?

3-5 The synchronous machine of Prob. 3-4 has an armature with 16 series turns per phase and a winding factor $k_w = 0.94$. For the flux condition of Prob. 3-4, calculate the rms generated voltage per phase.

3-6 What is the effect on the rotating mmf wave of a 3-phase winding if two of the phase connections are interchanged?

3-7 In a balanced 2-phase machine the two windings are displaced 90 electrical degrees in space, and the currents in the two windings are phase-displaced 90 electrical degrees in time. For such a machine, carry out the process leading up to an equation such as 3-40 for the rotating mmf wave.

3-8 The following statements are made in Art. 3-5 just after deriving and discussing Eq. 3-43: "In general it can be shown that a rotating field of constant amplitude will be produced by a q-phase winding excited by balanced q-phase currents when the respective phases are wound $2\pi/q$ electrical radians apart in space. The constant amplitude will be $q/2$ times the maximum contribution of any one phase, and the speed will be $\omega = 2\pi f$ electrical radians per second."

Prove these statements.

3-9 Figure 3-37 shows a 2-pole rotor revolving inside a smooth stator which carries a coil of 100 turns. The rotor produces a sinusoidal space distribution of flux at the stator surface, the peak value of the flux-density wave being

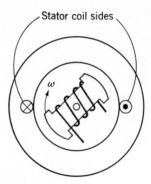

Fig. 3-37. Elementary generator, Prob. 3-9.

0.80 T when the current in the rotor is 10 A. The magnetic circuit is linear. The inside diameter of the stator is 0.10 m, and its axial length is 0.10 m. The rotor is driven at a speed of 60 r/s.

(a) The rotor is excited by a direct current of 10 A. Taking zero time as the instant when the axis of the rotor is vertical, find the expression for the instantaneous voltage generated in the open-circuited stator coil.

(b) The rotor is now excited by a 60-Hz sinusoidal alternating current whose rms value is 7.07 A. Consequently, the rotor current reverses every half revolution; it is timed to go through zero whenever the axis of the rotor is vertical. Taking zero time as the instant when the axis of the rotor is vertical, find the expression for the instantaneous voltage generated in the open-circuited stator coil. This scheme is sometimes suggested as a dc generator without a commutator, the thought being that, if alternate half cycles of the alternating voltage generated in part (a) are reversed by reversal of the polarity of the field (rotor) winding, a pulsating direct voltage will be generated in the stator. Explain whether this invention will work as described.

3-10 A 3-phase 2-pole winding is excited by balanced 3-phase 60-Hz currents as described by Eqs. 3-25 to 3-27. Although the winding distribution has been designed to minimize harmonics, there remains some third and fifth spatial harmonic. Thus the phase a mmf can be written as

$$\mathcal{F}_a = (A_1 \cos \theta + A_3 \cos 3\theta + A_5 \cos 5\theta)i_a$$

Similar expressions can be written for phases b (replace θ by $\theta - 120°$) and c (replace θ by $\theta + 120°$). Calculate the total 3-phase mmf. What is the angular velocity and rotational direction of each component of the mmf?

3-11 A small experimental 3-phase 4-pole alternator has the full-pitch concentrated Y-connected armature winding shown diagrammatically in **Fig.**

3-12b and c. Each coil (that represented by coil sides a and $-a$, for example) has two turns, and all the turns in any one phase are connected in series. The flux per pole is 0.25 Wb and is sinusoidally distributed in space. The rotor is driven at 1800 r/min.

Determine the rms generated voltage (a) to neutral and (b) between lines.

   (c) Consider an abc phase order and take zero time at the instant when the flux linkages with phase a are a maximum. Write a consistent set of time equations for the 3-phase voltages from terminals a, b, and c to neutral.
   (d) Under the conditions of part (c) write a consistent set of time equations for the three voltages between lines a and b, b and c, and c and a.

3-12 A 4-pole dc motor is designed to operate at 1500 r/min. It is proposed to increase the number of armature turns so that the machine will operate at 1200 r/min while maintaining the same terminal voltage and magnetic flux density. By what factor must the armature turns be increased?

3-13 The design of a 4-pole 3-phase 480-V 60-Hz induction motor is to be based upon a stator core length of 10 in and a stator inner diameter of 9 in. The stator winding distribution which has been chosen has a winding factor $k_w = 0.91$. The armature is to be Y-connected, and thus the rated voltage of each phase is $480/\sqrt{3}$ V.

   (a) The designer must pick the number of armature turns so that the flux density in the machine is large enough to make efficient use of the magnetic material without being so large that it results in excessive saturation. In this case, the optimal performance is achieved with a peak fundamental air-gap flux density of 1.1 T. Calculate the number of turns per phase.
   (b) For an air-gap length of 0.012 in, calculate the self-inductance of an armature phase based upon the results of part (a) using the inductance formulas of Appendix B. Neglect the reluctance of the rotor and stator iron and the armature leakage inductance.

3-14 A 2-pole 60-Hz 3-phase laboratory-size synchronous generator has a rotor radius of 2.25 in, an air-gap length of 0.010 in, and a rotor length of 7.1 in. The rotor field winding has 248 turns and a winding factor of 0.95. The armature winding consists of 60 turns per phase with a winding factor of 0.93.

   (a) Calculate the flux per pole and peak fundamental air-gap flux density for an armature voltage of 120 V rms/phase.
   (b) Calculate the dc field current in amperes required to achieve the conditions of part (a).

3-15 A 4-pole 60-Hz synchronous generator has a rotor length of 3.8 m, diameter of 1.12 m, and air-gap length of 6.3 cm. The rotor winding consists of 54 turns per pole with a winding factor of $k_r = 0.89$. The peak value of fundamental air-gap flux density is limited to 1.0 T and the rotor winding current to 2900 A. Calculate the maximum torque in newton-meters and power output in megawatts which can be supplied by the machine under these conditions.

3-16 Figure 3-38 shows in cross section a machine having a rotor winding ff and two identical stator windings aa and bb. The self-inductance of each stator winding is L_{aa} and of the rotor winding is L_{ff}. The air gap is uniform. The stator windings are in quadrature. The mutual inductance between a stator winding and the rotor winding depends on the angular position θ_0 of the rotor and may be assumed to be

$$M_{af} = M \cos \theta_0 \qquad M_{bf} = M \sin \theta_0$$

where M is the maximum value of the mutual inductance. The resistance of each stator winding is R_a.

(a) Derive a general expression for the torque T in terms of the angle θ_0, the inductance constants, and the instantaneous currents i_a, i_b, and i_f. Does this expression apply at standstill? When the rotor is revolving?

(b) Suppose the rotor is stationary and constant direct currents $I_a = 5$ A, $I_b = 5$ A, $I_f = 10$ A are supplied to the windings in the directions indicated by the dots and crosses in Fig. 3-38. If the rotor is allowed to move, will it rotate continuously or will it tend to come to rest? If the latter, at what value is θ_0?

(c) The rotor winding is now excited by a constant direct current I_f, and

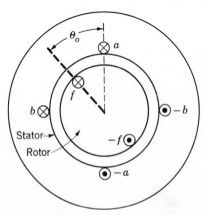

Fig. 3-38. Elementary cylindrical-rotor 2-phase synchronous machine, Prob. 3-16.

the stator windings carry balanced 2-phase currents

$$i_a = \sqrt{2}I_a \cos \omega t \qquad i_b = \sqrt{2}I_a \sin \omega t$$

The rotor is revolving at synchronous speed so that its instantaneous angular position θ_0 is given by $\theta_0 = \omega t - \delta$, where δ is a phase angle describing the position of the rotor at $t = 0$. The machine is an elementary 2-phase synchronous machine. Derive an expression for the torque under these conditions. Describe its nature.

(d) Under the conditions of part (c) derive an expression for the instantaneous terminal voltages of stator phases a and b.

3-17 Figure 3-39 shows in cross section a machine having two identical stator windings aa and bb arranged in quadrature on a laminated steel core. The salient-pole rotor is made of steel and carries a winding f connected to slip rings. The machine is an elementary 2-phase salient-pole synchronous machine.

Because of the nonuniform air gap, the self- and mutual inductances of the stator windings are functions of the angular position θ_0 of the rotor, as follows:

$$L_{aa} = L_0 + L_2 \cos 2\theta_0$$
$$L_{bb} = L_0 - L_2 \cos 2\theta_0$$
$$M_{ab} = L_2 \sin 2\theta_0$$

where L_0 and L_2 are positive constants. The mutual inductances between the rotor and the stator windings are functions of θ_0 as follows:

$$M_{af} = M \cos \theta_0 \qquad M_{bf} = M \sin \theta_0$$

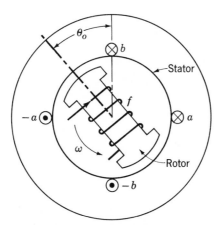

Fig. 3-39. Elementary salient-pole 2-phase synchronous machine, Prob. 3-17.

where M is a positive constant. The self-inductance L_{ff} of the rotor winding is constant, independent of θ_0.

The rotor (or field) winding f is excited with direct current I_f, and the stator windings are connected to a balanced 2-phase voltage source. The currents in the stator windings are

$$i_a = \sqrt{2}\, I_a \cos \omega t \qquad i_b = \sqrt{2}\, I_a \sin \omega t$$

The rotor is revolving at synchronous speed so that its instantaneous angular position is given by

$$\theta_0 = \omega t - \delta$$

where δ is a phase angle describing the position of the rotor at $t = 0$.

(a) Derive an expression for the electromagnetic torque acting on the rotor. Describe its nature.

(b) Can the machine be operated as a motor? As a generator? Explain.

(c) Will the machine continue to run if the field current I_f is reduced to zero? If so, give an expression for the torque and a physical explanation.

4

Engineering Aspects of Practical Electric-Machine Performance

In the previous chapter basic essential features of electric machinery were discussed; this material forms the basis for understanding the behavior of electric machinery. In this chapter our objective is to discuss some of the practical problems associated with engineering implementation of the machinery concepts which have been developed. Many of these problems are concerned with physical factors limiting the conditions under which a machine can operate successfully. Among them are saturation and the effects of losses on the rating and heating of the machine. These problems, common to all machine types, are discussed in this chapter; they will also be discussed in later chapters in the context of the various specific machine types.

4-1 MAGNETIC SATURATION

The characteristics of electric machines depend heavily upon the use of magnetic materials. These materials are required to form the magnetic circuit and

175

are used by the machine designer to obtain specific desired machine characteristics. As we have seen in Chap. 1, magnetic materials are less than ideal. As their magnetic flux is increased, they begin to saturate, with the result that their magnetic permeabilities begin to decrease, along with their effectiveness in contributing to the overall flux density in the machine.

Both electromagnetic torque and generated voltage in all machines depend on the winding flux linkages. For specific mmfs in the windings, the fluxes depend on the reluctances of the iron portions of the magnetic circuits and those of the air gaps. Saturation may therefore appreciably influence the characteristics of the machines. Another aspect of saturation, more subtle and more difficult to evaluate without experimental and theoretical comparisons, concerns its influence on the basic premises from which the analytic approach to machinery is developed. Specifically, all relations for mmf are based on negligible reluctance in the iron. When these relations are applied to practical machines with varying degrees of saturation in the iron, the actual machine must be, in effect, replaced for these considerations by an equivalent machine, one whose iron has negligible reluctance but whose air-gap length is increased by an amount sufficient to absorb the magnetic-potential drop in the iron of the actual machine. Incidentally, the effects of air-gap non-uniformities such as slots and ventilating ducts are also incorporated through the medium of an equivalent smooth gap, a replacement which, in contrast to that above, is made explicitly during magnetic-circuit computations for the machine structure. Thus, serious efforts must be made to reproduce the magnetic conditions at the air gap correctly, and the computed performance of machines is based largely on those conditions. Final assurance of the legitimacy of the approach must, of course, be the pragmatic one given by close experimental checks.

Magnetic-circuit data essential to handling saturation are given by the *open-circuit characteristic*, also called the *magnetization curve* or *saturation curve*. An example is shown in Fig. 4-1. Basically, this characteristic is the

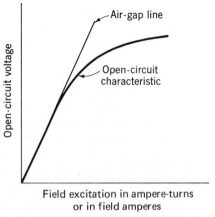

Fig. 4-1. Typical open-circuit characteristic and air-gap line.

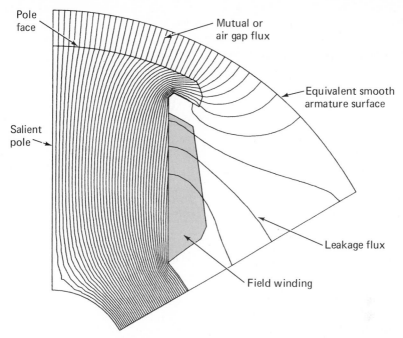

Fig. 4-2. Finite-element solution for flux distribution around a salient pole. (*General Electric Company.*)

magnetization curve for the particular iron and air geometry of the machine under consideration. Frequently, the abscissa is plotted in terms of field current or magnetizing current instead of mmf in ampere-turns. Also, the generated voltage with zero armature current is directly proportional to the flux when the speed is constant. For convenience in use, then, open-circuit terminal voltage is plotted on the ordinate scale rather than air-gap flux per pole, and the entire curve is drawn for a stated fixed speed, usually rated speed. The straight line tangent to the lower portion of the curve is the *air-gap line,* indicating very closely the mmf required to overcome the reluctance of the air gap. If it were not for the effects of saturation, the air-gap line and open-circuit characteristic would coincide, so that the departure of the curve from the air-gap line is an indication of the degree of saturation present. In typical machines the ratio at rated voltage of the total mmf to that required by the air gap alone usually is between 1.1 and 1.25.

The open-circuit characteristic can be calculated from design data by computer-implemented magnetic field solutions based upon techniques of finite-element or finite-difference analysis. These techniques can be used to obtain solutions for magnetic fields, including the effects of finite permeability. A typical finite-element solution for the flux distribution around the pole of a salient-pole machine is shown in Fig. 4-2. The distribution of air-gap flux found from this solution, together with the fundamental and third-harmonic components, is shown in Fig. 4-3.

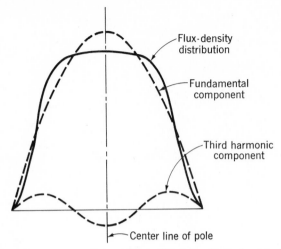

Fig. 4-3. Flux-density wave corresponding to Fig. 4-2, with its fundamental and third-harmonic components.

In addition to saturation effects, Fig. 4-2 clearly illustrates the effect of a nonuniform air gap. As expected, the flux density over the pole face, where the air gap is small, is much higher than that away from the pole. This type of detailed analysis is of great use to a designer in obtaining specific machine properties.

If the machine is an existing one, the magnetization curve is usually determined by operating the machine as an unloaded generator and reading the values of terminal voltage corresponding to a series of values of field current. For an induction motor the machine is operated at or close to synchronous speed, and values of the magnetizing current are obtained for a series of values of impressed stator voltage. It should be emphasized, however, that saturation in a fully loaded machine occurs as a result of the total mmf acting on the magnetic circuit. Since the flux distribution under load generally differs from that of no-load conditions, the details of the machine saturation characteristics may vary from the open-circuit curve of Fig. 4-1.

4–2 LEAKAGE FLUXES

In Art. 1-8 it was shown that in a 2-winding transformer the flux created by each winding can be separated into two components. One component consists of flux which links both windings, and the other consists of flux which links only the winding which is creating the flux. The first component, called *mutual flux,* is responsible for coupling between the two coils. The second, known as *leakage flux,* contributes only to the self-inductance of each coil.

Note that the concept of mutual and leakage flux is meaningful only in the context of a multiwinding system. For systems of three or more windings,

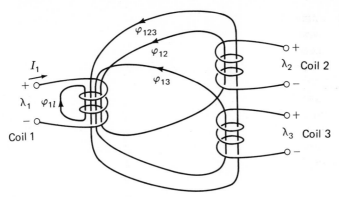

Fig. 4-4. Three-coil system showing components of mutual and leakage flux produced by current in coil 1.

the bookkeeping must be done very carefully. Consider, for example, the 3-winding system of Fig. 4-4. Shown schematically are the various components of flux created by a current in winding 1. Here φ_{123} is clearly mutual flux that links all three windings, and φ_{1l} is clearly leakage flux since it links only winding 1. However φ_{12} is mutual flux with respect to winding 2 yet is leakage flux with respect to winding 3, while φ_{13} is mutual flux with respect to winding 3 and leakage flux with respect to winding 2.

Electric machinery often contains systems of multiple windings requiring careful bookkeeping to account for the flux contributions of the various windings. Although the details of such analysis are beyond the scope of this book, it is useful to discuss these effects in a qualitative fashion and to describe how they affect the basic machine inductances.

Air-Gap Space-Harmonic Fluxes. In Chap. 3 we saw that although single distributed coils create air-gap flux with a significant amount of space-harmonic content, it is possible to so distribute these windings that the space-fundamental component is emphasized while the harmonic effects are greatly reduced. As a result, we can neglect harmonic effects and consider only space-fundamental fluxes in calculating the self- and mutual-inductance expressions of Eqs. B-27 and B-28.

Though often small, the space-harmonic components of air-gap flux do exist. In dc machines they are useful torque-producing fluxes and therefore can be counted as mutual flux between the rotor and stator windings. In ac machines, on the other hand, they may generate time-harmonic voltages or asynchronously rotating flux waves. These effects generally cannot be rigorously accounted for in most standard analyses. Nevertheless, it is consistent with the assumptions basic to these analyses to recognize that these fluxes form a part of the leakage flux of the individual windings which produce them.

Slot-Leakage Flux. Figure 4-5 shows the flux created by a single coil side in a slot. Notice that in addition to flux which crosses the air gap, contributing to

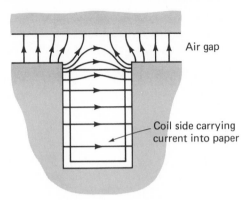

Fig. 4-5. Flux created by a single coil side in a slot.

the air-gap flux, there are also flux components which cross the slot. Since this flux links only the coil that is producing it, it too forms a component of the leakage inductance of the winding producing it.

End-Turn Fluxes. Figure 4-6 shows the stator end windings on an ac machine. The magnetic field distribution created by end turns is extremely complex. In general these fluxes do not contribute to useful rotor-to-stator mutual flux, and thus they too contribute to leakage inductance.

Fig. 4-6. End view of the stator of a 26-kV 908-MVA 3600-r/min turbine-generator with water-cooled windings. Hydraulic connections for coolant flow are provided for each winding end-turn. (*General Electric Company.*)

From this discussion we see that the self-inductance expression of Eq. B-27 must in general be modified by an additional term L_l, which represents the winding leakage inductance. Although leakage inductance is usually difficult to calculate analytically and must be determined by approximate or empirical techniques, it plays an important role in machine performance.

4–3 LOSSES

Consideration of machine losses is important for three reasons: (1) losses determine the efficiency of the machine and appreciably influence its operating cost; (2) losses determine the heating of the machine and hence the rating or power output that can be obtained without undue deterioration of the insulation; and (3) the voltage drops or current components associated with supplying the losses must be properly accounted for in a machine representation. Machine efficiency, like that of transformers or any energy-transforming device, is given by

$$\text{Efficiency} = \frac{\text{output}}{\text{input}} \qquad (4\text{-}1)$$

which can also be expressed as

$$\text{Efficiency} = \frac{\text{input} - \text{losses}}{\text{input}} = 1 - \frac{\text{losses}}{\text{input}} \qquad (4\text{-}2)$$

$$\text{Efficiency} = \frac{\text{output}}{\text{output} + \text{losses}} \qquad (4\text{-}3)$$

Rotating machines in general operate efficiently except at light loads. The full-load efficiency of average motors, for example, is in the neighborhood of 74 percent for 1-hp size, 89 percent for 50-hp, 93 percent for 500-hp, and 97 percent for 5000-hp. The efficiency of slow-speed motors is usually lower than that of high-speed motors, the total spread being 3 or 4 percent.

The forms given by Eqs. 4-2 and 4-3 are often used for electric machines, since their efficiency is most commonly determined by measurement of losses instead of by directly measuring the input and output under load. Loss measurements have the advantage of convenience and economy and of yielding more accurate and precise values of efficiency because a given percentage error in measuring losses causes only about one-tenth of that percentage error in the efficiency. Efficiencies determined from loss measurements can be used in comparing competing machines if exactly the same methods of measurement and computation are used in each case. For this reason the various losses and the conditions for their measurement are precisely defined by the American National Standards Institute, Inc. (ANSI), the Institute of Electrical and Electronic Engineers (IEEE), and the National Electrical Manufacturers Association (NEMA). The following discussion of individual losses incorporates many of these provisions as given in ANSI Standard C50, although no attempt is made to present all the details.

I^2R **Losses.** I^2R losses, are, of course, found in all the windings of the machine. By convention, these losses are computed on the basis of the dc resistances of the winding at 75°C. Actually the I^2R loss depends on the effective resistance of the winding under the operating frequency and flux conditions. The increment in loss represented by the difference between dc and effective resistances is included with stray load losses, discussed below. In the field circuits of synchronous and dc machines, only the losses in the field winding are charged against the machine; the losses in external sources supplying the excitation are charged against the plant of which the machine is a part. Closely associated with I^2R loss is the *brush-contact loss* at slip rings and commutators. By convention, this loss is normally neglected for induction and synchronous machines, and for industrial-type dc machines the voltage drop at the brushes is regarded as constant at 2 V total when carbon and graphite brushes with shunts (pigtails) are used.

Mechanical Losses. These losses consist of brush and bearing friction, windage, and the power required to circulate the air through the machine and ventilating system, if one is provided, whether by self-contained or external fans (except for the power required to force air through long or restricted ducts external to the machine). Friction and windage losses can be measured by determining the input to the machine running at the proper speed but unloaded and unexcited. Frequently they are lumped with core loss and determined at the same time.

Open-Circuit, or No-Load, Core Loss. Open-circuit core loss consists of the hysteresis and eddy-current losses arising from changing flux densities in the iron of the machine with only the main exciting winding energized. In dc and synchronous machines, these losses are confined largely to the armature iron, although the flux pulsations arising from slot openings will cause losses in the field iron as well, particularly in the pole shoes or surfaces of the field iron. In induction machines the losses are confined largely to the stator iron. Open-circuit core loss can be found by measuring the input to the machine when it is operating unloaded at rated speed or frequency and under the appropriate flux or voltage conditions and then deducting the friction and windage loss and, if the machine is self-driven during the test, the no-load armature I^2R loss (no-load stator I^2R loss for an induction motor). Usually, data are taken for a curve of core loss as a function of armature voltage in the neighborhood of rated voltage. The core loss under load is then considered to be the value at a voltage equal to rated voltage corrected for armature ohmic-resistance drop under load (a phasor correction for an ac machine). For induction motors, however, this correction is dispensed with, and the core loss at rated voltage is used. For efficiency determination alone, there is no need to segregate open-circuit core loss and friction and windage loss; the sum of these two losses is termed the *no-load rotational loss*.

Eddy-current loss is dependent on the squares of the flux density, frequency, and thickness of laminations. Under normal machine conditions it can be expressed to a sufficiently close approximation as

$$P_e = K_e(B_{max}f\tau)^2 \qquad (4\text{-}4)$$

where τ = lamination thickness
B_{max} = maximum flux density
f = frequency
K_e = proportionality constant

The value of K_e depends on the units used, the volume of iron, and the resistivity of the iron. Variation of hysteresis loss can be expressed in equation form only on an empirical basis. The most commonly used relation is

$$P_h = K_h f B_{max}^n \qquad (4\text{-}5)$$

where K_h is a proportionality constant dependent on the characteristics and volume of iron and the units used and the exponent n ranges from 1.5 to 2.5, a value of 2.0 often being used for estimating purposes in machines. In both Eqs. 4-4 and 4-5 frequency can be replaced by speed and flux density by the appropriate voltage, with the proportionality constants changed accordingly.

When the machine is loaded, the space distribution of flux density is significantly changed by the mmf of the load currents. The actual core losses may increase noticeably. For example, mmf harmonics cause appreciable losses in the iron near the air-gap surfaces. The total increment in core loss is classified as part of the stray load loss.

Stray Load Loss. Stray load loss consists of the losses arising from nonuniform current distribution in the copper and the additional core losses produced in the iron by distortion of the magnetic flux by the load current. It is a difficult loss to determine accurately. By convention it is taken as 1.0 percent of the output for dc machines. For synchronous and induction machines it can be found by test.

Study of the foregoing classification of the losses in a machine shows it to have a few features which, from a fundamental viewpoint, are somewhat artificial. Illustrations are offered by the division of iron losses into no-load core loss and an increment which appears under load, the division of I^2R losses into ohmic I^2R losses and an increment created by nonuniform current distribution, and the lumping of these two increments in the scavengerlike stray-load-loss category. These features are dictated by ease of testing. They are justified by the fact that the principal motivation is the determination of the total losses and efficiency nearly equal to the actual values as possible but also suitable for economic comparison of machines. Because of this seeming dominance of efficiency aspects, it may be appropriate to emphasize once more that losses play more than a bookkeeping role in machine operation.

In a generator, for example, components of mechanical input torque to the shaft are obviously required to supply I^2R and iron losses as well as friction and windage losses and the generator output. These losses may therefore be appreciable factors in the damping of electric and mechanical transients in the machine. Components of the stray load loss, although they may be individually only a fraction of a percent of the output, may be of first importance in the design of the machine. For example, rotor heating is usually a limiting factor in the design of large high-speed alternators, and the components of stray loss on the surface of the rotor structure are of great importance because they directly affect the dimensions of an alternator of given output. Of more direct concern in theoretical aspects is the influence of hysteresis and eddy currents in causing flux to lag behind mmf. There is a small angle of lag between the rotating mmf waves in a machine and the corresponding component flux-density waves. Associated with this influence is a torque on magnetic material in a rotating field, a torque proportional to the hysteresis and eddy-current losses in the material. Although the torque accompanying these losses is relatively small in normal machines, direct use of it is made in one type of small motor, the hysteresis motor.

4–4 RATING AND HEATING

One of the most common and important questions in the application of machines, transformers, and other electrical equipment is: What maximum output can be obtained? The answer, of course, depends on various factors, since the machine, while providing this output, must in general meet definite performance standards. A universal requirement is that the life of the machine not be unduly shortened by overheating. The temperature rise resulting from the losses considered in the previous article is therefore a major factor in the rating of a machine.

The operating temperature of a machine is closely associated with its life expectancy because deterioration of the insulation is a function of both time and temperature. Such deterioration is a chemical phenomenon involving slow oxidation and brittle hardening and leading to loss of mechanical durability and dielectric strength. In many cases the deterioration rate is such that the life of the insulation can be represented as an exponential

$$\text{Life} = A\varepsilon^{B/T} \tag{4-6}$$

where A and B are constants and T is the absolute temperature. Thus, according to Eq. 4-6, when life is plotted to a logarithmic scale against the reciprocal of absolute temperature on a uniform scale, a straight line should result. Such plots form valuable guides in the thermal evaluation of insulating materials and systems. A very rough idea of the life-temperature relation can be obtained from the old and more or less obsolete rule of thumb that the time to failure of organic insulation is halved for each 8 to 10°C rise.

The evaluation of insulating materials and complete systems of insulation (which may include widely different materials and techniques in combination) is to a large extent a functional one based on accelerated life tests. Both normal life expectancy and service conditions will vary widely for different classes of electric equipment. Life expectancy, for example, may be a matter of minutes in some military and missile applications, may be 500 to 1000 h in certain aircraft and electronic equipment, and may range from 10 to 30 years or more in large industrial equipment. The test procedures will accordingly vary with the type of equipment. Accelerated life tests on models, called *motorettes,* are commonly used in insulation evaluation. Such tests, however, cannot be easily applied to all equipment, especially the insulation systems of large machines.

Insulation life tests generally attempt to simulate service conditions. They usually include the following elements:

1 Thermal shock resulting from heating to the test temperature
2 Sustained heating at that temperature
3 Thermal shock resulting from cooling to room temperature or below
4 Vibration and mechanical stress such as may be encountered in actual service
5 Exposure to moisture
6 Dielectric testing to determine the condition of the insulation

Enough samples must be tested to permit statistical methods to be applied in analyzing the results. The life-temperature relations obtained from these tests lead to the classification of the insulation or insulating system in the appropriate temperature class.

For the allowable temperature limits of insulating systems used commercially, the latest standards of ANSI, IEEE, and NEMA should be consulted. The three NEMA insulation-system classes of chief interest for industrial machines are class B, class F, and class H. Class B insulation includes mica, glass fiber, asbestos, and similar materials with suitable bonding substances. Class F insulation also includes mica, glass fiber, and synthetic substances similar to those in class B, but the system must be capable of withstanding higher temperatures. Class H insulation, intended for still higher temperatures, may consist of materials such as silicone elastomer and combinations including mica, glass fiber, asbestos, etc., with bonding substances such as appropriate silicone resins. Experience and tests showing the material or system to be capable of operation at the recommended temperature form the important classifying criteria.

When the temperature class of the insulation is established, the permissible observable temperature rises for the various parts of industrial-type machines can be found by consulting the appropriate standards. Reasonably detailed distinctions are made with respect to type of machine, method of temperature measurement, machine part involved, whether the machine is enclosed or not, and the type of cooling (air-cooled, fan-cooled, hydrogen-cooled, etc.). Distinctions are also made between general-purpose machines and definite- or special-purpose machines. The term *general-purpose motor* refers to one of standard rating "up to 200 hp with standard operating characteristics and mechanical construction for use under usual service conditions without restriction to a particular application or type of application." In contrast a *special-purpose motor* is "designed with either operating characteristics or mechanical construction, or both, for a particular application." For the same class of insulation, the permissible rise of temperature is lower for a general-purpose motor than for a special-purpose motor, largely to allow a greater factor of safety where service conditions are unknown. Partially compensating the lower rise, however, is the fact that general-purpose motors are allowed a service factor of 1.15 when operated at rated voltage; *service factor* is a multiplier which, applied to the rated output, indicates a permissible loading which may be carried continuously under the conditions specified for that service factor.

Examples of allowable temperature rises can be seen from Table 4-1. The table applies to integral-horsepower induction motors, is based on 40°C ambient temperature, and assumes measurement of temperature rise by determining the increase of winding resistances.

The most common machine rating is the *continuous rating* defining the output (in kilowatts for dc generators, kilovoltamperes at a specified power factor for ac generators, and horsepower for motors) which can be carried indefinitely without exceeding established limitations. For intermittent, periodic, or varying duty a machine may be given a *short-time rating* defining the load which can be carried for a specific time. Standard periods for short-time ratings are 5, 15, 30, and 60 min. Speeds, voltages, and frequencies are also specified in machine ratings, and provision is made for possible variations in voltage and frequency. Motors, for example, must operate successfully at voltages 10 percent above and below rated voltage and, for ac motors, at

TABLE 4-1
ALLOWABLE TEMPERATURE RISE, °C†

Motor type	Class B	Class F	Class H
1.15 service factor	90	115	
1.00 service factor, encapsulated windings	85	110	
Totally enclosed, fan-cooled	80	105	125
Totally enclosed, nonventilated	85	110	135

†Excerpted from NEMA standards.

frequencies 5 percent above and below rated frequency; the combined variation of voltage and frequency may not exceed 10 percent. Other performance conditions are so established that reasonable short-time overloads can be carried. Thus, the user of a motor can expect to be able to apply for a short time an overload of, say, 25 percent at 90 percent of normal voltage with an ample margin of safety.

The converse problem to the rating of machinery, that of choosing the size of machine for a particular application, is a relatively simple one when the load requirements remain substantially constant. For many motor applications, however, the load requirements vary more or less cyclically and over a wide range. The duty cycle of a typical crane or hoist motor offers a good example. From the thermal viewpoint, the average heating of the motor must be found by detailed study of the motor losses during the various parts of the cycle. Account must be taken of changes in ventilation with motor speed for open and semiclosed motors. Judicious selection is based on a large number of experimental data and considerable experience with the motors involved. For estimating the required size of motors operating at substantially constant speeds it is sometimes assumed that the heating of the insulation varies as the square of the horsepower load, an assumption which obviously overemphasizes the role of armature I^2R loss at the expense of the core loss. The rms ordinate of the horsepower-time curve representing the duty cycle is obtained by the same technique used to find the rms value of periodically varying currents, and a motor rating is chosen on the basis of the result; i.e.,

$$\text{rms hp} = \sqrt{\frac{\Sigma(\text{hp})^2 \times \text{time}}{\text{running time} + (\text{standstill time}/k)}} \tag{4-7}$$

where the constant k accounts for the poorer ventilation at standstill and equals approximately 4 for an open motor. The time for a complete cycle must be short compared with the time for the motor to reach a steady temperature.

Although crude, the rms-horsepower method is used fairly often. The necessity for rounding off the result to a commercially available motor size† obviates the need for precise computations; if the rms horsepower were 87, for example, a 100-hp motor would be chosen. Special consideration must be given to motors that are frequently started or reversed, for such operations are thermally equivalent to heavy overloads. Consideration must also be given to duty cycles having such high torque peaks that motors with continuous ratings chosen on purely thermal bases would be unable to furnish the torques required. It is to such duty cycles that special-purpose motors with short-time ratings are often applied. Short-time-rated motors in general have better torque-producing ability than motors rated to produce the same power output continuously, although, of course, they have a lower thermal capacity.

†Commercially available motors are generally found in standard sizes as defined by NEMA. The NEMA Standards on Motors and Generators specify motor rating as well as the type and dimensions of the motor frame.

Both these properties follow from the fact that a short-time-rated motor is designed for high flux densities in the iron and high current densities in the copper. In general, the ratio of torque capacity to thermal capacity increases as the period of the short-time rating decreases. Higher temperature rises are allowed than for general-purpose motors. A motor with a 150-hp 1-h 50°C rating, for example, may have the torque ability of a 200-hp continuously rated motor; it will be able to carry only about 0.8 times its rated output, or 120 hp, continuously, however. In many cases it will be the economical solution for a drive requiring a continuous thermal capacity of 120 hp but having torque peaks which require the ability of a 200-hp continuously rated motor.

4-5 COOLING MEANS FOR ELECTRIC MACHINES

The cooling problem in electic apparatus in general increases in difficulty with increasing size. The surface area from which the heat must be carried away increases roughly as the square of the dimensions, whereas the heat developed by the losses is roughly proportional to the volume and therefore increases approximately as the cube of the dimensions. This problem is a particularly serious one in large turbine generators, where economy, mechanical requirements, shipping, and erection all demand compactness, especially for the rotor forging. Even in moderate sizes of machines, e.g., above a few thousand kilovoltamperes for generators, a closed ventilating system is commonly used. Rather elaborate systems of cooling ducts must be provided to ensure that the cooling medium will effectively remove the heat arising from the losses.

For turbine generators, hydrogen is commonly used as the cooling medium in the totally enclosed ventilating system. Hydrogen has the following properties which make it well suited to the purpose:

1 Its density is only about 0.07 times that of air at the same temperature and pressure, and therefore windage and ventilating losses are much less.
2 Its specific heat on an equal-weight basis is about 14.5 times that of air. This means that, for the same temperature and pressure, hydrogen and air are about equally effective in their heat-storing capacity per unit volume, but the heat transfer by forced convection between the hot parts of the machine and the cooling gas is considerably greater with hydrogen than with air.
3 The life of the insulation is increased and maintenance expenses decreased because of the absence of dirt, moisture, and oxygen.
4 The fire hazard is minimized. A hydrogen-air mixture will not explode if the hydrogen content is above about 70 percent.

The result of the first two properties is that for the same operating conditions the heat which must be dissipated is reduced and at the same time the ease with which it can be carried off is increased.

The machine and its water-cooled heat exchanger for cooling the hydro-

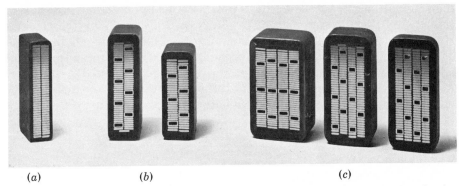

(a) (b) (c)

Fig. 4-7. Cross sections of bars for two-layer stator windings of turbine generators. Insulation system consists of synthetic resin with vacuum impregnation. (a) Indirectly cooled bar with tubular strands. (b) Water-cooled bars, two-wire-wide mixed strands. (c) Water-cooled bars, four-wire-wide mixed strands. (*Brown Boveri Corporation*.)

gen must be sealed in a gastight envelope. The crux of the problem is in sealing the bearings. The system is maintained at a slight pressure (at least $0.5 \, \mathrm{lb/in^2}$) above atmospheric so that gas leakage is outward and an explosive mixture cannot accumulate in the machine. At this pressure, the rating of the machine can be increased by about 30 percent above its air-cooled rating and the full-load efficiency increased by about 0.5 percent. The trend is toward the use of higher pressures (15 to $60 \, \mathrm{lb/in^2}$). Increasing the hydrogen pressure from 0.5 to $15 \, \mathrm{lb/in^2}$ increases the output for the same temperature rise by about 15 percent; a further increase to $30 \, \mathrm{lb/in^2}$ provides about an additional 10 percent.

An important step which has made it possible almost to double the output of a hydrogen-cooled turbine generator of given physical size is the development of *conductor cooling,* also called *inner cooling.* Here the coolant (liquid or gas) is forced through hollows or ducts inside the conductor or conductor strands. Examples of such conductors can be seen in Fig. 4-7. Thus,

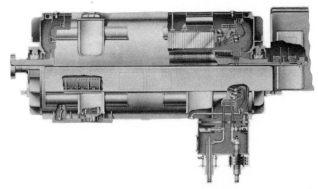

Fig. 4-8. Cutaway view of a 2-pole 3600-r/min turbine generator rated 500 MVA, 0.90 power factor, 22 kV, 60 Hz 45 $\mathrm{lb/in^2}$ gage H_2 pressure. Stator winding is water-cooled, rotor winding is hydrogen-cooled. (*General Electric Company*.)

the thermal barrier presented by the electric insulation is largely circumvented, and the conductor losses can be absorbed directly by the coolant. Hydrogen is usually the cooling medium for the rotor conductors. Either gas or liquid cooling may be used for the stator conductors. Hydrogen is the coolant in the former case, and transil oil or water is commonly used in the latter. A sectional view of a conductor-cooled turbine generator is given in Fig. 4-8. A large hydroelectric generator in which both stator and rotor are water-cooled is shown in Figs. 3-2 and 3-9.

4–6 EXCITATION SOURCES

The resultant flux in the magnetic circuit of a machine is established by the combined mmf of all the windings on the machine. For the conventional dc machine, the bulk of the effective mmf is furnished by the field windings. For the transformer, the net excitation may be furnished by either the primary or the secondary winding, or a portion may be furnished by each. A similar situation exists in ac machines. Furnishing excitation to ac machines has two different operational aspects which are of economic importance in the application of the machines.

a. Power Factor in AC Machines

The power factor at which ac machines operate is an economically important feature because of the cost of reactive kilovoltamperes. Low power factor adversely affects system operation in three principal ways. (1) Generators, transformers, and transmission equipment are rated in terms of kilovoltamperes rather than kilowatts because their losses and heating are very nearly determined by voltage and current regardless of power factor. The physical size and cost of ac apparatus is roughly proportional to its kVA rating. The investment in generators, transformers, and transmission equipment for supplying a given useful amount of active power therefore is roughly inversely proportional to the power factor. (2) Low power factor means more current and greater I^2R losses in the generating and transmitting equipment. (3) A further disadvantage is poor voltage regulation.

Factors influencing reactive-kVA requirements in motors can be visualized readily in terms of the relationship of these requirements to the establishment of magnetic flux. As in any electromagnetic device, the resultant flux necessary for motor operation must be established by a magnetizing component of current. It makes no difference either in the magnetic circuit or in the fundamental energy-conversion process whether this magnetizing current be carried by the rotor or stator winding, just as it makes no basic difference in a transformer which winding carries the exciting current. In some cases, part of it is supplied from each winding. If all or part of the magnetizing current is supplied to an ac winding, the input to that winding must include lagging reactive kilovoltamperes, because magnetizing current lags voltage drop by 90°. In effect, the lagging reactive kilovoltamperes set up flux in the motor.

The only possible source of excitation in an induction motor is the stator input. The induction motor therefore must operate at a lagging power factor. This power factor is very low at no load and increases to about 85 to 90 percent at full load, the improvement being caused by the increased active-power requirements with increasing load.

With a synchronous motor, there are two possible sources of excitation: alternating current in the armature or direct current in the field winding. If the field current is just sufficient to supply the necessary mmf, no magnetizing-current component or reactive kilovoltamperes are needed in the armature and the motor operates at unity power factor. If the field current is less, i.e., the motor is *underexcited,* the deficit in mmf must be made up by the armature and the motor operates at a lagging power factor. If the field current is greater, i.e., the motor is *overexcited,* the excess mmf must be counterbalanced in the armature and a leading component of current is present; the motor then operates at a leading power factor.

Because magnetizing current must be supplied to inductive loads such as transformers and induction motors, the ability of overexcited synchronous motors to supply lagging current is a highly desirable feature which may have considerable economic importance. In effect, overexcited synchronous motors act as generators of lagging reactive kilovoltamperes and thereby relieve the power source of the necessity for supplying this component. They thus may perform the same function as a local capacitor installation. Sometimes unloaded synchronous machines are installed in power systems solely for power-factor correction or for control of reactive-kVA flow. Such machines, called *synchronous condensers,* may be more economical in the larger sizes than static capacitors.

Both synchronous and induction machines may become self-excited when a sufficiently heavy capacitive load is present in their stator circuits. The capacitive current then furnishes the excitation and may cause serious overvoltage or excessive transient torques. Because of the inherent capacitance of transmission lines, the problem may arise when synchronous generators are energizing long unloaded or lightly loaded lines. The use of shunt reactors at the sending end of the line to compensate the capacitive current is sometimes necessary. For induction motors, it is normal practice to avoid self-excitation by limiting the size of any parallel capacitor when the motor and capacitor are switched as a unit.

b. Turbine-Generator Excitation Systems

As the available ratings of turbine-generators have increased, the problems of supplying the dc field excitation (amounting to 4000 A or more in the larger units) have grown progressively more difficult. The conventional excitation source is a dc generator whose output is supplied to the alternator field through brushes and slip rings. Cooling and maintenance problems are inevi-

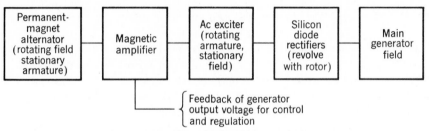

Fig. 4-9. Diagrammatic representation of a brushless excitation system.

tably associated with slip rings, commutators, and brushes. Many modern excitation systems have minimized these problems by minimizing the use of sliding contacts and brushes.

A schematic diagram of one such system is given in Fig. 4-9. At the heart of the system are silicon diode rectifiers, which are mounted on the same shaft as the generator field and which furnish dc excitation directly to the field. An ac exciter with a rotating armature feeds power along the shaft to the revolving rectifiers. The stationary field of the ac exciter is fed through a magnetic amplifier which controls and regulates the output voltage of the main generator. To make the system self-contained and free of sliding contacts, the excitation power for the magnetic amplifier is obtained from the stationary armature of a small permanent-magnet alternator also driven from the main shaft. The voltage and frequency of the ac exciter are chosen so as to optimize the performance and design of the overall system. Time delays in the response to a controlling signal are all short compared with the time constant of the main generator field. The system may have the additional advantage of doing away with need for spare exciters, generator-field circuit breakers, and field rheostats.

Another excitation system uses a shaft-driven alternator of conventional design as the main exciter. This alternator has a stationary armature and a rotating field winding. Its frequency may be 180 or 240 Hz. Its output is fed to a stationary solid-state rectifier, which in turn supplies the turbine-generator field through slip rings.

Excitation systems of the latest design are being built without any sort of rotating exciter-alternator. In these systems, the excitation power is obtained from a special auxiliary transformer fed from the local power system. Alternatively it may be obtained directly from the main generator terminals; in one system a special armature winding is included in the main generator to supply the excitation power. In each of these systems the power is rectified using phase-controlled silicon controlled rectifiers (SCRs). SCRs are similar to diode rectifiers, but they can be triggered by an external trigger signal so that the dc output voltage can be varied. These types of excitation system, which have been made possible by the development of reliable, high-power SCRs, are relatively simple in design and provide the fast response characteristics required in many modern applications.

4–7 ENERGY EFFICIENCY OF ELECTRIC MACHINERY

With increasing concern for both the supply and cost of energy comes a corresponding concern for efficiency in its use. Although electric energy can be converted into mechanical energy with great efficiency, achieving maximum efficiency requires both careful design of the electric machinery and proper matching of machine and intended application.

Clearly, one means to maximize the efficiency of an electric machine is to minimize its internal losses, such as those described in Art. 4-3. For example, the winding I^2R losses can be reduced by increasing the slot area so that more copper can be used, thus increasing the cross-sectional area of the windings and reducing the resistance.

Core loss can be reduced by decreasing the magnetic flux density in the iron of the machine. This can be done by increasing the volume of iron, but although the loss goes down in terms of watts per pound, the total volume of material (and hence the mass) is increased; depending on how the machine design is changed, there may be a point beyond which the losses actually begin to increase. Similarly, for a given flux density, eddy-current losses can be reduced by using thinner iron laminations.

One can see that there are tradeoffs involved here; machines of more efficient design generally require more material and thus are bigger and more costly. Users will generally choose the "lowest-cost" solution to a particular requirement; if the increased capital cost of a high-efficiency motor can be expected to be offset by energy savings over the expected lifetime of the machine, they will probably select the high-efficiency machine. If not, users are very unlikely to select this option in spite of the increased efficiency.

Similarly, some types of electric machines are inherently more efficient than others. For example, single-phase capacitor-start induction motors (Art. 11-1) are relatively inexpensive and highly reliable, finding use in all sorts of small appliances, e.g., refrigerators, air conditioners, and fans. Yet they are inherently less efficient than their three-phase counterparts. Modifications such as a capacitor-run feature can lead to greater efficiency in the single-phase induction motor, but they are expensive and often not economically justifiable.

To optimize the efficiency of use of electric machinery the machine must be properly matched to the application, both in terms of size and performance. Since typical induction motors tend to draw nearly constant reactive power, independent of load, and since this causes resistive losses in the supply lines, it is wise to pick the smallest-rating induction motor which can properly satisfy the requirements of a specific application. Alternatively, capacitive power-factor corrections may be used. Proper application of modern solid-state control technology (Arts. 6-6 and 10-5) can also play an important role in optimizing both performance and efficiency.

There are, of course, practical limitations which affect the selection of the motor for any particular application. Chief among them is that motors are

generally available only in certain standard sizes. For example, a typical man-
ufacturer might make fractional-horsepower ac motors rated at $\frac{1}{8}, \frac{1}{6}, \frac{1}{4}, \frac{1}{3}, \frac{1}{2}, \frac{3}{4}$,
and 1 hp (NEMA standard ratings). This discrete selection thus limits our
ability to fine tune our particular application; if our need is $\frac{7}{8}$ hp, we shall
undoubtedly end up buying a 1-hp device and settling for a somewhat lower
than optimum efficiency. A custom-designed and manufactured $\frac{7}{8}$-hp motor
can be economically justified only if it is needed in large quantities.

It is not the object of this book to delve into the details of machine design
or into a specific study of the optimization of machine performance. However,
a firm understanding of the principles of rotating machinery as provided by
the following chapters forms the basis of this type of analysis.

4–8 THE NATURE OF MACHINERY PROBLEMS

As we begin to converge on the theory of rotating machines, we need to reflect
momentarily on our objectives: What are the machine characteristics we need
to know in reasonably precise, quantitative form? The answer depends on
what, specifically, the machines are intended to do for us. The machine is one
component in an electromechanical energy-conversion system, and the machine characteristics often play the predominant part in the behavior of the
complete system.

In many applications of electric motors, the motor is supplied with electric power from a constant-voltage source. The motor drives a mechanical
load whose torque requirements vary with the speed at which it is driven. The
steady-state operating speed is then fixed by the point at which the torque the
motor can furnish electromagnetically is equal to the torque the load can
absorb mechanically. In Fig. 4-10, for instance, the solid curve is the speed-
torque characteristic of an induction motor. The dashed curve is a plot of the
torque input required by a fan for various operating speeds. When the fan and
motor are coupled, the steady-state operating point of the combination is at
the intersection of these two curves, i.e., where what the motor can give is the
same as what the fan can take.

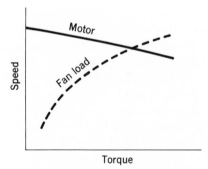

Fig. 4-10. Superposition of motor and load characteristics.

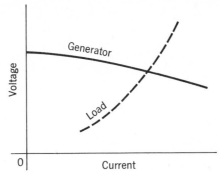

Fig. 4-11. Superposition of generator and load characteristics.

Motor power or torque requirements vary, of course, depending on conditions within the driven equipment. The requirements of some motor loads are satisfied by a speed which remains approximately constant as load varies; an ordinary hydraulic pump is an example. Others, like a phonograph turntable, require absolutely constant speed. Still others require a speed closely coordinated with another speed, e.g., raising both ends of a vertical-lift bridge. Some motor applications, such as cranes and many traction-type drives, inherently demand low speeds and heavy torques at one end of the range and relatively high speeds and light torques at the other, in other words, a varying-speed characteristic. Others may require an adjustable constant speed, e.g., some machine-tool drives in which the speed of operation may require adjustment over a wide range but must always be carefully predetermined, or an adjustable varying speed (again a crane is an example). In almost every application the torque which the motor is capable of supplying while starting, the maximum torque which it can furnish while running, and the current requirements are items of importance.

Many similar remarks can be made for generators. For example, the terminal voltage and power output of a generator are determined by the characteristics of both the generator and its load. Thus, the solid curve of Fig. 4-11 is a plot of the terminal voltage of a dc shunt generator as a function of its current output. The dashed curve is a plot of the voltampere characteristic of a load. When the load is connected to the generator terminals, the operating point of the combination is at the intersection of these two curves, where what the generator can give is the same as what the load can take. Often, as in the usual central station, the requirement is that terminal voltage remain substantially constant over a wide load range. Not infrequently, however, a motor is associated with its own individual generator in order to provide greater flexibility and more precise control. Then we may want the terminal voltage to vary with load in some particular fashion.

Among the features of outstanding importance, therefore, are the torque-speed characteristics of motors and the voltage-load characteristic of generators, together with knowledge of the limits between which these characteristics can be varied and ideas of how such variations can be obtained. Moreover, pertinent economic features are efficiency, power factor, comparative costs, and the effect of losses on the heating and rating of the machines. One of the objects of our analysis of machinery is accordingly to study and compare these features for the various machine types. Of course there are many important, interesting, and complex engineering problems associated with the design, development, and manufacturing of machines for which these studies are but the introduction; most such problems are beyond the scope of this book.

A knowledge of the steady-state characteristics is insufficient, however, for an understanding of the role played by rotating machines in modern technology. They are at the heart of a wide range of dynamic systems. Hand-held electronic calculators are widely available with small printers driven by miniature dc stepper motors. On the other hand, synchronous generators with ratings greater than 1000 MW are common for modern electric-power systems. Industrial processes use rotating electric machinery in a wide range of applications, often requiring precision control. The increased use of electric energy will include a significant growth of the use of electric machines in transportation systems. In each of these applications the emphasis is on the dynamic behavior of the complete electromechanical system of which the machine is one part.

Here the electromechanical transient behavior of the system as a whole is a major consideration; the system should respond accurately and rapidly to the control function, and oscillations should die out quickly. Not only the electrical characteristics but also the mechanical properties of the system, such as stiffness, inertia, and friction, must be considered and indeed may become the predominant factors.

In such studies, then, an adequate theory must be capable of treating the dynamic behavior of machines as system components. Since the analysis of a complete electromechanical system presents a complex problem, a theory of machines suitable for these purposes must be simplified as much as possible while still retaining the essential elements. The following chapters will develop such a theory. In addition to this basic theory, other factors which should be understood for a more complete and satisfying understanding of machine theory and which may have an important bearing on machine applications are discussed.

4-9 SUMMARY

As we go deeper into the study of rotating machines, detailed differences between the various machine types begin to appear. For example, commutation is a problem unique to dc machines. Magnetic saturation and salient poles may have significant effects on dc and synchronous machines. Speed control of induction motors presents problems. Understanding motor speed-control systems using solid-state rectifiers requires study of the rectifier and motor together. Important problems concerning transient and dynamic characteristics arise in many machine applications. Fractional-horsepower motors have their unique characteristics, not the least of which is a highly competitive market. Chapters 5 to 11 are concerned with some of these problems.

It should be recognized that the performance limitation of a machine is basically determined by the properties of the materials of which it is composed. Thus, much of the great progress in electric machinery over the years has stemmed from improvements in the quality and characteristics of steel and insulating material and in the cooling of the machines.

A group of interrelated problems common to all machine types is created by the losses in the machine and the need to dissipate the associated heat. The rating of the machine is closely connected with its ability to operate at temperatures compatible with reasonable life of the insulation and of the machine as a whole. Matters of rating, allowable temperatures rise, and determination of losses are all subjects of standardization by professional organizations such as the IEEE, NEMA, and ANSI. Such matters are obviously of great importance in the economics and engineering of a project involving machinery.

PROBLEMS

4-1 The following measurements have been made upon a 2-winding transformer whose nameplate has been lost. The primary self-inductance is 85 H, the secondary self-inductance is 0.76 mH, and with the primary open-circuited a secondary applied voltage of 120 V rms results in a primary voltage of 4.0 kV rms. Calculate (*a*) the primary-to-secondary mutual inductance in henrys and (*b*) the resultant open-circuit voltage in the secondary if a voltage of 4.0 kV is applied to the primary.

4-2 Assume that the rating of an induction motor is determined by the saturation flux density of the magnetic material and by the capabilities of the stator-cooling system. The saturation flux density determines the stator volts per turn, and the cooling system determines the stator ampere-turns.

Assume that the machine geometry and number of armature turns remain fixed. Discuss the effects on the machine rating (voltage, current, and power) of the following proposed design modifications.

(*a*) Replace the magnetic material with a material whose saturation flux density is 10 percent higher.

(*b*) Increase the cooling-system capability by 25 percent; that is 25 percent additional I^2R losses can be handled.

If more careful winding techniques permit achieving 20 percent more wire in the slots, discuss the effect on the machine rating if the cooling system capability and saturation flux density are unchanged and either

(c) the wire size remains the same and the number of turns increases by 20 percent or

(d) the number of turns remains the same and the wire cross-sectional area is increased by 20 percent.

4-3 Figure 4-12 shows a simplified block-diagram representation of a synchronous generator and its voltage regulation-excitation system. The terminal voltage V_t is compared with a reference voltage V_{Ref} and the error signal ϵ is

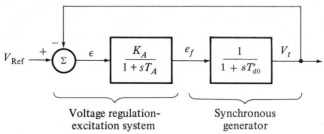

Fig. 4-12. Block diagram for Problem 4-3.

input to the exciter, represented by a gain K_A and time constant T_A. The exciter output e_f supplies the main generator, represented by time constant T'_{d0}.

(a) Write two first-order differential equations relating V_t to V_{Ref}. There will be one differential equation for the exciter output voltage de_f/dt and one differential equation for the terminal voltage dV_t/dt.

For $K_A = 50$, $T_A = 0.2$ s, and $T'_{d0} = 4.5$ s.

(b) Calculate the value of V_{Ref} required to obtain a steady-state terminal voltage of 1.0 per unit.

(c) Solve for the terminal voltage as a function of time if the reference voltage is suddenly increased to 1.05 per unit.

4-4 A production-line process operates at a load of 25 hp 10 percent of the time and 5 hp the remaining 90 percent of the time. Two induction motors, both rated at 20 hp, are being considered for this application.

Motor	Efficiency, %	
	At 25 hp	At 5 hp
1	92.0	85.0
2	91.0	86.5

(a) In the long run which of these two motors is the proper choice in terms of energy requirement?

(b) Assuming that the process runs nonstop for 1 year, calculate the energy bill for each motor for an average electric-energy cost of 7 cents per kilowatthour.

4-5 A dc compound motor is to be selected for the operation of a lift. The motor is to drive continuously a steel cable which runs over pulleys at the bottom and the top of the lift. When the load is descending, the motor becomes a generator and pumps power back into the line, the resulting torque supplying a braking action. The following operation cycle is repeated continuously throughout the day:

	Time, min	hp
Load going up	1	75
Loading period at top	2	5
Load going down	1	−60
Loading period at bottom	3	5

On the basis of heating, select the smallest motor suitable for this application. Motors are available in 25, 30, 40, 50, 60, 75, and 100 hp. What other factors, besides heating, should be considered?

4-6 In the design of a grab-bucket hoist for unloading coal from a barge into a bunker, a study is made of the mechanical requirements to determine the motor duty cycle. For an average cycle:

Part of cycle	Elapsed time, s	Required output, hp
Close bucket	6	40
Hoist	10	80
Open bucket	3	30
Lower bucket	10	45
Rest	16	0

Because of the conditions of service, a dustproof enclosed motor without forced ventilation is to be used, and the constant k associated with the standstill time may be taken as unity.

(a) Using the rms method, specify the continuous horsepower rating of the motor. Choose a commercially available motor size.

(b) Proposals to furnish this motor submitted by two manufacturers contain the following efficiency guarantees and prices:

Efficiency, %

Motor	$\frac{1}{4}$ load	$\frac{1}{2}$ load	$\frac{3}{4}$ load	1.0 load	$1\frac{1}{4}$ load	$1\frac{1}{2}$ load	Net price†
A	83.4	90.5	90.3	88.0	86.8	85.0	$3500
B	81.3	88.6	90.3	90.6	90.3	89.6	4200

†Both prices f.o.b. factory, freight allowed.

The average net cost of energy at this plant is 5.5 cents per kilowatthour. Assuming that the energy cost will remain constant, which of these motors would you recommend if the expected lifetime of each motor is 4 years? 8 years?

5

DC Machines: Steady State

Dc machines are characterized by their versatility. By means of various combinations of shunt-, series-, and separately excited field windings they can be designed to display a wide variety of volt-ampere or speed-torque characteristics for both dynamic and steady-state operation. Because of the ease with which they can be controlled, systems of dc machines are often used in applications requiring a wide range of motor speeds or precise control of motor output.

In this chapter the steady-state behavior of dc machines is investigated. Their dynamic behavior is discussed in Chap. 6.

5–1 INTRODUCTION TO DC MACHINES

The essential features of a dc machine are shown schematically in Fig. 5-1. The stator has salient poles and is excited by one or more field coils. The

201

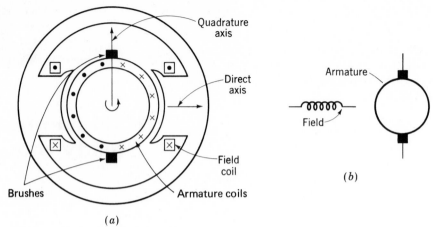

(a)

(b)

Fig. 5-1. Schematic representations of a dc machine.

air-gap flux distribution created by the field windings is symmetrical about the centerline of the field poles. This axis is called the *field axis* or *direct axis*.

As discussed in Art. 3-2c, the ac voltage generated in each rotating armature coil is converted to dc in the external armature terminals by means of a rotating commutator and stationary brushes to which the armature leads are connected. The commutator-brush combination forms a mechanical rectifier, resulting in a dc armature voltage as well as an armature mmf wave which is fixed in space. Commutator action is discussed in detail in Art. 5-2.

The brushes are located so that commutation occurs when the coil sides are in the neutral zone, midway between the field poles. The axis of the armature-mmf wave then is 90 electrical degrees from the axis of the field poles, i.e., in the *quadrature axis*. In the schematic representation of Fig. 5-1a the brushes are shown in the quadrature axis because this is the position of the coils to which they are connected. The armature-mmf wave then is along the brush axis, as shown. (The geometrical position of the brushes in an actual machine is approximately 90 electrical degrees from their position in the schematic diagram because of the shape of the end connections to the commutator. For example, see Fig. 5-7.) For simplicity, the circuit representation usually will be drawn as in Fig. 5-1b.

The magnetic torque and the speed voltage appearing at the brushes are independent of the spatial waveform of the flux distribution; for convenience we shall continue to assume a sinusoidal flux-density wave in the air gap. The torque can then be found from the magnetic field viewpoint of Art. 3-7b.

The torque can be expressed in terms of the interaction of the direct-axis air-gap flux per pole Φ_d and the space-fundamental component F_{a1} of the armature-mmf wave, in a form similar to Eq. 3-86. With the brushes in the quadrature axis the angle between these fields is 90 electrical degrees, and its sine equals unity. Substitution in Eq. 3-86 then gives for a P-pole machine

$$T = \frac{\pi}{2}\left(\frac{P}{2}\right)^2 \Phi_d F_{a1} \qquad (5\text{-}1)$$

in which the minus sign has been dropped because the positive direction of the torque can be determined from physical reasoning. The peak value of the sawtooth armature-mmf wave is given by Eq. 3-11, and its space fundamental F_{a1} is $8/\pi^2$ times its peak. Substitution in Eq. 5-1 then gives

$$T = \frac{PC_a}{2\pi m}\Phi_d i_a = K_a \Phi_d i_a \qquad \text{N} \cdot \text{m} \qquad (5\text{-}2)$$

where i_a = current in external armature circuit
$\quad\quad C_a$ = total number of conductors in armature winding
$\quad\quad m$ = number of parallel paths through winding

and

$$K_a = \frac{PC_a}{2\pi m} \qquad (5\text{-}3)$$

is a constant fixed by the design of the winding.

The rectified voltage generated in the armature has already been found in Art. 3-6*b* for an elementary single-coil armature, and its waveform is shown in Fig. 3-33. The effect of distributing the winding in several slots is shown in Fig. 5-2, in which each of the rectified sine waves is the voltage generated in one of the coils, commutation taking place at the moment when the coil sides are in the neutral zone. The generated voltage as observed from the brushes is the sum of the rectified voltages of all the coils in series between brushes and is shown by the rippling line labeled e_a in Fig. 5-2. With a dozen or so commutator segments per pole, the ripple becomes very small and the average generated voltage observed from the brushes equals the sum of the average values of the rectified coil voltages. From Eq. 3-59 the rectified voltage e_a between brushes, known also as the *speed voltage,* is

$$e_a = \frac{PC_a}{2\pi m}\Phi_d \omega_m = K_a \Phi_d \omega_m \qquad (5\text{-}4)$$

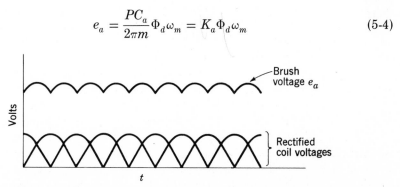

Fig. 5-2. Rectified coil voltages and resultant voltage between brushes in a dc machine.

where K_a is the design constant defined in Eq. 5-3. The rectified voltage of a distributed winding has the same average value as that of a concentrated coil. The difference is that the ripple is greatly reduced.

From Eqs. 5-2 and 5-4, with all variables expressed in SI units,

$$e_a i_a = T \omega_m \qquad (5\text{-}5)$$

This equation simply says that the instantaneous electric power associated with the speed voltage equals the instantaneous mechanical power associated with the magnetic torque, the direction of power flow being determined by whether the machine is acting as a motor or generator.

The direct-axis air-gap flux is produced by the combined mmf $\Sigma N_f i_f$ of the field windings, the flux-mmf characteristic being the *magnetization curve* for the particular iron geometry of the machine. A magnetization curve is shown in Fig. 5-3a, in which it is assumed that the armature mmf has no effect on the direct-axis flux because the axis of the armature-mmf wave is perpendicular to the field axis. It will be necessary to reexamine this assumption later in this chapter, where the effects of saturation are investigated more thoroughly. Because the armature emf is proportional to flux times speed, it is usually more convenient to express the magnetization curve in terms of the armature emf e_{a0} at a constant speed ω_{m0} as shown in Fig. 5-3b. The voltage e_a for a given flux at any other speed ω_m is proportional to the speed; i.e., from Eq. 5-4

$$\frac{e_a}{\omega_m} = K_a \Phi_d = \frac{e_{a0}}{\omega_{m0}} \qquad (5\text{-}6)$$

or

$$e_a = \frac{\omega_m}{\omega_{m0}} e_{a0} \qquad (5\text{-}7)$$

Figure 5-3c shows the magnetization curve with only one field winding excited. This curve can easily be obtained by test methods, no knowledge of any design details being required.

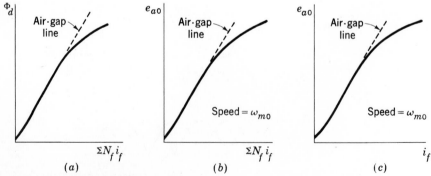

Fig. 5-3. Magnetization curves of a dc machine.

Over a fairly wide range of excitation the reluctance of the iron is negligible compared with that of the air gap. In this region the flux is linearly proportional to the total mmf of the field windings, the constant of proportionality being the *direct-axis air-gap permeance* $\mathcal{P}_d$; thus

$$\Phi_d = \mathcal{P}_d \Sigma N_f i_f \tag{5-8}$$

The dashed straight line through the origin coinciding with the straight portion of the magnetization curves in Fig. 5-3 is called the *air-gap line*.

The outstanding advantages of dc machines arise from the wide variety of operating characteristics which can be obtained by selection of the method of excitation of the field windings. The field windings may be *separately excited* from an external dc source, or they may be *self-excited*; i.e., the machine may supply its own excitation. Connection diagrams are shown in Fig. 5-4. The method of excitation profoundly influences not only the steady-state characteristics, which we shall describe briefly in this article, but also the dynamic behavior of the machine in control systems, discussed in Chap. 6.

The connection diagram of a separately excited generator is given in Fig. 5-4a. The required field current is a very small fraction of the rated armature current—of the order of 1 to 3 percent in the average generator. A small amount of power in the field circuit may control a relatively large amount of power in the armature circuit; i.e., the generator is a power amplifier. Separately excited generators are often used in feedback control systems when control of the armature voltage over a wide range is required.

The field windings of self-excited generators may be supplied in three different ways. The field may be connected in series with the armature (Fig. 5-4b), resulting in a *series generator*. The field may be connected in shunt

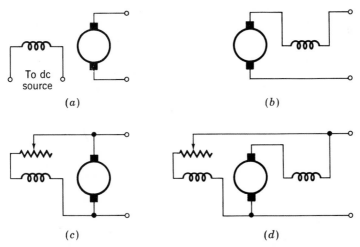

Fig. 5-4. Field-circuit connections of dc machines: (*a*) separate excitation; (*b*) series, (*c*) shunt, (*d*) compound.

with the armature (Fig. 5-4c), resulting in a *shunt generator,* or the field may be in two sections (Fig. 5-4d), one of which is connected in series and the other in shunt with the armature, resulting in a *compound generator*. With self-excited generators residual magnetism must be present in the machine iron to get the self-excitation process started.

Typical steady-state volt-ampere characteristics are shown in Fig. 5-5, constant-speed prime movers being assumed. The relation between the steady-state generated emf E_a and the terminal voltage V_t is

$$V_t = E_a - I_a R_a \qquad (5\text{-}9)$$

where I_a is the armature current output and R_a is the armature circuit resistance. In a generator, E_a is larger than V_t, and the electromagnetic torque T is a countertorque opposing rotation.

The terminal voltage of a separately excited generator decreases slightly with increase in the load current, principally because of the voltage drop in the armature resistance. The field current of a series generator is the same as the load current, so that the air-gap flux and hence the voltage vary widely with load. As a consequence, series generators are not often used. The voltage of shunt generators drops off somewhat with load but not in a manner which is objectionable for many purposes. Compound generators are normally connected so that the mmf of the series winding aids that of the shunt winding. The advantage is that through the action of the series winding the flux per pole can increase with load, resulting in a voltage output which is nearly constant or which even rises somewhat as load increases. The shunt winding usually contains many turns of relatively small wire. The series winding, wound on the outside, consists of a few turns of comparatively heavy conduc-

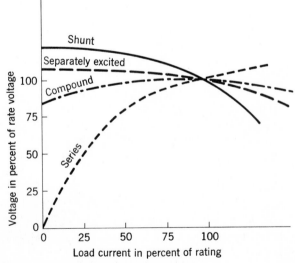

Fig. 5-5. Volt-ampere characteristics of dc generators.

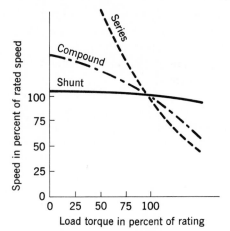

Fig. 5-6. Speed-torque characteristics of dc motors.

tor because it must carry the full armature current of the machine. The voltage of both shunt and compound generators can be controlled over reasonable limits by means of rheostats in the shunt field.

Any of the methods of excitation used for generators can also be used for motors. Typical steady-state speed-torque characteristics are shown in Fig. 5-6, in which it is assumed that the motor terminals are supplied from a constant-voltage source. In a motor the relation between the emf E_a generated in the armature and the terminal voltage V_t is

$$V_t = E_a + I_a R_a \qquad (5\text{-}10)$$

or

$$I_a = \frac{V_t - E_a}{R_a} \qquad (5\text{-}11)$$

where I_a is now the armature current input. The generated emf E_a is now smaller than the terminal voltage V_t, the armature current is in the opposite direction to that in a generator, and the electromagnetic torque is in the direction to sustain rotation of the armature.

In shunt and separately excited motors the field flux is nearly constant. Consequently, increased torque must be accompanied by a very nearly proportional increase in armature current and hence by a small decrease in counter emf to allow this increased current through the small armature resistance. Since counter emf is determined by flux and speed (Eq. 5-4), the speed must drop slightly. Like the squirrel-cage induction motor, the shunt motor is substantially a constant-speed motor having about 5 percent drop in speed from no load to full load. A typical speed-load characteristic is shown by the solid curve in Fig. 5-4. Starting torque and maximum torque are limited by the armature current that can be commutated successfully.

An outstanding advantage of the shunt motor is ease of speed control. With a rheostat in the shunt-field circuit, the field current and flux per pole can be varied at will, and variation of flux causes the inverse variation of speed to maintain counter emf approximately equal to the impressed terminal voltage. A maximum speed range of about 4 or 5 to 1 can be obtained by this method, the limitation again being commutating conditions. By variation of the impressed armature voltage, very wide speed ranges can be obtained.

In the series motor, increase in load is accompanied by increases in the armature current and mmf and the stator field flux (provided the iron is not completely saturated). Because flux increases with load, speed must drop in order to maintain the balance between impressed voltage and counter emf; moreover, the increase in armature current caused by increased torque is smaller than in the shunt motor because of the increased flux. The series motor is therefore a varying-speed motor with a markedly drooping speed-load characteristic of the type shown in Fig. 5-6. For applications requiring heavy torque overloads, this characteristic is particularly advantageous because the corresponding power overloads are held to more reasonable values by the associated speed drops. Very favorable starting characteristics also result from the increase in flux with increased armature current.

In the compound motor the series field may be connected either *cumulatively,* so that its mmf adds to that of the shunt field, or differentially, so that it opposes. The differential connection is very rarely used. As shown by the broken-dash curve in Fig. 5-6, a cumulatively compounded motor has speed-load characteristic intermediate between those of a shunt and a series motor, the drop of speed with load depending on the relative number of ampere-turns in the shunt and series fields. It does not have the disadvantage of very high light-load speed associated with a series motor, but it retains to a considerable degree the advantages of series excitation.

The application advantages of dc machines lie in the variety of performance characteristics offered by the possibilities of shunt, series, and compound excitation. Some of these characteristics have been touched upon briefly in this article. Still greater possibilities exist if additional sets of brushes are added so that other voltages can be obtained from the commutator. Thus the versatility of dc-machine systems and their adaptability to control, both manual and automatic, are their outstanding features. These characteristics will be discussed in this and the following chapter for both steady-state and dynamic operation.

5–2 COMMUTATOR ACTION

The dc machine differs in several respects from the ideal model of Art. 3-2. Although the basic concepts of Art. 3-2 are still valid, a reexamination of the assumptions and a modification of the model are desirable. The crux of the matter is the effect of the commutator shown in Figs. 3-1 and 3-16.

Figure 5-7 shows diagrammatically the armature winding of Fig. 3-22a with the addition of the commutator, brushes, and connections of the coils to

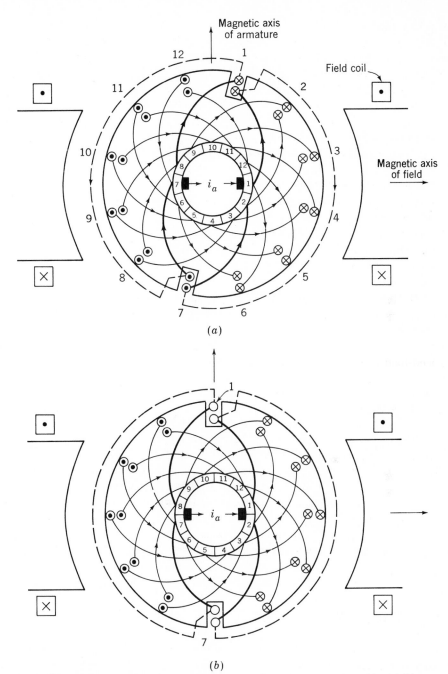

Fig. 5-7. DC-machine armature winding with commutator and brushes: (a) and (b) current directions for two positions of the armature.

the commutator segments. The commutator is represented by the ring of segments in the center of the figure. The segments are insulated from each other and from the shaft. Two stationary brushes are shown by the black rectangles inside the commutator. Actually the brushes usually contact the outer surface, as shown in Fig. 3-16. The coil sides in the slots are shown in cross section by the small circles with dots and crosses in them, indicating currents toward and away from the reader, respectively, as in Fig. 3-22. The connections of the coils to the commutator segments are shown by the circular arcs. The end connections at the back of the armature are shown dashed for the two coils in slots 1 and 7, and the connections of these coils to adjacent commutator segments are shown by the heavy arcs. All coils are identical. The back end connections of the other coils have been omitted, to avoid complicating the figure, but they can easily be traced by remembering that each coil has one side in the top of a slot and the other side in the bottom of the diametrically opposite slot.

In Fig. 5-7a the brushes are in contact with commutator segments 1 and 7. Current entering the right-hand brush divides equally between two parallel paths through the winding. The first path leads to the inner coil side in slot 1 and finally ends at the brush on segment 7. The second path leads to the outer coil side in slot 6 and also finally ends at the brush on segment 7. The current directions in Fig. 5-7a can readily be verified by tracing these two paths. They are the same as in Fig. 3-22. The effect is identical to that of a coil wrapped around the armature with its magnetic axis vertical, and a clockwise magnetic torque is exerted on the armature, tending to align its magnetic field with that of the field winding.

Now suppose the machine is acting as a generator driven in the counterclockwise direction by an applied mechanical torque. Figure 5-7b shows the situation after the armature has rotated through the angle subtended by half a commutator segment. The right-hand brush is now in contact with both segments 1 and 2, and the left-hand brush is in contact with both segments 7 and 8. The coils in slots 1 and 7 are now short-circuited by the brushes. The currents in the other coils are shown by the dots and crosses, and they produce a magnetic field whose axis again is vertical.

After further rotation, the brushes will be in contact with segments 2 and 8, and slots 1 and 7 will have rotated into the positions which were previously occupied by slots 12 and 6 in Fig. 5-7a. The current directions will be similar to those of Fig. 5-7a except that the currents in the coils in slots 1 and 7 will have reversed. The magnetic axis of the armature is still vertical.

During the time when the brushes are simultaneously in contact with two adjacent commutator segments, the coils connected to these segments are temporarily removed from the main circuits through the winding, short-circuited by the brushes, and the currents in them are reversed. Ideally, the current in the coils being commutated should reverse linearly with time. Serious departure from linear commutation will result in sparking at the brushes. Means for obtaining sparkless commutation are discussed in Art. 5-7. With

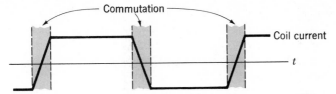

Fig. 5-8. Waveform of current in an armature coil with linear commutation.

linear commutation the waveform of the current in any coil as a function of time is trapezoidal, as shown in Fig. 5-8.

The winding of Fig. 5-7 is simpler than that used in most dc machines. Ordinarily more slots and commutator segments would be used, and except in small machines more than two poles are common. Nevertheless, the simple winding of Fig. 5-7 includes the essential features of more complicated windings.

5–3 EFFECT OF ARMATURE MMF

Armature mmf has definite effects on both the space distribution of the air-gap flux and the magnitude of the net flux per pole. The effect on flux distribution is important because the limits of successful commutation are directly influenced; the effect on flux magnitude is important because both the generated voltage and torque per unit of armature current are influenced thereby. These effects and the problems arising from them are described in this article.

It was shown in Art. 3-3*b* and Fig. 3-23 that the armature-mmf wave can be closely approximated by a sawtooth, corresponding to the wave produced by a finely distributed armature winding or current sheet. For a machine with brushes in the neutral position, the idealized mmf wave is again shown by the dashed sawtooth in Fig. 5-9, in which a positive mmf ordinate denotes flux lines leaving the armature surface. Current directions in all windings other than the main field are indicated by black and crosshatched bands. Because of the salient-pole field structure found in almost all dc machines, the associated space distribution of flux will not be triangular. The distribution of air-gap flux density with only the armature excited is given by the solid curve of Fig 5-9. As can readily be seen, it is appreciably decreased by the long air path in the interpolar space.

The axis of the armature mmf is fixed at 90 electrical degrees from the main-field axis by the brush position. The corresponding flux follows the paths shown in Fig. 5-10. The effect of the armature mmf is seen to be that of creating flux crossing the pole faces; thus its path in the pole shoes crosses the path of the main-field flux. For this reason, armature reaction of this type is called *cross-magnetizing armature reaction*. It evidently causes a decrease in the resultant air-gap flux density under one half of the pole and an increase under the other half.

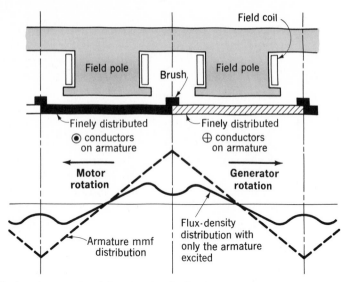

Fig. 5-9. Armature-mmf and flux-density distribution with brushes on neutral and only the armature excited.

When the armature and field windings are both excited, the resultant air-gap flux-density distribution is of the form given by the solid curve of Fig. 5-11. Superimposed on this figure are the flux distributions with only the armature excited (dashed curve) and only the field excited (dotted curve). The effect of cross-magnetizing armature reaction in decreasing the flux under one pole tip and increasing it under the other can be seen by comparing the solid and short-dash curves. In general, the solid curve is not the algebraic sum of the two dashed curves because of the nonlinearity of the iron magnetic circuit. Because of saturation of the iron, the flux density is decreased by a greater amount under one pole tip than it is increased under the other. Accordingly, the resultant flux per pole is lower than would be produced by the field winding alone, a consequence known as the *demagnetizing effect of cross-magnetizing armature reaction*. Since it is caused by saturation, its magnitude is a nonlinear function of both the field current and the armature current. For

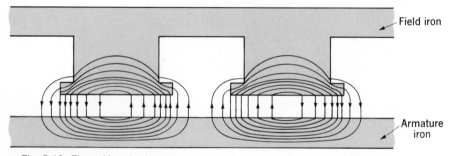

Fig. 5-10. Flux with only the armature excited and brushes on neutral.

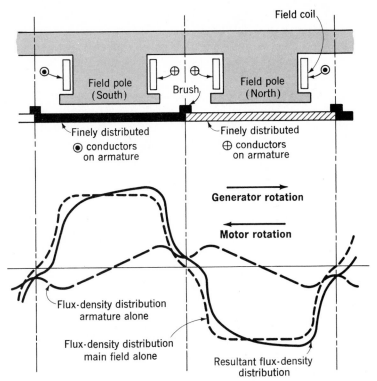

Fig. 5-11. Armature, main-field, and resultant flux-density distributions with brushes on neutral.

normal machine operation at the flux densities used commercially, the effect is usually significant, especially at heavy loads, and must often be taken into account in analyses of performance.

The distortion of the flux distribution caused by cross-magnetizing armature reaction may have a detrimental influence on the ability to commutate the current, especially if the distortion becomes excessive. In fact, this distortion is usually an important factor limiting the short-time overload of a dc machine. Tendency toward distortion of the flux distribution is most pronounced in a machine, such as a shunt motor, where the field excitation remains substantially constant while the armature mmf may reach very significant proportions at heavy loads. The tendency is least pronounced in a series-excited machine, such as the series motor, for both the field and armature mmf increase with load.

The effect of cross-magnetizing armature reaction can be limited in the design and construction of the machine. The mmf of the main field should exert predominating control on the air-gap flux, so that the condition of weak-field mmf and strong armature mmf can be avoided. The reluctance of the cross-flux path—essentially, the armature teeth, pole shoes, and the air gap, especially at the pole tips—can be increased by increasing the degree of saturation in the teeth and pole faces, by avoiding too small an air gap, and by using a chamfered or eccentric pole face, which increases the air gap at the pole tips. These expedients affect the path of the main flux as well, but the influence on the cross flux is much greater. The best but also the most expensive curative measure is to compensate the armature mmf by means of a winding embedded in the pole faces, a measure discussed in Art. 5-8.

If the brushes are not in the neutral position, the axis of the armature-mmf wave is not 90° from the main-field axis. The armature mmf then produces not only cross magnetization but also a direct demagnetizing or magnetizing effect, depending on the direction of brush shift. Shifting of the brushes from the neutral is usually inadvertent due to incorrect positioning of the brushes or a poor brush fit. Before the invention of interpoles, however, shifting the brushes was a common method of securing satisfactory commutation, the direction of the shift being such that demagnetizing action was produced. It can be shown that brush shift in the direction of rotation in a generator or against rotation in a motor produces a direct demagnetizing mmf which may result in unstable operation of a motor or excessive drop in voltage of a generator. Incorrectly placed brushes can be detected by a load test. If the brushes are on neutral, the terminal voltage of a generator or the speed of a motor should be the same for identical conditions of field excitation and armature current when the direction of rotation is reversed.

5–4 ANALYTICAL FUNDAMENTALS: ELECTRIC–CIRCUIT ASPECTS

From Eqs. 5-1 and 5-4 the electromagnetic torque and generated voltage of a dc machine are, respectively,

$$T = K_a \Phi_d I_a \tag{5-12}$$

and

$$E_a = K_a \Phi_d \omega_m \tag{5-13}$$

where

$$K_a = \frac{PC_a}{2\pi m} \tag{5-14}$$

Here the capital-letter symbols E_a for generated voltage and I_a for armature current are used to emphasize that we are primarily concerned with steady-state considerations in this chapter. The remaining symbols are as defined in Art. 5-1. These are basic equations for analysis of the machine. The quantity $E_a I_a$ is frequently referred to as the *electromagnetic power;* from Eqs. 5-12 and 5-13 it is related to electromagnetic torque by

$$T = \frac{E_a I_a}{\omega_m} \tag{5-15}$$

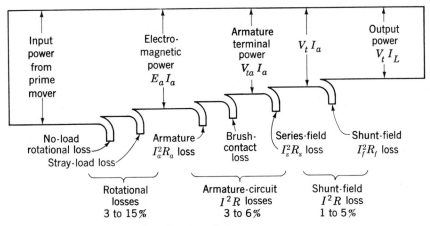

Fig. 5-12. Power division in a dc generator.

Figures 5-12 and 5-13 present in graphical form power balances for dc generators and motors, respectively, with both shunt and series fields. The connection diagram is given in Fig. 5-14. When either the shunt or the series field is not present in the machine, the associated entry is omitted from Figs. 5-12 to 5-14. In these diagrams V_t is the machine terminal voltage, V_{ta} the armature terminal voltage, I_L the line current, I_s the series-field current (equal to I_a for the connections shown in Fig. 5-14), I_f the shunt-field current, R_a the armature resistance, R_f the shunt-field resistance, and R_s the series-field resistance. Included in R_a is the resistance of any commutating and compensating winding. The armature-circuit I^2R losses, field-circuit I^2R losses, and rotational losses are those originally considered in Art. 4-3; typical full-load orders of magnitude of these losses, expressed in percent of the machine input, are quoted in Figs. 5-12 and 5-13 for general-purpose generators and motors in the 1- to 100-kW or 1- to 100-hp range, the smaller percentages applying to the larger ratings.

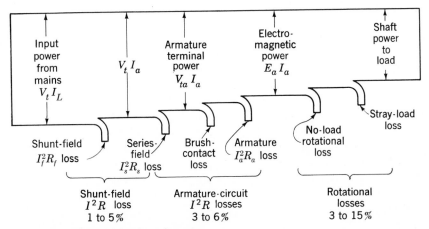

Fig. 5-13. Power division in a dc motor.

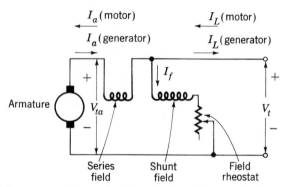

Fig. 5-14. Motor or generator connection diagram with current directions.

The electromagnetic power differs from the mechanical power at the machine shaft by the rotational losses and differs from the electric power at the machine terminals by the shunt-field and armature I^2R losses. The electromagnetic power is that measured at the points across which E_a exists; numerical addition of the rotational losses for generators and subtraction for motors yields the mechanical power at the shaft.

The interrelations between voltage and current are immediately evident from the connection diagram. Thus,

$$V_{ta} = E_a \pm I_a R_a \tag{5-16}$$

$$V_t = E_a \pm I_a(R_a + R_s) \tag{5-17}$$

and $$I_L = I_a \pm I_f \tag{5-18}$$

where the plus sign is used for a motor and the minus sign for a generator. Some of the terms in Eqs. 5-16 to 5-18 may be omitted when the machine connections are simpler than those shown in Fig. 5-14. The resistance R_a is to be interpreted as that of the armature plus brushes unless specifically stated otherwise. Sometimes R_a is taken as the resistance of the armature winding alone, and the brush-contact voltage drop is accounted for as a separate item, usually assumed to be 2 V.

For compound machines, another variation may occur. Figure 5-14 shows a so-called *long-shunt connection* in that the shunt field is connected directly across the line terminals with the series field between it and the armature. An alternative possibility is the *short-shunt connection,* illustrated in Fig. 5-15 for a compound generator, with the shunt field directly across the armature and the series field between it and the line terminals. The series-field current is then I_L instead of I_a, and the voltage equations are modified accordingly. There is so little practical difference between these two connections that the distinction can usually be ignored: unless otherwise stated, compound machines will be treated as though they were long-shunt-connected.

Although the difference between terminal voltage V_t and armature-gen-

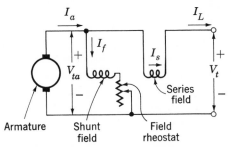

Fig. 5-15. Short-shunt compound-generator connections.

erated voltage E_a is comparatively small for normal operation, it has a definite bearing on performance characteristics. In effect, this difference, acting in conjunction with the circuit resistances and energy-conversion requirements, affects the value of armature current I_a and hence the rotor field strength. Complete determination of machine behavior requires a similar investigation of factors influencing the stator field strength or, more particularly, the net flux per pole Φ_d.

5-5 ANALYTICAL FUNDAMENTALS: MAGNETIC–CIRCUIT ASPECTS

The flux per pole is that resulting from the combined armature and field mmf's. The interdependence of armature-generated voltage E_a and magnetic-circuit conditions in the machine is accordingly a function of the sum of all the mmf's on the polar- or direct-axis flux path. We shall first consider the mmf purposely placed on the stator main poles to create the working flux, i.e., the *main-field mmf,* and then include armature-mmf effects.

a. Armature Reaction Neglected

With no load on the machine or with armature-reaction effects ignored, the resultant mmf is the algebraic sum of the mmf's on the main- or direct-axis poles. For the usual compound generator or motor having N_f shunt-field turns per pole and N_s series-field turns per pole,

$$\text{Main-field mmf} = N_f I_f \pm N_s I_s \qquad (5\text{-}19)$$

Additional terms will appear in this equation when there are additional field windings on the main poles and when, unlike the compensating windings of Art. 5-8, they are wound concentric with the normal field windings to permit specialized control. In Eq. 5-19 the plus sign is used when the two mmf's are aiding or when the two fields are cumulatively connected; the minus sign is used when the series field opposes the shunt field or for a differential connection. When either the series or the shunt field is absent, the corresponding term in Eq. 5-19 naturally is omitted.

Equation 5-19 thus sums up in ampere-turns per pole the gross mmf of the main-field windings acting on the main magnetic circuit. The magnetization curve for a dc machine is generally given in terms of current in only the principal field winding, which is almost invariably the shunt-field winding when one is present. The mmf units of such a magnetization curve and of Eq. 5-19 can be made the same by one of two rather obvious steps. The field current on the magnetization curve can be multiplied by the turns per pole in that winding, giving a curve in terms of ampere-turns per pole; or both sides of Eq. 5-19 can be divided by N_f, converting the units into the equivalent current in the N_f coil alone which produces the same mmf. Thus

$$\text{Gross mmf} = I_f \pm \frac{N_s}{N_f} I_s \qquad \text{equivalent shunt-field amperes} \qquad (5\text{-}20)$$

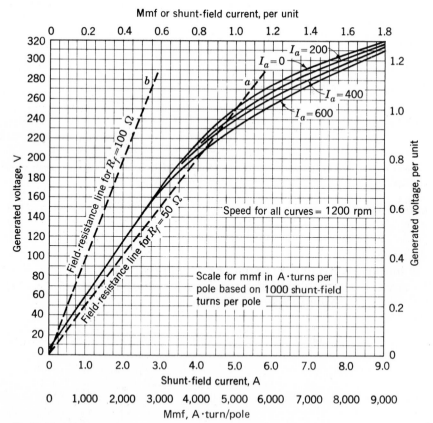

Fig. 5-16. Magnetization curves for a 250-V 1200-r/min dc machine.

The latter procedure is often the more convenient and the one more commonly adopted.

An example of a no-load magnetization characteristic is given by the curve for $I_a = 0$ in Fig. 5-16. The numerical scales on the left-hand and lower axes give representative values for a 100-kW 250-V 1200-r/min generator; the mmf scale is given in both shunt-field current and ampere-turns per pole, the latter being derived from the former on the basis of a 1000-turn-per-pole shunt field. The characteristic can also be presented in *normalized, or per unit, form,* as shown by the upper mmf and right-hand voltage scale. On these scales, 1.0 *per unit field current* or mmf is that required to produce rated voltage at rated speed when the machine is unloaded; similarly, 1.0 *per unit voltage* equals rated voltage.

Use of the magnetization curve with generated voltage rather than flux plotted on the vertical axis may be somewhat complicated by the fact that the speed of a dc machine need not remain constant and speed enters into the relation between flux and generated voltage. Hence generated-voltage ordinates correspond to a unique machine speed. The generated voltage E_a at any speed ω_m is, in accordance with Eq. 5-13, given by

$$E_a = E_{a0} \frac{\omega_m}{\omega_{m0}} \qquad (5\text{-}21)$$

where ω_{m0} is the magnetization-curve speed and E_{a0} the corresponding armature emf.

EXAMPLE 5-1

A 100-kW 250-V 400-A long-shunt compound generator has armature resistance (including brushes) of 0.025 Ω, a series-field resistance of 0.005 Ω, and the magnetization curve of Fig. 5-16. There are 1000 shunt-field turns per pole and 3 series-field turns per pole.

Compute the terminal voltage at rated current output when the shunt-field current is 4.7 A and the speed is 1150 r/min. Neglect armature reaction.

Solution

$I_s = I_a = I_L + I_f = 400 + 4.7 = 405$ A. From Eq. 5-20 the main-field gross mmf is

$$4.7 + \frac{3}{1000}(405) = 5.9 \text{ equivalent shunt-field amperes}$$

By entering the $I_a = 0$ curve of Fig. 5-16 with this current one reads 274 V.

Accordingly, the actual emf is

$$E_a = (274)\frac{1150}{1200} = 262 \text{ V}$$

Then

$$V_t = E_a - I_a(R_a + R_s) = 262 - 405(0.025 + 0.005) = 250 \text{ V}$$

b. Effect of Armature MMF

As described in Art. 5-3, excitation of the armature winding gives rise to a demagnetizing effect caused by a cross-magnetizing armature reaction. Analytic inclusion of this effect is not a straightforward task because of the nonlinearities involved. One common approach is to base the work on experimentally determined data for the machine involved or for one of similar design and frame size. Data are taken with both the field and armature excited, and the tests are conducted so that the effects on generated emf of varying both main-field excitation and the armature mmf can be noted.

One form of summarizing and correlating the results is illustrated in Fig. 5-16. Curves are plotted not only for the no-load characteristic ($I_a = 0$) but for a family of values of I_a. In the analysis of machine performance, then, the inclusion of armature reaction becomes simply a matter of using the magnetization curve corresponding to the armature current involved. Note that the ordinates of all these curves give values of armature-generated voltage E_a, not terminal voltage under load. Note also that all the curves tend to merge with the air-gap line as saturation of the iron decreases.

The load-saturation curves are displaced to the right of the no-load curve by an amount which is a function of I_a. The effect of armature reaction then is approximately the same as a demagnetizing mmf AR acting on the main-field axis. The *net* direct-axis mmf is then assumed to be

$$\text{Net mmf} = \text{gross mmf} - AR = N_f I_f \pm N_s I_s - AR \qquad (5\text{-}22)$$

The no-load magnetization curve can then be used as the relation between generated emf and net excitation under load with the armature reaction accounted for as a demagnetizing mmf. Over the normal operating range (about 240 to about 300 V for the machine of Fig. 5-16) the demagnetizing effect of armature reaction may be assumed to be approximately proportional to the armature current.

The amount of armature reaction present in Fig. 5-16 is chosen so that some of its disadvantageous effects will appear in a pronounced form in subsequent numerical examples and problems illustrating generator and motor performance features. It is definitely more than one would expect to find in a normal well-designed machine operating at normal currents.

5–6 ANALYSIS OF STEADY–STATE PERFORMANCE

Although exactly the same principles apply to analysis of a dc machine acting as a generator as to one acting as a motor, the general nature of the problems ordinarily encountered is somewhat different for the two methods of operation. For a generator, the speed is usually fixed by the prime mover, and problems often met are to determine the terminal voltage corresponding to a specified load and excitation or to find the excitation required for a specified load and terminal voltage. For a motor, on the other hand, problems frequently encountered are to determine the speed corresponding to a specific load and excitation or to find the excitation required for specified load and speed conditions; terminal voltage is often fixed at the value of the available supply mains. The routine techniques of applying the common basic principles therefore differ to the extent that the problems differ.

a. Generator Analysis

Since the main-field current is independent of the generator voltage, separately excited generators are the simplest to analyze. For a given load, the main-field excitation is given by Eq. 5-20, and the associated armature-generated voltage E_a is determined by the appropriate magnetization curve. This voltage, together with Eq. 5-16 or 5-17, fixes the terminal voltage.

In self-excited generators, the shunt-field excitation depends on the terminal voltage and the series-field excitation on the armature current. Dependence of shunt-field current on terminal voltage can be incorporated graphically in an analysis by drawing the *field-resistance line,* the line 0a in Fig. 5-16, on the magnetization curve. The field-resistance line 0a is simply a graphical representation of Ohm's law for the shunt field. It is the locus of the terminal voltage versus shunt-field-current operating point. Thus, the line 0a is drawn for $R_f = 50\ \Omega$ and hence passes through the origin and the point (1.0 A, 50 V).

One instance of the interdependence of magnetic- and electric-circuit conditions can be seen by examining the *buildup of voltage* for an unloaded shunt generator. When the field circuit is closed, the small voltage from residual magnetism (the 6-V intercept of the magnetization curve, Fig. 5-16) causes a small field current. If the flux produced by the resulting ampere-turns adds to the residual flux, progressively greater voltages and field currents are obtained. If the field ampere-turns oppose the residual magnetism, the shunt-field terminals must be reversed to obtain buildup. Buildup continues until the volt-ampere relations represented by the magnetization curve and the field-resistance line are simultaneously satisfied; i.e., at their intersection, 250 V for line 0a in Fig. 5-16. This statement ignores the extremely small voltage drop caused by the shunt-field current in the armature-circuit resistance. Notice that if the field resistance is too high, as shown by line 0b for $R_f = 100\ \Omega$, the intersection is at very low voltage and buildup is not obtained.

Notice also that if the field-resistance line is essentially tangent to the lower part of the magnetization curve, corresponding to 57 Ω in Fig. 5-16, the intersection may be anywhere from about 60 to 170 V, resulting in very unstable conditions. The corresponding resistance is the *critical field resistance,* beyond which buildup will not be obtained. The same buildup process and the same conclusions apply to compound generators; in a long-shunt compound generator, the series-field mmf created by the shunt-field current is entirely negligible.

This buildup of voltage is evidently a transient process in which, at any particular point, the vertical difference between the field-resistance line and the magnetization curve is the voltage serving to increase the current through the shunt-field inductance. The transient process is discussed in Art. 6-4*b*.

For a shunt generator, the magnetization curve for the appropriate value of I_a is the locus of E_a versus I_f. The field-resistance line is the locus V_t versus I_f. With steady-state operation and at any value of I_f, therefore, the vertical distance between the line and the curve must be the $I_a R_a$ drop at the load corresponding to that condition. Determination of the terminal voltage for a specified armature current is then simply a matter of finding where the line and curve are separated vertically by the proper amount; the ordinate of the field-resistance line at that field current is then the terminal voltage. For a compound generator, however, the series-field mmf causes corresponding points on the line and curve to be displaced horizontally as well as vertically. The horizontal displacement equals the series-field mmf measured in equivalent shunt-field amperes, and the vertical displacement is still the $I_a R_a$ drop.

Great precision is evidently not obtained from the foregoing computational process. The uncertainties caused by magnetic hysteresis in dc machines make high precision unattainable in any event. In general, the magnetization curve on which the machine operates on any given occasion may range from the rising to the falling part of the rather fat hysteresis loop for the magnetic circuit of the machine, depending essentially on the magnetic history of the iron just before. The curve used for analysis is usually the mean magnetization curve, and thus the results obtained are substantially correct on the average. Significant departures from the average may be encountered in the performance of any dc machine at a particular time, however.

EXAMPLE 5-2

A 100-kW 250-V 400-A 1200-r/min dc shunt generator has the magnetization curves (including armature-reaction effects) of Fig. 5-16. The armature-circuit resistance, including brushes, is 0.025 Ω. The generator is driven at a constant speed of 1200 r/min, and the excitation is adjusted to give rated voltage at no load.

(*a*) Determine the terminal voltage at an armature current of 400 A.

(*b*) A series field of 4 turns per pole having a resistance of 0.005 Ω is to be added. There are 1000 turns per pole in the shunt field. The generator is to be flat-compounded so that the full-load voltage is 250 V when the shunt-field rheostat is adjusted to give a no-load voltage of 250 V. Show how a resistance across the series field (a so-called *series-field diverter*) can be adjusted to produce the desired performance.

Solution

(a) The field-resistance line $0a$ (Fig. 5-16) passes through the 250-V 5.0-A point of the no-load magnetization curve. At $I_a = 400$ A

$$I_a R_a = 400(0.025) = 10 \text{ V}$$

A vertical distance of 10 V exists between the magnetization curve for $I_a = 400$ A and the field-resistance line at a field current of 4.1 A, corresponding to $V_t = 205$ V. The associated line current is

$$I_L = I_a - I_f = 400 - 4 = 396 \text{ A}$$

Note that a vertical distance of 10 V also exists at a field current of 1.2 A, corresponding to $V_t = 60$ V. The voltage-load curve is accordingly double-valued in this region. The point for which $V_t = 205$ is the normal operating point.

(b) For the no-load voltage to be 250 V, the shunt-field resistance must be 50 Ω, and the field-resistance line is $0a$ (Fig. 5-16). At full load, $I_f = 5.0$ A because $V_t = 250$ V. Then

$$I_a = 400 + 5.0 = 405 \text{ A}$$
and $$E_a = 250 + 405(0.025 + 0.005) = 262 \text{ V}$$

In the last equation, the effect of the diverter in reducing the series-field circuit resistance is ignored, a neglect which is permissible in view of the degree of precision warranted. From the 400-A magnetization curve, an E_a of 262 V requires a main-field excitation of 5.95 equivalent shunt-field amperes. (Strictly speaking, of course, a curve for $I_a = 405$ A should be used, but such a small distinction is obviously meaningless.) From Eq. 5-20

$$5.95 = 5.0 + \frac{4}{1000}I_s \qquad I_s = 238 \text{ A}$$

Hence, only 238 of the total 405 A of armature current must pass through the series field, a process requiring that the series field be shunted by a resistor of

$$\frac{238(0.005)}{405 - 238} = 0.0071 \text{ Ω}$$

b. Motor Analysis

Since the terminal voltage of motors is usually substantially constant at a specific value, there is no dependence of shunt-field excitation on a varying voltage as in shunt and compound generators. Hence, motor analysis most nearly resembles that for separately excited generators, although speed is now an important variable and often the one whose value is to be found. Analytical essentials include Eqs. 5-16 and 5-17 relating terminal voltage and generated voltage (counter emf), Eq. 5-20 for main-field excitation, the magnetization curve for the appropriate armature current as the graphical relation between counter emf and excitation, Eq. 5-12 showing the dependence of electromagnetic torque on flux and armature current, and Eq. 5-13 relating counter emf with flux and speed. The last two relations are particularly significant in motor analysis. The former is pertinent because the interdependence of torque and the stator and rotor field strengths must often be examined. The latter is the usual medium for determining motor speed from other specified operating conditions.

Motor speed corresponding to a given armature current I_a can be found by first computing the actual generated voltage E_a from Eq. 5-16 or 5-17. Next obtain the main-field excitation from Eq. 5-20. Since the magnetization curve will be plotted for a constant speed ω_{m0}, which in general will be different from the actual motor speed ω_m, the generated voltage read from the magnetization curve at the foregoing main-field excitation will correspond to the correct flux conditions but to the speed ω_{m0}. Substitution in Eq. 5-21 then yields the actual motor speed.

It will be noted that knowledge of the armature current is postulated at the start of this process. When, as is frequently the case, the speed at a stated shaft power or torque output is to be found, successive trials based on assumed values of I_a usually form the simplest procedure. Plotting the successive trials permits speedy determination of the correct armature current and speed at the desired output.

EXAMPLE 5-3

A 100-hp 250-V dc shunt motor has the magnetization curves (including armature-reaction effects) of Fig. 5-16. The armature circuit resistance, including brushes, is 0.025 Ω. No-load rotational losses are 2000 W, and stray load losses equal 1.0 percent of the output. The field rheostat is adjusted for a no-load speed of 1100 r/min.

(a) As an example of computing points on the speed-load characteristic, determine the speed in revolutions per minute and output in horsepower corresponding to an armature current of 400 A.

(b) Because the speed-load characteristic referred to in part (a) is considered undesirable, a *stabilizing winding* consisting of $1\frac{1}{2}$ cumulative series turns per pole is to be added. The resistance of this winding is negligible. There are 1000 turns per pole in the shunt field. Compute the speed corresponding to an armature current of 400 A.

Solution

(a) At no load, $E_a = 250$ V. The corresponding point on the 1200-r/min no-load saturation curve is

$$E_{a0} = 250 \times \frac{1200}{1100} = 273 \text{ V}$$

for which $I_f = 5.90$ A. The field current remains constant at this value.

At $I_a = 400$ A, the actual counter emf is

$$E_a = 250 - 400(0.025) = 240 \text{ V}$$

From Fig. 5-16 with $I_a = 400$ and $I_f = 5.90$, the value of E_a would be 261 V if the speed were 1200 r/min. The actual speed is then

$$n = \frac{240}{261}(1200) = 1100 \text{ r/min}$$

The electromagnetic power is

$$E_a I_a = 240(400) = 96{,}000 \text{ W}$$

Deduction of the rotational losses leaves 94,000 W. With stray load losses accounted for, the power output P_o is given by

$$94{,}000 - 0.01P_o = P_o$$

or

$$P_o = 93.1 \text{ kW} = 124.7 \text{ hp}$$

Note that the speed at this load is the same as at no load, indicating that armature-reaction effects have caused an essentially flat speed-load curve.

(b) With $I_f = 5.90$ A and $I_s = I_a = 400$ A, the main-field mmf in equivalent shunt-field amperes is

$$5.90 + \frac{1.5}{1000}(400) = 6.50$$

From Fig. 5-16 the corresponding value of E_a at 1200 r/min would be

273 V. Accordingly, the speed is now

$$n = \frac{240}{273}(1200) = 1055 \text{ r/min}$$

The power output is the same as in part (a). The speed-load curve is now drooping, due to the effect of the stabilizing winding.

EXAMPLE 5-4

To limit the starting current to the value which the motor can commutate successfully (see Art. 5-7), all except very small dc motors are started with external resistance in series with their armatures. This resistance is cut out either manually or automatically as the motor comes up to speed. In Fig. 5-17, for example, the contactors $1A$, $2A$, and $3A$ cut out successive steps R_1, R_2, and R_3 of the starting resistor.

    Consider that a motor is to be started with normal field flux. Armature reaction and armature inductance are to be ignored. During starting, the armature current and hence the electromagnetic torque are not to exceed twice the rated values, and a step of the starting resistor is to be cut out whenever the armature current drops to its rated value. Except in part (f) computations are to be made in the per unit system with magnitudes expressed as fractions of base values. (Base voltage equals rated line voltage, base armature current equals full-load armature current, and base resistance equals the ratio of base voltage to base current.)

    (a) What is the minimum per unit value of armature resistance which will permit these conditions to be met by a three-step starting resistor?

    (b) Above what per unit value of armature resistance will a two-step resistor suffice?

    (c) For the armature resistance of part (a), what are the per unit resistance values R_1, R_2, and R_3 of the starting resistor?

    (d) For a motor with the armature resistance of part (a), the contactors are to be closed by voltage-sensitive relays connected across the armature (called the *counter-emf method*). At what fractions of rated line voltage should the contactors close?

    (e) For a motor with the armature resistance of part (a), sketch approxi-

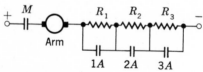

Fig. 5-17. Starting resistors and accelerating contactors for a dc motor.

mate curves of armature current, electromagnetic torque, and speed during the starting process, and label the ordinates with the appropriate per unit values at significant instants of time.

(f) For a 10-hp 230-V 500-r/min dc shunt motor having a full-load armature current of 37 A and fulfilling the conditions of part (a), list numerical values in their usual units for armature resistance, the results of parts (c) and (d), and the ordinate labelings of part (e).

Solution

(a) To prevent the armature current from exceeding 2.00 per unit at the instant main contactor M closes,

$$R_1 + R_2 + R_3 + R_a = \frac{V_t}{I_a} = \frac{1.00}{2.00} = 0.50$$

When the current has dropped to 1.00 per unit,

$$E_{a1} = V_t - I_a(R_1 + R_2 + R_3 + R_a)$$
$$= 1.00 - 1.00(0.50) = 0.50$$

At the instant the accelerating contactor $1A$ closes, short-circuiting R_1, the counter emf has attained this numerical value. Then, to prevent the allowable armature current from being exceeded

$$R_2 + R_3 + R_a = \frac{V_t - E_{a1}}{I_a} = \frac{1.00 - 0.50}{2.00} = 0.25$$

When the current has again dropped to 1.00 per unit,

$$E_{a2} = V_t - I_a(R_2 + R_3 + R_a) = 1.00 - 1.00(0.25) = 0.75$$

Repetition of this procedure for the closing of accelerating contactors $2A$ and $3A$ yields

$$R_3 + R_a = 0.125 \qquad E_{a3} = 0.875 \qquad R_a = 0.0625$$

and $\qquad$ Final E_a at full load $= 0.938$

The desired minimum per unit value of R_a is therefore 0.0625, because a lower value will allow the armature current to exceed twice the rated value when contactor $3A$ is closed.

(b) If a two-step resistor is to suffice, R_3 must be zero. Since, from part (a),

$$R_3 + R_a = 0.125$$

it follows that a three-step resistor is not required when R_a is equal to or greater than 0.125.

Under the specified starting conditions, a three-step resistor is appropriate for motors whose armature-circuit resistances are between 0.0625 and 0.125 per unit. For general-purpose continuously rated shunt motors, these values correspond to the lower integral-horsepower sizes. On the average, motor sizes up to about 10 hp will conform to these requirements, although the size limit will be lower for high-speed motors and higher for low-speed motors. For larger motors, either additional steps must be provided, or the limit on current and torque peaks must be relaxed. The results of this analysis are conservative because the armature resistance under transient conditions is higher than the static value.

(c) From the relations in part (a) the per unit starting resistances are

$$R_3 = 0.125 - 0.0625 = 0.0625$$
$$R_2 = 0.25 - 0.0625 - 0.0625 = 0.125$$
and $$R_1 = 0.50 - 0.0625 - 0.0625 - 0.125 = 0.25$$

(d) Just before contactor $1A$ closes,

$$V_{ta1} = E_{a1} + I_a R_a = 0.50 + 1.00(0.0625) = 0.563$$

In like manner,

$$V_{ta2} = 0.75 + 1.00(0.0625) = 0.813$$
and $$V_{ta3} = 0.875 + 1.00(0.0625) = 0.938$$

Acceleration contactors $1A$, $2A$, and $3A$, respectively, should pick up at these fractions of rated line voltage.

(e) Consider that main contactor M closes at $t = 0$ and that accelerating contactors $1A$, $2A$, and $3A$ close, respectively, at times t_1, t_2, and t_3. These values of time are not known (when armature and load inertias and torque-speed curve of the load are given, values of time can be computed by the methods of Chap. 6), so that only the general shapes of the current, electromagnetic torque, and speed curves can be given. They are indicated in Fig. 5-18.

The labeling of the speed curve follows from the fact that a counter emf $E_a = 0.938$ corresponds to rated speed at rated load and hence to unity speed. Other speeds are in proportion to E_a; thus, at t_1, t_2, and t_3, respectively,

$$n_1 = \frac{0.50}{0.938}(1.00) = 0.534$$

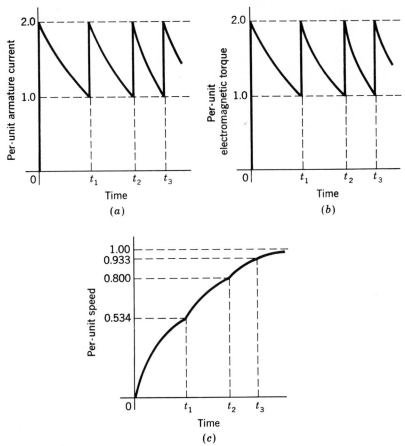

Fig. 5-18. (a) Armature current, (b) electromagnetic torque, and (c) speed during starting of a dc motor.

$$n_2 = \frac{0.75}{0.938}(1.00) = 0.800$$

and
$$n_3 = \frac{0.875}{0.938}(1.00) = 0.933$$

(f) Base quantities for this motor are as follows:

Base voltage = 230 V   Base armature current = 37 A

$$\text{Base armature-circuit resistance} = \frac{230}{37} = 6.22\ \Omega$$

Base speed = 500 r/min

TABLE 5-1
ABSOLUTE VALUES FOR PART (f) EXAMPLE 5-4

Part (c), Ω	Part (d), V	Part (e) scales of Fig. 5-18
$R_1 = 1.56$	Relay $1A = 129$	1.0 armature current = 37 A
$R_2 = 0.778$	Relay $2A = 187$	1.0 electromagnetic torque = 152 N · m
$R_3 = 0.389$	Relay $3A = 216$	1.0 speed = 500 r/min

$$\text{Base electromagnetic torque} = \frac{60}{2\pi n}E_a I_a$$

$$= \frac{60}{2\pi(500)}[230 - 37(0.0625)(6.22)](37)$$

$$= 152 \text{ N} \cdot \text{m}$$

Note that rated electromagnetic torque will be greater than rated shaft torque because of rotational and stray load losses.

The motor armature resistance is

$$R_a = 0.0625(6.22) = 0.389 \ \Omega$$

Values for the other quantities desired are listed in Table 5-1.

5–7 COMMUTATION AND INTERPOLES

One of the most important limiting factors on satisfactory operation of a dc machine is the ability to transfer the necessary armature current through the brush contact at the commutator without sparking and without excessive local losses and heating of the brushes and commutator. Sparking causes destructive blackening, pitting, and wear of both commutator and brushes, conditions which rapidly become worse and burn away the copper and carbon. It may be caused by faulty mechanical conditions, such as chattering of the brushes or a rough, unevenly worn commutator, or, as in any switching problem, by electrical conditions. The latter conditions are seriously influenced by the armature mmf and the resultant flux wave.

As indicated in Art. 5-2, a coil undergoing commutation is in transition between two groups of armature coils: at the end of the commutation period, the coil current must be equal but opposite to that at the beginning. Figure 5-7b shows the armature in an intermediate position during which the coils formed by inductors are being commutated. The commutated coils are short-circuited by the brushes. During this period the brushes must continue to conduct the armature current I_a from the armature winding to the external circuit. The short-circuited coil constitutes an inductive circuit with time-varying resistances at the brush contact, with, in general, rotational voltages induced in the coil and with both conductive and inductive coupling to the rest of the armature winding.

The attainment of good commutation is more an empirical art than a quantitative science. The principal obstacle to quantitative analysis lies in the electrical behavior of the carbon-copper contact film. Its resistance is nonlinear and is a function of current density, current direction, temperature, brush material, moisture, and atmospheric pressure. Its behavior in some respects is like that of an ionized gas or plasma. The most significant fact is that an unduly high current density in a portion of the brush surface (and hence an unduly high energy density in that part of the contact film) results in sparking and a breakdown of the film at that point. The boundary film also plays an important part in the mechanical behavior of the rubbing surfaces. At high altitudes, definite steps must be taken to preserve it, or extremely rapid brush wear takes place.

The empirical basis of securing sparkless commutation, then, is to avoid excessive current densities at any point in the copper-carbon contact. This basis, combined with the principle of utilizing all material to the fullest extent, indicates that optimum conditions are obtained when the current density is uniform over the brush surface during the entire commutation period. A linear change of current with time in the commutated coil, corresponding to linear commutation as shown in Fig. 5-8, brings about this condition and is accordingly the optimum.

The principal factors tending to produce linear commutation are changes in brush-contact resistance resulting from the linear decrease in area at the trailing brush edge and linear increase in area at the leading edge. Several electrical factors militate against linearity. Resistance in the commutated coil is one example. Usually, however, the voltage drop at the brush contacts is sufficiently large (of the order of 1.0 V) in comparison with the resistance drop in a single armature coil to permit the latter to be ignored. Coil inductance is a much more serious factor. Both the voltage of self-induction in the commutated coil and the voltage of mutual induction from other coils (particularly those in the same slot) undergoing commutation at the same time oppose changes in current in the commutated coil. The sum of these two voltages is often referred to as the *reactance voltage*. Its result is that current values in the short-circuited coil lag in time the values dictated by linear commutation. This condition is known as *undercommutation* or *delayed commutation*.

Armature inductance thus tends to produce high losses and sparking at the trailing brush tip. For best commutation, inductance must be held to a minimum by using the fewest possible number of turns per armature coil and by using a multipolar design with a short armature. The effect of a given reactance voltage in delaying commutation is minimized when the resistive brush-contact voltage drop is significant compared with it. This fact is one of the main reasons for the use of carbon brushes with their appreciable contact drop. When good commutation is secured by virtue of resistance drops, the process is referred to as *resistance commutation*. It is used today as the exclusive means only in fractional-horsepower machines.

Another important factor in the commutation process is the rotational voltage induced in the short-circuited coil. Depending on its sign, this voltage may hinder or aid commutation. In Fig. 5-11, for example, cross-magnetizing armature reaction creates a definite flux in the interpolar region. The direction of the corresponding rotational voltage in the commutated coil is the same as the current under the immediately preceding pole face. This voltage then encourages the continuance of current in the old direction and, like the reactance voltage, opposes its reversal. To aid commutation, the rotational voltage must oppose the reactance voltage. The general principle of producing in the coil undergoing commutation a rotational voltage which approximately compensates for the reactance voltage, a principle called *voltage commutation,* is used in almost all modern commutating machines. The appropriate flux density is introduced in the commutating zone by means of small, narrow poles located between the main poles. These auxiliary poles are called *interpoles* or *commutating poles.*

The general appearance of interpoles and an approximate map of the flux produced when they alone are excited are shown in Fig. 5-19. (The interpoles are the smaller poles between the larger main poles in Fig. 5-21.) The polarity of a commutating pole must be that of the main pole just ahead of it, i.e., in the direction of rotation for a generator, and just behind it for a motor. The interpole mmf must be sufficient to neutralize the cross-magnetizing armature mmf in the interpolar region and enough more to furnish the flux density required for the rotational voltage in the short-circuited armature coil to cancel the reactance voltage. Since both the armature mmf and the reactance voltage are proportional to the armature current, the commutating winding must be connected in series with the armature. To preserve the desired linearity, the commutating pole should operate at low saturations. By the use of

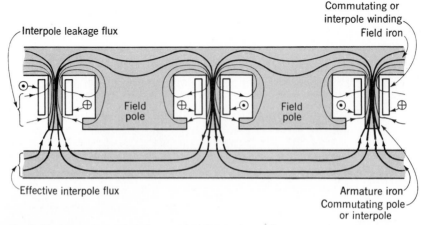

Fig. 5-19. Interpoles and their associated component flux.

commutating fields, then, sparkless commutation is secured over a wide range in modern machines. In accordance with the performance standards of the NEMA, general-purpose dc machines must be capable of carrying with successful commutation for 1 min loads of 150 percent of the current corresponding to the continuous rating with the field rheostat set for rated-load excitation.

5-8 COMPENSATING WINDINGS

For machines subjected to heavy overloads, rapidly changing loads, or operation with a weak main field, there is the possibility of trouble other than simply sparking at the brushes. At the instant when an armature coil is located at the peak of a badly distorted flux wave, the coil voltage may be high enough to break down the air between the adjacent segments to which the coil is connected and result in *flashover,* or arcing, between segments. The breakdown voltage here is not high, because the air near the commutator is in a condition favorable to breakdown. The maximum allowable voltage between segments is of the order of 30 to 40 V, a fact which limits the average voltage between segments to lower values and thus determines the minimum number of segments which can be used in a proposed design. Under transient conditions, high voltages between segments may result from the induced voltages associated with growth and decay of armature flux. Inspection of Fig. 5-10, for instance, may enable one to visualize very appreciable voltages of this nature being induced in a coil under the pole centers by the growth or decay of the armature flux shown in the sketch. Consideration of the sign of this induced voltage will show that it adds to the normal rotational emf when load is dropped from a generator or added to a motor. Flashing between segments may quickly spread around the entire commutator and, in addition to its possibly destructive effects on the commutator, constitutes a direct short circuit on the line. Even with interpoles present, therefore, armature reaction under the poles definitely limits the conditions under which a machine can operate.

These limitations can be considerably extended by compensating or neutralizing the armature mmf under the pole faces. Such compensation can be achieved by means of a *compensating,* or *pole-face, winding* (Fig. 5-20) embedded in slots in the pole face and having a polarity opposite to that of the adjoining armature winding. The physical appearance of such a winding can be seen in the stator of Fig. 5-21. Since the axis of the compensating winding is the same as that of the armature, it will almost completely neutralize the armature reaction of the armature conductors under the pole faces when it is given the proper number of turns. It must be connected in series with the armature in order to carry a proportional current. The net effect of the main field, armature, commutating winding, and compensating winding on the air-gap flux is that, except for the commutation zone, the resultant flux-density

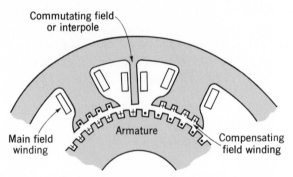

Fig. 5-20. Section of a dc machine showing compensating winding.

distribution is substantially the same as that produced by the main field alone (Fig. 5-11). Furthermore, the addition of a compensating winding improves the speed of response because it reduces the armature-circuit time constant.

The main disadvantage of pole-face windings is their expense. They are used in machines designed for heavy overloads or rapidly changing loads (steel-mill motors are a good example of machines subjected to severe duty cycles) or in motors intended to operate over wide speed ranges by shunt-field control. By way of a schematic summary, Fig. 5-22 shows the circuit diagram of a compound machine with a compensating winding. The relative position of the coils in this diagram indicates that the commutating and compensating

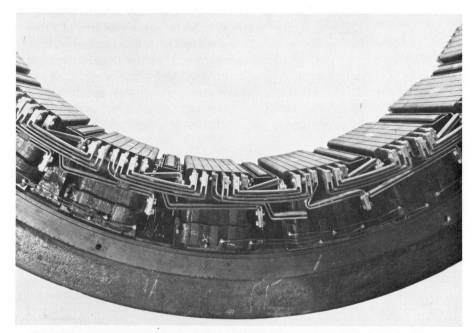

Fig. 5-21. Section of a dc motor stator or field showing shunt and series coils, interpoles, and pole-face compensating winding. (*Westinghouse Electric Company.*)

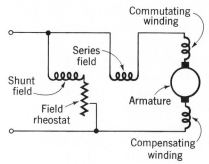

Fig. 5-22. Schematic connection diagram of a dc machine.

fields act along the armature axis and the shunt and series fields act along the axis of the main poles. Rather complete control of air-gap flux around the entire armature periphery is thus achieved.

5–9 SUMMARY; DC MACHINE APPLICATIONS

Discussion of dc machine applications involves recapitulation of the highlights of the machine's performance features, together with economic and technical evaluation of the machine's position with respect to competing energy-conversion devices. For dc machines in general, the outstanding advantage lies in their flexibility and versatility. The principal disadvantage is likely to be the initial investment concerned. Yet the advantages of dc motors are such that they retain a strong competitive position for industrial applications.

DC generators are the obvious answer to the problem of converting mechanical energy into electric energy in dc form. When the consumer of electric energy is geographically removed from the site of energy conversion by any appreciable distance, however, the advantages of ac generation, voltage transformation, and transmission are such that energy conversion and transmission in ac form are almost always adopted, ac-to-dc transformation taking place at or near the consumer. For ac-to-dc transformation the dc generator as part of an ac-to-dc motor-generator set must compete with semiconductor rectifier systems. When large-power rectification from ac to constant-voltage dc form is involved, the electronic methods usually offer determining economic advantages. The principal applications of dc generators, therefore, are to cases where the primary energy conversion occurs very near the point of consumption and cases where the ability to control output voltage in a prescribed manner is necessary, although solid-state controlled rectification using silicon controlled rectifiers (SCRs) and Triacs is finding widespread application.

Among dc generators themselves, separately excited and cumulatively compounded self-excited machines are the most common. Separately excited generators have the advantage of permitting a wide range of output voltages, whereas self-excited machines may produce unstable voltages in the lower ranges, where the field-resistance line becomes essentially tangent to the magnetization curve. Cumulatively compounded generators may produce a substantially flat voltage characteristic or one which rises with load, whereas shunt or separately excited generators (assuming no series field in the latter, which, of course, is not at all a practical restriction) produce a drooping voltage characteristic unless external regulating means are added. So far as the control potentialities of dc generators are concerned, the control-type generators (amplidynes and similar machines) discussed in Chap. 6 represent the results of a fuller exploration of the inherent possibilities.

Among dc motors, the outstanding characteristics of each type are as follows. The series motor operates with a decidedly drooping speed as load is added, the no-load speed usually being prohibitively high; the torque is proportional to almost the square of the current at low saturations and to some power between 1 and 2 as saturation increases. The shunt motor at constant field current operates at a slightly drooping but almost constant speed as load is added, the torque being almost proportional to armature current; equally important, however, is the fact that its speed can be controlled over wide ranges by shunt-field control, armature-voltage control, or a combination of both. Depending on the relative strengths of shunt and series field, the cumulatively compounded motor is intermediate between the other two and may be given essentially the advantages of one or the other.

By virtue of its ability to handle heavy torque overloads while cushioning the associated power overload with a speed drop, and by virtue of its ability to withstand severe starting duties, the series motor is best adapted to hoist, crane, and traction-type loads. Its ability is almost unrivaled in this respect. Speed changes are usually achieved by armature-resistance control. In some instances, the wound-rotor induction motor with rotor-resistance control competes with the series motor, but the principal argument concerns the availability and economics of a dc power supply rather than inherent motor characteristics.

Compound motors with a heavy series field have performance features approaching those of series motors except that the shunt field limits the no-load speed to safe values; the general remarks for series motors therefore apply. Compound motors with lighter series windings not infrequently find competition from squirrel-cage induction motors with high-resistance rotors—so-called *high-slip* motors (referred to in Chap. 9 as class D induction motors). Both motors provide a definitely drooping speed-load characteristic such as is desirable, for example, when flywheels are used as load equalizers to smooth out intermittent load peaks. Complete economic comparison of the two competing types must reflect both the usually higher initial cost of a compound-motor installation and the usually higher cost of losses in the high-slip induction motor.

Because of the comparative simplicity, cheapness, and ruggedness of the squirrel-cage induction motor, the shunt motor is not in a favorable competitive position for constant-speed service except at low speeds, where it becomes difficult and expensive to build high-performance induction motors with the requisite number of poles. The comparison at these low speeds is often likely to be between synchronous and dc motors. The outstanding feature of the shunt motor is its adaptability to adjustable-speed service as discussed in Art. 6-1, by means of armature-resistance control for speeds below the full-field speed, field-rheostat control for speeds above the full-field speed, and armature-voltage, or Ward Leonard, control for speeds below (and, at times, somewhat above) the normal-voltage full-field speed. The combination of armature-voltage control and shunt-field control, together with the possibility of additional field windings in either the motor or the associated generator to provide desirable inherent characteristics, gives the dc drives an enviable degree of flexibility. The use of solid-state motor drives, discussed in Chap. 6, reinforces the competitive position of dc machines where complete control of operation is important.

It should be emphasized that the choice of equipment for a significant engineering application to adjustable-speed drives is rarely a cut-and-dried matter or one to be decided from a mere verbal list of advantages and disadvantages. In general, specific, quantitative, economic, and technical comparison of all possibilities should be undertaken. Consideration must be given to the transient- and dynamic-response details of Chap. 6. Local conditions and the characteristics of the driven equipment (e.g., constant-horsepower, constant-torque, and variable-horsepower variable-torque requirements) invariably play an important role. One should also remember that comparative studies of motor cost and characteristics are based on the combination of motor and control equipment, for the latter plays an important part in determining motor performance under specific conditions and represents a by no means negligible portion of the total initial cost. Control equipment coupled with susceptibility to control makes dc machines the versatile energy-conversion devices that they are.

PROBLEMS

5-1 (*a*) Compare the effect on the speed of a dc shunt motor of varying the line voltage with that of varying only the armature terminal voltage, so that the field current remains fixed. (*b*) Compare both these effects with that of varying only the shunt-field current, the armature terminal voltage remaining fixed.

5-2 State approximately how the armature current and speed of a dc shunt motor would be affected by each of the following changes in the operating conditions:

   (*a*) Halving the armature terminal voltage while the field current and load torque remain constant

(b) Halving the armature terminal voltage while the field current and horsepower output remain constant

(c) Doubling the field flux while the armature terminal voltage and load torque remain constant

(d) Halving both the field flux and armature terminal voltage while the horsepower output remains constant

(e) Halving the armature terminal voltage while the field flux remains constant and the load torque varies as the square of the speed

Only brief quantitative statements of the order of magnitude of the changes are expected, e.g., "speed approximately doubled."

5-3 A 25-kW 250-V dc machine has an armature resistance of 0.10 Ω. Its magnetization curve at a constant speed of 1200 r/min is shown in Fig. 5-23. Its field is separately excited, and it is driven by a synchronous motor at a constant speed of 1200 r/min. Plot a family of curves of armature terminal voltage versus armature current for constant field currents of 2.5, 2.0, 1.5, and 1.0 A.

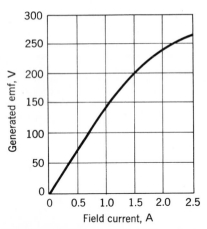

Fig. 5-23. Dc-machine magnetization curve at 1200 r/min for Probs. 5-3 and 5-4.

5-4 The dc machine of Prob. 5-3 is operated as a separately excited motor.

(a) For a constant field current of 2.0 A, plot a family of curves of speed in revolutions per minute versus torque in newton-meters for applied armature terminal voltages of 250, 200, 150, and 100 V.

(b) For a constant applied armature terminal voltage of 200 V, plot a family of curves of speed versus torque for field currents of 2.5, 2.0, 1.5, and 1.0 A.

5-5 An eddy-current brake of the type used for load tests on motors requires a torque of 100 lb·ft to drive it at a speed of 1000 r/min when the current in its magnetizing coils is 10.0 A. This brake is driven by a dc series motor rated to deliver 20 hp at 1000 r/min with an applied voltage of 230 V.

(a) Plot a family of five torque-speed curves for the brake at coil currents of 6, 8, 10, 12, and 14 A, respectively. For this problem assume that the flux is linearly proportional to the coil current and that the magnetic effect of the eddy currents generated in the brake disk is negligible. Also neglect windage and friction torques.

(b) On the same curve sheet plot a family of four torque-speed curves for the series motor at applied voltages of 120, 100, 80, and 60 percent of rated voltage. Assume that all losses are negligible, that the motor flux is linearly proportional to the motor field current, and that the resistive voltage drops in the field and armature windings are negligible.

(c) Plot curves of torque and speed against brake coil current when the motor is supplied with rated voltage.

(d) Plot curves of torque and speed against motor applied voltage when the brake coil current is constant at 10 A.

5-6 A 10-kW 230-V 1150-r/min shunt generator is driven by a prime mover whose speed is 1195 r/min when the generator delivers no load. The speed falls to 1150 r/min when the generator delivers 10 kW and may be assumed to decrease in proportion to the generator output. The generator is to be changed into a short-shunt compound generator by equipping it with a series field which will cause its voltage to rise from 230 V at no load to 250 V for a load of 43.5 A. It is estimated that the series field will have a resistance of 0.09 Ω. The armature resistance (including brushes) is 0.26 Ω. The shunt-field winding has 1800 turns per pole.

In order to determine the necessary series-field turns, the machine is run as a separately excited generator and the following load data obtained:

$$\text{Armature terminal voltage} = 254 \text{ V}$$
$$\text{Armature current} = 44.7 \text{ A}$$
$$\text{Field current} = 1.95 \text{ A}$$
$$\text{Speed} = 1145 \text{ r/min}$$

The magnetization curve at 1195 r/min is as follows:

E_a, V	230	240	250	260	270
I_f, A	1.05	1.13	1.26	1.46	1.67

Determine (a) the necessary number of series-field turns per pole and (b) the armature reaction in equivalent demagnetizing ampere-turns per pole for $I_a = 44.7$ A.

5-7 A small, lightweight dc shunt generator for use in aircraft has a rating of 9 kW, 30 V, 300 A. It is driven by one of the main engines of the airplane through an auxiliary power shaft. The generator speed is proportional to the main-engine speed and may have any value from 4500 to 8000 r/min. The terminal voltage of the generator is held constant at 30 V for all speeds and loads by means of a voltage regulator which automatically adjusts a carbon-pile field rheostat whose minimum resistance is 0.75 Ω. The resistance of the shunt-field, commutating, and armature (including brushes) windings are, respectively, 2.50, 0.0040, and 0.0120 Ω. Data for the magnetization curve at 4550 r/min are:

I_f, A	0	2.0	4.0	5.0	6.0	8.0	11.7
E_a, V	1.0	18.0	30.5	33.6	35.5	38.0	40.5

In a load test at 4550 r/min the field current required to maintain rated terminal voltage at rated load is 7.00 A. Determine the following characteristics of this generator: (*a*) maximum resistance required in the field rheostat, (*b*) maximum power dissipated in the field rheostat, (*c*) demagnetizing effect of armature reaction at rated load and 4550 r/min, expressed in terms of equivalent shunt-field current.

5-8 Assume that the demagnetizing effect of armature reaction in the aircraft generator of Prob. 5-7 is equivalent to a demagnetizing mmf proportional to the armature current.

(*a*) Plot the curve of terminal voltage as a function of line current for 4550 r/min and a rheostat setting held fixed at the value which results in 30 V at no load.

(*b*) When the airplane is on the ground and the engines are idling, the generator speed may be below 4500 r/min. Plot a curve of terminal voltage as a function of speed with minimum field-rheostat resistance and a constant line current of 300 A covering the subnormal speed range from 4500 to 3500 r/min. Estimate the minimum speed at which the generator is capable of delivering 300 A.

5-9 A dc series motor operates at 750 r/min with a line current of 80 A from the 230-V mains. Its armature-circuit resistance is 0.14 Ω, and its field resistance is 0.11 Ω.

Assuming that the flux corresponding to a current of 20 A is 40 percent of that corresponding to a current of 80 A, find the motor speed at a line current of 20 A at 230 V.

5-10 A certain series motor is so designed that flux densities in the iron part of the magnetic circuit are low enough to result in a linear relationship between field flux and field current throughout the normal range of operation. The rating of this motor is 50 hp, 190 A, 220 V, 600 r/min. Losses at full load in percentage of motor input are:

$$\text{Armature } I^2R \text{ loss (including brush loss)} = 3.7\%$$
$$\text{Field } I^2R \text{ loss} = 3.2\% \qquad \text{Rotational loss} = 2.8\%$$

Rotational loss may be assumed constant; armature reaction and stray load loss may be neglected. When this motor is operating from a 220-V supply with a current of half the rated value, find (*a*) the speed in revolutions per minute and (*b*) the shaft power output in horsepower.

5-11 A 150-hp 600-V 600-r/min dc series-wound railway motor has a combined field and armature resistance (including brushes) of 0.155 Ω. The full-load current at rated voltage and speed is 206 A. The magnetization curve at 400 r/min is as follows:

Generated emf, V	375	400	425	450	475
Field amperes	188	216	250	290	333

Determine the internal starting torque when the starting current is limited to 350 A. Assume armature reaction to be equivalent to a demagnetizing mmf which varies as the square of the current.

5-12 Following are the nameplate data of a certain dc motor: 230 V, 75.7 A, 20 hp, 900 r/min full-load, 50°C 1-h rating, series-wound. The field winding has 33 turns per pole and a hot resistance of 0.06 Ω; the hot armature-circuit resistance is 0.09 Ω (including brushes). Points on the magnetization curve at 900 r/min are as follows:

Ampere-turns per pole	500	1000	1500	2000	2500	3000
Generated voltage	95	150	188	212	229	243

To determine the fitness of this motor for driving a skip hoist, points on the motor speed-load curve are to be computed.

(*a*) For currents equal to $\frac{1}{3}$, $\frac{2}{3}$, 1, and $\frac{4}{3}$ times the nameplate value, compute the speed of the motor. Neglect armature reaction.

(*b*) For the same currents, compute the shaft-horsepower outputs. For this purpose, consider rotational losses to remain constant at the value determined by nameplate conditions.

(*c*) Compute the shaft torques in pound-feet corresponding to the values in part (b). Arrange the results of parts (a) to (c) in tabular form for convenience in checking.

(d) The maximum safe speed for the motor is 250 percent of full-load speed. What is the motor power input at this point?

(e) What value of resistance connected in series with the motor will enable the production of full-load electromagnetic torque at a speed of 500 r/min?

5-13 A 10-hp 230-V shunt motor has an armature-circuit resistance of 0.30 Ω and a field resistance of 170 Ω. At no load and rated voltage, the speed is 1200 r/min, and the armature current is 2.7 A. At full load and rated voltage, the line current is 38.4 A, and, because of armature reaction, the flux is 4 percent less than its no-load value. What is the full-load speed?

5-14 A 36-in axial-flow disk pressure fan is rated to deliver air at 27,120 ft^3/min against a static pressure of $\frac{1}{2}$ inH$_2$O when rotating at a speed of 1165 r/min. This fan has the following speed-load characteristics:

Speed, r/min	700	800	900	1000	1100	1200
Input, hp	2.9	3.9	5.2	6.7	8.6	11.1

It is proposed to drive the fan by a 10-hp 230-V 37.5 A 4-pole dc shunt motor. The motor has an armature winding with two parallel paths and $C_a = 666$ active conductors. Armature-circuit resistance is 0.267 Ω. The armature flux per pole is $\Phi_d = 10^{-2}$ Wb; armature reaction is negligible. No-load rotational losses (considered constant) are estimated at 600 W, a typical value for such a motor. Determine the shaft-horsepower output and the operating speed of the motor when it is connected to the fan load.

5-15 A 100-hp 250-V dc shunt motor has the magnetization curve of Fig. 5-16 and an armature resistance (including brushes) of 0.025 Ω. There are 1000 turns per pole on the shunt field. When the shunt-field rheostat is set for a motor speed of 1200 r/min at no load, the armature current is 8.0 A. How many series-field turns per pole must be added if the speed is to be 950 r/min for a load requiring an armature current of 250 A? Neglect the added resistance of the series field.

5-16 A shunt motor operating from a 230-V line draws a full-load armature current of 38.5 A and runs at a speed of 1200 r/min at both no load and full load. The following data are available on this motor:

$$\text{Armature-circuit resistance (including brushes)} = 0.21 \ \Omega$$
$$\text{Shunt-field turns per pole} = 2000 \text{ turns}$$

Magnetization curve taken as a generator at no load and 1200 r/min

E_a, V	180	200	220	240	250
I_f, A	0.74	0.86	1.10	1.45	1.70

(a) Determine the shunt-field current of this motor at no load and 1200 r/min when connected to a 230-V line. Assume negligible armature-circuit resistance drop and armature reaction at no load.

(b) Determine the effective armature reaction at full load in ampere-turns per pole.

(c) How many series-field turns should be added to make this machine into a long-shunt cumulatively compounded motor whose speed will be 1090 r/min when the armature current is 38.5 A and the applied voltage is 230 V? The series field will have a resistance of 0.052 Ω.

(d) If a series-field winding having 25 turns per pole and a resistance of 0.052 Ω is installed, determine the speed when the armature current is 38.5 A and the applied voltage is 230 V.

5-17 A 10-hp 230-V shunt motor has 2000 shunt-field turns per pole, an armature resistance (including brushes) of 0.20 Ω, and a commutating-field resistance of 0.041 Ω. The shunt-field resistance (exclusive of rheostat) is 235 Ω. When the motor is operated at no load with rated terminal voltage and varying field resistance, the following data are taken:

Speed, r/min	1110	1130	1160	1200	1240
I_f, A	0.932	0.880	0.830	0.770	0.725

The no-load armature current is negligible. When the motor is operated at full load and rated terminal voltage, the armature current is 37.5 A, the field current is 0.770 A, and the speed is 1180 r/min.

(a) Calculate the full-load armature reaction in equivalent demagnetizing ampere-turns per pole.

(b) Calculate the full-load electromagnetic torque.

(c) What starting torque will the motor exert with maximum field current if the starting armature current is limited to 75 A? The armature reaction under these conditions is 160 A·turns per pole.

(d) Design a series field to give a full-load speed of 1100 r/min when the no-load speed is 1200 r/min.

5-18 When operated at rated voltage, a 230-V shunt motor runs at 1600 r/min at full load and also at no load. The full-load armature current is 50.0 A. The shunt-field winding has 1000 turns per pole. The resistance of the armature circuit (including brushes and interpoles) is 0.20 Ω. The magnetization curve at 1600 r/min is:

E_a, V	200	210	220	230	240	250
I_f, A	0.80	0.88	0.97	1.10	1.22	1.43

(a) Compute the demagnetizing effect of armature reaction at full load in ampere-turns per pole.

(b) A long-shunt cumulative series-field winding have 5 turns per pole and a resistance of 0.05 Ω is added to the machine. Compute the speed at full-load current and rated voltage, with the same shunt-field circuit resistance as in part (a).

(c) With the series-field winding of part (b) installed, compute the internal starting torque in newton-meters if the starting armature current is limited to 100 A and the shunt-field current has its normal value. Assume that the corresponding demagnetizing effect of armature reaction is 260 A·turns per pole.

5-19 A weak shunt-field winding is to be added to a 50-hp 230-V 600-r/min series hoist motor for the purpose of preventing excessive speeds at very light loads. Its resistance will be 230 Ω. The combined resistance of the interpole and armature winding (including brushes) is 0.055 Ω. The series-field winding has 24 turns per pole with a total resistance of 0.021 Ω. In order to determine its design, the following test data were obtained before the shunt field was installed:

Load test as a series motor (output not measured):

$$V_t = 230 \text{ V} \qquad I_a = 184 \text{ A} \qquad n = 600 \text{ r/min}$$

No-load test with series field separately excited:

Voltage applied to armature, V	Speed, r/min	Armature current, A	Series-field current, A
230	1500	10.0	60
230	1200	9.2	74
230	900	8.0	103
215	700	7.7	135
215	600	7.5	175
215	550	7.2	201
215	525	7.1	225
215	500	7.0	264

(a) Determine the number of shunt-field turns per pole if the no-load speed at rated voltage is to be 1500 r/min. The armature, series-field, and interpole winding resistance drops are negligible at no load.

(b) Determine the speed after installation of the shunt field when the motor is operated at rated voltage with a load which results in a line current of 185 A. Assume that the demagnetizing mmf of armature reaction is unchanged by addition of the shunt field.

5-20 A 230-V dc shunt motor has an armature-circuit resistance of 0.1 Ω. This motor operates on the 230-V mains and takes an armature current of 100 A. An external resistance of 1.0 Ω is now inserted in series with the armature, and the electromagnetic torque and field-rheostat setting are unchanged. Give the percentage change in (a) the total current taken by the motor from the mains and (b) in the speed of the motor and state whether this will be an increase or a decrease.

5-21 A punch press is found to operate satisfactorily when driven by a 10-hp 230-V compound motor having a no-load speed of 1800 r/min and a full-load speed of 1200 r/min when the torque is 43.8 lb·ft. The motor is temporarily out of service, and the only available replacement is a compound motor with the following characteristics:

Rating = 230 V, 12.5 hp   No-load current = 4 A

No-load speed = 1820 r/min   Full-load speed = 1600 r/min

Full-load current = 57.0 A   Full-load torque = 43.8 lb·ft

Armature-circuit resistance = 0.2 Ω   Shunt-field current = 1.6 A

It is desired to use this motor as an emergency drive for the press without making any change in its field windings.

(a) How can it be made to have the desired speed regulation?
(b) Draw the pertinent circuit diagram and give complete specifications of the necessary apparatus.

5-22 A cumulatively compounded generator with interpoles and with its brushes on neutral is to be used as a compound motor. If no changes are made in the internal connections, will the motor be cumulatively or differentially compounded? Will the polarity of the interpoles be correct? Will the direction of rotation be the same as or opposite to the direction in which it was driven as a generator?

5-23 A self-excited dc machine with interpoles is adjusted for proper operation as an overcompounded generator. The machine is shut down, the connections to the shunt field are reversed, and the machine is then started with the direction of rotation reversed. The machine builds up normal terminal voltage. Answer the following questions, and give a brief explanation. Is the terminal-voltage polarity the same as before? Is the machine still cumulatively compounded? Do the interpoles have the proper polarity for good commutation?

DC Machine Dynamics and Control

In Chap. 5, the steady-state behavior of dc machines is described and models discussed. DC machine dynamic behavior is discussed in this chapter.

6-1 DC MOTOR SPEED CONTROL

DC machines are generally much more adaptable to adjustable-speed service than the ac machines associated with a constant-speed rotating field. Indeed, the ready susceptibility of dc motors to adjustment of their operating speed over wide ranges and by a variety of methods is one of the important reasons for the strong competitive position of dc machinery in modern industrial applications.

The three most common speed-control methods are adjustment of the flux, usually by means of field-current control, adjustment of the resistance

247

associated with the armature circuit, and adjustment of the armature terminal voltage.

Field-current control is the most common method and forms one of the outstanding advantages of shunt motors. The method is, of course, also applicable to compound motors. Adjustment of field current and hence the flux and speed by adjustment of the shunt-field circuit resistance or with a solid-state control when the field is separately excited is accomplished simply, inexpensively, and without much change in motor losses.

The lowest speed obtainable is that corresponding to maximum field current; the highest speed is limited electrically by the effects of armature reaction under weak-field conditions in causing motor instability or poor commutation. Addition of a *stabilizing winding* increases the speed range appreciably, and the alternative addition of a compensating winding still further increases the range. A stabilizing winding ensures attainment of a drooping speed-load characteristic even at weak shunt-field currents and heavy loads. It is used with adjustable-speed motors intended for operation over a wide speed range by shunt-field resistance control; its performance was illustrated in Example 5-3. With a compensating winding, the overall range may be as high as 8 to 1 for a small integral-horsepower motor. Economic factors limit the feasible range for very large motors to about 2 to 1, however, with 4 to 1 often being regarded as the limit for the average-sized motor.

To examine approximately the limitations on the allowable continuous motor output as the speed is changed, neglect the influence of changing ventilation and changing rotational losses on the allowable output. The maximum armature current I_a is then fixed at the nameplate value so that the motor will not overheat, and the speed voltage E_a remains constant because the effect of a speed change is compensated by the change of flux causing it. Then the $E_a I_a$ product and hence the allowable motor output remain substantially constant over the speed range. The dc motor with shunt-field-rheostat speed control is accordingly referred to as a *constant-horsepower drive*. Torque, on the other hand, varies directly with flux and therefore has its highest allowable value at the lowest speed. Field-current control is thus best suited to drives requiring increased torque at low speeds. When a motor so controlled is used with a load requiring constant torque over the speed range, the rating and size of the machine are determined by the product of the torque and the highest speed. Such a drive is inherently oversize at the lower speeds, which is the principal economic factor limiting the practical speed range of large motors.

Armature-circuit-resistance control consists of obtaining reduced speeds by the insertion of external series resistance in the armature circuit. It can be used with series, shunt, and compound motors; for the last two types, the series resistor must be connected between the shunt field and the armature, not between the line and the motor. It is the common method of speed control for series motors and is generally analogous in action to wound-rotor induction-motor control by series rotor resistance.

For a fixed value of series armature resistance, the speed will vary widely with load, since the speed depends on the voltage drop in this resistance and hence on the armature current demanded by the load. For example, a 1200 r/min shunt motor whose speed under load is reduced to 750 r/min by series armature resistance will return to almost 1200 r/min operation when the load is thrown off because the effect of the no-load current in the series resistance is insignificant. The disadvantage of poor speed regulation may not be important in a series motor, which is used only where varying speed service is required or satisfactory anyway.

Also, the power loss in the external resistor is large, especially when the speed is greatly reduced. In fact, for a constant-torque load, the power input to the motor plus resistor remains constant, while the power output to the load decreases in proportion to the speed. Operating costs are therefore comparatively high for long-time running at reduced speeds. Because of its low initial cost, however, the series-resistance method (or the variation of it discussed in the next paragraph) will often be attractive economically for short-time or intermittent slowdowns. Unlike shunt-field control, armature-resistance control offers a *constant-torque drive* because both flux and, to a first approximation, allowable armature current remain constant as speed changes.

A variation of this control scheme is given by the *shunted-armature method,* which may be applied to a series motor, as in Fig. 6-1a, or a shunt motor, as in Fig. 6-1b. In effect, resistors R_1 and R_2 act as a voltage divider applying a reduced voltage to the armature. Greater flexibility is possible because two resistors can now be adjusted to provide the desired performance. For series motors, the no-load speed can be adjusted to a finite, reasonable value, and the scheme is therefore applicable to the production of slow speeds at light loads. For shunt motors, the speed regulation in the low-speed range is appreciably improved because the no-load speed is definitely lower than the value with no controlling resistors.

Armature-terminal-voltage control utilizes the fact that a change in the armature terminal voltage of a shunt motor is accompanied in the steady

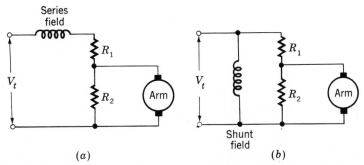

Fig. 6-1. Shunted-armature method of speed control applied to (a) series motor and (b) shunt motor.

state by a substantially equal change in the speed voltage and, with constant motor flux, a consequent proportional change in motor speed. Usually the power available is constant-voltage alternating current, so that auxiliary equipment in the form of a rectifier or a motor-generator set is required to provide the controlled armature voltage for the motor. The development of solid-state controlled rectifiers capable of handling many kilowatts has opened up a whole new field of applications where precise control of motor speed is required. These applications are of such importance that they are given special treatment in Arts. 6-5 and 6-6.

One common scheme, called the *Ward Leonard system,* shown schematically in Fig. 6-2, requires an individual motor-generator set to supply power to the armature of the motor whose speed is to be controlled. Control of the armature voltage of the main motor M is obtained by field-rheostat adjustment in the separately excited generator G, permitting close control of speed over a wide range. An obvious disadvantage is the initial investment in three full-size machines in contrast to that in a single motor. The speed-control equipment is located in low-power field circuits, however, rather than in the main power circuits. The smoothness and versatility of control are such that the method or one of its variants is often applied.

Frequently the control of generator voltage is combined with motor-field control, as indicated by the rheostat in the field of motor M in Fig. 6-2, in order to achieve the widest possible speed range. With such dual control, *base speed* can be defined as the normal-armature-voltage full-field speed of the motor. Speeds above base speed are obtained by motor-field control; speeds below base speed are obtained by armature-voltage control. As discussed in connection with field-current control, the range above base speed is that of a constant-horsepower drive. The range below base speed is that of a constant-torque drive because, as in armature-resistance control, the flux and the allowable armature current remain approximately constant. The overall output limitations are therefore as shown in Fig. 6-3a for approximate allowable torque and Fig. 6-3b for approximate allowable horsepower. The constant-torque characteristic is well suited to many applications in the machine-tool industry, where many loads consist largely in overcoming the friction of moving parts and hence have essentially constant torque requirements.

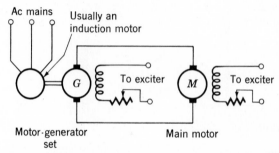

Fig. 6-2. Adjustable-armature-voltage, or Ward Leonard, method of speed control.

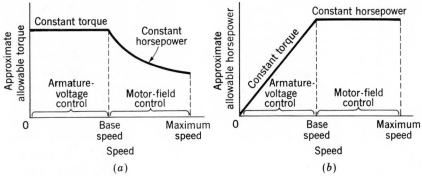

Fig. 6-3. (a) Torque and (b) power limitations of combined armature-voltage and field-rheostat methods of speed control.

The speed regulation and the limitations on the speed range above base speed are those already presented with reference to field-current control; the maximum speed thus does not ordinarily exceed 4 times base speed and preferably not twice base speed. In the region of armature-voltage control, the principal limitation in the basic system is residual magnetism in the generator, although considerations of speed regulation may also be determining. For conventional machines, the lower limit for reliable and stable operation is about one-tenth of base speed, corresponding to a total maximum-to-minimum range not exceeding 40 to 1. With armature reaction ignored, the decrease in speed from no-load to full-load torque is caused entirely by the full-load armature-resistance voltage drop in the dc generator and motor. This full-load armature-resistance voltage drop is constant over the voltage-control range, since full-load torque and hence full-load current are usually regarded as constant in that range. When measured in revolutions per minute, therefore, the speed decrease from no-load to full-load torque is a constant, independent of the no-load speed. The torque-speed curves accordingly are closely approximated by a series of parallel straight lines for the various generator-field adjustments. Note that a speed decrease of, say, 40 r/min from a no-load speed of 1200 r/min is often of little importance; a decrease of 40 r/min from a no-load speed of 120 r/min, however, may at times be of critical importance and require corrective steps in the layout of the system.

6-2 DYNAMIC EQUATIONS

Because of the complexity of dynamic-system problems, idealizing assumptions will be made. The usual assumptions are as follows:

1 The brushes are narrow, and commutation is linear, as in Fig. 5-8. The brushes are located so that commutation occurs when the coil sides are in

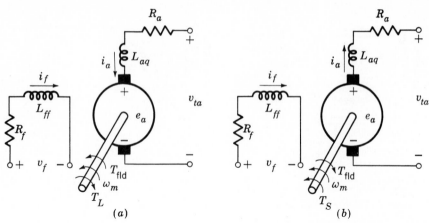

Fig. 6-4. Schematic representation of a dc machine showing (a) motor and (b) generator reference directions.

the neutral zone midway between the field poles. The axis of the armature-mmf wave then is fixed in space and lies along the *quadrature axis*.

2 The armature mmf is assumed to have no effect on the total direct-axis flux because the armature-mmf wave is perpendicular to the field axis. This assumption neglects the demagnetizing effect of armature reaction discussed in Chap. 5.

3 For most of the problems considered in this chapter the effects of magnetic saturation will be neglected. Superposition of magnetic fields can then be used, and inductances can be considered to be independent of the currents.

The schematic representation of the model is shown in Fig. 6-4. The convention will be adopted that the arrows represent the reference directions for both current and magnetic field. A consistent set of reference directions for all the other variables can then be adopted to conform with the flux-current reference directions. For example, in Fig. 6-4a the reference direction for magnetic torque is counterclockwise, tending to align the stator and rotor fields, as shown by the arrow labeled T_{fld}. If the machine is a motor, it will turn in the counterclockwise direction against the opposing torque T_L applied to the motor by the driven mechanical load, as shown by the arrows ω_m and T_L. The upper brush will be $+$, because electric power must be supplied to the motor. Figure 6-4b shows the reference directions for a generator, where T_S is the torque applied by the mechanical source. A consistent system of reference directions is especially important in dealing with the more complex cross-field machines discussed later in this chapter.

The electromechanical coupling terms are the magnetic torque T and the generated voltage e_a, already derived in Art. 5-1. From Eqs. 5-2 and 5-4

$$T = K_a \Phi_d i_a \tag{6-1}$$

$$e_a = K_a \Phi_d \omega_m \qquad (6\text{-}2)$$

where
$$K_a = \frac{PC_a}{2\pi m} \qquad (6\text{-}3)$$

The symbols have been defined in Eqs. 5-2 and 5-4. These equations, together with the differential equation of motion of the mechanical system, the volt-ampere equations for the armature and field circuits, and the magnetization curve, describe the system performance.

Consider first the ideal dc machine shown in Fig. 6-4 with one field winding and negligible magnetic saturation. The direct-axis air-gap flux Φ_d is then linearly proportional to the field current i_f, and Eqs. 6-1 and 6-2 can be expressed as

$$T = k_f i_f i_a \qquad (6\text{-}4)$$
$$e_a = k_f i_f \omega_m \qquad (6\text{-}5)$$

where k_f is a constant. With the brushes in the quadrature axis the mutual inductance between the field and armature circuits is zero, just as it would be for two coils whose axes are perpendicular. The voltage equation for the field circuit then is

$$v_f = L_{ff} p i_f + R_f i_f \qquad (6\text{-}6)$$

where v_f, i_f, R_f, and L_{ff} are the terminal voltage, current, resistance, and self-inductance of the field circuit, respectively, and p is the derivative operator d/dt.

For motor reference directions (Fig. 6-4a) the voltage equation for the armature circuit is

$$v_{ta} = e_a + L_{aq} p i_a + R_a i_a \qquad (6\text{-}7)$$
$$v_{ta} = k_f i_f \omega_m + L_{aq} p i_a + R_a i_a \qquad (6\text{-}8)$$

where v_{ta}, i_a, R_a, and L_{aq} are the terminal voltage, current, resistance, and self-inductance of the armature circuit, respectively. The subscript q is used with the inductance because the axis of the armature mmf is along the quadrature axis. The inductance L_{aq} includes the effect of any quadrature-axis stator windings in series with the armature, such as interpoles and pole-face compensating windings used on large machines to improve commutation, as described in Chap. 5. For a motor the dynamic equation for the mechanical system is

$$T = k_f i_f i_a = J p \omega_m + T_L \qquad (6\text{-}9)$$

where J is the moment of inertia and T_L is the mechanical load torque opposing rotation.

For generator reference directions (Fig. 6-4b), the armature voltage and torque equations become

$$v_{ta} = e_a - L_{aq}pi_a - R_a i_a \tag{6-10}$$

$$v_{ta} = k_f i_f \omega_m - L_{aq}pi_a - R_a i_a \tag{6-11}$$

and

$$T_S = Jp\omega_m + T = Jp\omega_m + k_f i_f i_a \tag{6-12}$$

where T_S now is the mechanical driving torque applied to the shaft in the direction of rotation.

Energy storage is associated with the magnetic fields produced by the field and armature currents and with the kinetic energy of the rotating parts. The state of a physical system can be described in terms of its stored energy. Accordingly, the field and armature currents and the speed are *state variables*. Equations 6-6 to 6-12 are first-order differential equations containing product nonlinearities $i_f \omega_m$ and $i_f i_a$ of these state variables. These equations, together with the Kirchhoff-law equations for the circuits connected to the field and armature terminals and the torque-speed characteristics of the mechanical system connected to the shaft, determine the system performance. Their application to specific cases will be illustrated in Arts. 6-3 and 6-7.

6–3 TRANSFER FUNCTIONS AND BLOCK DIAGRAMS OF DC MACHINES

The most difficult obstacle to overcome in analysis of dc machines is the inclusion of magnetic saturation. Linear analyses omitting saturation serve two useful purposes, however. (1) By virtue of the relatively simple linear differential equations which can then be written, a fuller appreciation of other factors affecting transient performance is possible and an approximate picture of the events is gained. (2) For system problems involving complex combinations of machines and other equipment, dynamic system studies can be made which otherwise would be practically prohibitive without resort to a computer.

In this article we shall consider separately excited dc machines. In part a we shall be concerned primarily with the electric transients in dc generators resulting from changes in excitation. The analysis will be made on a linear basis, with discussion of the effects of saturation postponed until Art. 6-4. In part b attention will be focused on the dynamics of dc motors with constant field excitation.

a. DC Generators; Linear Analysis

Consider the dc generator of Fig. 6-4b and assume that operation is restricted to the linear portion of the magnetization curve of Fig. 6-5. The inductance of

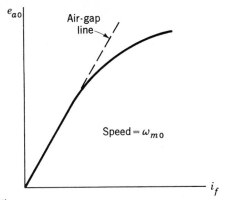

Fig. 6-5. Magnetization curve.

the field winding then is constant, and the voltage equation for the field circuit is

$$v_f = R_f i_f + L_{ff} p i_f = R_f (1 + \tau_f p) i_f \qquad (6\text{-}13)$$

where $\tau_f = L_{ff}/R_f$ is the time constant of the field circuit. At speed ω_{m0} corresponding to that of the magnetization curve and with operation restricted to the linear range, the speed voltage e_{a0} is

$$e_{a0} = K_g i_f \qquad (6\text{-}14)$$

where K_g is the slope of the air-gap line at speed ω_{m0}.

Rearrangement of Eq. 6-13 in state-variable form gives

$$p i_f = \frac{1}{\tau_f} \left(\frac{v_f}{R_f} - i_f \right) \qquad (6\text{-}15)$$

The block diagram with an integrator $1/p$ in the forward path is shown in Fig. 6-6a. Multiplication of the output i_f by K_g then gives the generated emf e_{a0} at magnetization-curve speed. Since generated emf is proportional to speed, the

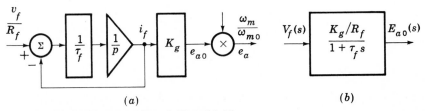

(a)   (b)

Fig. 6-6. Block diagram of Eqs. 6-15 and 6-17.

emf e_a at any other speed ω_m is

$$e_a = e_{a0} \frac{\omega_m}{\omega_{m0}} \tag{6-16}$$

as shown by the multiplier in the output, Fig. 6-6a. The corresponding transfer function in complex variables is obtained by replacing the derivative operator p in Eq. 6-13 by the complex frequency s. The variables $E_{a0}(s)$ and $V_f(s)$ then are the complex amplitudes of the corresponding time variables, and the equations become algebraic equations in s; thus

$$\frac{E_{a0}(s)}{V_f(s)} = \frac{K_g I_f}{V_f} = \frac{K_g/R_f}{1 + \tau_f s} \tag{6-17}$$

as shown in Fig. 6-6b.

The armature current i_a is determined by the generated emf e_a and the electric circuits connected to the armature terminals. The magnetic torque T is then determined by the direct-axis flux and armature current, as in Eq. 6-1. From Eq. 5-6 the torque can be expressed in terms of the magnetization curve as

$$T = \frac{e_{a0}}{\omega_{m0}} i_a \tag{6-18}$$

The relation between speed and torque is then given by Eq. 6-12 for a generator. The complete performance depends on the electric and mechanical systems connected to the machine.

EXAMPLE 6-1

A 200-kW 250-V dc generator has the following constants:

$$R_f = 33.7\ \Omega \qquad R_a = 0.0125\ \Omega \qquad L_{ff} = 25\ \text{H} \qquad L_{aq} = 0.008\ \text{H}$$

The slope of the air-gap line drawn on its magnetization curve at rated speed is

$$K_g = 38\ \text{V/field ampere}$$

The armature circuit is connected to a load having a resistance $R_L = 0.313\ \Omega$ and an inductance $L_L = 1.62$ H.

The generator is initially unexcited but rotating at rated speed. A 230-V dc source of negligible impedance is suddenly connected to the field terminals. Assume that as the terminal voltage builds up and the generator takes on load its speed does not change appreciably. Compute and plot a curve of the armature current $i_a(t)$ and investigate the possibilities of simplifying approximations.

Solution

The block diagram is shown in Fig. 6-7 in terms of complex-frequency variables. The first block represents the buildup of generated emf and the second the buildup of armature current, where R_a and L_a $(= R_a \tau_a)$ are the total resistance and inductance of the armature and load in series. Because the mutual inductance between armature and field is zero, the output of the second block does not influence the behavior of the first block. Each block represents an exponential term of the form $\varepsilon^{-t/\tau}$, where

$$\tau_f = \frac{L_{ff}}{R_f} = \frac{25}{33.7} = 0.74 \text{ s} \qquad\qquad \frac{1}{\tau_f} = 1.35$$

$$\tau_a = \frac{L_a}{R_a} = \frac{L_{aq} + L_L}{R_a + R_L} = \frac{1.63}{0.326} = 5 \text{ s} \qquad \frac{1}{\tau_a} = 0.2$$

The final steady-state value of the generated emf is

$$E_a = \frac{230(38)}{33.7} = 260 \text{ V}$$

and the final steady-state value of the armature current is

$$I_a = \frac{E_a}{R_a} = \frac{260}{0.326} = 800 \text{ A}$$

The equation for the armature current as a function of time is

$$i_a(t) = 800 + A\varepsilon^{-1.35t} + B\varepsilon^{-0.2t}$$

with initial conditions

$$i_a(0) = 0 \qquad \text{and} \qquad \frac{di_a}{dt}(0) = 0$$

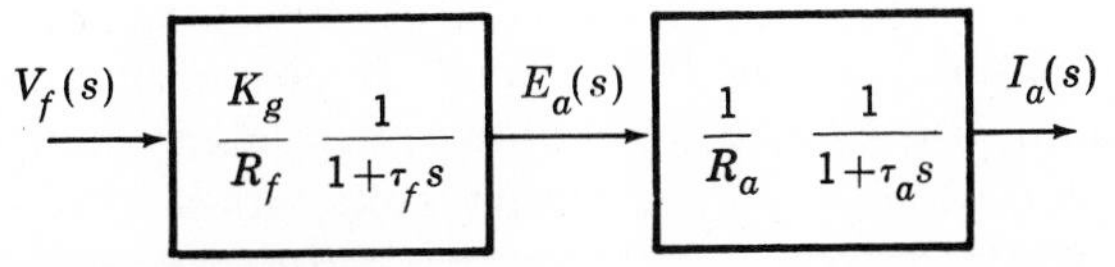

Fig. 6-7. Block diagram, Example 6-1.

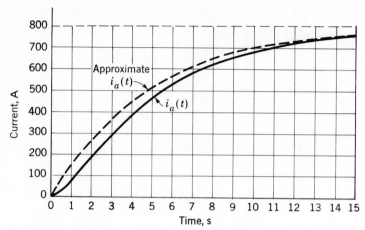

Fig. 6-8. Generator armature-current buildup, Example 6-1.

whence

$$i_a(t) = 800 + 139\varepsilon^{-1.35t} - 939\varepsilon^{-0.2t}$$

A plot of the current buildup is shown by the solid curve in Fig. 6-8. If the smaller of the two time constants were ignored, the current buildup would be given by

$$i_a(t) = 800 - 800\varepsilon^{-0.2t}$$

which is shown by the dashed curve in Fig. 6-8. Comparison of these two curves shows that the shorter of two time lags in series often can be ignored when its time constant is less than about one-quarter of the longer one.
The terminal voltage $v_{ta}(t)$ is

$$v_{ta}(t) = (R_L + L_L p)i_a(t) = 250 - 260\varepsilon^{-1.35t} + 10\varepsilon^{-0.2t}$$

Compare with the equation for generated emf given by

$$e_a(t) = 260 - 260\varepsilon^{-1.35t}$$

An idea of the influence of armature resistance and inductance of this machine, which is typical of dc generators used as exciters for synchronous generators, can now be gained. Obviously the influence of armature inductance is very small. The principal effect of armature resistance is to reduce the final voltage from 260 V to the 250 V actually obtained. When this effect is taken into account, it is evidently possible to base many engineering analyses on the assumption of negligible armature inductance.

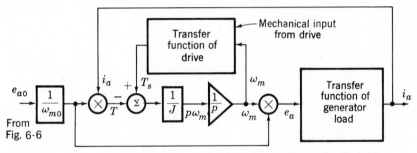

Fig. 6-9. Schematic block diagram for a dc generator and its mechanical drive.

In Example 6-1 the generator speed is assumed to be constant. The dynamics of the mechanical drive do not enter into the problem; it is assumed that the drive is capable of delivering whatever mechanical torque is required to hold constant speed. Figure 6-9 shows schematically the components which would have to be added to the block diagram of Fig. 6-6a or b to take into account the dynamics of the mechanical drive. In Fig. 6-9 the first multiplier (reading from left to right) represents Eq. 6-18 for the magnetic torque T. Rearrangement of Eq. 6-12 gives

$$p\omega_m = \frac{T_s - T}{J} \tag{6-19}$$

as shown by the summation and the coefficient multiplier $1/J$, where J is the combined inertia of the generator and drive and T_s is obtained from the transfer function of the drive. Integration then gives ω_m, which becomes an input to the transfer function of the drive and also to the multiplier representing Eq. 6-16 for e_a. Finally i_a, obtained from the transfer function of the electric load on the generator, is fed back into the first multiplier.

As more and more refinements are added, it can readily be appreciated that a complete system problem can soon become too complex for an analytical solution. For example, in Fig. 6-9 the transfer function of the mechanical drive may involve the transfer function of a mechanical speed governor. Often simplifying approximations can be made. Laying out a block diagram to take into account as many refinements as desired is not especially difficult, however, because it can be constructed piece by piece.

b. Separately Excited DC Motors

DC motors are often used in applications requiring precise control of speed and torque output over a wide range. One of the common ways of control is the use of a separately excited motor with constant field excitation. The speed is controlled by variation of the voltage applied to the armature terminals.

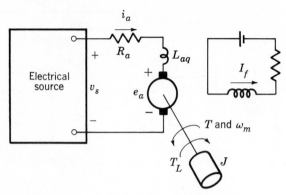

Fig. 6-10. Schematic diagram of a separately excited dc motor.

The analysis then involves the electic transients in the armature circuit and the dynamics of the mechanical load driven by the motor.

A separately excited motor is shown in Fig. 6-10. At constant field current I_f, the magnetic torque and generated voltage are given by

$$T = K_m i_a \qquad \text{N} \cdot \text{m} \tag{6-20}$$

$$e_a = K_m \omega_m \qquad \text{V} \tag{6-21}$$

where $K_m = k_f I_f$ is a constant. In terms of the magnetization curve

$$K_m = \frac{e_{a0}}{\omega_{m0}} \tag{6-22}$$

with e_{a0} the generated voltage corresponding to the field current I_f at the speed ω_{m0} rad/s. In SI units the constant K_m in newton-meters per ampere (Eq. 6-20) equals the constant K_m in volt-seconds per radian (Eqs. 6-21 and 6-22).

The response of the motor to changes in source voltage and the effects of load torque will now be investigated.

From Eq. 6-7, after rearrangement of the terms and division by R_a, the differential equation for the armature current i_a is

$$\frac{L_{aq}}{R_a} p i_a = \tau_a p i_a = \frac{v_s - e_a}{R_a} - i_a \tag{6-23}$$

where v_s = source voltage
e_a = speed voltage (Eq. 6-21)
$\tau_a = L_{aq}/R_a$ = *electrical time constant* of armature circuit

R_a and L_{aq} include the series resistance and inductance of the source and armature circuit. The magnetic torque T is given by Eq. 6-20, and from Eq. 6-9 the acceleration is

$$p\omega_m = \frac{T}{J} - \frac{T_L}{J} = \frac{K_m i_a}{J} - \frac{T_L}{J} \tag{6-24}$$

where J is the moment of inertia including that of the load and T_L is the load torque opposing rotation.

The block diagram representing Eqs. 6-20 to 6-24 is shown in Fig. 6-11a in terms of the state variables i_a and ω_m with v_s and T_L/J as inputs. Division of the input v_s by K_m and combination of the constants in the forward path yields the simpler form shown in Fig. 6-11b, where

$$\tau_m = \frac{JR_a}{K_m^2} \tag{6-25}$$

is the *inertial time constant*. Physically interpreted, v_s/K_m is the steady-state no-load speed corresponding to a constant dc input voltage V_{dc}.

In general, the load torque is a function of speed. It is sometimes assumed that the load torque is proportional to the speed; thus

$$T_L = B\omega_m \qquad \text{or} \qquad \frac{T_L}{J} = \frac{B\omega_m}{J} \tag{6-26}$$

where B is the slope of the torque-speed curve at the operating point and may be assumed to be constant for small changes. The parameter J/B is the *load time constant* τ_L describing the rate at which the motor coasts when its arma-

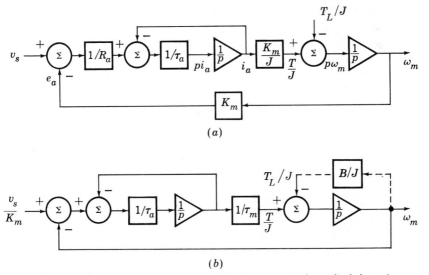

(a)

(b)

Fig. 6-11. Block diagrams of Eqs. 6-20 to 6-24 for a separately excited dc motor.

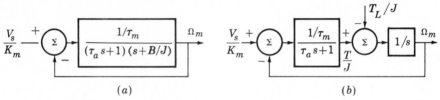

Fig. 6-12. Block diagrams of a separately excited dc motor in terms of complex variables.

ture circuit is open, and B/J is the corresponding damping factor. It varies over a wide range from no load to full load, but its effect usually is not very important with integral-horsepower motors. The effect of the load damping is shown in Fig. 6-11b by the feedback B/J around the second integrator.

The block diagram in terms of complex frequency s is shown in Fig. 6-12, where now Ω_m, V_s, and T_L are the complex amplitudes of the corresponding time variables.† The functional notation "of s" will be omitted for the sake of simplicity in the rest of this chapter with the understanding that capital letters such as Ω_m and V_s are complex amplitudes. [Although the same symbol is used for torque $T(t)$ and torque $T(s)$, the meaning should be clear from the context.] The first integrator in Fig. 6-11b becomes the algebraic term $1/(\tau_a s + 1)$ in Fig. 6-12. The second integrator in Fig. 6-11b with damping B/J becomes the algebraic term $1/(s + B/J)$ in Fig. 6-12a. The block diagram with mechanical damping neglected and T_L assumed to be an independent variable is shown in Fig. 6-12b. The transfer function relating speed to input voltage, found by elimination of the negative feedback in Fig. 6-12a, is

$$\frac{\Omega_m}{V_s/K_m} = \frac{1}{\tau_m(\tau_a s + 1)(s + B/J) + 1} \tag{6-27}$$

With mechanical damping neglected

$$\frac{\Omega_m}{V_s/K_m} = \frac{1}{\tau_m s(\tau_a s + 1) + 1} \tag{6-28}$$

and the transfer function relating speed to load torque then is

$$\frac{\Omega_m}{T_L} = -\frac{R_a}{K_m^2} \frac{1}{\tau_m s(\tau_a s + 1) + 1} \tag{6-29}$$

The natural frequencies s of the system are given by the poles of the transfer function, Eq. 6-27, or the roots of the equation

$$\left(s + \frac{1}{\tau_a}\right)\left(s + \frac{B}{J}\right) + \frac{1}{\tau_a \tau_m} = 0 \tag{6-30}$$

†The reader is cautioned to note the distinction between roman "s," which is the abbreviation for seconds, and italic "s," which is the complex frequency and has units of inverse seconds.

$$s^2 + \left(\frac{1}{\tau_a} + \frac{B}{J}\right)s + \frac{1}{\tau_a}\left(\frac{1}{\tau_m} + \frac{B}{J}\right) = 0 \qquad (6\text{-}31)$$

Comparison with the standard form for a second-order equation,

$$s^2 + 2\alpha s + \omega_n^2 = 0 \qquad (6\text{-}32)$$

shows that the undamped natural frequency ω_n is

$$\omega_n = \sqrt{\frac{1}{\tau_a}\left(\frac{1}{\tau_m} + \frac{B}{J}\right)} \qquad (6\text{-}33)$$

and the damping factor α is

$$\alpha = \frac{1}{2}\left(\frac{1}{\tau_a} + \frac{B}{J}\right) \qquad (6\text{-}34)$$

The relative damping factor or damping ratio ζ is

$$\zeta = \frac{\alpha}{\omega_n} \qquad (6\text{-}35)$$

The roots are given by the well-known solution

$$s_1, s_2 = -\zeta\omega_n \pm \omega_n \sqrt{\zeta^2 - 1} = -\zeta\omega_n \pm j\omega_n \sqrt{1 - \zeta^2} \qquad (6\text{-}36)$$

where the first form, $\zeta > 1$, gives two exponential terms with negative real exponents, and the second form, $\zeta < 1$, gives a damped sinusoid. The mechanical load usually has only a small effect on ω_n and α, although of course it does affect the steady-state speed. If B/J is neglected,

$$\omega_n = \sqrt{\frac{1}{\tau_a \tau_m}} \qquad (6\text{-}37)$$

$$\alpha = \frac{1}{2\tau_a} \qquad (6\text{-}38)$$

$$\zeta = \frac{1}{2}\sqrt{\frac{\tau_m}{\tau_a}} \qquad (6\text{-}39)$$

The solutions of a second-order system with a step input and initial rest conditions are shown in normalized form by the family of curves in Fig. 6-13. The ordinates are the ratio of the output X to its final steady-state value X_∞. These curves can be used to find the change in speed $\Delta\omega_m(t)$ resulting from a step change Δv_s in source voltage. The ordinates are then interpreted as the ratio $\Delta\omega_m(t)/\Delta\omega_m(\infty)$, where $\Delta\omega_m(\infty)$ is the final value of the change in speed.

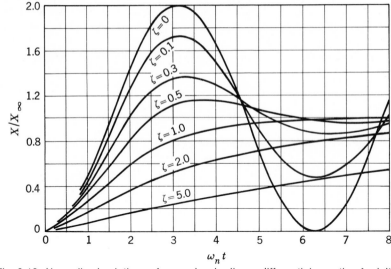

Fig. 6-13. Normalized solutions of second-order linear differential equation for initial-rest conditions.

From Eq. 6-27, with $s = 0$

$$\frac{\Delta\omega_m(\infty)}{\Delta v_s/K_m} = \frac{1}{(\tau_m B/J) + 1} = \frac{1}{(R_a B/K_m^2) + 1} \tag{6-40}$$

From Eq. 6-29, with the load torque assumed to be independent of speed

$$\frac{\Delta\omega_m(\infty)}{\Delta T_L} = -\frac{R_a}{K_m^2} \tag{6-41}$$

EXAMPLE 6-2

The following constants are given for two typical high-performance compensated dc motors.†

	Motor 1 1 hp, 500 r/min, 240 V	Motor 2 100 hp, 1750 r/min, 240 V
R_a	7.56 Ω	0.0144 Ω
L_{aq}	0.055 H	0.0011 H
K_m	4.23 V · s/rad	1.27 V · s/rad
J	0.050 lb · ft · s²	1.34 lb · ft · s²

Assume that the resistance and inductance of the source equal the resistance and inductance of the motor, and that the load inertia equals the motor

---

†A table of dc motor constants is given by A. Kusko, "Solid-State DC Motor Drives," The M.I.T. Press, Cambridge, Mass., 1969, pp. 22–23.

inertia. Also assume that the torque-speed characteristic of the load is a straight line through the origin and the rated-load point. Neglect the rotational losses in the motors.

Find the undamped natural frequency ω_n and the damping factor ζ for each motor. Discuss the effect of approximations.

Solution

Convert the motor rating and inertia into SI units.

	Motor 1 1 hp = 746 W 500 r/min = 52.3 rad/s	Motor 2 100 hp = 74,600 W 1750 r/min = 183 rad/s
Rated torque $T =$	$\dfrac{746}{52.3} = 14.3\ \mathrm{N \cdot m}$	$\dfrac{74,600}{183} = 407\ \mathrm{N \cdot m}$
$B = \dfrac{T}{\omega} =$	$0.273\ \mathrm{N \cdot m \cdot s/rad}$	$2.23\ \mathrm{N \cdot m \cdot s/rad}$
$J =$	$\dfrac{2(0.050)}{0.738} = 0.136\ \mathrm{kg \cdot m^2}$	$\dfrac{2(1.34)}{0.738} = 3.64\ \mathrm{kg \cdot m^2}$
$\dfrac{B}{J} =$	2.0	0.61
$\tau_a =$	$\dfrac{0.110}{15.1} = 0.0073\ \mathrm{s}$	$\dfrac{0.0022}{0.0288} = 0.0765\ \mathrm{s}$
$\dfrac{1}{\tau_a} =$	1.37	13.1
$\tau_m =$	$\dfrac{0.136(15.1)}{(4.23)^2} = 0.115\ \mathrm{s}$	$\dfrac{3.64(0.0288)}{(1.27)^2} = 0.065\ \mathrm{s}$
$\dfrac{1}{\tau_m} =$	8.7	15.4

For the assumed source and load parameters the effect of the load damping B/J is small, especially with the larger motor. From Eqs. 6-37 to 6-39 the results are:

	Motor 1	Motor 2
$\omega_n =$	$\sqrt{137(8.7)} = 34.5$	$\sqrt{13.1(15.4)} = 14.2$
$\zeta =$	$\dfrac{1}{2}\sqrt{\dfrac{0.115}{0.0073}} = 1.98$	$\dfrac{1}{2}\sqrt{\dfrac{0.065}{0.0765}} = 0.46$

From the curves of Fig. 6-13 the transient response of the 100-hp motor to a step change in source voltage is a damped sinusoid with an overshoot to about 1.2 at $\omega_n t = 3.5$, or $t = 3.5/14.2 = 0.25$ s. The response of the 1-hp motor is overdamped. Its transient is 0.8 of its final value at $\omega_n t = 6.5$, or $t = 6.5/34.5 = 0.19$ s. If its armature-circuit time constant is neglected, the expression for its response reduces to

$$\frac{\Delta\omega_m(t)}{\Delta\omega_m(\infty)} = 1 - \varepsilon^{-t/\tau_m}$$

The approximate expression gives a value of 0.8 when $\varepsilon^{-t/\tau_m} = 0.2$, or $t/\tau_m = 1.61, t = 1.61(0.115) = 0.185$ s. This approximate value is substantially the same as the value of 0.19 s from the curves. The armature-circuit time constant then has very little effect on the transient behavior of the 1-hp motor but a significant effect on the 100-hp motor.

In general, armature-circuit inductance may be neglected for damping ratios ζ greater than about 1.5, corresponding to time-constant ratios τ_a/τ_m less than about $\frac{1}{9}$. Critical damping corresponds to the ratio $\tau_a/\tau_m = \frac{1}{4}$. As a general trend for the motor alone, τ_a increases with increasing frame size and τ_m decreases slightly. Large low-speed motors, above 10 hp at speeds of 1150 r/min and below, are underdamped. Of course, the source impedance affects τ_a and the load inertia affects τ_m. The damping B/J usually is negligible.

For a system consisting of a dc motor, its drive system, and load, analysis depends upon specific details of the electric source and mechanical load. An example of this type of analysis is found in Art. 6-7.

6–4 EFFECTS OF SATURATION: SELF–EXCITED GENERATORS

In Art. 6-3a the transient response of a dc generator was studied with magnetic saturation neglected. Often the critical portion of the transient response takes place in substantially the linear region, or the response to small disturbances can be treated on an incrementally linear basis. For more comprehensive studies, however, transient investigations may require a nonlinear analysis because of saturation. For example, the response of an exciter driven into saturation by the demands of the voltage-regulating system during major system disturbances or the analysis of self-excited shunt generators definitely require that saturation be taken into account.

The purpose of this article is to show how magnetic saturation can be included by a modification of the block diagram of Fig. 6-6a in a form adaptable to computer solution of system problems. We shall also investigate the voltage buildup of a shunt generator as an example of a nonlinear problem that can be analyzed by relatively simple graphical means.

a. The Block Diagram with Saturation

Consider a dc generator driven at constant speed ω_{m0} with a voltage v_f applied to its field terminals. With saturation included, the relation between generated voltage e_{a0} and field current i_f is the magnetization curve. Furthermore, the inductance of the field winding is no longer constant. It is more convenient then to express the voltage induced in the field winding in terms of the field flux linkages; thus

$$v_f - R_f i_f = N_f p \Phi_f \qquad (6\text{-}42)$$

where R_f = field-circuit resistance
$\quad i_f$ = field current
$\quad \Phi_f$ = field flux per pole
$\quad N_f$ = total number of turns in field winding (all poles assumed connected in series)
$\quad p$ = derivative operator d/dt

The field flux Φ_f is somewhat greater than the direct-axis air-gap flux Φ_d because of field leakage flux. The increase can be included approximately by the use of a *coefficient of dispersion* σ, usually about 1.15, which we shall assume to be constant. Thus

$$\Phi_f = \sigma \Phi_d \qquad (6\text{-}43)$$

The air-gap flux Φ_d is related to the generated voltage e_{a0} by Eq. 6-2; thus

$$\Phi_d = \frac{e_{a0}}{K_a \omega_{m0}} \qquad (6\text{-}44)$$

Substitution of Eq. 6-44 in Eq. 6-43, differentiation, and substitution of the result in Eq. 6-42 yields

$$\frac{N_f \sigma}{K_a \omega_{m0}} p e_{a0} = v_f - R_f i_f \qquad (6\text{-}45)$$

The coefficient of the left-hand side of Eq. 6-45 can now be expressed in terms of recognizable and easily determinable constants if its numerator and denominator are multiplied by $N_f \mathcal{P}_{ag}$, where $\mathcal{P}_{ag}$ is the permeance of the air gap; thus

$$\frac{N_f \sigma}{K_a \omega_{m0}} = \frac{N_f^2 \sigma \mathcal{P}_{ag}}{K_a \omega_{m0} \mathcal{P}_{ag} N_f} \qquad (6\text{-}46)$$

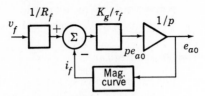

Fig. 6-14. Block diagram with saturation included.

Now, $N_{f}^2 \sigma \mathcal{P}_{\text{ag}}$ is the unsaturated value of the field inductance L_{ff}, and $K_a \omega_{m0} \mathcal{P}_{\text{ag}} N_f$ is the slope K_g of the air-gap line in generated volts per field ampere. Both these quantities are *constants*. They can easily be determined by tests taken under conditions of negligible saturation. Substitution of these constants in Eq. 6-45 gives

$$\frac{L_{ff}}{K_g} pe_{a0} = v_f - R_f i_f \tag{6-47}$$

or, after division by R_f and solving for pe_{a0}

$$pe_{a0} = \frac{K_g}{\tau_f}\left(\frac{v_f}{R_f} - i_f\right) \tag{6-48}$$

where $\tau_f = L_{ff}/R_f$ is the unsaturated value of the field-circuit time constant.

The block diagram representing Eq. 6-48 is shown in Fig. 6-14. It is a relatively simple modification of the linearized block diagram of Fig. 6-6a. Saturation is taken care of by feeding back the output e_{a0} through the magnetization curve to obtain i_f. This operation can readily be carried out on a computer.

b. Shunt-Generator Voltage Buildup

The buildup of voltage of a self-excited shunt generator, described qualitatively in Art. 5-8a, is obviously a process inherently dependent on saturation. A shunt generator driven at constant speed ω_{m0} is shown in Fig. 6-15 and its magnetization curve in Fig. 6-16. A small residual flux is assumed, corresponding to a small generated voltage at zero excitation. The straight line $0a$ in Fig.

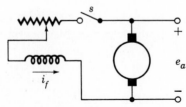

Fig. 6-15. Shunt generator.

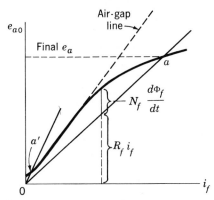

Fig. 6-16. Magnetization curve.

6-16, called the *field-resistance line,* is a plot of the relation

$$v_f = R_f i_f \qquad (6\text{-}49)$$

where v_f, i_f, and R_f are the voltage, current, and resistance of the field circuit. The slope of line $0a$ is adjustable by means of the field rheostat. In Fig. 6-16 the slope of $0a$ is less than that of the air-gap line, and point a is the intersection of the field-resistance line with the magnetization curve.

Now let the field switch S be closed at $t = 0$. The small voltage generated in the armature by the residual flux is then applied to the shunt-field circuit, and the shunt-field current starts to build up. If the connections of the shunt field to the armature terminals are such that the resulting field current increases the flux, positive feedback results and the voltage continues to build up until it is finally limited by magnetic saturation at point a, where the generated voltage supplies just enough field current to sustain itself. This statement ignores the extremely small voltage drop caused by the shunt-field current in the armature-circuit resistance. The normal operating point of a shunt generator is well up on the magnetization curve.

Figure 6-17 shows a curve of the armature terminal voltage as a function of time following closure of the field switch. This curve can be calculated by the following method. Since the generated voltage e_a is applied to the field

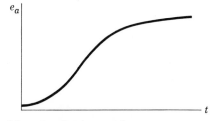

Fig. 6-17. Voltage buildup of a shunt generator.

circuit, the voltage equation for the field circuit is

$$N_f \frac{d\Phi_f}{dt} = e_a - R_f i_f \qquad (6\text{-}50)$$

where Φ_f is the field flux per pole and N_f is the number of turns in the field winding. Thus, the rate of change of field flux, and consequently of generated voltage, is proportional to the vertical difference between the magnetization curve and the field-resistance line.

The curve of e_a as a function of time can be computed by separation of the variables and graphical integration. From Eq. 6-47 with e_{a0} substituted for v_f and the variables separated

$$dt = \frac{L_{ff}}{K_g} \frac{de_{a0}}{e_{a0} - R_f i_f} \qquad (6\text{-}51)$$

The time required for the voltage to change from an initial value e_r to the value e_a is then

$$t = \frac{L_{ff}}{K_g} \int_{e_r}^{e_a} \frac{1}{e_{a0} - R_f i_f} de_{a0} \qquad (6\text{-}52)$$

This integral can be evaluated graphically by finding the area on a plot of $1/(e_{a0} - R_f i_f)$ as a function of e_{a0}. Alternatively, the magnetization curve can be expressed in a functional or tabular form and the integration can be performed on a computer. The response is rather slow, because only relatively small voltage differences act to build up the flux.

The graphical process can be applied to many first-order equations where the variables can be separated in this manner. Example 10-3 shows an application of this method to the starting transient of an induction motor.

6-5 INTRODUCTION TO RECTIFIER CIRCUITS

As discussed in Art. 6-1, the torque-speed characteristics of a dc motor can be controlled by adjusting the armature voltage, by adjusting the field current, and by inserting resistance into the armature circuit. Solid-state motor controls are designed to use each of these modes for particular purposes.

Adjustable-armature-voltage systems use phase-controlled thyristor rectifier circuits to provide dc power for the dc motors. The rectifier circuits operate from single- and 3-phase ac lines, and in various configurations, depending upon horsepower rating, reversibility, and braking requirements. A commercial system is shown in Fig. 6-18. The armature voltage is adjusted by controlling the electrical angle within the ac wave at which the gate signal is applied to each thyristor.

Fig. 6-18. Commercial thyristor dc motor-drive system. (*Electric Regulator Corporation*.)

Field current is supplied from the main rectifier when the field current is not controlled. When the field current must be reduced to obtain speeds above the base speed, the current is supplied from an auxiliary controlled-rectifier circuit, which is actuated by a crossover circuit in the speed-control reference circuit.

The equivalent of armature-resistance control is used for speed control of dc series motors operating from dc sources, such as a battery in an electric-drive vehicle or a third rail in a rapid-transit system. A thyristor is made to switch on and off at a fast rate so that the applied voltage divides between the armature and the switch, resulting in a controllable average armature voltage. The thyristor switch circuit is called a *chopper* and acts as a lossless armature resistance.

The rectifier is the principal item in the solid-state control system for a dc motor. The rectifier is a circuit assembled from diodes and thyristors which supplies direct current to the armature and field of the motor from the ac supply line. The dc voltage of the rectifier, which is controlled by the thyristors, controls in turn the speed or torque of the motor. Several circuits with

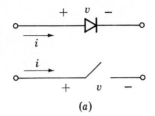

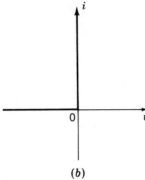

Fig. 6-19. Diode representation: (a) symbol for diode (upper) and circuit model switch (lower); (b) volt-ampere characteristic of circuit model.

passive loads will be described in this article; motor loads will be discussed in Art. 6-6.†

The diode and the thyristor used in rectifier circuits are highly nonlinear devices. In order to analyze circuits containing these devices, we require circuit models and the technique for using these models in the repetitive operation that occurs.‡

Although the actual volt-ampere characteristics of a diode are nonlinear and depend upon temperature, for many applications the diode can be represented as a switch which conducts current when forward-biased and blocks current when reverse-biased. Its circuit model and electrical characteristics are shown in Fig. 6-19. When forward voltage is impressed on the model, the switch is closed and the diode carries current with zero voltage drop. When reverse voltage is impressed, the switch is opened and the diode carries zero reverse current.

A thyristor is a three-terminal device which is similar to a diode except that it must be turned on by a signal to its gate lead before it will begin conduction of current when it is forward-biased. The circuit model is a switch as shown in Fig. 6-20a, and the characteristic for the model is shown in Fig.

†B. Dewan and A. Straughen, "Power Semiconductor Circuits," Wiley, New York, 1975.
‡A discussion of modeling issues and an extensive bibliography can be found in J. G. Kassakian, Simulating Power Electronic Systems—A New Approach, *Proc. IEEE,* **67**(10):1428–1439, (October 1979).

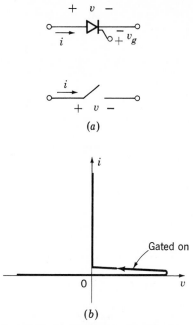

Fig. 6-20. Thyristor representation: (*a*) symbol for thyristor (*upper*) and circuit model switch (*lower*); (*b*) volt-ampere characteristic of circuit model.

6-20*b*. The switch is open for either direction of current flow before a gate signal is applied. When forward voltage and a gate signal are applied, the switch is closed and the thyristor model carries forward current with no voltage drop.

a. Single-Phase Half-Wave Diode Rectifier

The simplest rectifier circuit consists of a single diode operating into a resistance load, as shown in Fig. 6-21*a*. The line voltage is a sine wave, as shown in Fig. 6-21*b*. According to the diode model of Fig. 6-19, the diode acts like an open switch when the diode voltage is negative and conducts zero current; this occurs typically from $\omega t = \pi$ to 2π. The diode acts like a closed switch when the diode voltage is positive, typically from $\omega t = 0$ to π, and conducts current i_n. The load voltage v_n is equal to the line voltage v_0 during the conducting interval because the diode model is lossless.

b. Single-Phase Half-Wave Thyristor

The simplest controlled-rectifier circuit consists of a single thyristor operating into a resistance load, as shown in Fig. 6-22*a*. The circuit for applying gating pulses to the thyristor is not shown. According to the thyristor model of Fig. 6-20, the thyristor can conduct current only when its voltage is positive, typically during the intervals $\omega t = 0$ to π, 2π to 3π, etc. In addition, the

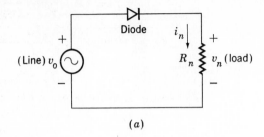

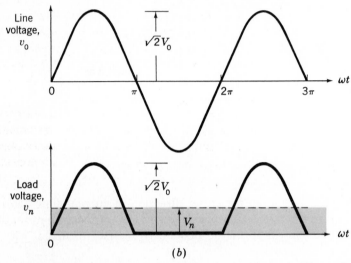

Fig. 6-21. Operation for half-wave diode-rectifier circuit: (a) circuit; (b) waveforms of line voltage v_0 and load voltage, v_n, where V_0 is the rms value of the sine-wave line voltage and V_n is the average value of the load voltage, shown shaded.

thyristor must also receive a gating pulse, which we assume is applied at the firing angle α. By convention, the firing angle α is measured from the angle that produces the largest load voltage, in this case, from $\omega t = 0$, 2π, etc. As shown in Fig. 6-22b, the thyristor conducts from $\omega t = \alpha$ to π, $2\pi + \alpha$ to 3π, etc. During the conduction intervals the load voltage v_n is equal to the line voltage v_0. As the firing angle α is shifted by the control circuit from zero to π, the average value V_n of the load voltage decreases as well.

EXAMPLE 6-3

Find the expression for the average load voltage V_n as a function of the firing angle α for the circuit of Fig. 6-22a.

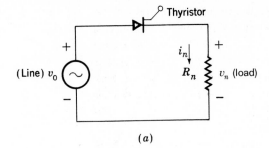

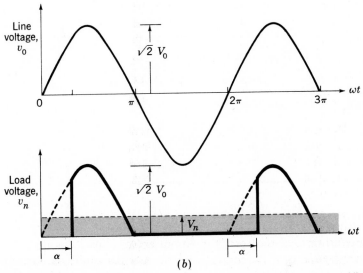

Fig. 6-22. Operation of half-wave thyristor circuit: (a) circuit, (b) waveforms of line voltage v_0 and load voltage v_n, where V_0 is the rms value of the sine-wave line voltage and V_n is the average value of the load voltage, shown shaded.

Solution

The average voltage is given by

$$V_n = \frac{1}{2\pi} \int_{\alpha}^{\pi} v_n \, d(\omega t) = \frac{1}{2\pi} \int_{\alpha}^{\pi} \sqrt{2} \, V_0 \sin \omega t \, d(\omega t)$$

$$= \frac{-1}{\sqrt{2}\pi} [V_0 \cos \omega t]_{\alpha}^{\pi} = 0.225 \, V_0 (1 + \cos \alpha)$$

The relationship is shown in Fig. 6-23. The maximum value of V_n, corresponding to the half-wave diode-rectifier operation of Fig. 6-21, occurs for $\alpha = 0$ and

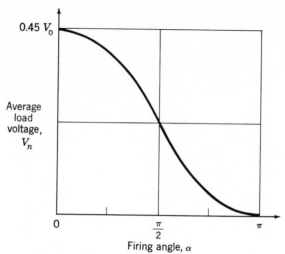

Fig. 6-23. Plot of average voltage V_n of half-wave thyristor circuit as a function of firing angle α.

is

$$V_n = 0.450 \, V_0$$

c. Single-Phase Rectifier, Reactive Load

Actual dc motor loads on rectifiers do not act as resistance alone. Armature circuits have resistance and inductance; the armature and load inertia act like capacitance. Field circuits are highly inductive. The operation of the diode and thyristor elements is different from that with resistance load, but the elements must still behave in accordance with the constraints of their models.

A half-wave diode rectifier with RL load is shown in Fig. 6-24a. The waveforms of line voltage v_0, resistance voltage v_r, and inductance voltage v_l are shown in Fig. 6-24b. At $\omega t = 0$, the line voltage v_0 becomes positive and the diode starts to conduct the current i_n. The inductance forces the current to lag the voltage until the current reaches its peak value at angle ωt_1. During this period, the flux linkage λ in the inductance increases by

$$\Delta\lambda = \int_0^{t_1} v_l \, dt = \frac{1}{\omega} \int_0^{\omega t_1} v_l \, d(\omega t) \tag{6-53}$$

From angle ωt_1 to β, the current declines until the flux linkage λ returns to its value at $\omega t = 0$, or

$$\Delta\lambda = \int_{t_1}^{\beta/\omega} v_l \, dt = \frac{1}{\omega} \int_{\omega t_1}^{\beta} v_l \, d(\omega t) \tag{6-54}$$

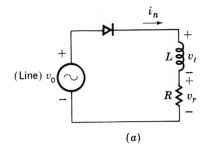

(a)

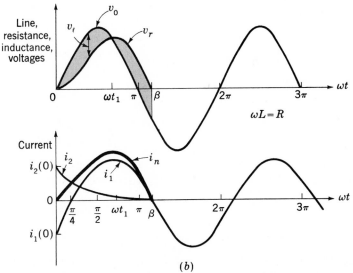

(b)

Fig. 6-24. Operation of half-wave diode rectifier with reactive RL load: (a) circuit, (b) waveforms of load-voltage components of v_r and v_l and load current i_n.

The integrals of Eqs. 6-53 and 6-54 are represented by the areas shown shaded in Fig. 6-24b. The shaded areas, frequently termed *volt-time areas*, on both sides of the current peak at ωt_1 must be equal.

EXAMPLE 6-4

For the circuit of Fig. 6-24a find the current i_n for the conduction interval $\omega t = 0$ to β. Assume that $R = \omega L$.

Solution

The current i_n consists of a steady-state component i_1 and a transient component i_2. The boundary condition at $\omega t = 0$ is

$$i_n(0) = i_1(0) + i_2(0) = 0$$

The steady-state component i_1 is given by

$$i_1 = \frac{\sqrt{2}\, V_0}{[R^2 + (\omega L)^2]^{1/2}} \sin(\omega t - \phi)$$

where $\phi = \tan^{-1}(\omega L/R)$. The transient component is

$$i_2 = I_2 \varepsilon^{-t/\tau}$$

where $\tau = L/R$. The boundary condition requires that

$$I_2 = \frac{\sqrt{2}\, V_0}{[R^2 + (\omega L)^2]^{1/2}} \sin\phi$$

For $R = \omega L$, the components are

$$i_1 = \frac{V_0}{R} \sin\left(\omega t - \frac{\pi}{4}\right) \qquad i_2 = 0.707 \frac{V_0}{R} \varepsilon^{-t/\tau}$$

The components and total current i_n for the example are shown in the wave-forms of Fig. 6-24b. The peak current occurs at approximately $\omega t_1 = 3\pi/4 = 135°$ and the extinction angle at $\beta = 5\pi/4 = 225°$.

During the period $\omega t = 0$ to ωt_1 in Fig. 6-24b the line supplies energy to the resistance R at the rate $v_r i_n$ and to the inductance L at the rate $v_l i_n$. From $\omega t = \omega t_1$ to π the line continues to supply energy to the resistance R in addition to the energy extracted from the inductance L at the rate $(-)\,v_l i_n$. From $\omega t = \pi$ to β the energy extracted from the inductance L is not only supplied to the resistance R but is returned to the line as well, forcing the conduction period of the current to extend beyond π to the extinction angle β.

d. Three-Phase Half-Wave Rectifier

The simplest 3-phase rectifier circuit is the half-wave configuration shown in Fig. 6-25a. The circuit consists of three diodes connected to a common resistance load R; the load current is returned to the neutral n of the 3-phase supply line. The 3-phase supply voltages, v_{an}, v_{bn}, and v_{cn} and the load voltage are shown in Fig. 6-25b and c.

The diodes conduct one at a time for 120° each, in sequence. The cathode ends of the three diodes are connected to a common point on the load resistance, while the anode ends are connected to the respective supply voltages. In accordance with the model of Fig. 6-19, each diode will conduct only when its voltage is positive and block when it is negative. At any time, the conducting

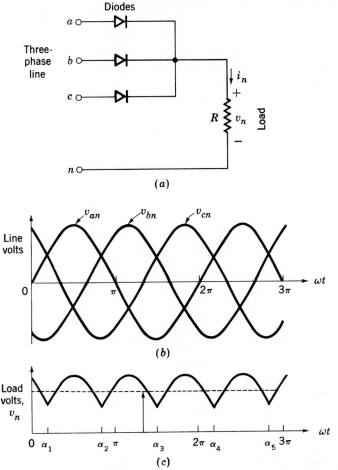

Fig. 6-25. Operation of three-phase half-wave rectifier circuit: (a) circuit, (b) 3-phase line-to-neutral voltages; (c) load voltage v_n.

diode is the one connected to the highest instantaneous supply voltage. The common terminal of the cathodes is also raised to the same voltage, so that the blocking diodes are *reverse-biased*. As shown in Fig. 6-25c, the load voltage follows the envelope of the highest instantaneous supply voltages.

The load voltage v_n is continuous and has an average value V_n, given by

$$V_n = \frac{3}{2\pi} \int_{\alpha_1}^{\alpha_2} \sqrt{2}\, V_0 \sin(\omega t)\, d(\omega t) = \frac{-3}{\sqrt{2}\pi} [\, V_0 \cos(\omega t)\,]_{\alpha_1}^{\alpha_2} \qquad (6\text{-}55)$$

For $\alpha_1 = \pi/6$ and $\alpha_2 = 5\pi/6$,

$$V_n = 1.17 V_0 \qquad (6\text{-}56)$$

There are numerous 3-phase rectifier circuits using both diodes and thyristors. In the circuits using diodes, the dc load voltage is directly proportional to the ac source voltage. In the circuits using thyristors, the load voltage can be controlled independently of the ac source voltage. Rectifier circuits are selected to meet the requirements of the application and to minimize the cost. The cost of diodes and thyristors increases with both current and voltage rating; the use of fewer costly elements must be balanced against a greater number of less costly elements to achieve the best rectifier design.

6-6 SOLID–STATE DC MOTOR–DRIVE SYSTEMS

We shall consider various dc motor-drive circuits corresponding to the rectifier circuits described in Art. 6-5 for resistance loads.† The rectifiers will be used to supply the armature circuit of a dc motor. For simplicity, the armature circuit will be modeled as an armature-generated voltage e_a and an armature-circuit inductance L_a; the resistance R_a and the effect of the brushes will be neglected. The field flux Φ_f is assumed constant. The armature will be assumed to have an inertia J and a mechanical-load torque $B\omega_m$ proportional to velocity.

a. Singe-Phase Half-Wave Diode Motor-Drive System

The simplest dc motor-drive circuit operating from an ac line consists of a single diode supplying the armature circuit. The field winding requires a separate source of excitation, or the motor can use a permanent-magnet field. The drive system has no means for armature-voltage control; unless the field current is adjusted, it will operate on one torque-speed characteristic. However, the circuit will be used to show the basic principles of armature-circuit operation from a rectifier. In the next section, we shall replace the diode with a thyristor and consider a controlled drive system.

The waveforms of voltage, current, and speed are shown in Fig. 6-26 for two levels of torque at the same average speed. The armature receives pulses of current i_a once per cycle of the line voltage v_0. As shown in Fig. 6-26a, the current i_a starts at $\omega t = \alpha$, when the line voltage v_0 attempts to rise above the armature emf e_a, and the diode conducts. The current i_a continues until $\omega t = \beta$ for a conduction angle γ. Just as for the passive RL circuit of Fig. 6-24, the conduction period ends at $\omega t = \beta$, when the shaded volt-time areas of the induction voltage balance. The end of current conduction indicates that the armature inductance has returned its stored energy to the circuit, or the armature-circuit flux linkages have returned to their starting point at $\omega t = \alpha$.

During the armature-current conduction period, electric energy flows into the armature circuit; the interaction of the current and field flux results in positive electromagnetic torque. The motor accelerates during the conduc-

---

†A. Kusko, "Solid State DC Motor Drives," M.I.T. Cambridge, MA, 1969; also S. B. Dewan and A. Straughen, "Power Semiconductor Circuits," Wiley, New York, 1975.

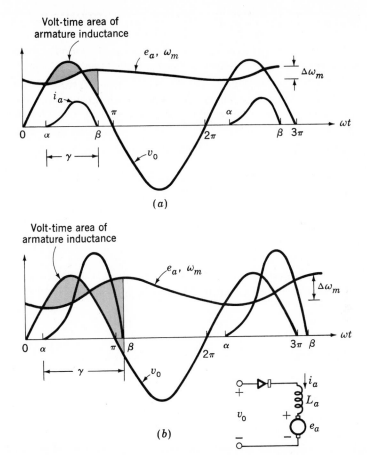

Fig. 6-26. DC motor operating from half-wave diode-rectifier circuit: (a) waveforms of speed ω_m, current i_a, and emf e_a at light load; (b) some waveforms at heavy load and same average speed.

tion period by the speed rise $\Delta\omega_m$, as shown in Fig. 6-26a. From the end of the conduction period at $\omega t = \beta$ of one cycle and the start of the next conduction period at $\omega t = 2\pi + \alpha$, the motor coasts down by $\Delta\omega_m$. The motor supplies the load energy during the coasting period from its own kinetic energy; the torque is $T_m = J\,d\omega_m/dt$. The armature inductance acts as a reservoir of electric energy during the conduction period; the armature and load inertia act as a reservoir of mechanical energy during the coasting period.

The waveforms of Fig. 6-26b show the operation when the load torque is increased but the speed readjusted to the same average value. The speed dip $\Delta\omega_m$ during the coasting period increases because more torque must be supplied than for Fig. 6-26a and the diode starts conducting earlier during its half cycle. Furthermore, the amplitude and average value of the current pulse i_a is greater, the volt-time areas are greater, and the conduction period γ increases. At low speeds and high torques, the conduction period will exceed the angle π and will approach 2π.

EXAMPLE 6-5

A 1-hp 500-r/min dc motor having armature inertia $J = 0.050$ lb $\cdot$ ft $\cdot$ s^2 and a rated torque of 20.3 N $\cdot$ m is driven by a 60-Hz half-wave diode circuit. Assume that the coasting period of the motor is π rad between conduction periods, as shown in Fig. 6-26b. Find the speed dip between conduction periods when the motor is delivering rated torque at rated speed.

Solution

During the coasting period, the air-gap torque is zero; the differential equation describing the velocity of the motor is given by

$$J\frac{d\omega_m}{dt} + B\omega_m = 0$$

where we have assumed that the load torque is proportional to velocity. The motor velocity in terms of the initial velocity ω_{m0} is thus

$$\omega_m = \omega_{m0}\varepsilon^{-t/\tau_L}$$

The mechanical time constant of the motor is given by

$$\tau_L = \frac{J}{B}$$

The values of J and B are

$$J = (0.050 \text{ lb} \cdot \text{ft} \cdot \text{s}^2)\frac{1}{0.738} = 0.0678 \text{ kg} \cdot \text{m}^2$$

$$B = \frac{20.3}{500(2\pi)(1/60)} = 0.388 \text{ N} \cdot \text{m} \cdot \text{s}$$

and
$$\tau_L = \frac{0.0678}{0.388} = 0.174 \text{ s}$$

The mechanical time constant τ_L is large compared with the time Δt of π/ω 8.3 ms. Hence, the speed dip is given approximately by the linearized expression

$$\Delta\omega_m = \frac{d\omega_m}{dt}\bigg|_{t=0}\Delta t^-$$

where
$$\frac{d\omega_m}{dt}\bigg|_{t=0} = \frac{d}{dt}(\omega_{m0}\varepsilon^{-t/\tau_L})\bigg|_{t=0} = -\frac{\omega_{m0}}{\tau_L}$$

and thus

$$\Delta\omega_m = -\frac{\omega_{m0}}{\tau_L}\Delta t$$

Similarly, in terms of motor speed n in revolutions per minute

$$\Delta n = -\frac{n_0}{\tau_L}\Delta t = -\frac{500}{0.174}(8.3 \times 10^{-3}) = -23.8\ \text{r/min}$$

The motor-speed fluctuation is almost 5 percent of the base speed, so that the motor can supply the load torque between the conduction periods without stalling.

---

b. Single-Phase Half-Wave Thyristor Drive System

The half-wave diode drive system shown in Art. 6-6a is suitable for field control over a typical 3-to-1 range of speed. The circuit and waveforms for a thyristor-drive system are shown in Figs. 6-27 and 6-28. The diode of Fig. 6-26 has been replaced by a thyristor; by control of the firing angle the voltage applied to the armature circuit and the speed of the motor can be adjusted over a wide range. The operation of the circuit is analogous to the RL load case of Art. 6-5c and also to the previous case of the half-wave diode circuit.

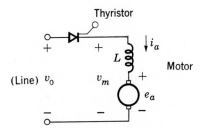

Fig. 6-27. Circuit of half-wave thyristor dc motor drive.

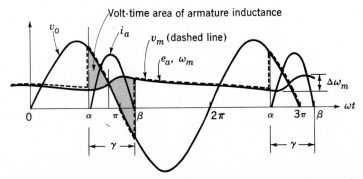

Fig. 6-28. Waveforms of speed ω_m, current i_a, motor voltage v_m, and speed voltage e_a for circuit of Fig. 6-27.

As shown in Fig. 6-28, the pulse of armature current i_a starts at firing angle α when the line voltage v_0 exceeds the motor voltage v_m and a gate pulse is applied to the thyristor. The current pulse continues for conduction angle γ until it reaches zero at $\omega t = \beta$; the thyristor then blocks until it receives its next gate pulse at $\omega t = 2\pi + \alpha$. The volt-time area of the armature inductance voltage v_1 is shown by the shaded area. The positive and negative areas balance when the current i_a reaches zero, designating the return of the flux linkage to its initial value and the return of the magnetic field energy to the circuit. For larger torques at the same speed, the angle α must be advanced; the current pulse increases in amplitude and conduction angle and the speed dip is more pronounced. For higher speed at the same torque, the angle α must be advanced so that the conduction period takes place closer to the peak of the line voltage v_0 wave.

The equation for the armature circuit during the conduction period is

$$v_0 = i_a R_a + L_a \frac{di_a}{dt} + e_a \tag{6-57}$$

The equation can be integrated over the conduction period

$$\int_{\alpha/\omega}^{\beta/\omega} v_0 dt = R_a \int_{\alpha/\omega}^{\beta/\omega} i_a \, dt + L_a \int di_a + \int_{\alpha/\omega}^{\beta/\omega} e_a \, dt \tag{6-58}$$

The interpretation of the terms of Eq. 6-58 must be done carefully. The inductance term is zero because the current i_a returns to its initial value at the end of the period of integration. The left-hand term can be evaluated only during the conduction period, when the thyristor is conducting and the voltage v_0 is a segment of the line voltage. Each term can be divided by the conduction period γ/ω and interpreted as

$$V'_m = I'_a R_a + E'_a \tag{6-59}$$

where V'_m = average voltage applied to motor over conduction period, a segment of line voltage
I'_a = average current over conduction period
E'_a = average speed voltage over conduction period

If we wish to consider the average values over the period of a line-voltage cycle, we can write the unprimed expression

$$V_m = I_a R_a + E_a \tag{6-60}$$

where V_m = average voltage at motor terminals
I_a = average armature current
E_a = average armature-generated voltage = $K_m \Omega_m$
Ω_m = average motor speed

The primed and unprimed variables can be related as follows:

$$I_a = \frac{\omega}{2\pi} \int_0^{2\pi} i_a \, dt = \frac{\omega}{2\pi} \int_{\alpha/\omega}^{\beta/\omega} i_a \, dt = \frac{\gamma}{2\pi} I'_a \tag{6-61}$$

and since the average power input per period of the line voltage is equal to $\gamma/2\pi$ times the average power input during the conduction period,

$$E_a I_a = \frac{\gamma}{2\pi} E'_a I'_a \tag{6-62}$$

Thus (using Eq. 6-61) we have

$$E_a = E'_a \tag{6-63}$$

The relationships given in Art. 5-1 for the magnetic torque T and generated voltage e_a can be expressed in terms of an electromechanical constant K_m for the condition of constant field flux such that $T = K_m i_a$ and $e_a = K_m \omega_m$.
The equation for the mechanical system is

$$T = K_m i_a = T_L + J \frac{d\omega_m}{dt} \tag{6-64}$$

The equation can be integrated over a period of the line voltage

$$K_m \int_0^{2\pi/\omega} i_a \, dt = \int_0^{2\pi/\omega} T_L \, dt + J \int d\omega_m \tag{6-65}$$

to yield
$$K_m I_a = T_L \tag{6-66}$$

Equations 6-60 and 6-66 show that the relationship between the average speed Ω_m and average load torque T_L is given in terms of average motor voltage V_m and average current I_a, just as for a dc motor operated from a fixed dc voltage source. One must be careful in using the equations to predict performance because the average motor voltage V_m is not an independent variable; the firing angle α is usually the independent variable, and the average motor voltage is a function of the angle α and the conduction angle γ. The conduction angle γ changes with the armature current. Example 6-6 will show how the equations are used.

EXAMPLE 6-6

The 1-hp motor of Example 6-5 is operating at 38.4 percent magnetic torque in a half-wave thyristor circuit. The firing angle α is observed to be 90° and the extinction angle β is 210°. The motor constants are $R_a = 7.56 \, \Omega$,

$L_a = 0.55$ H, $K_m = 4.23$ N · m/A or 4.23 V · s/rad. The rms line voltage is 120 V. Find the average speed Ω_m.

Solution

The average motor voltage over the conduction period $\gamma = 120° = 2\pi/3$ is

$$V_m' = \frac{3\sqrt{2}V_0}{2\pi} \int_{\pi/2}^{7\pi/6} \sin \omega t \, d(\omega t) = \frac{-3\sqrt{2}V_0}{2\pi}[\cos \omega t]_{\pi/2}^{7\pi/6}$$

$$= \frac{3\sqrt{2}(120)(0.866)}{2\pi} = 69.6 \text{ V}$$

The average armature current for 38.4 percent of rated torque is

$$I_a = \frac{T_L}{K_m} = \frac{7.80}{4.23} = 1.84 \text{ A}$$

The average armature current over the conduction period is

$$I_a' = 3I_a = 5.52 \text{ A}$$

The average speed Ω_m is given by Eq. 6-59 as

$$\Omega_m = \frac{E_a'}{K_m} = \frac{V_m' - I_a' R_a}{K_m} = \frac{69.6 - 5.52(7.56)}{4.23} = 6.6 \text{ rad/s}$$

$$\text{Average speed} = \frac{6.6 \text{ rad}}{1 \text{ s}} \frac{1 \text{ r}}{2\pi \text{ rad}} \frac{60 \text{ s}}{1 \text{ min}} = 63 \text{ r/min}$$

Using the results of Example 6-5, we see that the speed dip for a coasting period of about $3\pi/2$ rad (1.5 times that of Example 6-5) and 38.4 percent torque is $0.384(23.8)(1.5) = 13.7$ r/min. This operating point at 63 and 13.7 r/min speed dip is obviously not one that yields smooth steady-speed operation.

The half-wave drive system is inexpensive because it uses only a single thyristor, but it has these disadvantages. (1) The armature current flows in relatively short pulses, one pulse per cycle of line voltage. The current has a high rms-to-average ratio, so that the armature heating for the same torque is higher than for continuous current. The motor must be either force-cooled to obtain its rated horsepower or derated. (2) The motor coasts between conduction periods; at high torques and low speeds the speed fluctuation is very pronounced. (3) The half-wave circuits all introduce a dc component into the supply line which can saturate supply transformers and cause other difficulties. One solution is to use single-phase full-wave circuits, which produce two current pulses per cycle, thus reducing the coasting time, speed dip, and the rms-to-average ratio of the current. Another solution is to use 3-phase circuits.

c. Three-Phase Half-Wave Thyristor-Drive System

Three-phase supply voltages are generally used for dc motor-drive systems of about 5 hp and larger. Three-phase rectifier circuits provide more voltage pulses per cycle of line frequency, thus assuring armature current over a larger portion of the cycle, increasing the ratio of average-to-rms current and thereby reducing the heating of the armature. Furthermore, the power is taken from a 3-phase system which generally has more capability of supplying the power than a single-phase system. Three configurations of thyristors and diodes are used for such drives: 3-phase bridge using six thyristors; 3-phase incomplete bridge using three thyristors and three diodes; and 3-phase half-wave circuit using three thyristors. The half-wave circuit is described in Art. 6-5d for resistive loads.

The diagram for the half-wave drive circuit is shown in Fig. 6-29. It consists of three thyristors connected so that when I is fired, a segment of voltage v_{an} is applied to the motor; when II is fired, a segment of voltage v_{bn} is applied; and when III is fired, a segment of voltage v_{cn} is applied. Two thyristors cannot in general conduct simultaneously because the one connected to the instantaneously highest line voltage will apply a negative anode-to-cathode voltage to the other and turn it off. However, at the instant a thyristor is gated on in normal operation, there is a short interval when two thyristors conduct simultaneously to allow the current in the inductance associated

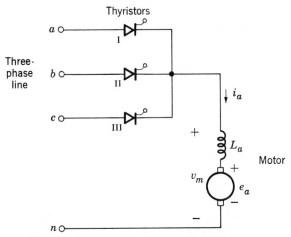

Fig. 6-29. Circuit diagram for 3-phase half-wave drive system.

with each thyristor branch to decline to zero. This interval is called the *commutation interval* and is treated in detail in texts on rectifier circuits.

A pair of 3-phase half-wave circuits is frequently used for a reversing drive; one set of thyristors is connected to apply positive voltage to the motor; the other set is connected to apply negative voltage. By allocating the gating signals for the thyristors speed control of the motor in either direction is obtained.

d. Choppers

A special class of solid-state circuits has been developed for controlling dc motors which are supplied from fixed voltage sources. These circuits are used where the source is a battery, as in industrial electrically powered trucks, and where it is a third rail or overhead trolley wire, as in rapid-transit cars. The circuits, called *choppers,* are used to replace switched series armature-circuit resistors; the advantage is higher efficiency, continuous control, and the ability to operate the motor in a regenerative braking mode.† A simplified diagram of the chopper circuit with a series motor is shown in Fig. 6-30. The chopper operates from a fixed dc voltage V_0 and controls the average motor voltage V_m from zero to V_0. The thyristor acts as a switch which is closed and opened at the rate of several hundred hertz. The relative on-to-off time of the thyristor determines the average motor voltage. The thyristor is unable to turn itself off when carrying current; it requires a commutating circuit which impresses a negative voltage on the thyristor for a short period of typically $40\,\mu s$ to turn it off. The commutating circuit is represented by a switch.

The waveforms of the circuit are shown in Figs. 6-31 and 6-32. In Fig. 6-31 the chopper is operating at about $0.2V_0$ motor voltage. When the thyristor is turned on by a gating signal at $t = 0$, the armature current i_a is delivered from the battery and rises as the circuit inductance absorbs the volt-time area of the difference between V_0 and the armature emf e_a. When the thyristor is

†S. B. Dewan and A. Straughen, "Power Semiconductor Circuits," Wiley, New York, 1975, chap. 6.

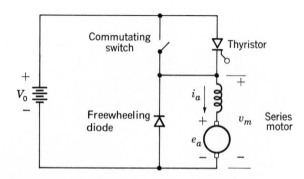

Fig. 6-30. Simplified circuit of chopper speed-control system.

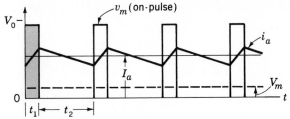

Fig. 6-31. Waveforms of motor voltage and current at low speed.

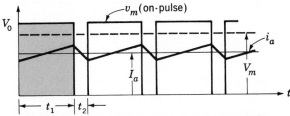

Fig. 6-32. Waveforms of motor voltage and current at high speed.

turned off after the time t_1, the armature current declines through the freewheeling diode as the energy stored in the circuit inductance is applied to the armature. The diode serves as a path to maintain armature current which can no longer flow through the external circuit. As the ratio of on-to-off time t_1/t_2 is increased, the average motor voltage V_m rises, as shown in Fig. 6-32.

The chopper circuit requires inductance to store energy. Usually a series motor is employed, and the field winding serves as the inductance. However, shunt motors can be employed with external inductors and the normal field winding supplied from the dc source. The chopper can be controlled by pulse width or pulse frequency. The motor operates as though it were subjected to voltage control, not to resistance control, because the average motor voltage V_m is independent of the armature current.

EXAMPLE 6-7

A 100-hp series motor rated 180 A is operating in a chopper circuit from a 500-V dc source. The armature and field inductance is 0.060 H. At the minimum ratio $t_1/(t_1 + t_2)$ of 0.20, as shown in Fig. 6-31, find the pulse frequency to limit the amplitude of armature-current excursion to 10 A.

Solution

For a pulse ratio of 0.20 the average armature voltage is

$$0.2(500) = 100 \text{ V}$$

The volt-time area applied to the inductance is $(500 - 100)t_1$ V · s. Thus the rise of current is

$$\Delta i_a = \frac{400t_1}{0.060} = 10 \text{ A} \qquad t_1 = \frac{0.6}{400} = 1.5 \text{ ms}$$

$$t_1 + t_2 = \frac{1.5}{0.2} = 7.5 \text{ ms}$$

$$\text{Pulse frequency} = \frac{1}{7.5 \times 10^{-3}} = 133 \text{ pulses per second}$$

6-7 AN ELEMENTARY MOTOR–SPEED REGULATOR

When high precision and freedom from the effects of external disturbances are important specifications, a feedback control system must be employed. The theory of feedback control has been developed to a high degree of accuracy, and numerous textbooks and technical papers are available covering the broad subject.† We shall assume that the reader is familiar with the basic concepts. An example of a speed-control system will be discussed in this article.

The schematic diagram of a speed-control system using a separately ex-cited dc motor is shown in Fig. 6-33a. The motor speed is measured by means of a dc tachometer generator and its voltage e_t compared with a reference voltage E_R. The error voltage ϵ is amplified and controls the output voltage of the power-conversion equipment, so as to maintain substantially constant speed at the value set by the reference voltage.

The details of the error-processing amplifier A and the power-conversion equipment P depend on the type of system. For example, P may be a solid-state controlled rectifier, one of the types described in Art. 6-6, and A may be a phase-shifting circuit for controlling the firing angle of the rectifiers. Or P may be a motor-generator set with A an amplifier controlling the field current of the dc generator, as in the Ward Leonard system described in Art. 6-1. We shall assume that the combination of A and P is equivalent to a linear con-trolled voltage source $v_s = K_A \epsilon$ with negligible time lag and gain K_A. (With solid-state rectifiers the time lags are about $\frac{1}{2}$ cycle of the ac supply and will be neglected. With the Ward Leonard system the time lag in the dc generator field may be significant, however.) We shall assume that the load torque T_L is independent of the speed; i.e., the damping B/J is zero.

---

†J. J. D'Azzo and C. H. Houpis, "Linear Control System Analysis and Design," 2d ed., McGraw-Hill, New York, 1981; R. C. Dorf, "Modern Control Systems," 3d ed., Addison-Wesley, Reading, Mass., 1980; B. C. Kuo, "Automatic Control Systems," 4th ed., Prentice-Hall, Englewood Cliffs, N.J., 1982.

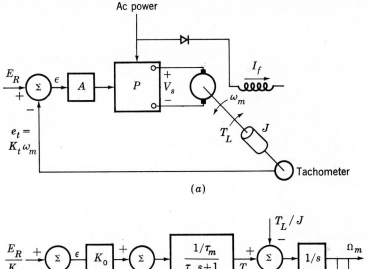

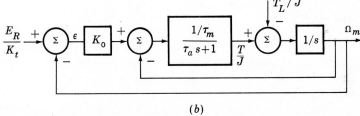

Fig. 6-33. Feedback speed-control system: (a) schematic diagram and (b) block diagram.

The block diagram is shown in Fig. 6-33b, where E_R/K_t is the steady-state no-load speed setting, K_t is the tachometer speed-voltage constant, and

$$K_0 = \frac{K_t K_A}{K_m} \tag{6-67}$$

This block diagram is found by addition of the tachometer feedback to the block diagram of the motor, Fig. 6-12b. The response to changes in reference voltage E_R and load torque T_L will now be investigated.

With $T_L = 0$, reduction of the block diagram gives the transfer function

$$\frac{\Omega_m}{E_R} = \frac{K_0}{K_t} \frac{1}{\tau_m s(\tau_a s + 1) + 1 + K_0} \tag{6-68}$$

Similarly, with $E_R = 0$

$$\frac{\Omega_m}{T_L} = -\frac{R_a}{K_m^2} \frac{\tau_a s + 1}{\tau_m s(\tau_a s + 1) + 1 + K_0} \tag{6-69}$$

The natural frequencies s_1, s_2 of the closed-loop system are given by the poles of the transfer function. The undamped natural frequency ω_n is

$$\omega_n = \sqrt{\frac{1 + K_0}{\tau_a \tau_m}} \tag{6-70}$$

and the damping factor α is

$$\alpha = \frac{1}{2\tau_a} \tag{6-71}$$

The damping ratio ζ is

$$\zeta = \frac{\alpha}{\omega_n} = \frac{1}{2}\sqrt{\frac{\tau_m}{\tau_a}\frac{1}{1 + K_0}} \tag{6-72}$$

The natural frequencies s_1 and s_2 are then given by Eq. 6-36. If the armature-circuit inductance is neglected, the response reduces to a single natural frequency $s_1 = -(1 + K_0)/\tau_m$ describing a single exponential term with a time constant $\tau'_m = \tau_m/(1 + K_0)$.

Examination of the above equations shows several important facts concerning the transient behavior of the system and its design for satisfactory performance. For a step input ΔE_R the final steady-state response $\Delta\omega_m(\infty)$ is, from Eq. 6-68 with $s = 0$,

$$\frac{\Delta\omega_m(\infty)}{\Delta E_R} = \frac{1}{K_t}\frac{K_0}{1 + K_0} \tag{6-73}$$

For initial rest conditions the step-input response is shown in normalized form by the curves in Fig. 6-13. For a step input ΔT_L of load torque, from Eq. 6-69,

$$\frac{\Delta\omega_m(\infty)}{\Delta T_L} = -\frac{R_a}{K_m^2}\frac{1}{1 + K_0} \tag{6-74}$$

For best performance, the system should be insensitive to load-torque disturbances. This requirement means that R_a should be as small as possible and K_m as large as possible; i.e., the motor should be operated at the maximum permissible flux. Also, K_0 should be as large as possible. Recall that K_0 is proportional to the amplifier gain. Increasing the amplifier gain therefore stiffens the system against the effect of load disturbances.

Increasing the amplifier gain has undesirable effects on the dynamic behavior, however. From Eqs. 6-70 and 6-72, increasing the amplifier gain increases the natural frequency ω_n and decreases the relative damping factor ζ;

that is, the system oscillates rapidly and through wide extremes. The relative stability is poor. The components may wear out rapidly. In more complicated systems with three or more time lags, too much gain may lead to absolute instability, i.e., exponentially increasing oscillations. In practice it has been found that amplifier gains resulting in damping ratios lying in the range between 0.4 and 0.7 usually give satisfactory results. If straight amplification will not give a satisfactory system, various forms of compensation can be added. The techniques of compensation by use of corrective networks are discussed in texts devoted to feedback theory.

EXAMPLE 6-8

A 5-hp 240-V 1750-r/min dc motor is used in the speed-control system shown in Fig. 6-33a. The armature is supplied from a solid-state controlled rectifier. The armature-circuit resistance and inductance including the rectifier are

$$R_a = 1.20 \ \Omega \qquad L_a = 0.010 \ \text{H}$$

The motor field is supplied with constant field current provided by a separate rectifier. The speed-voltage constant of the motor is

$$K_m = 1.21 \ \text{V} \cdot \text{s/rad}$$

The motor + load inertia is

$$J = 0.068 + 0.140 = 0.208 \ \text{kg} \cdot \text{m}^2$$

The tachometer speed-voltage constant is

$$K_t = 0.1 \ \text{V/(r/min)} \qquad \text{or} \qquad 0.96 \ \text{V} \cdot \text{s/rad}$$

The voltage gain of the error-detector amplifier and rectifier is

$$K_A = 10 \ \text{V/V}$$

The reference voltage E_R is adjusted for a no-load speed of 1800 r/min.

    (a) Solve for E_R.
    (b) Find the steady-state speed drop resulting from an applied torque of 20 N · m (approximately rated load).
    (c) Find ω_n, α, and ζ. Comment on the system performance.

Work in SI units. Neglect no-load rotational losses.

Solution

(a) $1800 \, \text{r/min} = 188 \, \text{rad/s}$

$$\text{Motor speed voltage } E_a = K_m \omega_m = 1.21(188) = 228 \text{ V}$$
$$\text{Input to amplifier} = E_a/K_A = 22.8 \text{ V}$$
$$\text{Reference voltage } E_R = 180 + 22.8 = 202.8 \text{ V}$$

(b) From Eq. 6-67

$$K_0 = \frac{0.96(10)}{1.21} = 7.9$$

and from Eq. 6-74

$$\frac{\Delta\omega_m(\infty)}{\Delta T_L} = -\frac{1.20}{(1.21)^2}\frac{1}{8.9} = -0.092$$

Speed drop $= 20(0.092) = 1.84 \, \text{rad/s} = 1.0\%$ of no-load speed

(c) The time constants are

$$\tau_a = \frac{0.010}{1.20} = 0.00833 \text{ s}$$

$$\frac{R_a}{K_m^2} = \frac{1.20}{(1.21)^2} = 0.82 \qquad \tau_m = 0.208(0.82) = 0.171 \text{ s}$$

From Eqs. 6-70 to 6-72

$$\omega_n = \sqrt{\frac{8.9}{0.0083(0.171)}} = 79 \text{ rad/s}$$

$$\alpha = \frac{1}{2(0.00833)} = 60 \text{ rad/s} \qquad \zeta = \frac{60}{79} = 0.76$$

The steady-state speed regulation and the damping are within satisfactory limits for most industrial applications.

Many industrial applications of adjustable-speed drives require the coordinated control of several motors in tandem drives for processing continuous strips of material passed through a succession of rollers. Papermaking machines and some steel-mill drives are examples. These drives require careful coordination of the motor controls to avoid a disastrous tug-of-war between the individual drive motors.†

· The transient analysis of dynamic systems in terms of the time response to disturbances shows what may happen to a system when it is subjected to disturbances such as step changes in inputs. However, the transient analysis becomes unwieldy when the system contains more than two energy-storage elements. Dynamic analysis by frequency-response methods is an important complement to transient analysis and is treated thoroughly in texts devoted to feedback theory.

6–8 METADYNES AND AMPLIDYNES

So far, we have considered dc machines with brushes located only in the quadrature axis. The purpose of this article is to examine the effects of additional brushes located in the direct axis. By these means the armature mmf can be used to provide most of the excitation, and high-power gains can be achieved. Machines with more than two brush sets per pair of poles are called *metadynes*. This article is concerned with metadyne generators, with emphasis on the most common form, the amplidyne.‡

a. Basic Metadyne Generators

A modification of the basic dc machine is shown in Fig. 6-34. The stator has a control-field winding f on the direct axis. Brushes qq' are located on the commutator so that commutation takes place along the quadrature axis, as in the normal dc generator. With the generator driven at constant speed ω_{m0} and with magnetic saturation neglected, the voltage e_{aq} generated in the armature

---

†W. K. Boice, Controlling Speed in Multidrive Systems, *Mach. Des.* **42**(2):130–134 (1970).
‡For a discussion of the steady-state theory and descriptions of a number of applications, see J. M. Pestarini, "Metadyne Statics," MIT-Wiley, New York, 1952. For discussions of the transient theory, see M. Riaz, Transient Analysis of the Metadyne Generator, *Trans. AIEE,* **72**(III):52–62 (1953); K. A. Fegley, Metadyne Transients, *Trans. AIEE,* **74**(III):1179–1188 (1955).

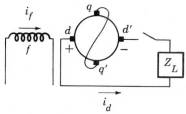

Fig. 6-34. Basic metadyne.

between the quadrature-axis brushes is

$$e_{aq} = K_{qf}i_f \tag{6-75}$$

where K_{qf} is a constant and i_f is the field current.

Now reduce the field current to a small value and short-circuit the quadrature-axis brushes, as shown in Fig. 6-34. Since the impedance of the short-circuited armature is small, a weak control-field current will produce a relatively much larger quadrature-axis armature current and a corresponding flux-density wave centered on the quadrature axis. By commutator action this magnetic field is stationary in space. Its effect is similar to that of a fictitious stator winding on the quadrature axis.

If brushes dd' are now placed on the commutator in the direct axis, as shown in Fig. 6-34, an emf e_{ad} generated in the armature by its rotation in the quadrature-axis flux will appear across these brushes. With the continued assumption of constant speed and negligible saturation

$$e_{ad} = K_{dq}i_q \tag{6-76}$$

where i_q is the quadrature-axis armature current and K_{dq} is a constant.

Now connect a load Z_L to the direct-axis brushes. The direct-axis armature current i_d produces an mmf which *opposes* the control-field mmf. Each stage of voltage generation results in a current whose magnetic field is spatially 90° ahead of the flux wave producing the voltage. With two stages of voltage generation, the mmf of the direct-axis output current is shifted 90° twice and therefore opposes the original field excitation. The quadrature-axis-generated emf now is

$$e_{aq} = K_{qf}i_f - K_{qd}i_d \tag{6-77}$$

where K_{qd} is a constant under the assumed conditions of constant speed and negligible saturation.

The metadyne generator of Fig. 6-34 is therefore a two-stage power amplifier with strong negative current feedback from the final output stage to the input. For a fixed value of field current it maintains very nearly constant output current i_d over a wide range of load impedance. Its power amplification, however, is reduced by the effect of the negative feedback.

b. Amplidynes

The commonest version of the metadyne, the *amplidyne,* consists of the basic metadyne generator plus a cumulative winding on the direct axis connected in series with the direct-axis load current, as shown by the winding labeled

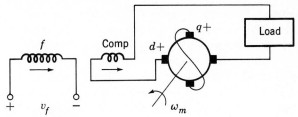

Fig. 6-35. Basic amplidyne.

"Comp" in the schematic diagram of Fig. 6-35. This winding, called a *compensating winding,* is carefully designed to provide a flux as nearly as possible equal and opposite to the flux produced by the direct-axis armature current. The negative-feedback effect of the load current is thereby canceled, and the control-field winding has almost complete control over the direct-axis flux. Very little control-field power input is required to produce a large current in the short-circuited quadrature axis of the armature. The quadrature-axis current then produces the principal magnetic field. The power required to sustain the quadrature-axis current and the load is supplied mechanically by the motor driving the amplidyne. Power amplification of the order of 20,000 to 1 can easily be obtained. This power amplification may be compared with values in the range from about 20 to 1 and 100 to 1 for conventional generators.

If we assume perfect compensation, negligible saturation, and constant speed, the transfer function relating direct-axis generated emf E_{ad} to control-field applied voltage V_f is

$$\frac{E_{ad}}{V_f} = \left(\frac{K_{qf}/R_f}{\tau_f s + 1}\right)\left(\frac{K_{dq}/R_{aq}}{\tau_{aq} s + 1}\right) \qquad (6\text{-}78)$$

where R_f and τ_f are the resistance and time constant of the control field and R_{aq} and τ_{aq} are the resistance and time constant of the quadrature-axis armature circuit. The resistance of the direct-axis armature circuit is considered to be lumped with the load. The direct-axis armature inductance usually is neglected because the compensating winding very nearly eliminates the flux produced by the direct-axis armature current. The principal time lag is that produced by the quadrature-axis time constant τ_{aq} and is in the range from 0.02 to 0.25 s.

Various auxiliary or control-field windings can be added to either axis of an amplidyne to improve performance characteristics. For example, a cumulative series field can be wound on the quadrature axis and connected in series with the quadrature-axis current. This field decreases the quadrature-axis current for a specified direct-axis voltage output. Quadrature-axis commutation is thereby improved.

Amplidynes are used to provide the power amplification in a variety of feedback control systems requiring controlled power output in the range from about 1 to 50 kW. For example, they are used as the voltage-regulating unit in the excitation systems of large ac generators to insert a buck-or-boost voltage in series with the field winding of the main exciter.† Or the main exciter may be an amplidyne when the excitation requirements are within the range where amplidynes are competitive with other types of excitation systems. An amplidyne can be used as the generator in a Ward Leonard speed or position control system if the power requirements of the regulated motor do not exceed a few kilowatts. Position-control servomechanisms are treated in textbooks devoted to feedback theory.

6-9 SUMMARY

In this chapter models are developed describing the electromechanical dynamic behavior of dc machines. From these models we wish to find the dynamic performance not only of the machine itself as either a generator or motor but also of combinations of dc machines and simple control elements.

For the basic dc machine the dynamic equations are relatively simple and are easily established. They are given in Art. 6-2, the basic one in the chapter. The articles which follow it are devoted to examples of applying the equations, to the development of techniques for solving them, and to illustrating the adaptability and versatility of dc machines as control devices.

When the dc machine comprises a single field winding and an armature circuit, the basic equations show that there are three sources of time lag in the dynamic response of the machine. One is created by the field inductance, one by the armature-circuit inductance, and one by the mechanical equipment connected to the machine shaft, including the inertia of the armature itself. Analysis can often be simplified, however, by virtue of one or more time constants being small compared with others. Such simplification is of special value when one is concerned with analysis of a system of machines rather than a single machine.

Since the more important and more interesting problems do deal with systems of machines, we have illustrated the use of several techniques for expediting their analysis. A common system of machines for wide-range and precise control of speed is based on a dc motor whose armature is supplied from an adjustable-voltage solid-state rectifier or dc generator (the so-called Ward Leonard system). Among the techniques of analysis, in addition to the classical solution of differential equations, are the use of block diagrams, transfer functions, equivalent circuits, and frequency-response methods. These techniques, of course, are broadly applicable to system problems in general.

One result of using these techniques is to view the machine as more than

†H. C. Barnes, J. A. Oliver, A. S. Rubenstein, and M. Temoshok, Alternator-Rectifier Exciter for Cardinal Plant 724-MVA Generator, *IEEE Trans. Power Appar. Syst.*, **87**(4):1189–1198 (1968).

simply a brute-force energy-conversion device. The separately excited generator, for example, is looked upon as a power amplifier—a gain and one or more
time constants, together with an increase in power level. The shunt generator
is seen to be similar to a feedback oscillator. From such broader viewpoints
one can more fully assess the control possibilities of more complex dc machines. The additions to the basic machine may include brushes in the direct
as well as in the quadrature axis (as in the amplidyne and metadyne). When
combined with feedback through external circuits, much can be done to tailor
the system characteristics to meet performance specifications. Ideally, the
objectives are to increase the sensitivity or gain, to decrease the effective time
constants or response time, and to decrease the sensitivity of the system to
uncontrolled external disturbances. Not all these ideals are compatible. For
example, the addition of negative feedback to a system has the beneficial
effect of decreasing the response time and stiffening the system against the
effects of disturbances; but the gain is decreased. To obtain the same power
output with the same control power input, the power amplification must accordingly be increased. The cost of equipment with the increased power amplification must be balanced against the improvements in system performance.

Throughout all these analyses, then, it should always be borne in mind
that the machines must have adequate capability to handle the voltage, current, and power surges demanded by the control signals. In other words, the
signal-flow characteristics of the ideal machine must be coordinated with the
limitations imposed by the properties of the materials making up the realistic
machine and the cost of equipment to give the required power amplification.

DC machines have seen widespread use in electromechanical systems
because of the relative ease with which their dynamics can be controlled by
variable levels of dc voltage applied to their armature and/or field terminals.
However, solid-state technology now permits the generation of variable-frequency ac voltages of significant power level, resulting in the use of synchronous and induction machines in applications once considered almost exclu-
sively the domain of dc machines. Here there is a tradeoff between machine
and drive-system cost and simplicity. In any event, the analysis techniques
developed in this chapter can readily be modified to apply to these new situations, although this fact will not be stressed in this book.

PROBLEMS

6-1 (*a*) A 230-V dc shunt-wound motor is used as an adjustable-speed drive
over the range from 0 to 1000 r/min. Speeds from 0 to 500 r/min are
obtained by adjusting the armature terminal voltage from 0 to 230 V
with the field current kept constant. Speeds from 500 to 1000 r/min
are obtained by decreasing the field current with the armature terminal voltage maintained at 230 V. Over the entire speed range the

torque required by the load remains constant. Show the general form of the curve of armature current versus speed over the entire range. Ignore machine losses and armature-reaction effects.

(b) Suppose that, instead of keeping the load torque constant, the armature current is not to exceed a specified value. Show the general form of the curve of allowable load torque versus speed. Conditions otherwise areas in part (a).

6-2 Two adjustable-speed dc shunt motors have maximum speeds of 1650 r/min and minimum speeds of 450 r/min. Speed adjustment is obtained by field-rheostat control. Motor A drives a load requiring constant horsepower over the speed range; motor B drives one requiring constant torque. All losses and armature reaction may be neglected.

(a) If the horsepower outputs are equal at 1650 r/min and the armature currents are each 100 A, what will the armature currents be at 450 r/min?

(b) If the horsepower outputs are equal at 450 r/min and the armature currents are each 100 A, what will the armature currents be at 1650 r/min?

(c) Answer parts (a) and (b) for speed adjustment by armature-voltage control with conditions otherwise the same.

6-3 Consider a dc shunt motor connected to constant-voltage mains and driving a load requiring constant electromagnetic torque. Show that if $E_a > 0.5 V_t$ (the normal situation), increasing the resultant air-gap flux decreases the speed, whereas if $E_a < 0.5 V_t$ (as might be brought about by inserting a relatively high resistance in series with the armature), increasing the resultant air-gap flux increases the speed.

6-4 Two identical 5-hp 230-V 17-A dc shunt machines are to be used as the generator and motor, respectively, in a Ward Leonard system. The generator is driven by a synchronous motor whose speed is constant at 1200 r/min. The armature-circuit resistance of each machine is 0.47 Ω (including brushes). Armature reaction is negligible. Data for the magnetization curve of each machine at 1200 r/min are as follows:

I_f, A	0.2	0.4	0.6	0.8	1.0	1.2
E_a, V	108	183	230	254	267	276

(a) Compute the maximum and minimum values of generator-field current needed to give the motor a speed range from 300 to 1500 r/min at full-load armature current (17.0 A) with the motor-field current held constant at 0.50 A.

(b) Compute the speed regulation of the motor for the conditions of maximum speed and minimum speed found in part (a).

(c) Compute the maximum motor speed obtainable at full-load armature current if the motor-field current is reduced to 0.20 A and the generator-field current is not allowed to exceed 1.10 A.

6-5 One of the commonest industrial applications of dc series motors is for crane and hoist drives. This problem relates to the computation of selected motor performance characteristics for such a drive. The specific motor concerned is a series-wound 230-V totally enclosed motor having a $\frac{1}{2}$-h crane rating of 65 hp with a 75°C temperature rise. The performance characteristics of the motor alone on 230 V as taken from the manufacturer's catalog are listed in Table 6-1. The resistance of the armature (including brushes) plus commutating field is 0.090 Ω, and that of the series-field winding is 0.040 Ω. Armature reaction should be ignored.

TABLE 6-1

Line current, A	Shaft torque, lb·ft	Speed, r/min
50	80	940
100	210	630
150	380	530
200	545	475
250	730	438
300	910	407
350	1105	385
400	1265	370

The motor is to be connected as in Fig. 6-36a for hoisting and Fig. 6-36b for lowering. The former connection is simply one for series-resistance control. The latter connection is one for lowering by dynamic braking with the field reconnected in shunt and having an adjustable resistance in series with it.

A few samples of the torque-speed curves determining the suitability of the motor and control for its particular application are to be plotted. Plot all these curves on the same sheet, torque horizontally and speed vertically, covering about the torque-magnitude range embraced in Table 6-1. Provide for both positive and negative values of speed, corresponding, respectively, to

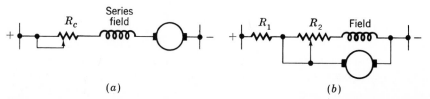

(a)   (b)

Fig. 6-36. Series crane motor, Prob. 6-5: (a) hoisting connection, (b) lowering connection.

hoisting and lowering; provide also for both positive and negative values of torque, corresponding, respectively, to torque in the direction of raising the load and torque in the direction of lowering the load; thus, use all four quadrants of the conventional rectangular coordinate system.

(a) For the hoisting connection, plot torque-speed curves for the control resistor R_c set at 0, 0.65, and 1.30 Ω. If any of these curves extend into the fourth quadrant within the range of torques covered, plot them in that region, and interpret physically what operation there means.

(b) For the lowering connection, plot a torque-speed curve for $R_1 = 0.65$ Ω and R_2 set at 0.65 Ω. The most important portion of this curve is in the fourth quadrant, but if it extends into the third quadrant, that region should also be plotted and interpreted physically.

(c) In part (b) what is the lowering speed corresponding to rated torque?

(d) How is the speed in part (c) affected by decreasing R_2? Why?

(e) How would the speed of part (c) be affected by adding resistance in series with the motor armature? Why?

6-6 An automatic starter is to be designed for a 15-hp 230-V shunt motor. The resistance of the armature circuit is 0.162 Ω. When operated at rated voltage and loaded until its armature current is 32 A, the motor runs at a speed of 1100 r/min with a field-circuit resistance of 115 Ω. When the motor is delivering rated output, the armature current is 56 A. The motor is to be started with a load which requires a torque proportional to speed and which under running conditions requires 15 hp. The field winding is connected across the 230-V mains, and the resistance in series with the armature is to be adjusted automatically so that during the starting period the armature current does not exceed 200 percent of rated value or fall below rated value. That is, the machine is to start with 200 percent of rated armature current, and as soon as the current falls to rated value, sufficient series resistance is to be cut out to restore current to 200 percent. This process is repeated until all the series resistance has been cut out.

(a) What should the total resistance of the starter be?

(b) How much resistance should be cut out at each step in the starting operation?

6-7 Figure 6-37 shows schematically the connections of a Ward Leonard system using a generator with three field windings. The separately excited motor M drives the forward motion of the scoop on a very large power shovel used for open-pit strip mining of coal. The motor has a commutating winding C and a separately excited field winding F_m. Its armature is connected to the armature of a generator G having a commutating winding C and three field windings: a separately excited control field F_1, a self-excited shunt field F_2, and a differential series field S. The purpose of field S is to limit the armature current if the motor should be stalled. The exciter E supplies a constant

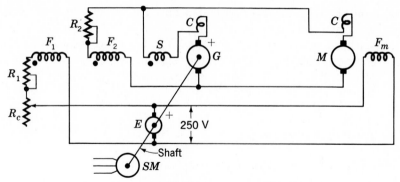

Fig. 6-37. Ward Leonard system with three-field generator, Prob. 6-7.

voltage of 250 V to the motor field F_m and to the generator control field F_1. The generator G and exciter E are driven by a 2300-V 3-phase synchronous motor SM.

Main generator G (rating 500 kW, 500 V):

S: 1 turn per pole, total resistance = 0.001 Ω

F_1: 200 turns per pole, resistance = 25 Ω

F_2: 100 turns per pole, resistance = 12 Ω

R_1 and R_2 are fixed resistors

R_c is the controller

Resistance of armature plus commutating winding = 0.009 Ω

Armature reaction negligible

Data for the generator magnetization curve are as follows:

Field excitation, ampere-turns/pole	500	1000	1500	2000	2500	3000
Generated volts	250	450	540	585	615	640

Motor M (rating 500 V, 1000 A):

F_m: Excited at 250 V

With 500 V applied to motor armature, no-load speed = 600 r/min

Resistance of armature plus commutating winding plus cables = 0.015 Ω

Armature reaction negligible

(a) Find the resistance of R_2 which makes the excitation line of F_2 coincide with the magnetization-curve air-gap line.

(b) Find the resistance of R_1 which limits the stalled torque of the motor of 1.5 per unit with $R_c = 0$.

(c) With the above settings of R_1 and R_2, plot the generator volt-ampere characteristic and the motor speed-torque characteristic with $R_c = 0$. Use per unit values of the variables, with 600 r/min, 500 V, 1000 A, and motor torque at 1000 A as base values.

6-8 A separately excited dc generator has the following constants:

Field-winding resistance $R_f = 100\ \Omega$

Field-winding inductance $L_{ff} = 50\ \mathrm{H}$

Armature resistance $R_a = 0.05\ \Omega$

Armature inductance $L_a = 0.5\ \mathrm{mH}$

Generated emf constant $K_g = 100\ \mathrm{V/field\ ampere}$ at 1200 r/min

The generator is driven at a constant speed of 1200 r/min. Its field and armature circuits are initially open.

(a) At $t = 0$ a constant-voltage source of 250 V is suddenly applied to the terminals of its field winding. Find the equation for the armature terminal voltage as a function of time.

(b) After steady-state conditions have been established in part (a), the armature is suddenly connected to a load of resistance $1.20\ \Omega$ and inductance 1.5 mH in series. Find the equations for (1) the armature current and (2) the armature terminal voltage as functions of time. Include the effect of the armature inductance and resistance.

(c) Find the magnetic torque as a function of time.

6-9 A dc motor M has its armature permanently connected to a source S as shown in Fig. 6-38a. The volt-ampere characteristic of the source is shown in Fig. 6-38b. The motor field f is separately excited from a voltage source E_f as shown.

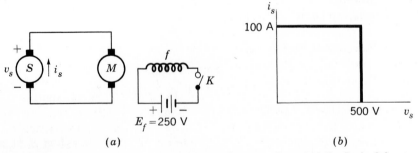

Fig. 6-38. (a) Circuit diagram and (b) idealized source characteristic, Prob. 6-9.

Motor armature resistance $R_a = 0.5\,\Omega$

Motor armature inductance negligible

Motor field resistance $R_f = 50\,\Omega$

Motor field inductance $L_{ff} = 50\,\text{H}$

Motor torque at 5-A field current and 100-A armature current = 200 N · m

Motor load is pure inertia. Total moment of inertia J of load and armature is $10\,\text{kg}\cdot\text{m}^2$

Neglect magnetic-saturation effects in the motor

Motor field switch K closed at $t = 0$

Derive an expression, with numerical values, for the speed in radians per second as a function of time in seconds. Sketch this curve roughly to scale. Indicate on it the final steady-state speed and the speed and time at which the breakpoint in the source characteristic is reached. Compute the time required for the motor to reach approximately 96 percent of its final speed.

6-10 A separately excited dc motor drives a pure-inertia load. The combined inertia of motor and load is J. The armature resistance is R_a. Neglect armature inductance. The motor torque constant is K_m. With the motor initially at standstill and the field excited in the steady state, a constant voltage V_t is suddenly applied to the armature. Find the total heat dissipated in the motor armature in bringing the motor up to its final steady-state speed. Compare with the kinetic energy stored in the rotating masses.

6-11 A dc shunt motor is driving a pure-inertia load. The armature and field are supplied from a source of constant direct voltage. The motor is initially operating in the steady state. Neglect all rotational losses and armature reaction. Assume that the flux is directly proportional to the field current and that armature inductance is negligible.

The field rheostat is suddenly short-circuited. Develop the differential equation for the speed of the motor following this disturbance, using the ordinary symbols for the various quantities. Indicate all initial conditions. It is not necessary to solve the equation.

6-12 In continuous rolling mills the stands, or rolls, through which the bar passes in the rolling process are arranged in tandem, with the majority of the stands driven by separate motors. It is common to use dc motors supplied with power from one or several generators. The transient changes of the motor speed under suddenly applied loads as the bar enters one stand after another may seriously affect the quality of the product. In particular, the *impact speed drop* which occurs at the maximum of the transient oscillation is of major importance.

Consider a single motor M supplied by a generator G, each with separate and constant field excitation. The internal voltage E of the generator may be considered constant, and the armature reaction of both machines may be considered negligible. With the motor running without external load and the system in the steady state, a bar enters the stand at $t = 0$, causing the load torque to be increased suddenly from zero to T. The following numerical values apply:

Internal voltage E of $G = 387$ V

Motor-plus-generator armature inductance $L = 0.00768$ H

Motor-plus-generator armature resistance $R = 0.0353\ \Omega$

Moment of inertia of motor armature and connected rolls, all referred to motor speed, $J = 42.2\ \text{kg} \cdot \text{m}^2$

Electromechanical conversion constant for motor $K_m = 4.23\ \text{N} \cdot \text{m/A}$

No-load armature current $i_0 = 35$ A

Suddenly applied torque $T = 2040\ \text{N} \cdot \text{m}$

Determine (a) the undamped angular frequency of the transient speed oscillations, (b) the damping ratio of the system, (c) the initial speed in revolutions per minute, (d) the initial acceleration in (r/min)/s, (e) the ultimate speed drop in revolutions per minute, (f) the impact speed drop in revolutions per minute.

6-13 Figure 6-39 shows a dc generator whose field current is supplied from an exciter. The generator and exciter are driven at constant speed. The machine constants are:

	Exciter	Main generator
Field inductance	$L_1 = 125$ H	$L_2 = 100$ H
Field resistance	$R_1 = 250\ \Omega$	$R_2 = 100\ \Omega$
Generated voltage	1000 V/field ampere	250 V/field ampere
Armature resistance	Negligible	Negligible

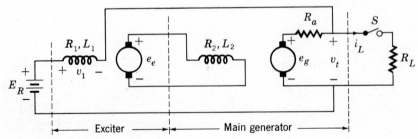

Fig. 6-39. Voltage-regulating system, Prob. 6-13.

Neglect the effect of the exciter-field current on the voltage drop in the armature of the main generator.

(a) With switch S closed the reference voltage E_R is adjusted until $V_t = 250$ V with $R_L = 10\,\Omega$. Find E_R.

(b) With the system in the steady state as in part (a), switch S is opened at $t = 0$. Find $v_t(t)$.

6-14 This problem concerns one of the two motor-generator sets in the Francis Bitter National Magnet Laboratory. Each set consists of the following machines, all mechanically coupled on one shaft:

One 6000-hp 360-r/min synchronous motor

One 600-hp 18-pole wound-rotor induction motor

One 84-ton flywheel

Two dc generators each with the following capability:

	V	kA	r/min
Continuous	250	10	360
5-s pulse	200	40	385/300

The dc generators are used to supply direct current to water-cooled air-core magnets for experimental work in very high magnetic fields. The dc generators can be connected in parallel to supply 20 kA at 250 V continuously, or a maximum current of 80 kA at 200 V for 5 s.

Three-phase power is supplied to the drive motors at 4160 V, 60 Hz. The induction motor is used to start the set. The starting current is held constant at 90 A by an amplidyne servosystem which adjusts liquid resistors in the rotor circuits. It takes about 15 min to get up to 360 r/min. For continuous loads the synchronous motor is then synchronized and the induction motor is disconnected. For pulsed loads the synchronous motor is disconnected, the induction motor drives the set at 385 r/min initially, and most of the energy is supplied by the flywheel. The dc generators have interpoles, pole-face compensating windings, and series fields to improve commutation, speed of response, and load sharing.

The excitation system consists of an exciter motor-generator set comprising a 200-hp induction motor driving two 75-kW 250-V fast-response amplidynes for excitation for the two main dc generator fields. The amplidynes are controlled by feedback amplifiers to regulate the main generator outputs.

With the main generators driven at 360 r/min and no load, the response of the control system is as shown in Fig. 6-40, where v_f is the voltage applied to the main-generator field terminals in per unit. Here 1.0 per unit is the field voltage that will result in 250 V generated emf in the main generator at 360 r/min, no load.

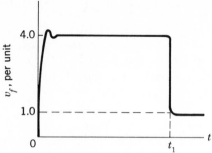

Fig. 6-40. Excitation system response, Prob. 6-14.

Estimated data are as follows:

Moment of inertia $J = 4 \times 10^5$ kg · m²

Main-generator field time constant $\tau_f = 1.3$ s

Main-generator armature-circuit resistance (including all series fields) = 0.02 per unit on continuous rating

(a) Estimate the time t_1 (Fig. 6-40) for the generator output voltage to reach 250 V at 360 r/min, no load. Sketch the curve.

(b) With the two generators in parallel supplying the magnet, estimate the speed at the end of a 5-s 80-kA pulse.

(c) With the load of part (b) estimate the net field excitation in per unit required to maintain 200 V at 80-kA output.

6-15 The power source in Fig. 6-22a is an inverter which delivers a square-wave voltage of amplitude 100 V at 200 Hz. Find the relationship of average load voltage V_n to firing angle α.

6-16 The load in the circuit of Prob. 6-15 is replaced with a series-connected combination of $R = 10\,\Omega$ and $L = 25$ mH. The thyristor is fired at $\alpha = 10°$. Find an expression for the current $i_n(t)$ over one cycle.

6-17 The motor of Example 6-5 is operated with a mechanical load that imposes rated torque at rated speed and a load inertia of 0.10 lb · ft · s². The motor is supplied from a 60-Hz single-phase full-wave rectifier. The coasting period is estimated to be $\pi/2$ rad between conduction periods. Find the speed dip between conduction periods.

6-18 The 100-hp motor of Example 6-7 is operated at a firing angle $\alpha = 0$. Find the average motor speed at no load and full load for a line voltage of 480 V line to line.

6-19 The speed-control system described in Art. 6-7 is used with a 10-hp 1750-r/min 240-V separately excited motor.

Motor + source resistance $R_a = 1.20\ \Omega$

Motor + source inductance $L_a = 0.008\ H$

Motor + load inertia $J = 0.20\ kg \cdot m^2$

Motor speed-voltage constant $K_m = 1.27\ V \cdot s/rad$

Tachometer constant $= 0.1\ V/(r/min)$

The reference voltage E_R is adjusted for 1800 r/min at no load. The voltage gain is adjusted so that the damping ratio ζ is 0.50.

(a) Compute the corresponding values of the amplifier gain K_A, the undamped natural frequency ω_n, and the damping factor α.

With the system initially in the steady state at no load, a step of load torque $T_L = 40\ N \cdot m$ (approximately full load) is suddenly applied to the motor shaft. For (1) the motor speed and (2) the motor armature current find (b) the final steady-state values, (c) the initial values ($t = 0+$), and (d) the initial rates of change.

(e) Find the equations for the speed and the armature current as functions of time. Sketch the curves. Are the normalized curves of Fig. 6-13 applicable? Estimate the minimum speed and the maximum armature current.

6-20 A 2-kW 200-V metadyne generator of the type shown in Fig. 6-34 is driven by a synchronous motor at 1800 r/min and has the following constants:

Control-field resistance $R_f = 20\ \Omega$

Control-field inductance $L_{ff} = 2\ H$

Voltage constant $K_{qf} = 240\ V/field\ amperes$

Armature resistances $R_{aq} = R_{ad} = 4\ \Omega$

Armature inductance $L_{aq} = 1.0\ H$

Voltage constants $K_{dq} = K_{qd} = 60\ V/A$

The metadyne supplies a 20-Ω resistive load at a voltage of 200 V. Find the power input to the control field and the power amplification.

6-21 A compensating winding is added to the metadyne of Prob. 6-20, converting it into an amplidyne. The amplidyne supplies 200 V to a 20-Ω load. Find the power input to the control field, and compare with the result of Prob. 6-20. Neglect the resistance of the compensating winding.

6-22 An amplidyne exciter supplies field current to a 10-kW 125-V dc generator.

Amplidyne data:

Control-field turns $N_c = 400$
Control-field resistance $R_c = 40\,\Omega$
Control-field inductance $L_c = 1.6\,\mathrm{H}$
Quadrature-axis time constant $\tau_q = 0.10\,\mathrm{s}$
Direct-axis generated voltage $= 5.0\,\mathrm{V/control\text{-}field\ A\cdot turn}$
Armature resistance $= 5.0\,\Omega$

Generator data:

Field resistance $R_{fg} = 35\,\Omega$
Field inductance $L_{fg} = 10\,\mathrm{H}$
Generated voltage constant $= 50\,\mathrm{V/field\ ampere}$
Armature resistance $= 0.075\,\Omega$

(a) Compute the open-loop transfer function $G(j\omega)H(j\omega)$ relating generator output voltage at no load to amplidyne control-field voltage.
(b) Compute the complex value of $G(j\omega)H(j\omega)$ at $\omega = 20\,\mathrm{rad/s}$.
(c) The system is now connected as a closed-loop voltage-regulating system. Will the system be stable without any antihunt feedback?
(d) Assume that the system is inherently stable or has been stabilized by an antihunt feedback circuit. What must the value of the constant reference voltage be to give a generator output voltage of 125 V at no load?
(e) With the reference voltage held constant as in part (d) what will the generator terminal voltage be when it is delivering an armature current of 80 A?

6-23 A metadyne having no stator windings is driven at a constant speed. Its armature resistance $R_a = 0.10\,\Omega$ measured between either pair of brushes. Its armature inductance $L_a = 0.01\,\mathrm{H}$ measured between either pair of brushes. A test taken with the direct axis open and a constant voltage of 20 V applied to the quadrature-axis brushes gives a steady-state direct-axis open-circuit voltage of 600 V. A similar test with the direct and quadrature axes interchanged gives similar results. Magnetic saturation is negligible.

This machine is used as a constant-voltage–to–constant-current transformer to supply substantially constant current to a variable-resistance load. The constant source voltage of 600 V is applied to the quadrature-axis brushes, and the resistive load is connected to the direct-axis brushes. Compute the steady-state load current and source current when the load voltage is 600 V.

7

Synchronous Machines:
Steady State

A synchronous machine is an ac machine whose speed under steady-state conditions is proportional to the frequency of the current in its armature. The magnetic field created by the armature currents rotates at the same speed as that created by field current on the rotor (which is rotating at synchronous speed), and a steady torque results. An elementary picture of how a synchronous machine works is given in Art. 3-2a with emphasis on torque production in terms of the interactions between its magnetic fields.

Analytical methods of examining the steady-state performance of polyphase synchronous machines will be developed in this chapter. Initial consideration will be given to cylindrical-rotor machines; the effects of salient poles are taken up in Arts. 7-6 and 7-7.

311

7-1 INTRODUCTION TO POLYPHASE
SYNCHRONOUS MACHINES

As indicated in Art. 3-2a, a synchronous machine is one in which alternating current flows in the armature winding and dc excitation is supplied to the field winding. The armature winding is almost invariably on the stator and is usually a 3-phase winding, as described in Chap. 3. The field winding is on the rotor. The cylindrical-rotor construction shown in Figs. 3-10 and 3-11 is used for 2- and 4-pole turbine generators. The salient-pole construction shown in Fig. 3-9 is best adapted to multipolar slow-speed hydroelectric generators and most synchronous motors. The dc power required for excitation—approximately 1 to a few percent of the rating of the synchronous machine—usually is supplied through slip rings from a dc generator called an *exciter,* which is often mounted on the same shaft as the synchronous machine. Various excitation systems using ac exciters and solid-state rectifiers are used with large turbine generators. One of these systems is described in Art. 4-6.

A single synchronous generator supplying power to an impedance load acts as a voltage source whose frequency is determined by its prime-mover speed, as in Eq. 3-2. The current and power factor are then determined by the generator field excitation and the impedance of the generator and load.

Synchronous generators can readily be operated in parallel, and, in fact, the electricity supply systems of industrialized countries may have scores or even hundreds of alternators operating in parallel, interconnected by hundreds of miles of transmission lines, and supplying electric energy to loads scattered over areas of hundreds of thousands of square miles. These huge systems have grown in spite of the necessity for designing the system so that synchronism will be maintained following disturbances and the problems, both technical and administrative, which must be solved to coordinate the operation of such a complex system of machines and personnel. The principal reasons for these interconnected systems are continuity of service and economies in plant investment and operating costs.

When a synchronous generator is connected to a large interconnected system containing many other synchronous generators, the voltage and frequency at its armature terminals are substantially fixed by the system. As a result, armature currents will produce a component of the air-gap magnetic field which rotates at synchronous speed (Eq. 3-42 or 3-43) as determined by the system frequency. For the production of a steady unidirectional electromagnetic torque the fields of the stator and rotor must rotate at the same speed, and therefore the rotor must turn at precisely synchronous speed. Because any individual generator is a small fraction of the total system generation, it cannot significantly affect the system voltage or frequency. It is often useful for the purposes of analysis to represent the system as a constant-frequency, constant-voltage source known as an *infinite bus.* Many important features of synchronous-machine behavior can be understood from analysis of a single machine connected to an infinite bus.

The steady-state behavior of a synchronous machines can be visualized in terms of the torque equation. From Eq. 3-86, with changes in notation appropriate to synchronous-machine theory,

$$T = \frac{\pi}{2}\left(\frac{\text{poles}}{2}\right)^{2} \Phi_R F_f \sin \delta_{RF} \qquad (7\text{-}1)$$

where Φ_R = resultant air-gap flux per pole
F_f = mmf of dc field winding
δ_{RF} = phase angle between magnetic axes of Φ_R and F_f

The minus sign of Eq. 3-86 has been omitted with the understanding that the electromagnetic torque acts in the direction to bring the interacting fields into alignment. In normal steady-state operation the electromagnetic torque balances the mechanical torque applied to the shaft. In a generator, the prime-mover torque acts in the direction of rotation of the rotor, pushing the rotor mmf wave ahead of the resultant air-gap flux. The electromagnetic torque then opposes rotation. The opposite situation exists in a synchronous motor, where the electromagnetic torque is in the direction of rotation, in opposition to the retarding torque of the mechanical load on the shaft.

Variations in the electromagnetic torque result in corresponding variations in the torque angle δ_{RF}, as seen from Eq. 7-1. The relationship is shown in the form of a torque-angle curve in Fig. 7-1, where field current (rotor mmf) and resultant air-gap flux are assumed constant. Positive values of torque represent generator action, and positive values of δ_{RF} represent angles of lead of the rotor mmf wave with respect to the resultant air-gap flux.

As the prime-mover torque is increased, the magnitude of δ_{RF} must increase until the electromagnetic torque balances the shaft torque. The readjustment process is actually a dynamic one, accompanied by a temporary change in the instantaneous mechanical speed of the rotor and a damped mechanical oscillation, called *hunting,* of the rotor about its new steady-state torque angle. In a practical machine some changes in the amplitudes of the

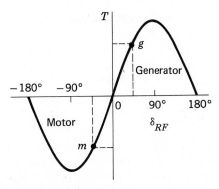

Fig. 7-1. Torque-angle characteristic.

resultant flux-density and mmf waves may also occur because of factors, e.g., saturation and leakage impedance, neglected in the present argument. The adjustment of the rotor to its new angular position following a load change can be observed experimentally in the laboratory by viewing the machine rotor with stroboscopic light having a flashing frequency which causes the rotor to appear stationary when it is turning at its normal synchronous speed. Synchronous-machine transient behavior is discussed in Chap. 8.

As the prime-mover torque is further increased, δ_{RF} increases. When δ_{RF} becomes 90°, the electromagnetic torque reaches its maximum value, known as the *pull-out torque*. Any further increase in prime-mover torque cannot be balanced by a corresponding increase in synchronous electromagnetic torque, with the result that the rotor will speed up and synchronous operation will not be maintained. This phenomenon is known as *losing synchronism* or *pulling out of step*. Under these conditions, the generator is usually disconnected from the line by automatic circuit breakers, and the prime mover is quickly shut down to prevent dangerous overspeed. Note from Eq. 7-1 that the value of pull-out torque can be increased by increasing either the field current or the resultant air-gap flux.

As seen from Fig. 7-1, a similar situation occurs in a synchronous motor for which an increase in the shaft load beyond the pull-out torque will cause the rotor to lose synchronism and thus to slow down.

Since a synchronous motor per se has no net starting torque, means must be provided for bringing it up to synchronous speed by induction-motor action, as described briefly at the end of Art. 9-1.

7-2 SYNCHRONOUS–MACHINE INDUCTANCES; EQUIVALENT CIRCUITS

In Art. 7-1 we described synchronous-machine torque-angle characteristics in terms of the interacting air-gap flux and mmf waves. Our purpose now is to derive an equivalent circuit which represents the steady-state terminal volt-ampere characteristics.

A cross-sectional sketch of a 3-phase cylindrical-rotor synchronous machine is shown schematically in Fig. 7-2. The figure shows a 2-pole machine; alternatively, this can be considered as two poles of a *P*-pole machine. The 3-phase armature winding on the stator is of the same type used in the discussion of rotating magnetic fields in Art. 3-5. The coils *aa′*, *bb′*, and *cc′* represent distributed windings producing sinusoidal mmf and flux-density waves in the air gap. The reference directions for the currents are shown by dots and crosses. The field winding *ff′* on the rotor also represents a distributed winding which produces a sinusoidal mmf and flux-density wave centered on its magnetic axis and rotating with the rotor.

When the flux linkages with the armature phases *a, b, c* and field winding *f* are expressed in terms of the inductances and currents as follows, the in-

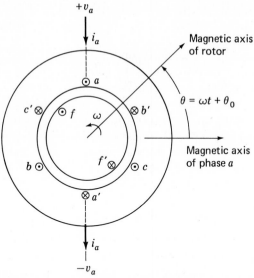

Fig. 7-2. Schematic diagram of a 3-phase cylindrical-rotor synchronous machine.

duced voltages can be found from Faraday's law

$$\lambda_a = \mathcal{L}_{aa}i_a + \mathcal{L}_{ab}i_b + \mathcal{L}_{ac}i_c + \mathcal{L}_{af}i_f \qquad (7\text{-}2)$$
$$\lambda_b = \mathcal{L}_{ba}i_a + \mathcal{L}_{bb}i_b + \mathcal{L}_{bc}i_c + \mathcal{L}_{bf}i_f \qquad (7\text{-}3)$$
$$\lambda_c = \mathcal{L}_{ca}i_a + \mathcal{L}_{cb}i_b + \mathcal{L}_{cc}i_c + \mathcal{L}_{cf}i_f \qquad (7\text{-}4)$$
$$\lambda_f = \mathcal{L}_{fa}i_a + \mathcal{L}_{fb}i_b + \mathcal{L}_{fc}i_c + \mathcal{L}_{ff}i_f \qquad (7\text{-}5)$$

Here two like subscripts denote a self-inductance and two unlike subscripts denote a mutual inductance between the two windings. The script $\mathcal{L}$ is used for inductance in order to conform with the notation in Chap. 8, where the effects of salient poles and transients are analyzed.

Before proceeding farther it is useful to investigate the nature of the various inductances. All these inductances can be expressed in terms of constants which can be computed from design data or measured by tests on an existing machine.

a. Rotor Self-Inductance

With a cylindrical stator, the self-inductance of the field winding is independent of rotor position θ when the harmonic effects of stator slot openings are neglected. Hence

$$\mathcal{L}_{ff} = L_{ff} = L_{ff0} + L_{fl} \qquad (7\text{-}6)$$

where the italic L is used for an inductance which is independent of θ. The component L_{ff0} takes into account the space-fundamental component of air-gap flux. This component can be computed from air-gap dimensions and winding data, as shown in Appendix B. The additional component L_{fl} accounts for field-winding leakage flux.

Under transient or unbalanced conditions, the flux linkages with the field winding, Eq. 7-5, vary with time, and the voltages induced in the rotor circuits have an important effect on machine performance, as shown in Chap. 8. With balanced 3-phase armature currents, however, the constant-amplitude magnetic field of the armature currents rotates in synchronism with the rotor. The flux linkages with the field winding produced by the armature currents do not vary with time, and the voltage induced in the field winding is therefore zero. With constant dc voltage V_f applied to the field-winding terminals the dc field current I_f is given by Ohm's law, $I_f = V_f/R_f$.

b. Stator-to-Rotor Mutual Inductances

The stator-to-rotor mutual inductances vary periodically with the angle θ between the magnetic axes of the field winding and the armature phase a (Fig. 7-2). With space-mmf and air-gap flux distribution assumed sinusoidal, the mutual inductance between the field winding f and phase a varies as $\cos \theta$; thus

$$\mathcal{L}_{af} = \mathcal{L}_{fa} = L_{af} \cos \theta \qquad (7\text{-}7)$$

Similar expressions apply to phases b and c, with θ replaced by $\theta - 120°$ and $\theta + 120°$, respectively. Attention will be focused on phase a. The inductance L_{af} can be calculated as indicated in Appendix B.

With the rotor rotating at synchronous speed ω

$$\theta = \omega t + \theta_0 \qquad (7\text{-}8)$$

where θ_0 is the angle of the rotor at time $t = 0$. With dc excitation I_f in the field winding, the flux linkage λ_{af} of phase a is

$$\lambda_{af} = L_{af} I_f \cos (\omega t + \theta_0) \qquad (7\text{-}9)$$

c. Stator Inductances; Synchronous Inductance

With a cylindrical rotor, the air gap is independent of θ if the effects of rotor slots are neglected. The stator self-inductances then are constant; thus

$$\mathcal{L}_{aa} = \mathcal{L}_{bb} = \mathcal{L}_{cc} = L_{aa} = L_{aa0} + L_{al} \qquad (7\text{-}10)$$

where L_{aa0} is the component of self-inductance due to space-fundamental air-gap flux (Appendix B) and L_{al} is the additional component due to armature-leakage flux (Art. 4-2).

The armature phase-to-phase mutual inductances can be found on the assumption that the mutual inductance is due solely to space-fundamental air-gap flux. Because the armature phases are displaced by 120° and $\cos(\pm120°) = -\frac{1}{2}$, the mutual inductances between armature phases are

$$\mathcal{L}_{ab} = \mathcal{L}_{ba} = \mathcal{L}_{ac} = \mathcal{L}_{ca} = \mathcal{L}_{bc} = \mathcal{L}_{cb} = -\tfrac{1}{2}L_{aa0} \qquad (7\text{-}11)$$

The phase a flux linkages (Eq. 7-2) can be written as

$$\lambda_a = (L_{aa0} + L_{al})i_a - \tfrac{1}{2}L_{aa0}(i_b + i_c) + \lambda_{af} \qquad (7\text{-}12)$$

where λ_{af} is given by Eq. 7-9.

With balanced 3-phase armature currents

$$i_a + i_b + i_c = 0 \qquad \text{or} \qquad i_b + i_c = -i_a \qquad (7\text{-}13)$$

Substitution of Eq. 7-13 in Eq. 7-12 results in

$$\lambda_a = (L_{aa0} + L_{al})i_a + \tfrac{1}{2}L_{aa0}i_a + \lambda_{af} = (\tfrac{3}{2}L_{aa0} + L_{al})i_a + \lambda_{af} \qquad (7\text{-}14)$$

It is useful to define the *synchronous inductance* L_s as

$$L_s = \tfrac{3}{2}L_{aa0} + L_{al} \qquad (7\text{-}15)$$

and thus

$$\lambda_a = L_s i_a + \lambda_{af} \qquad (7\text{-}16)$$

The synchronous inductance L_s is the effective inductance seen by phase a under the balanced 3-phase conditions of normal machine operation. It is made up of three components. The first, L_{aa0}, is due to the space-fundamental air-gap component of phase a self-flux linkages. The second, L_{al}, is due to the leakage component of phase a flux linkages. The third component, $\frac{1}{2}L_{aa0}$, is due to the phase a flux linkages from the space-fundamental component of air-gap flux produced by currents in phases b and c. Under balanced 3-phase conditions, the phase b and c currents are related to the current in phase a by Eq. 7-13; thus the synchronous inductance is an apparent inductance which accounts for the flux linkages of phase a in terms of the current in phase a even though this is not truly the self-inductance of phase a alone.

The significance of the synchronous inductance can be further appreciated with reference to the discussion of rotating magnetic fields in Art. 3-5, where it was shown that under balanced 3-phase conditions, the armature currents create a rotating magnetic flux wave in the air gap $\frac{3}{2}$ times the magnitude of that due to phase a alone, the additional component being due to the phase b and c currents. This corresponds directly to the $\frac{3}{2}L_{aa0}$ component of the synchronous inductance in Eq. 7-15; this component of the synchronous inductance accounts for the total space-fundamental air-gap component of phase a flux linkages produced by the three armature currents.

The phase a terminal voltage is the sum of the armature-resistance voltage drop $R_a i_a$ and the induced voltages. From the derivative of Eq. 7-16

$$V_{ta} = R_a i_a + \frac{d\lambda_a}{dt} = R_a i_a + L_s \frac{di_a}{dt} + \frac{d\lambda_{af}}{dt} \qquad (7\text{-}17)$$

From Eq. 7-9 the corresponding voltage e_{af} is

$$e_{af} = \frac{d\lambda_{af}}{dt} = -\omega L_{af} I_f \sin(\omega t + \theta_0) \qquad (7\text{-}18)$$

It is the voltage generated by the flux produced by the rotating field winding and is defined as the *excitation voltage*. From the trigonometric identity

$$\sin \alpha = -\cos\left(\alpha + \frac{\pi}{2}\right)$$

$$e_{af} = +\omega L_{af} I_f \cos\left(\omega t + \theta_0 + \frac{\pi}{2}\right) \qquad (7\text{-}19)$$

The excitation voltage e_{af} leads the flux linkage λ_{af} by $90°$. Its rms value E_{af} is

$$E_{af} = \frac{\omega L_{af} I_f}{\sqrt{2}} \qquad (7\text{-}20)$$

The equation for the rms voltage generated in the armature phases by a rotating flux wave (Eq. 3-56) is repeated here for convenience; thus

$$E_{af} = 4.44 f k_w N_{ph} \Phi_{af} \qquad (7\text{-}21)$$

where Φ_{af} is the flux per pole produced by the field winding and the other symbols are defined in Chap. 3. It can be shown that Eqs. 7-20 and 7-21 are identical.

In the analysis of steady-state sinusoids it is convenient to use phasor symbolism. Thus Eq. 7-19 describes a voltage whose rms value is given by Eq. 7-20 and which leads the flux linkages $\widehat{\lambda}_{af}$ by $\pi/2 = 90°$. Similarly the phase a terminal voltage $\widehat{V}_{ta}$ and armature current $\widehat{I}_a$ can be represented by phasors. The complex impedance equivalent of Eq. 7-17 then is

$$\widehat{V}_{ta} = R_a \widehat{I}_a + j X_s \widehat{I}_a + \widehat{E}_{af} \qquad (7\text{-}22)$$

where $X_s = \omega L_s$ is the *synchronous reactance*. An equivalent circuit in complex form is shown in Fig. 7-3a. The reader should note that Eq. 7-22 and Fig.

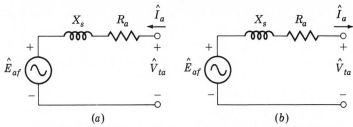

Fig. 7-3. Synchronous-machine equivalent circuits: (a) motor reference direction and (b) generator reference direction.

7-3a assume the motor reference direction for $\hat{I}_a$, with $\hat{I}_a$ defined as positive into the + terminals of the machine. Alternatively, the generator reference direction ($\hat{I}_a$ positive out of the + armature terminals) can be chosen; for this choice Eq. 7-22 becomes

$$\hat{V}_{ta} = -R_a\hat{I}_a - jX_s\hat{I}_a + \hat{E}_{af} \qquad (7\text{-}23)$$

and the equivalent circuit is that of Fig. 7-3b. Although these two representations are equivalent, the generator reference direction is more common and will generally be used from this point on in the text.

Figure 7-4 shows an alternative form of the equivalent circuit in which the synchronous reactance is shown in terms of its components. From Eq. 7-15

$$X_s = \omega L_s = \omega L_{al} + \omega(\tfrac{3}{2}L_{aa0}) = X_{al} + X_A \qquad (7\text{-}24)$$

where $X_{al} = \omega L_{al}$ is the *armature leakage reactance* and $X_A = \omega(\tfrac{3}{2}L_{aa0})$ is reactance corresponding to the rotating space-fundamental air-gap flux produced by the three armature currents. The reactance X_A is defined as the *reactance of armature reaction*. The voltage $\hat{E}_R$ is the internal voltage generated by the resultant air-gap flux and is usually referred to as the *air-gap voltage* or the voltage "behind" leakage reactance. Its rms value E_R is related to the resultant air-gap flux Φ_R, as in Eq. 3-56. As a phasor, $\hat{E}_R$ leads the resultant flux $\hat{\Phi}_R$ by 90°.

It is helpful to have a rough idea of the order of magnitude of the imped-

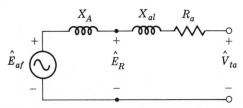

Fig. 7-4. Synchronous-machine equivalent circuit showing air-gap and leakage components of synchronous reactance and air-gap voltage.

ance components. For machines with ratings above a few hundred kilovolt-amperes the armature-resistance voltage drop at rated current usually is less than 0.01 times rated voltage; i.e., the armature resistance usually is less than 0.01 per unit on the machine rating as a base. (The per unit system is described in Art. 1-11.) The armature leakage reactance usually is in the range from 0.1 to 0.2 per unit, and the synchronous reactance is in the vicinity of 1.0 to 2.0 per unit. In general, the per unit armature resistance increases and the per unit synchronous reactance decreases with decreasing size of the machine. In small machines, such as those in educational laboratories, the armature resistance may be in the vicinity of 0.05 per unit and the synchronous reactance in the vicinity of 0.5 per unit. In all but small machines the armature resistance usually is neglected except insofar as its effect on losses and heating is concerned.

7–3 OPEN– AND SHORT–CIRCUIT CHARACTERISTICS

Two basic sets of characteristic curves for a synchronous machine are involved in the inclusion of saturation effects and in the determination of the appropriate machine constants. These sets are discussed here. Except for a few remarks on the degree of validity of certain assumptions, the discussions apply to both cylindrical-rotor and salient-pole machines.

a. Open-Circuit Characteristic and No-Load Rotational Losses

Like the magnetization curve for a dc machine, the open-circuit characteristic of a synchronous machine is a curve of the armature terminal voltage on open circuit as a function of the field excitation when the machine is running at synchronous speed, as shown by the curve *occ* in Fig. 7-5*a*. The curve is often plotted in per unit terms, as in Fig. 7-5*b*, where unity voltage is the rated

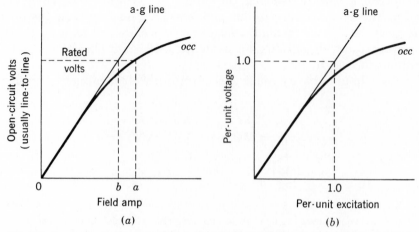

Fig. 7-5. Open-circuit characteristic: (*a*) in terms of volts and field amperes and (*b*) in per unit.

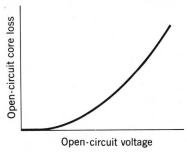

Fig. 7-6. Open-circuit core-loss curve.

voltage and unity field current is the excitation corresponding to rated voltage on the air-gap line. Essentially, the open-circuit characteristic represents the relation between the space-fundamental component of the air-gap flux and the mmf on the magnetic circuit when the field winding constitutes the only mmf source. When the machine is an existing one, the open-circuit characteristic is usually determined experimentally by driving it mechanically at synchronous speed with its armature terminals on open circuit and reading the terminal voltage corresponding to a series of values of field current. If the mechanical power required to drive the synchronous machine during the open-circuit test is measured, the no-load rotational losses can be obtained. These losses comprise friction, windage, and core loss corresponding to the flux in the machine at no load. The friction and windage losses at synchronous speed are constant, while the open-circuit core loss is a function of the flux, which in turn is proportional to the open-circuit voltage.

The mechanical power required to drive the machine at synchronous speed and unexcited is its friction and windage loss. When the field is excited, the mechanical power equals the sum of the friction, windage, and open-circuit core loss. The open-circuit core loss therefore can be found from the difference between these two values of mechanical power. A curve of open-circuit core loss as a function of open-circuit voltage is shown in Fig. 7-6.

b. Short-Circuit Characteristic and Load Loss

If the armature terminals of a synchronous machine which is being driven as a generator at synchronous speed are short-circuited through suitable ammeters, as shown in Fig. 7-7a, and the field current is gradually increased until the armature current has reached a maximum safe value (perhaps twice rated current), data can be obtained from which the short-circuit armature current can be plotted against the field current. This relation is known as the *short-circuit characteristic*. An open-circuit characteristic *occ* and a short-circuit characteristic *scc* are shown in Fig. 7-7b.

The phasor relation between the excitation voltage $\widehat{E}_{af}$ and the steady-state armature current $\widehat{I}_a$ under polyphase short-circuit conditions is

$$\widehat{E}_{af} = \widehat{I}_a(R_a + jX_s) \qquad (7\text{-}25)$$

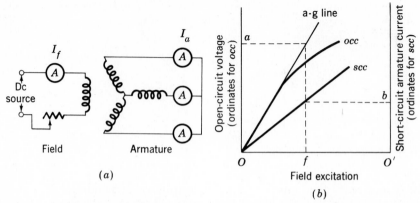

Fig. 7-7. (*a*) Connections for short-circuit test; (*b*) open- and short-circuit characteristics.

The phasor diagram is shown in Fig. 7-8. Because the resistance is much smaller than the synchronous reactance, the armature current lags the excitation voltage by very nearly 90°. Consequently the armature-reaction-mmf wave is very nearly in line with the axis of the field poles and in opposition to the field mmf, as shown by the phasors $\hat{A}$ and $\hat{F}$ representing the space waves of armature-reaction and field mmf, respectively.

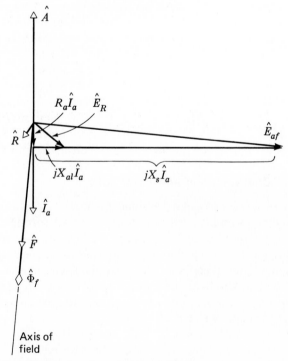

Fig. 7-8. Phasor diagram for short-circuit conditions.

The resultant mmf creates the resultant air-gap flux wave which generates the air-gap voltage $\widehat{E}_R$ equal to the voltage consumed in armature resistance R_a and leakage reactance X_{al}; as an equation,

$$\widehat{E}_R = I_a(R_a + jX_{al}) \tag{7-26}$$

In most synchronous machines the armature resistance is negligible, and the leakage reactance is between 0.10 and 0.20 per unit; a representative value is about 0.15 per unit. That is, at rated armature current the leakage-reactance voltage drop is about 0.15 per unit. From Eq. 7-26, therefore, the air-gap voltage at rated armature current on short circuit is about 0.15 per unit; i.e., the resultant air-gap flux is only about 0.15 times its normal-voltage value. Consequently, the machine is operating in an unsaturated condition. The short-circuit armature current therefore is directly proportional to the field current over the range from zero to well above rated armature current.

The unsaturated synchronous reactance can be found from the open- and short-circuit data. At any convenient field excitation, such as Of in Fig. 7-7b, the armature current on short circuit is $O'b$, and the excitation voltage for the same field current corresponds to Oa read from the air-gap line. Note that the voltage on the air-gap line should be used because the machine is operating on short circuit in an unsaturated condition. If the voltage per phase corresponding to Oa is $E_{af,ag}$ and the armature current per phase corresponding to $O'b$ is $I_{a,sc}$, then from Eq. 7-25, with armature resistance neglected, the unsaturated value $X_{s,ag}$ of the synchronous reactance is

$$X_{s,ag} = \frac{E_{af,ag}}{I_{a,sc}} \tag{7-27}$$

where the subscripts ag indicate air-gap-line conditions. If $E_{af,ag}$ and $I_{a,sc}$ are expressed in per unit, the synchronous reactance will be in per unit. If $E_{af,ag}$ and $I_{a,sc}$ are expressed in volts per phase and amperes per phase, respectively, the synchronous reactance will be ohms per phase.

Note that the synchronous reactance in ohms is calculated using the phase or line-neutral voltage. Often the open-circuit saturation curve is given in terms of the line-line voltage, in which case the voltage must be converted into the line-neutral value by dividing by $\sqrt{3}$.

For operation at or near rated terminal voltage, it is sometimes assumed that the machine is equivalent to an unsaturated one whose magnetization curve is a straight line through the origin and the rated-voltage point on the open-circuit characteristic, as shown by the dashed line Op in Fig. 7-9. According to this approximation, the saturated value of the synchronous reactance at rated voltage V_{ta} is

$$X_s = \frac{V_{ta}}{I'_{a,sc}} \tag{7-28}$$

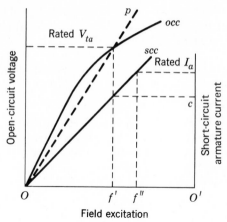

Fig. 7-9. Open- and short-circuit characteristics.

where $I'_{a,\text{sc}}$ is the armature current $O'c$ read from the short-circuit characteristic at the field current Of' corresponding to V_{ta} on the open-circuit characteristic, as shown in Fig. 7-9. This method of handling the effects of saturation usually gives satisfactory results when great accuracy is not required.

The *short-circuit ratio* (SCR) is defined as the ratio of the field current required for rated voltage on open circuit to the field current required for rated armature current on short circuit. That is, in Fig. 7-9 the SCR is

$$\text{SCR} = \frac{Of'}{Of''} \tag{7-29}$$

It can be shown that the SCR is the reciprocal of the per unit value of the saturated synchronous reactance given by Eq. 7-28.

EXAMPLE 7-1

The following data are taken from the open- and short-circuit characteristics of a 45-kVA 3-phase Y-connected 220-V (line-to-line) 6-pole 60-Hz synchronous machine. From the open-circuit characteristic

Line-to-line voltage = 220 V   Field current = 2.84 A

From the short-circuit characteristic

Armature current, A	118	152
Field current, A	2.20	2.84

From the air-gap line

Field current = 2.20 A   Line-to-line voltage = 202 V

Compute the unsaturated value of the synchronous reactance, its saturated value at rated voltage in accordance with Eq. 7-28, and the short-circuit ratio. Express the synchronous reactance in ohms per phase and in per unit on the machine rating as a base.

Solution

At a field current of 2.20 A the voltage to neutral on the air-gap line is

$$E_{af,ag} = \frac{202}{\sqrt{3}} = 116.7 \text{ V}$$

and for the same field current the armature current on short circuit is

$$I_{a,sc} = 118 \text{ A}$$

From Eq. 7-27

$$X_{s,ag} = \frac{116.7}{118} = 0.987 \text{ } \Omega/\text{phase}$$

Note that rated armature current is $45,000/\sqrt{3}$ (220) = 118 A. Therefore, $I_{a,sc} = 1.00$ per unit. The corresponding air-gap-line voltage is

$$E_{af,ag} = \frac{202}{220} = 0.92 \text{ per unit}$$

From Eq. 7-27 in per unit

$$X_{s,ag} = \frac{0.92}{1.00} = 0.92 \text{ per unit}$$

From the open- and short-circuit characteristics and Eq. 7-28

$$X_s = \frac{220}{\sqrt{3}(152)} = 0.836 \text{ } \Omega/\text{phase}$$

In per unit $I'_{a,sc} = \frac{152}{118} = 1.29$, and from Eq. 7-28

$$X_s = \frac{1.00}{1.29} = 0.775 \text{ per unit}$$

From the open- and short-circuit characteristics and Eq. 7-29

$$\text{SCR} = \frac{2.84}{2.20} = 1.29$$

If the mechanical power required to drive the machine is measured while the short-circuit test is being made, information can be obtained regarding the losses caused by the armature current. The mechanical power required to drive the synchronous machine during the short-circuit test equals the sum of friction and windage plus losses caused by the armature current. The losses caused by the armature current can then be found by subtracting friction and windage from the driving power. The losses caused by the short-circuit armature current are known collectively as the *short-circuit load loss*. A curve of short-circuit load loss plotted against armature current is shown in Fig. 7-10. It is approximately parabolic.

The short-circuit load loss comprises I^2R loss in the armature winding, local core losses caused by the armature leakage flux, and a very small core loss caused by the resultant flux. The dc resistance loss can be computed if the dc resistance is measured and corrected, when necessary, for the temperature of the windings during the short-circuit test. For copper conductors

$$\frac{R_T}{R_t} = \frac{234.5 + T}{234.5 + t} \tag{7-30}$$

where R_T and R_t are the resistances at Celsius temperatures T and t, respectively. If this dc resistance loss is subtracted from the short-circuit load loss, the difference will be the loss due to skin effect and eddy currents in the armature conductors plus the local core losses caused by the armature leakage flux. (The core loss caused by the resultant flux on short circuit is customarily neglected.) This difference between the short-circuit load loss and the dc resistance loss is the additional loss caused by the alternating current in the armature. It is the stray load loss described in Art. 4-3, commonly considered

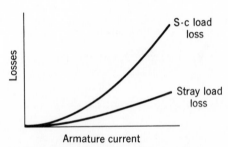

Fig. 7-10. Short-circuit load-loss and stray load-loss curves.

to have the same value under normal load conditions as on short circuit. It is a function of the armature current, as shown by the curve in Fig. 7-10.

As with any ac device, the effective resistance of the armature is the power loss attributable to the armature current divided by the square of the current. On the assumption that the stray load loss is a function of only the armature current, the effective resistance $R_{a,\text{eff}}$ of the armature can be determined from the short-circuit load loss

$$R_{a,\text{eff}} = \frac{\text{short-circuit load loss}}{(\text{short-circuit armature current})^2} \qquad (7\text{-}31)$$

If the short-circuit load loss and armature current are in per unit, the effective resistance will be in per unit. If they are in watts per phase and amperes per phase, respectively, the effective resistance will be in ohms per phase. Usually it is sufficiently accurate to find the value of $R_{a,\text{eff}}$ at rated current and then to assume it to be constant.

EXAMPLE 7-2

For the 45-kVA 3-phase Y-connected synchronous machine of Example 7-1, at rated armature current (118 A) the short-circuit load loss (total for three phases) is 1.80 kW at a temperature of 25°C. The dc resistance of the armature at this temperature is 0.0335 Ω/phase. Compute the armature effective resistance in per unit and in ohms per phase at 25°C.

Solution

In per unit the short-circuit load loss is

$$\frac{1.80}{45} = 0.040$$

at $I_a = 1.00$ per unit. Therefore,

$$R_{a,\text{eff}} = \frac{0.040}{(1.00)^2} = 0.040 \text{ per unit}$$

On a per phase basis the short-circuit load loss is

$$\frac{1800}{3} \text{ W/phase}$$

and consequently the effective resistance is

$$R_{a,\text{eff}} = \frac{1800}{3(118)^2} = 0.043 \text{ Ω/phase}$$

The ratio of ac to dc resistance is

$$\frac{R_{a,\text{eff}}}{R_{a,\text{dc}}} = \frac{0.043}{0.0335} = 1.28$$

Because this is a small machine, its per unit resistance is relatively high. The armature resistance of machines with ratings above a few hundred kilovoltamperes usually is less than 0.01 per unit.

7-4 STEADY-STATE POWER–ANGLE CHARACTERISTICS

The maximum power a synchronous machine can deliver is determined by the maximum torque which can be applied without loss of synchronism with the external system to which it is connected. The purpose of this article is to derive expressions for the steady-state power limits of simple situations in which the external system can be represented as an impedance in series with a voltage source.

Since the machine can be represented by a simple impedance, the study of power limits becomes merely a special case of the more general problem of the limitations on power flow through an inductive impedance. The impedance can include that of a line and transformer bank as well as the synchronous impedance of the machine.

Consider the simple circuit of Fig. 7-11a, comprising two ac voltages $\widehat{E}_1$ and $\widehat{E}_2$ connected by an impedance Z through which the current is $\widehat{I}$. The phasor diagram is shown in Fig. 7-11b. The power P_2 delivered through the impedance to the load end $\widehat{E}_2$ is

$$P_2 = E_2 I \cos \phi_2 \tag{7-32}$$

where ϕ_2 is the phase angle of $\widehat{I}$ with respect to $\widehat{E}_2$. The phasor current is

$$\widehat{I} = \frac{\widehat{E}_1 - \widehat{E}_2}{Z} \tag{7-33}$$

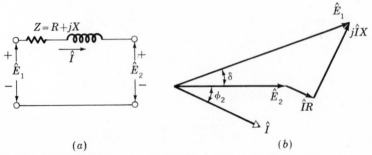

(a)   (b)

Fig. 7-11. (a) Impedance interconnecting two voltages; (b) phasor diagram.

If the phasor voltages and the impedance are expressed in polar form,

$$\widehat{I} = \frac{E_1\underline{/\delta} - E_2\underline{/0°}}{|Z|\underline{/\phi_z}} = \frac{E_1}{|Z|}\underline{/\delta - \phi_z} - \frac{E_2}{|Z|}\underline{/-\phi_z} \qquad (7\text{-}34)$$

where E_1, E_2 = magnitudes of voltages
$\qquad \delta$ = phase angle by which $\widehat{E}_1$ leads $\widehat{E}_2$
$\qquad |Z|$ = magnitude of impedance
$\qquad \phi_z$ = angle of impedance in polar form

The real part of the phasor equation 7-34 is the component of $\widehat{I}$ in phase in $\widehat{E}_2$, whence

$$I \cos \phi_2 = \frac{E_1}{|Z|} \cos (\delta - \phi_z) - \frac{E_2}{|Z|} \cos (-\phi_z) \qquad (7\text{-}35)$$

when it is noted that

$$\cos (-\phi_z) = \cos \phi_z = \frac{R}{|Z|}$$

substitution of Eq. 7-35 in Eq. 7-32 gives

$$P_2 = \frac{E_1 E_2}{|Z|} \cos (\delta - \phi_z) - \frac{E_2^2 R}{|Z|^2} \qquad (7\text{-}36)$$

and

$$P_2 = \frac{E_1 E_2}{|Z|} \sin (\delta + \alpha_z) - \frac{E_2^2 R}{|Z|^2} \qquad (7\text{-}37)$$

where

$$\alpha_z = 90° - \phi_z = \tan^{-1}\frac{R}{X} \qquad (7\text{-}38)$$

usually is a small angle.

Similarly the power P_1 at source end $\widehat{E}_1$ of the impedance can be expressed as

$$P_1 = \frac{E_1 E_2}{|Z|} \sin (\delta - \alpha_z) + \frac{E_1^2 R}{|Z|^2} \qquad (7\text{-}39)$$

If, as is frequently the case, the resistance is negligible,

$$P_1 = P_2 = \frac{E_1 E_2}{X} \sin \delta \qquad (7\text{-}40)$$

Equation 7-40 is commonly referred to as the *power-angle characteristic*

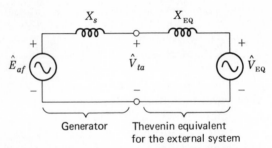

Fig. 7-12. Representation of a synchronous machine connected to an external system.

for a synchronous machine and the angle δ is known as the *power angle*. If the resistance is negligible and the voltages are constant, the maximum power

$$P_{1,\text{max}} = P_{2,\text{max}} = \frac{E_1 E_2}{X} \tag{7-41}$$

occurs when $\delta = 90°$.

Equation 7-40 is valid for any voltage sources $\widehat{E}_1$ and $\widehat{E}_2$ separated by a reactive impedance jX. Thus for a synchronous machine connected to a system whose Thevenin equivalent is a voltage source $\widehat{V}_{\text{EQ}}$ in series with a reactive impedance jX_{EQ}, as shown in Fig. 7-12, the power-angle characteristic can be written

$$P = \frac{E_{af} V_{\text{EQ}}}{X_s + X_{\text{EQ}}} \sin \delta \tag{7-42}$$

where P is the power transferred from the synchronous machine to the system and δ is the phase angle of $\widehat{E}_{af}$ with respect to $\widehat{V}_{\text{EQ}}$.

In a similar fashion it is possible to write a power-angle characteristic in terms of X_s, E_{af}, the terminal voltage V_{ta}, and the relative angle between them or alternatively using X_{EQ}, V_{ta}, and V_{EQ}. Although both expressions are equally valid, they are not equally useful. For example, while E_{af} and V_{EQ} remain constant as P is varied, V_{ta} will not. Thus, while Eq. 7-42 gives an easily solved relation between P and δ, a power-angle characteristic based upon V_{ta} cannot be solved without an additional expression relating V_{ta} to P.

It should be emphasized that the derivation of Eqs. 7-32 to 7-42 is based upon a single-phase ac circuit. For a balanced 3-phase system, if E_1 and E_2 are the line-neutral voltages, the results must be multiplied by 3 to get the total 3-phase power; alternatively E_1 and E_2 can be expressed in terms of the line-to-line voltage (equal to $\sqrt{3}$ times the line-neutral voltage), in which case their result gives 3-phase power directly.

When Eq. 7-40 is compared with Eq. 7-1 for torque in terms of interacting flux and mmf waves, they are seen to be of the same form. This is no coincidence. First remember that torque and power are linearly proportional when, as here, speed is constant. What we are really saying is that Eq. 7-1 applied specifically to the idealized cylindrical-rotor machine and translated to circuit terms becomes Eq. 7-40. A quick mental review of the background of each relation should show that they stem from the same fundamental considerations.

From Eq. 7-42 we see that the maximum power transfer associated with synchronous-machine operation is proportional to the magnitude of the system voltage, corresponding to V_{EQ}, and the generator internal voltage E_{af}. Thus for constant system voltage the maximum power transfer can be increased by increasing the synchronous-machine field current and thus the internal voltage. Of course, this cannot be done without limit; neither the field current nor the machine fluxes can be raised past the point where cooling requirements can no longer be met.

In general, stability considerations dictate that a synchronous machine achieve steady-state operation for a power angle considerably less than 90°. Thus, for a given system configuration it is necessary to ensure that the machine will be able to achieve its rated operation and that this operating condition will be within acceptable operating limits for both the machine and the system.

EXAMPLE 7-3

A 2000-hp 1.0-power-factor 3-phase Y-connected 2300-V 30-pole 60-Hz synchronous motor has a synchronous reactance of 1.95 Ω/phase. For this problem all losses may be neglected.

(a) Compute the maximum torque in pound-feet which this motor can deliver if it is supplied with power from a constant-frequency source, commonly called an *infinite bus,* and if its field excitation is constant at the value which would result in 1.00 power factor at rated load.

(b) Instead of the infinite bus of part (a) suppose that the motor is supplied with power from a 3-phase Y-connected 2300-V 1750-kVA 2-pole 3600-r/min turbine generator whose synchronous reactance is 2.65 Ω/phase. The generator is driven at rated speed, and the field excitations of generator and motor are adjusted so that the motor runs at 1.00 power factor and rated terminal voltage at full load. The field excitations of both machines are then held constant, and the mechanical load on the synchronous motor is gradually increased. Compute the maximum motor torque under these conditions and the terminal voltage when the motor is delivering its maximum torque.

Solution

Although this machine is undoubtedly of the salient-pole type, we shall solve the problem by simple cylindrical-rotor theory. The solution accordingly ne-

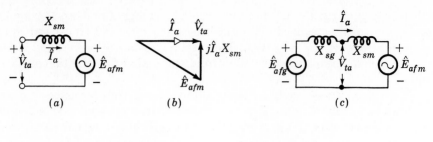

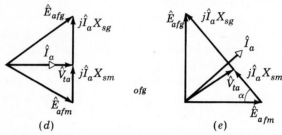

Fig. 7-13. Equivalent circuits and phasor diagrams for Example 7-3.

glects reluctance torque. The machine actually would develop a maximum torque somewhat greater than our computed value, as discussed in Art. 7-7.

(a) The equivalent circuit is shown in Fig. 7-13a and the phasor diagram at full load in Fig. 7-13b, where $\hat{E}_{afm}$ is the excitation voltage of the motor and X_{sm} is its synchronous reactance. From the motor rating with losses neglected,

$$\text{Rated kVA} = 2000(0.746) = 1492 \text{ kVA, 3-phase}$$
$$= 497 \text{ kVA/phase}$$

$$\text{Rated voltage} = \frac{2300}{\sqrt{3}} = 1330 \text{ V to neutral}$$

$$\text{Rated current} = \frac{497{,}000}{1330} = 374 \text{ A/phase Y}$$

The synchronous-reactance voltage drop in the motor is

$$I_a X_{sm} = 374(1.95) = 730 \text{ V/phase}$$

From the phasor diagram at full load

$$E_{afm} = \sqrt{V_{ta}^2 + (I_a X_{sm})^2} = 1515 \text{ V}$$

When the power source is an infinite bus and the field excitation is constant, V_{ta} and E_{afm} are constant. Substitution of V_{ta} for E_1, E_{afm}

for E_2, and X_{sm} for X in Eq. 7-41 then gives

$$P_{max} = \frac{V_{ta}E_{afm}}{X_{sm}} = \frac{1330(1515)}{1.95} = 1030 \text{ kW/phase}$$

$$= 3090 \text{ kW for 3 phases}$$

(In per unit, $P_{max} = 3090/1492 = 2.07$.) With 30 poles at 60 Hz, synchronous speed $= 4 \text{ r/s}$.

$$T_{max} = \frac{P_{max}}{\omega_s} = \frac{3090 \times 10^3}{2\pi(4)} = 123,000 \text{ N} \cdot \text{m}$$

$$= 0.738(123,000) = 90,800 \text{ lb} \cdot \text{ft}$$

(b) When the power source is the turbine generator, the equivalent circuit becomes that shown in Fig. 7-13c, where $\widehat{E}_{afg}$ is the excitation voltage of the generator and X_{sg} is its synchronous reactance. The phasor diagram at full motor load, 1.00 power factor, is shown in Fig. 7-13d. As before

$$V_{ta} = 1330 \text{ V at full load} \qquad E_{afm} = 1515 \text{ V}$$

The synchronous-reactance voltage drop in the generator is

$$I_a X_{sg} = 374(2.65) = 991 \text{ V}$$

and from the phasor diagram

$$E_{afg} = \sqrt{V_{ta}^2 + (I_a X_{sg})^2} = 1655 \text{ V}$$

Since the field excitations and speeds of both machines are constant, E_{afg} and E_{afm} are constant. Substitution of E_{afg} for E_1, E_{afm} for E_2, and $X_{sg} + X_{sm}$ for X in Eq. 7-41 then gives

$$P_{max} = \frac{E_{afg}E_{afm}}{X_{sg} + X_{sm}} = \frac{1655(1515)}{4.60} = 545 \text{ kW/phase}$$

$$= 1635 \text{ kW for 3 phases}$$

(In per unit, $P_{max} = 1635/1492 = 1.095$.)

$$T_{max} = \frac{P_{max}}{\omega_s} = \frac{1635 \times 10^3}{2\pi(4)} = 65,000 \text{ N} \cdot \text{m} = 48,000 \text{ lb} \cdot \text{ft}$$

Synchronism would be lost if a load torque greater than this value were applied to the motor shaft. The motor would stall, the generator would tend to overspeed, and the circuit would be opened by circuit-breaker action.

With fixed excitations, maximum power occurs when $\widehat{E}_{afg}$ leads $\widehat{E}_{afm}$ by 90°, as shown in Fig. 7-13e. From this phasor diagram

$$I_a(X_{sg} + X_{sm}) = \sqrt{E_{afg}^2 + E_{afm}^2} = 2240 \text{ V}$$

$$I_a = \frac{2240}{4.60} = 488 \text{ A}$$

$$I_a X_{sm} = 488(1.95) = 951 \text{ V}$$

$$\cos \alpha = \frac{E_{afm}}{I_a(X_{sg} + X_{sm})} = \frac{1515}{2240} = 0.676$$

$$\sin \alpha = \frac{E_{afg}}{I_a(X_{sg} + X_{sm})} = \frac{1655}{2240} = 0.739$$

The phasor equation for the terminal voltage is

$$\widehat{V}_{ta} = \widehat{E}_{afm} + j\widehat{I}_a X_{sm} = \widehat{E}_{afm} - I_a X_{sm} \cos \alpha + jI_a X_{sm} \sin \alpha$$
$$= 1515 - 643 + j703 = 872 + j703$$

The magnitude of $\widehat{V}_{ta}$ is

$$V_{ta} = 1120 \text{ V to neutral} = 1940 \text{ V line to line}$$

When the source is the turbine generator, as in part (b), the effect of its impedance causes the terminal voltage to decrease with increasing load, thereby reducing the maximum power from 3090 kW in part (a) to 1635 kW in part (b).

7-5 STEADY-STATE OPERATING CHARACTERISTICS

The principal steady-state operating characteristics are the interrelations between terminal voltage, field current, armature current, power factor, and the efficiency. A selection of performance curves of importance in practical application of the machines is presented here. All of them can be computed for application studies by the methods presented in this chapter.

Consider a synchronous generator delivering power at constant frequency to a load whose power factor is constant. The curve showing the field current required to maintain rated terminal voltage as the constant-power-factor load is varied is known as a *compounding curve*. Three compounding curves at various constant power factors are shown in Fig. 7-14.

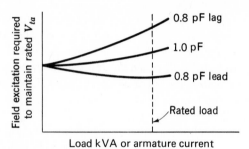

Fig. 7-14. Generator compounding curves.

Synchronous generators are usually rated in terms of the maximum kVA load at a specific voltage and power factor (often 80, 85, or 90 percent lagging) which they can carry continuously without overheating. The active power output of the generator is usually limited to a value within the kVA rating by the capability of its prime mover. By virtue of its voltage-regulating system, the machine normally operates at a constant voltage whose value is within ± 5 percent of rated voltage. When the active-power loading and voltage are fixed, the allowable reactive-power loading is limited by either armature or field heating. A typical set of reactive-power capability curves for a large turbine generator is shown in Fig. 7-15. They give the maximum reactive-

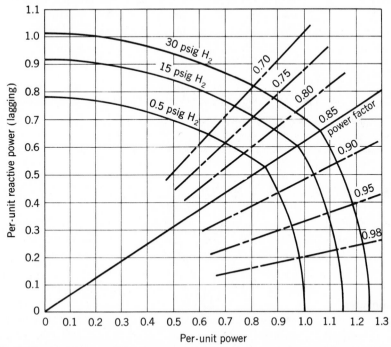

Fig. 7-15. Reactive-capability curves of hydrogen-cooled turbine generator, 0.85 power factor, 0.80 short-circuit ratio. Base kVA is rated kVA at 0.5 lb/in² hydrogen.

power loadings corresponding to various power loadings with operation at rated voltage. Armature heating is the limiting factor in the region from unity to rated power factor (0.85 in Fig. 7-15). For lower power factors, field heating is limiting. Such a set of curves forms a valuable guide in planning and operating the system of which the generator is a part. Also shown in Fig. 7-15 is the effect of increased hydrogen pressure (resulting in increased cooling) on allowable machine loadings.

The power factor at which a synchronous motor operates, and hence its armature current, can be controlled by adjusting its field excitation. The curve showing the relation between armature current and field current at a constant terminal voltage and with a constant shaft load is known as a *V curve* because of its characteristic shape. A family of V curves is shown in Fig. 7-16. For constant power output, armature current is, of course, a minimum at unity power factor and increases as power factor decreases. The dashed lines are loci of constant power factor. They are the synchronous-motor compounding curves showing how the field current must be varied as load is changed in order to maintain constant power factor. Points to the right of the unity-power-factor compounding curve correspond to overexcitation and leading current input; points to the left correspond to underexcitation and lagging current input. The synchronous-motor compounding curves are very similar to the generator compounding curves of Fig. 7-14. (Note the interchange of armature-current and field-current axes when comparing Figs. 7-14 and 7-16.) In fact, if it were not for the small effects of armature resistance, the motor and generator compounding curves would be identical except that the lagging- and leading-power-factor curves would be interchanged.

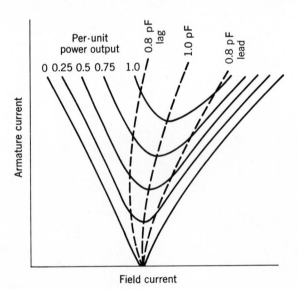

Fig. 7-16. Synchronous-machine V curves.

As in all electromagnetic machines, the losses in synchronous machines comprise I^2R losses in the windings, core losses, and mechanical losses. The conventional efficiency is computed in accordance with a set of rules agreed upon by the ANSI. The general principles upon which these rules are based are described in Art. 4-3. The purpose of the following example is to show how these rules are applied specifically to synchronous machines.

EXAMPLE 7-4

Data are given in Fig. 7-17 with respect to the losses of the 45-kVA synchronous machine of Examples 7-1 and 7-2. Compute its efficiency when running as a synchronous motor at a terminal voltage of 230 V and with a power input to its armature of 45 kW at 0.80 power factor, leading current. The field current measured in a load test taken under these conditions is I_f (test) = 5.50 A.

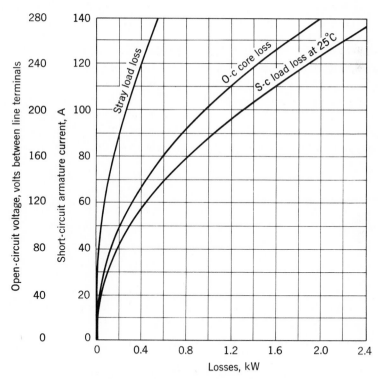

Friction and windage loss = 0.91 kW
Armature dc resistance at 25°C = 0.0335 Ω per phase
Field-winding resistance at 25°C = 29.8 Ω

Fig. 7-17. Losses in 3-phase 45-kVA Y-connected 220-V 60-Hz 6-pole synchronous machine, Example 7-4.

Solution

For the specified operating conditions, the armature current is

$$I_a = \frac{45,000}{\sqrt{3}(230)(0.80)} = 141 \text{ A}$$

The I^2R losses are to be computed on the basis of the dc resistances of the windings at 75°C. Correcting the winding resistances by means of Eq. 7-30 gives

$$\text{Field-winding resistance } R_f \text{ at } 75°\text{C} = 35.5 \, \Omega$$
$$\text{Armature dc resistance } R_a \text{ at } 75°\text{C} = 0.0399 \, \Omega/\text{phase}$$

The field I^2R loss is

$$I_f^2 R_f = (5.50)^2(35.5) = 1.07 \text{ kW}$$

According to the ANSI standards, field-rheostat and exciter losses are not charged against the machine. The armature I^2R loss is

$$3I_a^2 R_a = 3(141)^2(0.0399) = 2.38 \text{ kW}$$

and from Fig. 7-17 at $I_a = 141$ A, stray load loss $= 0.56$ kW. According to the ANSI standards, no temperature correction is to be applied to the stray load loss.

The core loss is read from the open-circuit core-loss curve at a voltage equal to the internal voltage behind the resistance of the machine. The stray load loss is considered to account for the losses caused by the armature leakage flux. For motor action this internal voltage is, as a phasor,

$$\widehat{V}_{ta} - \widehat{I}_a R_a = \frac{230}{\sqrt{3}} - 141(0.80 + j0.60)(0.0399) = 128.4 - j3.4$$

The magnitude is 128.4 V/phase, or 222 V between line terminals. From Fig. 7-17 open-circuit core loss is 1.20 kW. Also, friction and windage loss is 0.91 kW. All losses have now been found:

$$\text{Total losses} = 1.07 + 2.38 + 0.56 + 1.20 + 0.91 = 6.12 \text{ kW}$$

The power input is the sum of the ac input to the armature and the dc input to the field, or

$$\text{Input} = 46.07 \text{ kW}$$

Therefore

$$\text{Efficiency} = 1 - \frac{\text{losses}}{\text{input}} = 1 - \frac{6.12}{46.1} = 0.867 = 86.7\%$$

7-6 EFFECTS OF SALIENT POLES; INTRODUCTION TO DIRECT- AND QUADRATURE-AXIS THEORY

The essential features of salient-pole machines are developed in this article based upon physical reasoning. A mathematical treatment, based upon an inductance formulation like that presented in Art. 7-2, is given in Arts. 8-2 and 8-3, where the dq0 transformation is developed.

a. Flux and MMF Waves

The flux produced by an mmf wave in the uniform-air-gap machine is independent of the spatial alignment of the wave with respect to the field poles. The salient-pole machine, on the other hand, has a preferred direction of magnetization determined by the protruding field poles. The permeance along the *polar,* or *direct, axis* is appreciably greater than that along the *interpolar,* or *quadrature, axis.*

We have seen that the armature-reaction flux wave lags the field flux wave by a space angle of $90° + \phi_{\text{lag}}$, where ϕ_{lag} is the time-phase angle by which the armature current in the direction of the excitation emf lags the excitation emf. If the armature current $\widehat{I}_a$ lags the excitation emf $\widehat{E}_f$ by $90°$, the armature-reaction flux wave $\widehat{\Phi}_{ar}$ is directly opposite the field poles and in the opposite direction to the field flux $\widehat{\Phi}_f$, as shown in the phasor diagram of Fig. 7-18a. The corresponding component flux-density waves at the armature surface produced by the field current and by the synchronously rotating space-fundamental component of armature-reaction mmf are shown in Fig. 7-18b, in which the effects of slots are neglected. The waves consist of a space fundamental and a family of odd-harmonic components. The harmonic effects usually are small (see Art. 3-3a). Accordingly only the space-fundamental components will be considered. It is the fundamental components which are represented by the flux per pole phasors $\widehat{\Phi}_f$ and $\widehat{\Phi}_{ar}$ in Fig. 7-18a.

Conditions are quite different when the armature current is in phase with the excitation emf, as shown in the phasor diagram of Fig. 7-19a. The axis of the armature-reaction wave then is opposite an interpolar space, as shown in Fig. 7-19b. The armature-reaction flux wave is badly distorted, comprising principally a fundamental and a prominent third space harmonic. The third-harmonic flux wave generates third-harmonic emf's in the armature phases, but these voltages do not appear between the line terminals.

Because of the high reluctance of the air gap between poles, the space-

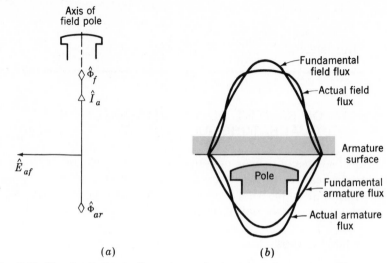

Fig. 7-18. Direct-axis air-gap fluxes in a salient-pole synchronous machine.

fundamental armature-reaction flux when the armature reaction is in quadrature with the field poles (Fig. 7-19) is less than the space-fundamental armature-reaction flux which would be created by the same armature current if the armature flux wave were directly opposite the field poles (Fig. 7-18). Hence, the magnetizing reactance is less when the armature current is in time phase with the excitation emf (Fig. 7-19) than when it is in time quadrature with respect to the excitation emf (Fig. 7-18a).

The effects of salient poles can be taken into account by resolving the armature current $\hat{I}_a$ into two components, one in time quadrature with, and the other in time phase with, the excitation voltage $\hat{E}_{af}$, as shown in the

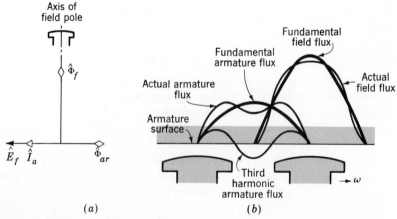

Fig. 7-19. Quadrature-axis air-gap fluxes in a salient-pole synchronous machine.

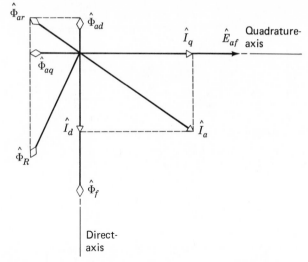

Fig. 7-20. Phasor diagram of a salient-pole synchronous generator.

phasor diagram of Fig. 7-20. This diagram is drawn for an unsaturated salient-pole generator operating at a lagging power factor. The component $\hat{I}_d$ of
the armature current, in time quadrature with the excitation voltage, produces a component fundamental armature-reaction flux $\hat{\Phi}_{ad}$ along the axes of
the field poles, as in Fig. 7-18. The component $\hat{I}_q$ in phase with the excitation
voltage, produces a component fundamental armature-reaction flux $\hat{\Phi}_{aq}$ in
space quadrature with the field poles, as in Fig. 7-19. The subscripts d and q
refer to the space phase of the armature-reaction fluxes, and not to the time
phase of the component currents producing them. Thus a *direct-axis quantity*
is one whose magnetic effect is centered on the axes of the field poles. Direct-
axis mmf's act on the main magnetic circuit. A *quadrature-axis quantity* is
one whose magnetic effect is centered on the interpolar space. For an unsaturated machine, the armature-reaction flux $\hat{\Phi}_{ar}$ is the sum of the components
$\hat{\Phi}_{ad}$ and $\hat{\Phi}_{aq}$. The resultant flux $\hat{\Phi}_R$ is the sum of $\hat{\Phi}_{ar}$ and the main-field flux $\hat{\Phi}_f$.

b. Phasor Diagrams for Salient-Pole Machines

With each of the component currents $\hat{I}_d$ and $\hat{I}_q$ there is associated a component synchronous-reactance voltage drop, $j\hat{I}_d X_d$ and $j\hat{I}_q X_q$, respectively. The
reactances X_d and X_q are, respectively, the *direct-* and *quadrature-axis synchronous reactances*; they account for the inductive effects of all the funda-
mental-frequency-generating fluxes created by the armature currents, along
the direct- and quadrature axes including both armature-leakage and armature-reaction fluxes. Thus, the inductive effects of the direct- and quadrature-axis armature-reaction flux waves can be accounted for by *direct-* and
quadrature-axis magnetizing reactances $X_{\varphi d}$ and $X_{\varphi q}$, respectively, similar to
the magnetizing reactance X_φ of cylindrical-rotor theory. The direct- and

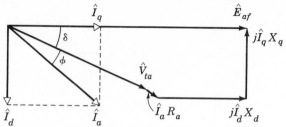

Fig. 7-21. Phasor diagram for synchronous generator.

quadrature-axis synchronous reactances then are

$$X_d = X_{al} + X_{\varphi d} \tag{7-43}$$

$$X_q = X_{al} + X_{\varphi q} \tag{7-44}$$

where X_{al} is the armature leakage reactance, assumed to be the same for direct- and quadrature-axis currents. Compare Eqs. 7-43 and 7-44 with Eq. 7-24 for the non-salient-pole case. As shown in the generator phasor diagram (Fig. 7-21), the excitation voltage $\widehat{E}_{af}$ equals the phasor sum of the terminal voltage $\widehat{V}_{ta}$ plus the armature-resistance drop $\widehat{I}_a R_a$ and the component synchronous-reactance drops $j\widehat{I}_d X_d + j\widehat{I}_q X_q$.

The reactance X_q is less than the reactance X_d because of the greater reluctance of the air gap in the quadrature axis. Usually, X_q is between $0.6X_d$ and $0.7X_d$. Typical values are given in Table 8-1. Note that a small salient-pole effect is present in turboalternators, even though they are cylindrical-rotor machines, because of the effect of the rotor slots on the quadrature-axis reluctance.

In using the phasor diagram of Fig. 7-21 the armature current must be resolved into its direct- and quadrature-axis components. This resolution assumes that the phase angle $\phi + \delta$ of the armature current with respect to the excitation voltage is known. Often, however, the power-factor angle ϕ at the machine terminals is explicitly known, rather than the internal power-factor angle $\phi + \delta$. The phasor diagram of Fig. 7-21 is repeated by the solid-line phasors in Fig. 7-22. Study of this phasor diagram shows that the dashed phasor $o'a'$, perpendicular to $\widehat{I}_a$, equals $j\widehat{I}_a X_q$. This result follows geometrically from the fact that triangles $o'a'b'$ and oab are similar, because their corresponding sides are perpendicular. Thus

$$\frac{o'a'}{oa} = \frac{b'a'}{ba} \tag{7-45}$$

or

$$o'a' = \frac{b'a'}{ba}oa = \frac{j\widehat{I}_q X_q}{\widehat{I}_q}\widehat{I}_a = jX_q\widehat{I}_a \tag{7-46}$$

The phasor sum $\widehat{V}_{ta} + \widehat{I}_a R_a + j\widehat{I}_a X_q$ then locates the angular position of the excitation voltage $\widehat{E}_{af}$ and therefore the direct and quadrature axes. Physi-

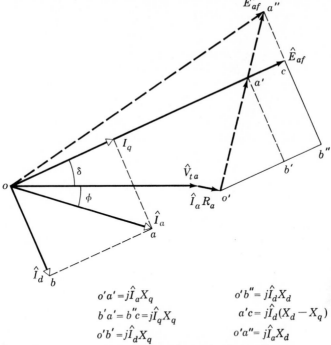

$$o'a' = j\hat{I}_a X_q \qquad\qquad o'b'' = j\hat{I}_d X_d$$
$$b'a' = b''c = j\hat{I}_q X_q \qquad a'c = j\hat{I}_d(X_d - X_q)$$
$$o'b' = j\hat{I}_d X_q \qquad\qquad o'a'' = j\hat{I}_a X_d$$

Fig. 7-22. Relations between component voltages in phasor diagram.

cally this must be so, because all the field excitation in a normal machine is in the direct axis. One use of these relations in determining the excitation requirements for specified operating conditions at the terminals of a salient-pole machine is illustrated in Example 7-5.

EXAMPLE 7-5

The reactances X_d and X_q of a salient-pole synchronous generator are 1.00 and 0.60 per unit, respectively. The armature resistance is negligible. Compute the excitation voltage when the generator delivers rated kilovoltamperes at 0.80 power factor, lagging current, and rated terminal voltage.

Solution

First, the phase of $\hat{E}_{af}$ must be found so that $\hat{I}_a$ can be resolved into its direct- and quadrature-axis components. The phasor diagram is shown in Fig. 7-23.

$$\hat{I}_a = 0.80 - j0.60 = 1.00\underline{/-36.9^\circ}$$
$$j\hat{I}_a X_q = j(0.80 - j0.60)(0.60) = 0.36 + j0.48$$
$$\hat{V}_{ta} = \text{reference phasor} = 1.00 + j0$$
$$\text{Phasor sum} = \hat{E}' = 1.36 + j0.48 = 1.44\underline{/19.4^\circ}$$

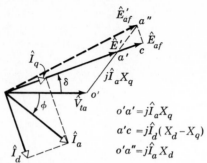

Fig. 7-23. Generator phasor diagram for Example 7-5.

The angle $\delta = 19.4°$, and the phase angle between $\widehat{E}_{af}$ and $\widehat{I}_a$ is $36.9° + 19.4° = 56.3°$.

The armature current can now be resolved into its direct- and quadrature-axis components. Their magnitudes are

$$I_d = 1.00 \sin 56.3° = 0.832 \qquad I_q = 1.00 \cos 56.3° = 0.555$$

As phasors,

$$\widehat{I}_d = 0.832\underline{/-90° + 19.4°} = 0.832\underline{/-70.6°}$$

and
$$\widehat{I}_q = 0.555\underline{/19.4°}$$

We can now find E_{af} by adding the length

$$a'c = I_d(X_d - X_q)$$

numerically to the magnitude of $\widehat{E}'$; thus, the magnitude of the excitation voltage is the algebraic sum

$$E_{af} = E' + I_d(X_d - X_q) = 1.44 + 0.832(0.40) = 1.77 \text{ per unit}$$

As a phasor, $\widehat{E}_{af} = 1.77\underline{/19.4°}$

In the simplified theory of Art. 7-2 the synchronous machine is assumed to be representable by a single reactance, the synchronous reactance of Eq. 7-24. The question naturally arises: How serious an approximation is involved if a salient-pole machine is treated in this simple fashion. Suppose the salient-pole machine of Figs. 7-22 and 7-23 were treated by cylindrical-rotor theory as if it had a single synchronous reactance equal to its direct-axis value X_d. For the same conditions at its terminals, the synchronous-reactance drop $j\widehat{I}_a X_d$ would be the phasor $o'a''$, and the equivalent excitation voltage would be $\widehat{E}'_{af}$, as shown in these figures. Because ca'' is perpendicular to $\widehat{E}_{af}$, there is little difference in magnitude between the correct value $\widehat{E}_{af}$ and the approximate value $\widehat{E}'_{af}$ for a normally excited machine. Recomputation of the excitation voltage on this basis for Example 7-5 gives a value of $1.79\underline{/26.6°}$.

Insofar as the interrelations between terminal voltage, armature current, power, and excitation over the normal operating range are concerned, the effects of salient poles usually are of minor importance, and such characteristics of a salient-pole machine usually can be computed with satisfactory accuracy by the simple cylindrical-rotor theory. Only at small excitations will the differences between cylindrical-rotor and salient-pole theory become important.

There is, however, considerable difference in the phase angles of $\widehat{E}_{af}$ and $\widehat{E}'_{af}$ in Figs. 7-22 and 7-23. This difference is caused by the reluctance torque in a salient-pole machine. Its effect is investigated in the following article.

7-7 POWER–ANGLE CHARACTERISTICS OF SALIENT–POLE MACHINES

We shall limit the discussion to the simple system shown in the schematic diagram of Fig. 7-24a, comprising a salient-pole synchronous machine SM connected to an infinite bus of voltage $\widehat{E}_e$ through a series impedance of reactance X_e per phase. Resistance will be neglected because it usually is small. Consider that the synchronous machine is acting as a generator. The phasor diagram is shown by the solid-line phasors in Fig. 7-24b. The dashed phasors show the external reactance drop resolved into components due to $\widehat{I}_d$ and $\widehat{I}_q$. The effect of the external impedance is merely to add its reactance to the reactances of the machine; i.e., the total values of reactance interposed between the excitation voltage $\widehat{E}_{af}$ and the bus voltage $\widehat{E}_e$ are

$$X_{dT} = X_d + X_e \tag{7-47}$$

$$X_{qT} = X_q + X_e \tag{7-48}$$

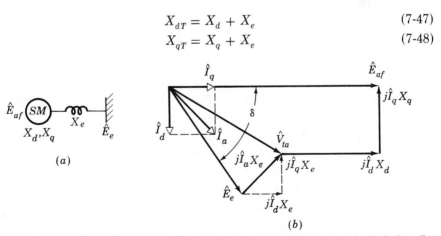

Fig. 7-24. Salient-pole synchronous machine and series impedance: (a) single-line diagram; (b) phasor diagram.

If the bus voltage $\widehat{E}_e$ is resolved into components $E_e \sin \delta$ and $E_e \cos \delta$ in phase with $\widehat{I}_d$ and $\widehat{I}_q$, respectively, the power P delivered to the bus per phase is

$$P = I_d E_e \sin \delta + I_q E_e \cos \delta \tag{7-49}$$

Also, from Fig. 7-24b,

$$I_d = \frac{E_{af} - E_e \cos \delta}{X_{dT}} \tag{7-50}$$

and

$$I_q = \frac{E_e \sin \delta}{X_{qT}} \tag{7-51}$$

Substitution of Eqs. 7-50 and 7-51 in Eq. 7-49 gives

$$P = \frac{E_{af}E_e}{X_{dT}} \sin \delta + E_e^2 \frac{X_{dT} - X_{qT}}{2X_{dT}X_{qT}} \sin 2\delta \tag{7-52}$$

As discussed in Art. 7-4, Eq. 7-52 gives power per phase when E_{af} and E_e are expressed as line-neutral voltages. The result must be multiplied by 3 to get 3-phase power; alternatively, expressing E_{af} and E_e as line-to-line voltages will result in 3-phase power directly.

This power-angle characteristic is shown in Fig. 7-25. The first term is the same as the expression obtained for a cylindrical-rotor machine. This term is merely an extension of the basic concepts of Chap. 3 to include the effects of series reactance. The second term introduces the effect of salient poles. It

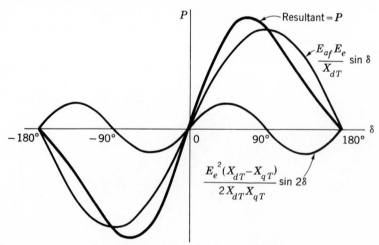

Fig. 7-25. Power-angle characteristic of a salient-pole synchronous machine showing fundamental component due to field excitation and second-harmonic component due to reluctance torque.

represents the fact that the air-gap flux wave creates torque tending to align the field poles in the position of minimum reluctance. This term is the power corresponding to the *reluctance torque* and is of the same general nature as the reluctance torque discussed in Art. 2-5. Note that the reluctance torque is independent of field excitation. Note, also, that if $X_{dT} = X_{qT}$, as in a uniform-air-gap machine, there is no preferential direction of magnetization, the reluctance torque is zero, and Eq. 7-52 reduces to the power-angle equation for a cylindrical-rotor machine whose synchronous reactance is X_{dT} (see Eq. 7-40).

Notice that the characteristic for negative values of δ is the same except for a reversal in the sign of P. That is, the generator and motor regions are alike if the effects of resistance are negligible. For generator action $\widehat{E}_{af}$ leads $\widehat{E}_e$; for motor action $\widehat{E}_{af}$ lags $\widehat{E}_e$. Steady-state operation is stable over the range where the slope of the power-angle characteristic is positive. Because of the reluctance torque, a salient-pole machine is "stiffer" than one with a cylindrical rotor; i.e., for equal voltages and equal values of X_{dT}, a salient-pole machine develops a given torque at a smaller value of δ, and the maximum torque which can be developed is somewhat greater.

EXAMPLE 7-6

The 2000-hp 1.0-power-factor 3-phase Y-connected 2300-V synchronous motor of Example 7-3 has reactances of $X_d = 1.95$ and $X_q = 1.40 \, \Omega$/phase. Neglecting all losses, compute the maximum mechanical power in kilowatts which this motor can deliver if it is supplied with electric power from an infinite bus (Fig. 7-26a) at rated voltage and frequency and if its field excitation is constant at the value which would result in 1.00 power factor at rated load. The shaft load is assumed to be increased gradually so that transient swings are negligible and the steady-state power limit applies. Include the effects of salient poles.

Solution

The first step is to compute the synchronous-motor excitation at rated voltage, full load, 1.0 power factor. As in Example 7-4, the full-load terminal

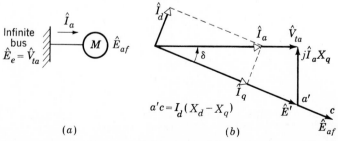

Fig. 7-26. (a) Single-line diagram and (b) phasor diagram for motor of Example 7-6.

voltage and current are 1330 V to neutral and 374 A/phase Y. The phasor diagram for the specified full-load conditions is shown in Fig. 7-26b. The only essential difference between this phasor diagram and the generator phasor diagram of Fig. 7-23 is that $\widehat{I}_a$ in Fig. 7-26 represents motor *input* current; i.e., we have switched to the motor reference direction for $\widehat{I}_a$. The phasor voltage equation then becomes

$$\widehat{E}_{af} = \widehat{V}_{ta} - j\widehat{I}_d X_d - j\widehat{I}_q X_q$$

In Fig. 7-26b

$$\widehat{E}' = \widehat{V}_{ta} - j\widehat{I}_a X_q = 1330 + j0 - j374(1.40) = 1429\underline{/-21.5°}$$

That is, the angle δ is 21.5°, with $\widehat{E}_{af}$ lagging $\widehat{V}_{ta}$. The magnitude of $\widehat{I}_d$ is

$$I_d = I_a \sin δ = 374(0.367) = 137 \text{ A}$$

The magnitude of $\widehat{E}_{af}$ can now be found by adding the length $a'c = I_d(X_d - X_q)$ numerically to the magnitude of $\widehat{E}'$; thus

$$E_{af} = E' + I_d(X_d - X_q) = 1429 + 137(0.55) = 1504 \text{ V to neutral}$$

From Eq. 7-52 the power-angle characteristic for this motor is

$$P = \frac{E_{af}E_e}{X_d} \sin δ + E_e^2 \frac{X_d - X_q}{2X_d X_q} \sin 2δ$$

$$= 1030 \sin δ + 178 \sin 2δ \qquad \text{kW/phase}$$

The maximum power occurs when $dP/dδ = 0$

$$\frac{dP}{dδ} = 1030 \cos δ + 356 \cos 2δ$$

Setting this equal to zero and using the trigonometric identity

$$\cos 2α = (2 \cos^2 α) - 1$$

permits us to solve for the angle δ at which the maximum power occurs:

$$δ = 73.2°$$

Therefore the maximum power is

$$P_{\max} = 1080 \text{ kW/phase} = 3240 \text{ kW for 3 phases}$$

Compare with $P_{max} = 3090$ kW found in Example 7-3, where the effects of salient poles were neglected. The error caused by neglecting saliency is slightly less than 5 percent in this case.

The effect of salient poles on the power limit increases as the excitation voltage is decreased, as can be seen from Eq. 7-52. For a normally excited machine the effect of salient poles usually amounts to a few percent at most. Only at small excitations does the reluctance torque become important. Thus, except at small excitations or when very accurate results are required, a salient-pole machine usually can be adequately treated by simple cylindrical-rotor theory.

7–8 INTERCONNECTED SYNCHRONOUS GENERATORS

As discussed in Art. 7-1, electric-power systems consist of the interconnection of large numbers of synchronous generators operating in parallel, interconnected by transmission lines, and supplying large numbers of widely distributed loads. To illustrate the basic features of parallel operation on a very simplified scale, consider an elementary system comprising two identical 3-phase generators G_1 and G_2 with their prime movers PM_1 and PM_2 supplying power to a load L, as shown in the single-line diagram of Fig. 7-27. Suppose generator G_1 is supplying the load at rated voltage and frequency, with generator G_2 disconnected. Generator G_2 can be paralleled with G_1 by driving it at synchronous speed and adjusting its field excitation so that its terminal voltage equals that of the bus. If the frequency of the incoming machine is not exactly equal to that of the bus, the phase relation between its voltage and that of the bus will vary at a frequency equal to the difference between the frequencies of the two voltages—perhaps a fraction of a cycle per second. The switch S_2 should be closed when the two voltages are momentarily in phase and the voltage across the switch is zero. A device for indicating the appropriate moment is known as a *synchroscope*. After G_2 has been synchronized in this manner, each machine can be made to take its share of the active- and reactive-power load by appropriate adjustments of the prime-mover throttles and field excitation.

In contrast with dc generators, paralleled synchronous generators must run at exactly the same steady-state speed (for the same number of poles).

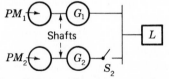

Fig. 7-27. Parallel operation of two synchronous generators.

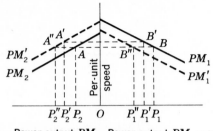

Fig. 7-28. Prime-mover speed-power characteristics.

Consequently, how the active power divides between them depends almost wholly on the speed-power characteristics of their prime movers. In Fig. 7-28 the sloping solid lines PM_1 and PM_2 represent the speed-power characteristics of the two prime movers for constant throttle openings. All practical prime movers have drooping speed-power characteristics. The total load P_L is shown by the solid horizontal line AB, and the generator power outputs are P_1 and P_2 (losses being neglected). Now suppose the throttle opening of PM_2 is increased, translating its speed-power curve upward to the dashed line PM_2'. The dashed line $A'B'$ now represents the load power. Note that the power output of generator 2 has now increased from P_2 to P_2' while that of generator 1 has decreased from P_1 to P_1'. At the same time, the system frequency has increased. The frequency can be restored to normal with a further load shift from generator 1 to generator 2 by closing the throttle on generator 1, lowering its speed-power curve to the dashed line PM_1'. The load power is now represented by $A''B''$, and the power outputs of the generators are P_1'' and P_2''. Thus, the system frequency and the division of active power between the generators can be controlled by the prime-mover throttles.

Changes in excitation affect the terminal voltage and reactive-power distribution. For example, let the two identical generators of Fig. 7-27 be adjusted to share the active and reactive loads equally. The phasor diagram is shown by the solid lines in Fig. 7-29, where $\widehat{V}_{ta}$ is the terminal voltage, $\widehat{I}_L$ is the load current, $\widehat{I}_a$ is the armature current in each generator, and $\widehat{E}_{af}$ is the excitation voltage. The synchronous-reactance voltage drop in each generator is $j\widehat{I}_a X_s$, and the resistance drops are neglected. Now suppose the excitation of generator 1 is increased. The bus voltage $\widehat{V}_{ta}$ will increase. It can then be restored to normal by decreasing the excitation of generator 2. The final condition is shown by the dashed phasors in Fig. 7-29. The terminal voltage, load current, and load power factor have been unchanged. Since the prime-mover throttles have not been touched, the power output and in-phase components of the generator armature currents have not been changed. The excitation voltages $\widehat{E}_{af1}$ and $\widehat{E}_{af2}$ have shifted in phase so that $\widehat{E}_{af} \sin \delta$ remains constant. The generator with the increased excitation has now taken on more of the lagging reactive-kVA load. For the condition shown by the dashed phasors in Fig. 7-29, generator 1 is supplying all the reactive kilovoltamperes, and gener-

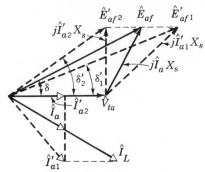

Fig. 7-29. Effects of changing excitations in two paralleled synchronous generators.

ator 2 is operating at unity power factor. Thus the terminal voltage and reactive-power distribution between the generators can be controlled by means of the field excitation.

Usually the prime-mover throttles are controlled by governors and automatic frequency regulators so that system frequency is maintained very nearly constant and power is divided properly among the generators. Voltage and reactive-power flow often are automatically regulated by voltage regulators acting on the field circuits of the generators and by transformers with automatic tap-changing devices.

7–9 SUMMARY

The physical picture of the internal workings of a synchronous machine in terms of rotating magnetic fields is rather simple. It is that of Art. 3-7: interaction of component fields of rotor and stator when the two are stationary with respect to each other. For both round-rotor and salient-pole machines, the component fields and mmf's, together with the associated voltages and currents, can be represented on phasor diagrams like Fig. 7-20. The phasor diagrams in turn lead to the concept of the synchronous reactances X_s, X_d, and X_q. These reactances are derived by replacing the effect of the rotating armature-reaction wave by magnetizing reactions X_φ or $X_{\varphi d}$ and $X_{\varphi q}$.

The unsaturated synchronous reactance X_s or X_d can be found from the results of an open-circuit and a short-circuit test. These test methods are a variation of a testing technique applicable not only to synchronous machines but also to anything whose behavior can be approximated by a linear equivalent circuit to which Thevenin's theorem applies. From the Thevenin-theorem viewpoint, an open-circuit test gives the internal voltage, and a short-circuit test gives information regarding the internal impedance. From the more specific viewpoint of electromagnetic machinery, an open-circuit test gives information regarding excitation requirements, core losses, and (for rotating machines) friction and windage losses; a short-circuit test gives infor-

mation regarding the magnetic reactions of the load current, leakage impedances, and losses associated with the load current such as I^2R and stray load
losses. The only real complication arises from the effects of magnetic nonlinearity, effects which can be taken into account approximately by considering
the machine to be equivalent to an unsaturated one whose magnetization
curve is the straight line Op of Fig. 7-9 and whose synchronous reactance is
empirically adjusted for saturation as in Eq. 7-28.

Prediction of the steady-state synchronous-machine characteristics then
becomes merely a study of power flow through a simple impedance with constant or easily determinable voltages at its ends. Study of the maximum-
power limits for short-time overloads is simply a special case of the limitations
on power flow through an inductive impedance. The power flow through such
an impedance can be expressed conveniently in terms of the voltages at its
sending and receiving ends and the phase angles associated with these voltages, as in Eq. 7-40 for a cylindrical-rotor machine and 7-52 for a salient-pole
machine. These analyses show that saliency has relatively little effect on the
interrelations between field excitation, terminal voltage, armature current,
and power; but the power-angle characteristics are affected by the presence of
a reluctance-torque component. Because of the reluctance torque, a salient-
pole machine is stiffer than one with a cylindrical rotor.

In the next chapter we extend our analysis of synchronous machines to
include transient conditions. This includes both the transient behavior of the
currents and fluxes in the machine and the transient interactions of the machine with the system with which it is connected.

PROBLEMS

7-1 The full-load torque angle δ_{RF} of a synchronous motor at rated voltage
and frequency is 30 electrical degrees. Neglect the effects of armature resistance and leakage reactance. If the field current is constant, how would the
torque angle be affected by the following changes in operating conditions?

(a) Frequency reduced 10 percent, load torque constant
(b) Frequency reduced 10 percent, load power constant
(c) Both frequency and applied voltage reduced 10 percent, load torque
    constant
(d) Both frequency and applied voltage reduced 10 percent, load power
    constant

7-2 A synchronous generator has a synchronous reactance of 1.6 per unit,
negligible armature resistance, and an armature leakage reactance of 0.13 per
unit. The generator is supplying 0.85 per unit power at unity power factor
directly to an infinite bus of unity voltage magnitude. Use the infinite-bus
voltage $\widehat{V}_\infty$ as a reference.

(a) Calculate the excitation voltage $\widehat{E}_{af}$ and the air-gap voltage $\widehat{E}_R$.

(b) Draw a phasor diagram showing the voltage $\widehat{V}_\infty$, $\widehat{E}_{af}$, and $\widehat{E}_R$ and the armature current $\widehat{I}_a$.

(c) On the same phasor diagram indicate the field flux $\widehat{\Phi}_{af}$, the resultant air-gap flux $\widehat{\Phi}_R$, and the flux of armature reaction $\widehat{\Phi}_{ar}$, equal to the phasor difference of the air-gap flux minus the field flux.

7-3 Design calculations show the following parameters for a round-rotor synchronous generator.

$$\text{Phase } a \text{ self-inductance } L_{aa} = 2.65 \text{ mH}$$
$$\text{Armature leakage inductance } L_{al} = 0.255 \text{ mH}$$

Calculate the phase-phase mutual inductance and the machine synchronous inductance.

7-4 A 2-pole 60-Hz synchronous generator develops an rms excitation voltage of 2.4 kV line-neutral with an applied field current of 45 A dc. What is the armature-to-field mutual inductance L_{af} in henrys?

7-5 A 50-MVA 13.8-kV synchronous machine is operating as a synchronous condenser, as discussed in Art. 4-6a. The generator short-circuit ratio is 0.625, and the field current at rated voltage, no load, is 250 A. Assume the generator to be connected directly to a 13.8-kV voltage source.

(a) What is the saturated synchronous reactance of the generator in per unit and in ohms per phase?

The generator field current is adjusted to 200 A.

(b) Draw a phasor diagram indicating the terminal voltage, the excitation voltage, and the armature current.

(c) Calculate the armature current magnitude in per unit and in amperes.

(d) What is the phase angle of the armature current with respect to the terminal voltage?

(e) Under these conditions, does the synchronous condenser appear inductive or capacitive to the 13.8-kV system?

(f) Repeat parts (b) to (e) for a field current of 400 A.

7-6 Figure 7-30 shows the single-phase equivalent circuit of a 3-phase 60-Hz 3600-r/min synchronous generator connected to a resistive load R_L. The synchronous reactance is 45 Ω/phase, and the field current required to achieve an rms excitation voltage of 2.4 kV (line-to-neutral) is 60 A. The field current and prime-mover torque are adjusted so that the 3-phase power delivered to the resistive load is 433 kW at 60 Hz.

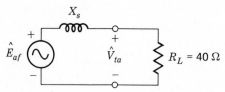

Fig. 7-30. Single-phase equivalent circuit of a synchronous generator connected to a resistive load.

(a) Calculate the line-to-line terminal voltage.

(b) Calculate the line-to-neutral excitation voltage and the field current.

(c) With the field current held constant, the prime-mover torque is reduced until the generator speed is 3000 r/min. Calculate the resultant terminal voltage and 3-phase load power.

7-7 A synchronous generator is connected directly to a 13.8-kV system. The generator synchronous reactance is 7.36 Ω/phase, and the armature resistance is negligible. The generator real power output is measured as 23 MW, and the reactive-power output is 10.3 MVA.

(a) Calculate the magnitude of the rms excitation voltage $\widehat{E}_{af}$ and its angle with respect to the generator terminal voltage.

(b) Draw a phasor diagram indicating the terminal voltage, the excitation voltage, the armature current, and the voltage drop across the synchronous reactance.

7-8 The following readings are taken from the results of an open- and a short-circuit test on a 9375-kVA 3-phase Y-connected 13,800-V (line-to-line) 2-pole 60-Hz turbine generator driven at synchronous speed:

Field current, A	169	192
Armature current, short-circuit test, A	392	446
Line voltage, open-circuit characteristic, V	13,000	13,800
Line voltage, air-gap line, V	15,400	17,500

The armature resistance is 0.064 Ω/phase. The armature leakage reactance is 0.10 per unit on the generator rating as a base. Find (a) the unsaturated value of the synchronous reactance in ohms per phase and per unit, (b) the short-circuit ratio, and (c) the saturated synchronous reactance in per unit and Ω/phase.

7-9 (a) Compute the field current required in the generator of Prob. 7-8 at rated voltage, rated-kVA load, 0.80 power factor lagging. Account for saturation under load by the method described in the paragraph relating to Eq. 7-28.

(b) In addition to the data given in Prob. 7-8, more points on the open-circuit characteristic are given below:

Field current, A	200	250	300	350
Line voltage, V	14,100	15,200	16,000	16,600

Find the voltage regulation for the load of part (a). *Voltage regulation* is defined as the rise in voltage when load is removed, the speed and field excitation being held constant. It is usually expressed as a percentage of the voltage under load.

7-10 Loss data for the generator of Prob. 7-8 are as follows:

Open-circuit core loss at 13,800 V = 68 kW

Short-circuit load loss at 392 A, 75°C = 50 kW

Friction and windage = 87 kW

Field-winding resistance at 75°C = 0.285 Ω

Compute the efficiency at rated load, 0.80 power factor lagging.

7-11 A 3-phase synchronous generator is rated 12,000 kVA, 13,800 V, 0.80 power factor, 60 Hz. What should be its kVA and voltage rating at 0.80 power factor and 50 Hz if the field and armature I^2R losses are to be the same as at 60 Hz? If its voltage regulation at rated load and 60 Hz is 18 percent, what will the value of the voltage regulation be at its rated load for 50 Hz operation? The effect of armature-resistance voltage drop on regulation may be neglected.

7-12 A synchronous motor is operating at half load. An increase in its field excitation causes a decrease in armature current. Before the increase, was the motor delivering or absorbing lagging reactive kilovoltamperes?

7-13 A dc shunt motor is mechanically coupled to a 3-phase cylindrical-rotor synchronous generator. The dc motor is connected to a 230-V constant-voltage dc supply, and the ac generator is connected to a 230-V (line-to-line) constant-voltage constant-frequency 3-phase supply. The 4-pole Y-connected synchronous machine is rated 25 kVA, 230 V, and has a synchronous reactance of 1.60 Ω/phase. The 4-pole dc machine is rated 25 kW, 230 V. All losses are to be neglected.

(a) If the two machines act as a motor-generator set receiving power from the dc mains and delivering power to the ac mains, what is the excitation voltage of the ac machine in volts per phase (line to neutral) when it delivers rated kilovoltamperes at 1.00 power factor?

(b) Leaving the field current of the ac machine as in part (a), what adjustment can be made to reduce the power transfer (between ac and dc) to zero? Under this condition of zero transfer, what is the armature current of the dc machine? What is the armature current of the ac machine?

(c) Leaving the field current of the ac machine as in parts (a) and (b), what adjustment can be made to cause 25 kW to be taken from the ac mains and delivered to the dc mains? Under these conditions, what is the armature current of the dc machine? What are the magnitude and phase of the current of the ac machine?

7-14 A 150-hp 0.8-power-factor 2300-V 38.0-A 60-Hz 3-phase synchronous motor has a direct-connected exciter to supply its field current. Assume the efficiency of the exciter to be constant at a value of 80 percent. The synchronous motor is run at no load from a 2300-V 60-Hz circuit, with its field current supplied by its exciter, and the following readings are taken:

Armature voltage between terminals = 2300 V

Armature current = 38.0 A/terminal

3-phase power input = 13.7 kW

Field current = 20.0 A

Voltage applied to field = 300 V

When the synchronous motor is loaded so that its input is 38.0 A at 0.80 power factor and 2300 V between terminals, its field current is 17.3 A. Under these conditions, what is the efficiency of the synchronous motor exclusive of the losses in its exciter? What is the useful mechanical-power output in horsepower?

7-15 From the phasor diagram of a synchronous machine with constant synchronous reactance X_s operating at constant terminal voltage V_{ta} and constant excitation voltage E_{af}, show that the locus of the tip of the armature-current phasor is a circle. On a phasor diagram with terminal voltage chosen as the reference phasor indicate the position of the center of this circle and its radius. Express the coordinates of the center and the radius of the circle in terms of V_{ta}, E_{af}, and X_s.

7-16 Superconducting synchronous generators are designed with superconducting field windings which can carry large current densities and create large magnetic flux densities. Since typical operating magnetic flux densities exceed the saturation flux densities of iron, these machines are designed without iron in the magnetic circuit; as a result, these machines exhibit no saturation effects.

A 2-pole 60-Hz 13.8-kV 10-MVA superconducting generator achieves rated open-circuit armature voltage at a field current of 842 A. It achieves rated armature current into a terminal 3-phase short circuit with a field current of 226 A. Consider this generator to be connected to a 13.8-kV distribution feeder of negligible impedance with an output power of 8.5 MW at 0.85 power factor lagging. Calculate (a) the per unit synchronous inductance; (b) the field current in amperes and rotor angle for this loading; and (c) the resultant rotor angle and reactive-power output in megavoltamperes if the field current is reduced to 842 A while the prime-mover torque is held constant.

7-17 A synchronous generator is connected to an infinite bus through two parallel 3-phase transmission circuits each having a reactance of 0.60 per unit including step-up and step-down transformers at each end. The synchronous reactance of the generator (which may be handled on a cylindrical-rotor basis) is 0.90 per unit. All resistances are negligible, and reactances are expressed on the generator rating as a base. The infinite-bus voltage is 1.00 per unit.

(a) The power output and excitation of the generator are adjusted so that it delivers rated current at 1.0 power factor at its terminals in the steady state. Compute the generator terminal and excitation voltages, the power output, and the reactive power delivered to the infinite bus.

(b) The throttle of the prime mover is now adjusted so that there is no power transfer between the generator and the infinite bus. The field current of the generator is adjusted until 0.50 per unit lagging reactive kilovoltamperes are delivered to the infinite bus. Under these conditions, compute the terminal and excitation voltages of the generator.

(c) The system is then returned to the operating conditions of part (a). One of the two parallel transmission circuits is disconnected by tripping the circuit breakers at its ends. The generator excitation is kept constant. Will the generator remain in synchronism? After comparing the desired power transfer with the maximum under these conditions, give an opinion regarding the adequacy of the transmission system.

7-18 A 1000-hp 2300-V Y-connected 3-phase 60-Hz 20-pole synchronous motor has a synchronous reactance of 4.00 Ω/phase. In this problem cylindrical-rotor theory may be used, and all losses may be neglected.

(a) This motor is operated from an infinite bus supplying rated voltage at rated frequency, and its field excitation is adjusted so that the power factor is unity when the shaft load is such as to require an input of 800 kW. If the shaft load is slowly increased with the field excitation held constant, determine the maximum torque (in pound-feet) that the motor can deliver.

(b) Instead of the infinite bus of part (a) suppose that the power supply is a 1000-kVA 2300-V Y-connected synchronous generator whose synchronous reactance is also 4.00 Ω/phase. The frequency is held constant by a governor, and the field excitations of motor and generator are held constant at the values which result in rated terminal voltage when the motor absorbs 800 kW at unity power factor. If the shaft load on the synchronous motor is slowly increased, determine the maximum torque (in pound-feet). Also determine the armature current, terminal voltage, and power factor at the terminals corresponding to this maximum load.

(c) Determine the maximum motor torque if, instead of remaining constant as in part (b), the field currents of the generator and motor are slowly increased so that rated terminal voltage and unity power factor are always maintained while the shaft load is increased.

7-19 Draw the steady-state direct-quadrature phasor diagram for an overexcited synchronous motor, i.e., one whose field current is high enough for lagging reactive kVA to be delivered to the supply system. From this phasor diagram show that the torque angle δ between the excitation and terminal voltage phasors is given by

$$\tan \delta = \frac{I_a X_q \cos \phi + I_a R_a \sin \phi}{V_{ta} - I_a X_q \sin \phi + I_a R_a \cos \phi}$$

Consider ϕ to be negative when $\widehat{I}_a$ lags $\widehat{V}_{ta}$.

7-20 What percent of its rated output will a salient-pole synchronous motor deliver without loss of synchronism when the applied voltage is normal and the field excitation is zero if $X_d = 0.80$ per unit and $X_q = 0.50$ per unit? Compute the per unit armature current at maximum power.

7-21 A salient-pole synchronous motor has $X_d = 0.80$ and $X_q = 0.50$ per unit. It is running from an infinite bus of $V_\infty = 1.00$ per unit. Neglect all losses. What is the minimum per unit excitation for which the machine will stay in synchronism with full-load torque?

7-22 A salient-pole synchronous generator with saturated synchronous inductances $X_d = 1.85$ per unit and $X_q = 1.65$ per unit is connected to an infinite bus through an external reactance $X_e = 0.15$ per unit. The generator is supplying its rated output power, unity power factor, as measured at the infinite bus.

(a) Draw a phasor diagram indicating the bus voltage, the armature current, the generator terminal voltage, the excitation voltage, and the rotor angle.

(b) Calculate the rotor angle in degrees.

(c) Calculate the per unit terminal and excitation voltages.

(d) Repeat parts (a) to (c) for the generator supplying its rated output power, unity power factor, as measured at the generator terminals.

8

Synchronous Machines: Transient Performance

In Chap. 7 the steady-state performance of synchronous machines was described. This steady-state behavior is based upon the interaction between the armature flux wave, which rotates at synchronous speed, and the synchronously rotating rotor flux wave. The steady-state torque level and the magnitude of the flux waves determine the constant angular separation between these flux waves.

During transient conditions, various disturbances can cause these fluxes to change magnitude and angular displacement as the rotor deviates from synchronous speed. Since electrical windings and solid-iron structures tend to oppose changes in flux linkages, these disturbances induce transient currents in the rotor and stator windings as well as in the rotor body of solid-rotor machines. The transient analysis of synchronous machines is thus concerned with determining transient fluxes and currents and their influence upon the electrical and electromechanical behavior of the machine.

361

In Art. 7-6 the concept of direct- and quadrature-axis analysis was introduced. This technique, based upon viewing the machine from a reference frame attached to the rotor, is extremely valuable in transient analysis of synchronous machines and will be developed further in this chapter.

8-1 SYNCHRONOUS–MACHINE TRANSIENTS

The inherent complexity of synchronous-machine transient phenomena can be appreciated by inspecting the main structure details of the machine with the object of discussing the significant circuits when transient conditions prevail. A salient-pole structure is shown in the schematic diagram of Fig. 8-1. Damper bars are used in such machines to provide induction-motor-type torques, which aid in damping electromechanical rotor oscillations following transient disturbances; the damper bars and field collar are all shorted together at the end of the rotor, forming a structure which appears much like the squirrel cage in an induction motor and whose function is quite similar.

Although solid cylindrical-rotor machines are generally not constructed with damper windings, following transients, currents which are induced directly in the rotor body play the same role as damper currents in a salient-pole machine. The rotors of salient-pole machines are normally made of stacks of thin insulated laminations, and hence rotor body currents are not induced in these machines. In fact, in the analysis of solid-rotor machines these rotor body currents are often referred to as damper currents. The salient-pole structure also emphasizes the difference between the polar, or direct, axis and the interpolar, or quadrature, axis; similar differences exist be-

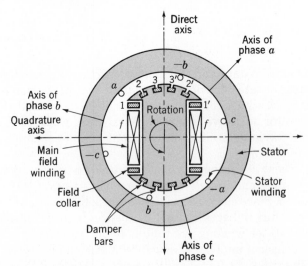

Fig. 8-1. Schematic diagram for synchronous machine showing significant circuits for transients.

tween the two axes of cylindrical-rotor machines when considering transient analysis, so that the two types of machines are treated in similar fashion in most transient analyses.

Under balanced steady-state conditions the mmf wave of the stator winding and the associated flux wave revolve at the same speed as the rotor and are of essentially fixed waveform. Then the flux linkages with the rotor circuits do not change with time, and no voltages are induced in these circuits. Thus the main field winding is the only rotor circuit which carries current, and its excitation is determined by the voltage applied to the field terminals.

Under transient conditions the magnitudes of the rotor and stator fluxes change, and often the rotor speed changes, causing the rotor to change its alignment with respect to the synchronously rotating stator flux. As a consequence, the flux linkages with all the rotor circuits change with time, and currents are induced in each of them, in the damper windings, in the rotor body, and in the field winding. As a consequence the net rotor mmf is produced by all the windings and the resultant electrical and electromechanical behavior of the machine is determined by the transient behavior of each of these currents. Transient analysis is most often accomplished by representing the machine as a set of mutually coupled circuits.

In Fig. 8-1 there is a stator circuit for each of the three phases, *a, b* and *c,* and there are rotor circuits corresponding to the field winding and to bars 2-2′ and 3-3′ and the conducting field collars 1-1′. An additional rotor circuit may be formed by the bolts and iron of the rotor structure. Currents induced in the rotors of cylindrical-rotor machines can also be represented in terms of current induced in equivalent rotor circuits although the exact current paths are not easily identifiable.

All these circuits have their own resistance, self-inductance, and mutual inductances with respect to every other circuit. Basic analysis of synchronous-machine transient behavior can thus be accomplished by the solution of a set of simultaneous coupled-circuit differential equations. This analysis is often linear in the sense that saturation effects are often ignored or at best represented by changing the magnetizing inductance as a function of net flux. Nevertheless, the solution of these equations can be a formidable task. The solution is expedited appreciably by a linear transformation of variables (known as the dq0 transformation) in which the stator currents, voltages, and fluxes are replaced by equivalent quantities rotating at rotor speed. For example, the three stator-phase currents i_a, i_b, and i_c are replaced by three transformed quantities, the *direct-axis component* i_d, the *quadrature-axis component* i_q, and the *zero-sequence component* i_0, which is zero under balanced terminal conditions of the sort considered in this chapter.

The mathematical formalization of this transformation (Art. 8-2) permits machine analysis to be performed in a reference frame fixed with respect to the rotor; the transformed variables are the equivalent armature quantities as seen by an observer in this reference frame. The physical significance of the direct- and quadrature-axis quantities is that given in Art. 7-6; direct-axis quantities lie along the axis of the field winding, whereas quadrature-axis quantities lie along an axis perpendicular to the field winding. The idea of making this transformation arose from an extension of the physical picture corresponding to the steady-state analysis.

This change of variables permits the simultaneous equations to be separated into two sets of equations (one for each axis) based upon inductances which are constant in time. This can easily be seen by recognizing that the actual mutual inductances between the stator-phase windings and the rotor circuits change with rotor position and hence with time whereas the transformed stator variables remain fixed in the rotor frame and hence the mutual inductances remain unchanged with time. Furthermore, this type of analysis quickly leads to the identification of various reactances and time constants which have proved quite useful in the analysis and characterization of synchronous machines and which are discussed at various points throughout this chapter.

The most common situation for which synchronous-machine transient analysis is required is that of a synchronous machine interconnected into a system consisting of many machines, loads, and a large transmission network. For some sorts of analyses it is sufficient to represent the system by a Thevenin-equivalent series impedance and voltage source (infinite bus), as discussed in Art. 7-1. In these cases, the emphasis is on the behavior of the single machine under the assumption that the transient disturbance can be considered localized to that particular unit.

In most cases, however, the disturbance is not so localized, and the analysis must involve a number of machines (perhaps even up to a few hundred units). In addition, the dynamics of turbine-governor and excitation systems may play an important role in the transient behavior and must be modeled. Although it is in theory possible to represent each machine by a complex model with a large number of differential equations to be solved, as a practical matter such a procedure often leads to formulations which are at best very slow and at worst too large to be implemented in any form. Thus, many levels of representation are used, ranging from a complex, multiwinding model down to a model in which the machine is represented as a constant voltage in series with a transient reactance.

This chapter cannot begin to cover the range of issues involved in the transient analysis of synchronous machines. It is intended rather to provide a basic understanding of the transient phenomena and as an introduction to the various models and analytical tools which are available.

8-2 TRANSFORMATION TO DIRECT- AND QUADRATURE-AXIS VARIABLES

In Art. 7-6 the concept of resolving armature quantities into two rotating components, one aligned with the field-winding axis, the direct-axis compo-

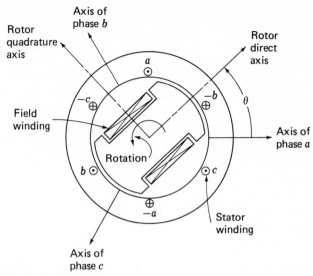

Fig. 8-2. Idealized synchronous machine.

nent, and one in quadrature with the field-winding axis, the quadrature-axis component, was introduced as a means of facilitating analysis in salient-pole machines. The usefulness of this concept stems from the fact that although each of the stator phases sees a time-varying inductance due to the saliency of the rotor, the transformed quantities rotate with the rotor and hence see constant magnetic paths. Additional saliency effects are present under transient conditions, due to the different conducting paths in the rotor, rendering the concept of this transformation all the more useful.

The idea behind the transformation is an old one, stemming from the work of André Blondel in France, and the technique is sometimes referred to as the *Blondel two-reaction method*. Much of the development in the form used here was carried out by R. E. Doherty, C. A. Nickle, R. H. Park, and their associates in the United States. The transformation itself (known as the dq0 transformation) can be represented in a straightforward fashion in terms of the electrical angle θ (equal to poles/2 times the spatial angle) between the rotor direct axis and the stator phase-a axis, as shown in Fig. 8-2. Letting S represent the stator quantity to be transformed (current, voltage, or flux), we can write the transformation in matrix form as

$$\begin{bmatrix} S_d \\ S_q \\ S_0 \end{bmatrix} = \frac{2}{3} \begin{bmatrix} \cos \theta & \cos (\theta - 120°) & \cos (\theta + 120°) \\ -\sin \theta & -\sin (\theta - 120°) & -\sin (\theta + 120°) \\ \frac{1}{2} & \frac{1}{2} & \frac{1}{2} \end{bmatrix} \begin{bmatrix} S_a \\ S_b \\ S_c \end{bmatrix} \qquad (8-1)$$

and the inverse transformation as

$$\begin{bmatrix} S_a \\ S_b \\ S_c \end{bmatrix} = \begin{bmatrix} \cos\theta & -\sin\theta & 1 \\ \cos(\theta - 120°) & -\sin(\theta - 120°) & 1 \\ \cos(\theta + 120°) & -\sin(\theta + 120°) & 1 \end{bmatrix} \begin{bmatrix} S_d \\ S_q \\ S_0 \end{bmatrix} \tag{8-2}$$

Here the subscripts d and q represent the direct and quadrature axes, respectively. A third component, the *zero-sequence component,* indicated by the subscript 0, is also included. This component is required to yield a unique transformation of the three stator-phase quantities; it corresponds to components of armature current which produce no net air-gap flux and hence no net flux linking the rotor circuits. As can be seen from Eq. 8-1, under balanced 3-phase conditions, there are no zero-sequence components. Only balanced 3-phase conditions are considered in this book, and hence zero-sequence components will not be discussed in any detail.

EXAMPLE 8-1

A 2-pole synchronous machine is carrying balanced 3-phase armature currents

$$i_a = I_a \cos \omega t \qquad i_b = I_a \cos(\omega t - 120°) \qquad i_c = I_a \cos(\omega t + 120°)$$

The rotor is rotating at synchronous speed ω, and the rotor direct axis is aligned with the stator phase-a axis at $t = 0$. Find the direct- and quadrature-axis current components.

Solution

The angle between the rotor direct axis and the stator phase-a axis can be expressed as

$$\theta = \omega t$$

From Eq. 8-1

$$i_d = \tfrac{2}{3}[i_a \cos \omega t + i_b \cos(\omega t - 120°) + i_c \cos(\omega t + 120°)]$$
$$= \tfrac{2}{3}I_a[\cos^2 \omega t + \cos^2(\omega t - 120°) + \cos^2(\omega t + 120°)]$$

Using the trigonometric identity $\cos^2 \alpha = \tfrac{1}{2}(1 + \cos 2\alpha)$ gives

$$i_d = I_a$$

Similarly,

$$i_q = -\tfrac{2}{3}[i_a \sin \omega t + i_b \sin (\omega t - 120°) + i_c \sin (\omega t + 120°)]$$
$$= -\tfrac{2}{3}I_a[\cos \omega t \sin \omega t + \cos (\omega t - 120°) \sin (\omega t - 120°)$$
$$+ \cos (\omega t + 120°) \sin (\omega t + 120°)]$$

and using the trigonometric identity $\cos \alpha \sin \alpha = \tfrac{1}{2} \sin 2\alpha$ gives

$$i_q = 0$$

This result corresponds directly to our physical picture of the dq transformation. From the discussion of Art. 3-5 we recognize that the balanced 3-phase currents correspond to a synchronously rotating current (or mmf) wave which is aligned with the stator phase-a axis at $t = 0$. This stator-current wave is thus aligned with the rotor direct axis at $t = 0$ and remains so since the rotor is rotating at the same speed. Hence the stator current consists only of a direct-axis component.

8-3 BASIC MACHINE RELATIONS IN dq0 VARIABLES

Equation 7-9 gives the flux-current relationships for a synchronous machine consisting of a field winding and three stator phase windings. This simple machine is sufficient to demonstrate the basic features of the machine representation in dq0 variables; the effects of additional rotor circuits can be introduced in a straightforward fashion.

The flux-current relationships in terms of phase variables (Eqs. 7-2 to 7-5) are repeated here for convenience

$$\begin{bmatrix} \lambda_a \\ \lambda_b \\ \lambda_c \\ \lambda_f \end{bmatrix} = \begin{bmatrix} \mathcal{L}_{aa} & \mathcal{L}_{ab} & \mathcal{L}_{ac} & \mathcal{L}_{af} \\ \mathcal{L}_{ba} & \mathcal{L}_{bb} & \mathcal{L}_{bc} & \mathcal{L}_{bf} \\ \mathcal{L}_{ca} & \mathcal{L}_{cb} & \mathcal{L}_{cc} & \mathcal{L}_{cf} \\ \mathcal{L}_{fa} & \mathcal{L}_{fb} & \mathcal{L}_{fc} & \mathcal{L}_{ff} \end{bmatrix} \begin{bmatrix} -i_a \\ -i_b \\ -i_c \\ i_f \end{bmatrix} \qquad (8\text{-}3)$$

where the negative signs have been added consistent with a choice of generator reference for the armature currents. Unlike the analysis of Art. 7-2, this analysis will include the effects of saliency, which cause the stator self- and mutual inductances to vary with rotor position.

For the purposes of this analysis the idealized synchronous machine of Fig. 8-2 is assumed to satisfy two conditions: (1) the air-gap permeance has a constant component as well as a smaller component which varies cosinusoidally with rotor angle as measured from the direct axis and (2) the effects of space harmonics in the air-gap flux can be ignored. Although these approximations may appear somewhat restrictive, they form the basis of classical dq0 machine analysis and have been found to give excellent results in a wide variety of applications. Essentially they involve neglecting effects which result in time-harmonic stator voltages and currents and are thus consistent with our previous assumptions neglecting harmonics produced by discrete windings.

The various machine inductances can then be written in terms of the electrical rotor angle θ (between the rotor direct axis and the stator phase-a axis), using the notation of Art. 7-2, as follows. For the stator self-inductances

$$\mathcal{L}_{aa} = L_{aa0} + L_{al} + L_{g2} \cos 2\theta \tag{8-4}$$

$$\mathcal{L}_{bb} = L_{aa0} + L_{al} + L_{g2} \cos (2\theta + 120°) \tag{8-5}$$

$$\mathcal{L}_{cc} = L_{aa0} + L_{al} + L_{g2} \cos (2\theta - 120°) \tag{8-6}$$

For the stator-to-stator mutual inductances

$$\mathcal{L}_{ab} = \mathcal{L}_{ba} = -\tfrac{1}{2}L_{aa0} + L_{g2} \cos (2\theta - 120°) \tag{8-7}$$

$$\mathcal{L}_{bc} = \mathcal{L}_{cb} = -\tfrac{1}{2}L_{aa0} + L_{g2} \cos 2\theta \tag{8-8}$$

$$\mathcal{L}_{ac} = \mathcal{L}_{ca} = -\tfrac{1}{2}L_{aa0} + L_{g2} \cos (2\theta + 120°) \tag{8-9}$$

For the field-winding self-inductance

$$\mathcal{L}_{ff} = L_{ff} \tag{8-10}$$

and for the stator-to-rotor mutual inductances

$$\mathcal{L}_{af} = \mathcal{L}_{fa} = L_{af} \cos \theta \tag{8-11}$$

$$\mathcal{L}_{bf} = \mathcal{L}_{fb} = L_{af} \cos (\theta - 120°) \tag{8-12}$$

$$\mathcal{L}_{cf} = \mathcal{L}_{fc} = L_{af} \cos (\theta + 120°) \tag{8-13}$$

Comparison with Art. 7-2 shows that the effects of saliency appear only in the stator self- and mutual-inductance terms as an inductance term which varies with 2θ. This twice-angle variation can be understood with reference to Fig. 8-2, where it can be seen that rotation of the the rotor through 180° reproduces the original geometry of the magnetic circuit. Notice that the self-inductance of each stator phase is a maximum when the rotor direct axis is aligned with the axis of that phase and that the phase-phase mutual inductance is maximum when the rotor direct axis is aligned with the midpoint of the two phases. This is the expected result since the rotor direct axis is the path of lowest reluctance (maximum permeance) for air-gap flux.

The flux-linkage expressions of Eq. 8-3 become much simpler when expressed in terms of dq0 variables. This can be done by application of the transformation of Eq. 8-1 to both the flux linkages and the currents of Eq. 8-3. The manipulations are somewhat laborious and will be omitted here because they are simply algebraic. The results are

$$\lambda_d = -L_d i_d + L_{af} i_f \qquad (8\text{-}14)$$
$$\lambda_q = -L_q i_q \qquad (8\text{-}15)$$
$$\lambda_f = -\tfrac{3}{2} L_{af} i_d + L_{ff} i_f \qquad (8\text{-}16)$$
$$\lambda_0 = -L_0 i_0 \qquad (8\text{-}17)$$

In these equations, new inductance terms appear. They are

$$L_d = L_{al} + \tfrac{3}{2}(L_{aa0} + L_{g2}) \qquad (8\text{-}18)$$
$$L_q = L_{al} + \tfrac{3}{2}(L_{aa0} - L_{g2}) \qquad (8\text{-}19)$$
$$L_0 = L_{al} \qquad (8\text{-}20)$$

The quantities L_d and L_q are the *direct-axis* and *quadrature-axis synchronous inductances,* as discussed in Art. 7-6. The inductance L_0 is the *zero-sequence inductance.* Notice that the transformed flux-current relationship expressed in Eqs. 8-14 to 8-17 no longer contain inductances which are functions of rotor position. It is this feature which is responsible for the usefulness of the dq0 transformation.

Transformation of the voltage equations (where $p = d/dt$)

$$v_a = -R_a i_a + p\lambda_a \qquad (8\text{-}21)$$
$$v_b = -R_a i_b + p\lambda_b \qquad (8\text{-}22)$$
$$v_c = -R_a i_c + p\lambda_c \qquad (8\text{-}23)$$
$$v_f = R_f i_f + p\lambda_f \qquad (8\text{-}24)$$

results in

$$v_d = -R_a i_d + p\lambda_d - \omega\lambda_q \qquad (8\text{-}25)$$
$$v_q = -R_a i_q + p\lambda_q + \omega\lambda_d \qquad (8\text{-}26)$$
$$v_f = R_f i_f + p\lambda_f \qquad (8\text{-}27)$$
$$v_0 = -R_a i_0 + p\lambda_0 \qquad (8\text{-}28)$$

(algebraic details again omitted), where $\omega = p\theta = d\theta/dt$ is the rotor electrical angular velocity.

In Eqs. 8-25 and 8-26 the terms $\omega\lambda_q$ and $\omega\lambda_d$ are speed-voltage terms which come as a result of the fact that we have chosen to define our variables in a reference frame rotating at angular velocity ω. These speed-voltage terms are directly analogous to the speed-voltage terms found in the dc machine analysis of Chap. 5. In the dc machine the commutator-brush system performs the transformation of transferring armature (rotor) voltages to the field-winding (stator) reference frame.

We now have the basic relations for analysis of our simple synchronous machine. They consist of the flux-current equations 8-14 to 8-17, the voltage equations 8-25 to 8-28, and the transformation equations 8-1 and 8-2. When the machine speed ω is constant, the differential equations are linear with constant coefficients. In addition, the transformer terms $p\lambda_d$ and $p\lambda_q$ in Eqs. 8-25 and 8-26 are often negligible with respect to the speed-voltage terms $\omega\lambda_q$ and $\omega\lambda_d$, providing further simplification. As illustrated later, omission of these terms corresponds to neglecting the harmonics and dc component in the transient solution for stator voltages and currents. In any case, the transformed equations are generally much easier to solve, both analytically and by computer simulation, than the equations expressed directly in terms of phase variables.

In using these equations and the corresponding equations in the machinery literature, careful note should be made of the sign convention and units employed. Here, we have chosen the generator convention for armature currents, with positive armature current flowing out of the machine. Also, standard SI units (volts, amperes, ohms, henrys, etc.) are used here; often in the literature one of several per unit systems is used to provide numerical simplification.†

To complete the basic array, relations for power and torque are needed. The instantaneous power output from the 3-phase stator is

$$p_s = v_a i_a + v_b i_b + v_c i_c \tag{8-29}$$

Phase quantities can be eliminated from Eq. 8-29 by using Eq. 8-2 written for voltages and currents. The result is

$$p_s = \tfrac{3}{2}(v_d i_d + v_q i_q + 2v_0 i_0) \tag{8-30}$$

The electromagnetic torque developed, which acts to decelerate the shaft, is readily obtained as the power output corresponding to the speed voltages divided by the shaft speed (in mechanical radians per second). From Eq. 8-30 with the speed-voltage terms from 8-25 and 8-26, by recognizing ω as the speed in electrical radians per second we get

$$T = \frac{3}{2}\frac{\text{poles}}{2}(\lambda_d i_q - \lambda_q i_d) \tag{8-31}$$

The torque of Eq. 8-31 is positive for generator action. It can be seen that this result is in general conformity with torque production from interacting magnetic fields as expressed in Eq. 3-86. In Eq. 8-31 we are superimposing the

†See A. W. Rankin, Per-Unit Impedances of Synchronous Machines, *Trans. AIEE,* **64**:569–573, 839–841(1945).

interaction of components: direct-axis magnetic field with quadrature-axis mmf, and quadrature-axis magnetic field with direct-axis mmf. For both interactions the sine of the space-displacement angle has unity magnitude. Again the use of the dq transformation leads to a compact result which can be related to our earlier work.

8–4 ANALYSIS OF A SUDDEN THREE–PHASE SHORT CIRCUIT

In order to demonstrate the application of the dq0 transformation and to begin to develop an understanding of the transient behavior of synchronous machines it is useful to analyze the transient following a sudden 3-phase short circuit at the armature terminals. The machine is assumed to be initially unloaded and to continue operating at synchronous speed ω after the short circuit occurs. Since the short circuit is balanced, zero-sequence quantities do not appear. For the purposes of this analysis, the only rotor circuit to be considered will be a field winding, and hence the machine can be described by Eqs. 8-14 to 8-16 and 8-25 to 8-27.

a. Physical Description of the Transient

Before investigating the analytical solution to this problem, it is useful to start with a physical description of the transient. Since the machine is initially unloaded, the only predisturbance current flowing in the machine is the field current. Each armature phase sees a resultant time-varying flux linkage as the rotor rotates.

The situation changes radically upon application of the 3-phase short circuit. Currents now flow in the armature windings in such a fashion as to maintain their flux linkages at the value they had at the time of the short circuit. There are two components of these currents. One is an ac component, corresponding to the armature current required to oppose the time-varying flux produced by the field winding as it rotates. The other is a dc component, corresponding to the initial flux linkage which existed at the time of the short. Thus, the net result of these currents is an armature flux linkage which is fixed in space, each armature phase holding constant its own portion of the initial armature flux linkage.

A similar situation occurs on the field winding. It is now rotating in a trapped armature-flux wave and hence must respond with an alternating component of field current to oppose the tendency of this trapped armature flux to change its own flux linkages. In addition, there must be an induced dc component of field current to oppose the synchronously rotating component of flux created by the ac components of armature current, which, in order to oppose the tendency of the rotating-field flux to change the armature flux linkages, in turn create a rotating flux which attempts to demagnetize (reduce) the flux created by the field winding.

Quite clearly, all these things happen simultaneously, and it is difficult to give a description in terms of cause and effect. However, it is possible to recognize that during this transient, ac currents in the stator correspond to dc currents on the rotor and vice versa. It should also be clear that these conditions do not remain indefinitely. Resistance in the armature and field windings means that the transient dc currents tend to decay away since there is no source to drive them. As a result, the dc armature current and the corresponding ac field current decay with a time constant determined by the armature resistance R_a. Similarly, the transient dc field current and corresponding transient ac component of armature current decay with a time constant determined by the field resistance R_f.

Figure 8-3 shows the waveforms of the 3-phase currents and the field

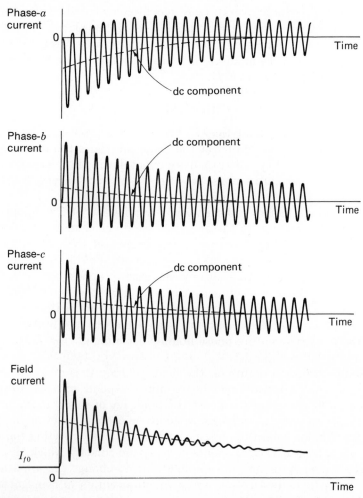

Fig. 8-3. Armature and field currents following a sudden short circuit on an initially unloaded synchronous machine.

current following a sudden 3-phase short circuit. In the armature-current waveform, the amount of dc offset in each phase is determined by the flux in that phase at the time of the short. The armature currents are balanced and thus, like the ac components, the dc components sum to zero at any given instant. The ac component of field current is due to the dc component of armature current and decays at the same rate determined by the armature resistance. The transient dc component of field current and the corresponding transient ac component of armature current decay together at a rate determined by the field-winding resistance.

b. Armature and Field Resistances Neglected

An analysis based upon the assumption that the armature and field resistances are negligibly small will result in a transient similar to that discussed above. However, the dc components of armature and field current will not decay with time. The basic machine equations under these conditions become

$$\lambda_f = L_{ff}i_f - \tfrac{3}{2}L_{af}i_d \tag{8-32}$$

$$\lambda_d = L_{af}i_f - L_d i_d \tag{8-33}$$

$$\lambda_q = -L_q i_q \tag{8-34}$$

$$v_d = p\lambda_d - \lambda_q \omega \tag{8-35}$$

$$v_q = p\lambda_q + \lambda_d \omega \tag{8-36}$$

$$v_f = p\lambda_f \tag{8-37}$$

The currents i_d and i_q are initially zero, and i_f is initially I_{f0}.

It is now necessary to represent the short circuit in mathematical form. This can be done by noting its effect on the machine voltages. Both before and after the short circuit

$$v_d = 0 \tag{8-38}$$

Hence the short circuit has no effect on v_d. The voltage v_q, however, is $\omega L_{af}I_{f0}$ before the short circuit and zero afterward. Since the short circuit causes v_q to vanish suddenly, its effect can be found by suddenly impressing on the machine the voltage

$$v_q = -\omega L_{af}I_{f0} \tag{8-39}$$

The currents found in this way must be added to the initial values of the currents existing before the short circuit in order to obtain the resultant currents after the short circuit.

The response to the voltage of Eq. 8-39 must be the response for this voltage alone, not the combined effect of this voltage and the initial excitation. Accordingly, the response must be evaluated with the field winding considered closed and initially unexcited, i.e., with

$$v_f = 0 \qquad (8\text{-}40)$$

The subscript t (for transient) will be added to the symbols $i_f, \lambda_f, \lambda_d,$ and λ_q to denote the values under these conditions. Then the true initial values are added, as in

$$i_f = I_{f0} + i_{ft} \qquad (8\text{-}41)$$

to find the resultant quantities after the short circuit, thereby superimposing the effect of the true initial excitation. No added subscripts are necessary for i_d and i_q because their initial values are obviously zero for an unloaded machine.

From Eqs. 8-37 and 8-40 it can be seen that the field linkages λ_{ft} must be zero. In other words, with R_f neglected, the flux linkages λ_f with the field winding cannot change. This is a specific case of what is sometimes called the *principle of constant flux linkages,* which simply states that the flux linkages with a closed circuit having zero resistance and no voltage source cannot change. To maintain field linkages constant, a current must be induced in the field winding to counteract the magnetic effects of i_d. From Eq. 8-32, with $\lambda_f = 0$, this current is

$$i_{ft} = +\frac{3}{2}\frac{L_{af}}{L_{ff}}i_d \qquad (8\text{-}42)$$

Upon substitution of Eq. 8-42 in 8-33,

$$\lambda_{dt} = -\left(L_d - \frac{3}{2}\frac{L_{af}^2}{L_{ff}}\right)i_d \qquad (8\text{-}43)$$

$$= -L_d' i_d \qquad (8\text{-}44)$$

where $\qquad\qquad L_d' = L_d - \frac{3}{2}\frac{L_{af}^2}{L_{ff}} \qquad (8\text{-}45)$

is the *direct-axis transient inductance.* As we shall see, it is an important constant in machine-transient theory.

The specific values of the voltages v_d and v_q are now introduced. From Eqs. 8-34 and 8-35, with $v_d = 0$,

$$0 = p\lambda_{dt} + \omega L_q i_q \qquad (8\text{-}46)$$

or, after using Eq. 8-44 to obtain $p\lambda_{dt}$,

$$0 = -L_d' p i_d + \omega L_q i_q \qquad (8\text{-}47)$$

Similarly, from Eqs. 8-36 and 8-39

$$-\omega L_{af}I_{f0} = p\lambda_q + \lambda_d\omega \qquad (8\text{-}48)$$

which, with Eqs. 8-34 and 8-44, becomes

$$\omega L_{af}I_{f0} = L_q p i_q + \omega L_d' i_d \qquad (8\text{-}49)$$

Equations 8-47 and 8-49 are two simultaneous equations in i_d and i_q. With i_q eliminated and the resulting equation written in differential-equation form, we have

$$\frac{d^2 i_d}{dt^2} + \omega^2 i_d = \frac{\omega^2}{L_d'}L_{af}I_{f0} \qquad (8\text{-}50)$$

The solution of Eq. 8-50 is

$$i_d = \frac{L_{af}I_{f0}}{L_d'}(1 - \cos\omega t) \qquad (8\text{-}51)$$

From Eq. 8-47

$$i_q = \frac{L_d'}{\omega L_q}\frac{d i_d}{dt} = \frac{L_{af}I_{f0}}{L_q}\sin\omega t \qquad (8\text{-}52)$$

From Eqs. 8-41 and 8-42

$$i_f = I_{f0} + \frac{3}{2}\frac{L_{af}}{L_{ff}}\frac{L_{af}I_{f0}}{L_d'}(1 - \cos\omega t) \qquad (8\text{-}53)$$

The waveforms of i_d, i_q, and i_f as given by these results are sketched in Fig. 8-4.
Upon substitution in the dq0 transformation (Eq. 8-2), with θ taken as $\omega t + \theta_0$, the phase current is

$$i_a = \frac{L_{af}I_{f0}}{L_d'}\cos(\omega t + \theta_0) - \frac{L_{af}I_{f0}}{2}\left(\frac{1}{L_d'} + \frac{1}{L_q}\right)\cos\theta_0$$

$$- \frac{L_{af}I_{f0}}{2}\left(\frac{1}{L_d'} - \frac{1}{L_q}\right)\cos(2\omega t - \theta_0) \qquad (8\text{-}54)$$

From Eqs. 8-3 and 8-21 the peak value of prefault armature-voltage mag-

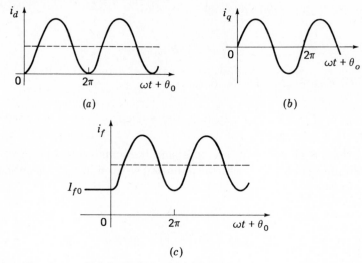

Fig. 8-4. Waveforms of (a) i_d, (b) i_q, and (c) i_f following short circuit. Machine resistances are ignored.

nitude is $\sqrt{2}\,E_{af0} = \omega L_{af} I_{f0}$. Thus Eq. 8-54 can be rewritten

$$i_a = \frac{\sqrt{2}E_{af0}}{X_d'} \cos\,(\omega t + \theta_0) - \frac{\sqrt{2}E_{af0}}{2}\left(\frac{1}{X_d'} + \frac{1}{X_q}\right)\cos\theta_0$$

$$- \frac{\sqrt{2}E_{af0}}{2}\left(\frac{1}{X_d'} - \frac{1}{X_q}\right)\cos\,(2\omega t + \theta_0) \tag{8-55}$$

where
$$X_d' = \omega L_d' \tag{8-56}$$

is the *direct-axis transient reactance.*

The phase current is seen to consist of a fundamental-frequency term, a dc component, and a second harmonic. The fundamental is the most important. Its magnitude depends on the prefault excitation and the transient reactance. The dc term depends on θ_0 and hence on the instant when the short circuit appears. The rotor position θ_0 at the time of the short circuit determines the initial flux linkages of each armature phase; the dc currents flow to maintain these flux linkages. If, for example, the fault occurs at the instant when the direct axis is 90° from the axis of phase a, no dc component appears in the phase-a current. Even then, dc components do appear in the phase-b and phase-c currents because of the 120° displacement between phases. The second-harmonic term depends on so-called *transient saliency,* the differences in transient circuitry in the two rotor axes as represented by the quantity $(1/X_d') - (1/X_q)$. The second-harmonic amplitude is usually small and often neglected.

Notice that as a result of the transformation from dq0 to phase variables, the dc component of i_d corresponds to the ac component of i_a and the ac components of i_d and i_q correspond to dc and second-harmonic components of i_a. This is the expected result of a transformation from rotating to fixed-frame variables.

Because the field and armature resistances were neglected, none of the terms in Eqs. 8-51 to 8-55 dies away with time. Actually, exponential decays are encountered, as they are in any inductive circuit when resistance is present. This fact is illustrated in Fig. 8-3, which shows the phase currents with their dc components after a short circuit. Before investigating the rates of decay, however, we wish to simplify our basic relations somewhat further. This will be done by showing that neglecting the transformer-voltage terms $p\lambda_d$ and $p\lambda_q$ in Eqs. 8-35 and 8-36 corresponds to dropping out the dc and second-harmonic terms in the above solutions.

c. Resistances and Transformer Voltages Neglected

When the $p\lambda_d$ and $p\lambda_q$ terms are omitted from Eqs. 8-35 and 8-36, very simple expressions for the machine currents are obtained. They are

$$i_d = \frac{L_{af}I_{f0}}{L_d'} \tag{8-57}$$

$$i_q = 0 \tag{8-58}$$

$$i_f = I_{f0} + i_{ft} = I_{f0} + \frac{3}{2}\frac{L_{af}}{L_{ff}}\frac{L_{af}I_{f0}}{L_d'} \tag{8-59}$$

Substitution in the dq0 transformation (Eq. 8-2) and use of Eq. 8-56 give

$$i_a = \frac{\sqrt{2}E_{af0}}{X_d'}\cos(\omega t + \theta_0) \tag{8-60}$$

The dc and second-harmonic terms have now dropped out of the phase current, as have the fundamental-frequency terms in i_d, i_q, and i_f. The currents i_d and i_f now follow the dashed lines in Fig. 8-4a and c; they have a sudden step at $t = 0$.

Neglect of the second-harmonic and dc components in the phase currents is very common in machine analysis. The former is usually quite small, and the latter dies away very rapidly. Neither has a significant influence on the average torque of the machine for an appreciable time, so that their influence on dynamic performance is relatively small. Accordingly, many machine analyses are carried out with the $p\lambda_d$ and $p\lambda_q$ terms omitted. The advantage is that the resulting equations are much simpler.

d. Field Resistance Included, Transformer Voltages Neglected

We now wish to investigate the decay of the currents i_d and i_{ft} of Eqs. 8-57 and 8-59. To do so, we follow substantially the approach of part b of this article except that the $p\lambda_d$ and $p\lambda_q$ terms are omitted but the $R_f i_f$ term is retained in Eq. 8-27. Again,

$$i_q = 0 \tag{8-61}$$

The equation followed by the induced field current i_{ft} becomes

$$\frac{L_{ff}}{L_d}\left(L_d - \frac{3}{2}\frac{L_{af}^2}{L_{ff}}\right)\frac{di_{ft}}{dt} + R_f i_{ft} = 0 \tag{8-62}$$

or, by using Eq. 8-45 and dividing through by R_f,

$$\frac{L_{ff}}{R_f}\frac{L_d'}{L_d}\frac{di_{ft}}{dt} + i_{ft} = 0 \tag{8-63}$$

Now the fraction

$$\frac{L_{ff}}{R_f} = T_{d0}' \tag{8-64}$$

is the open-circuit time constant of the field winding. (The symbol T is used for time constant in this chapter to conform with standard notation. Elsewhere the symbol τ is used; however, there is practically no possibility of confusion with torque here.) It is called the *direct-axis open-circuit transient time constant*. It characterizes the decay of transients with the armature open-circuited. With the armature short-circuited, however, the apparent field inductance is lower because of coupling to the armature winding. From Eq. 8-63 the time constant is then

$$T_d' = T_{d0}'\frac{L_d'}{L_d} = T_{d0}'\frac{X_d'}{X_d} \tag{8-65}$$

and is called the *direct-axis short-circuit transient time constant*. Both T_{d0}' and T_d' are important constants of the synchronous machine. Typical values for large machines are about 5 s for T_{d0}' and 1 to 2 s for T_d'.

Equation 8-63 can then be written

$$T_d'\frac{di_{ft}}{dt} + i_{ft} = 0 \tag{8-66}$$

Similarly, for the current i_d one obtains

$$T_d'\frac{di_d}{dt} + i_d = \frac{L_{af}I_{f0}}{L_d} \tag{8-67}$$

The currents i_d and i_f become, subject to their being given by Eqs. 8-57 and 8-59 at $t = 0+$,

$$i_d = \frac{L_{af}I_{f0}}{L_d} + L_{af}I_{f0}\left(\frac{1}{L'_d} - \frac{1}{L_d}\right)\varepsilon^{-t/T'_d} \tag{8-68}$$

$$i_f = I_{f0} + \frac{3}{2}\frac{L_{af}}{L_{ff}}\frac{L_{af}I_{f0}}{L'_d}\varepsilon^{-t/T'_d} \tag{8-69}$$

Finally, the phase current is

$$i_a = \frac{\sqrt{2}E_{af0}}{X_d}\cos(\omega t + \theta_0) + \sqrt{2}E_{af0}\left(\frac{1}{X'_d} - \frac{1}{X_d}\right)\varepsilon^{-t/T'_d}\cos(\omega t + \theta_0) \tag{8-70}$$

The first term on the right in Eq. 8-70 is the steady-state short-circuit current. The second is the transient term. The phase current is a damped sinusoid. Its initial amplitude is limited by the transient reactance X'_d and its final amplitude by the synchronous reactance X_d. It passes from one to the other exponentially as determined by the time constant T'_d.

The machine-current solutions obtained in this article can readily be adapted to include external reactance X_e between the machine terminals and the short circuit. Such external reactance merely plays the same role as the armature leakage reactance ωL_{al}, which is a component of X_d, X'_d, and X_q. All that need be done is to add X_e to X_d, X'_d, and X_q wherever they appear (or the corresponding inductance L_e to L_d, L'_d, and L_q). The results are thus made more widely applicable.

To summarize the results of this article, note first that we avoid a frontal attack on the basic machine equations with all terms retained. We do so to avoid complex and cumbersome analysis in which it would be difficult to distinguish between the important and the unimportant. By comparison of the results of parts b and c of the article we find neglect of the $p\lambda_d$ and $p\lambda_q$ terms to be an admissible approximation in most cases. In part d we adopt this approximation and investigate how the transients die away. The results are quite simple, and we have established valuable guides to the conduct of other machine-transient analyses. At no time should it be forgotten, however, that comparison with actual machine performance can be the only ultimate guide.

8-5 TRANSIENT POWER–ANGLE CHARACTERISTICS

It can be seen from Art. 8-4 that in the first instant following a transient disturbance in the armature circuit of a synchronous machine the total field linkage λ_f remains constant at its value before the disturbance. Thereafter, in the absence of voltage-regulator action, λ_f decays to a new steady-state value. The time constant of the decay is an appropriately adjusted value of T'_{d0}. For periods of time which are short compared with this time constant the decrement in field linkages may be neglected. This approximation leads to a simplified representation of the machine which is especially useful in investigating dynamic problems.

The field and direct-axis armature linkages are, from Eqs. 8-14 and 8-16,

$$\lambda_f = L_{ff}i_f - \tfrac{3}{2}L_{af}i_d \tag{8-71}$$

$$\lambda_d = L_{af}i_f - L_d i_d \tag{8-72}$$

Solving 8-72 for i_f and substituting in 8-71 give

$$\lambda_f = \frac{L_{ff}}{L_{af}}(\lambda_d + L_d i_d) - \tfrac{3}{2}L_{af}i_f \tag{8-73}$$

or

$$\frac{L_{af}}{L_{ff}}\lambda_f = \lambda_d + \left(L_d - \frac{3}{2}\frac{L_{af}^2}{L_{ff}}\right)i_d \tag{8-74}$$

$$\frac{L_{af}}{L_{ff}}\lambda_f = \lambda_d + L'_d i_d \tag{8-75}$$

where Eq. 8-45 is used for the transient inductance L'_d. When both sides of Eq. 8-75 are multiplied by ω, we have

$$\omega\frac{L_{af}}{L_{ff}}\lambda_f = \omega\lambda_d + X'_d l_d \tag{8-76}$$

In terms of rms values

$$E'_q = V_q + X'_d i_d \tag{8-77}$$

where

$$E'_q = \omega\frac{L_{af}}{L_{ff}}\frac{\lambda_f}{\sqrt{2}} \tag{8-78}$$

and V_q is written by virtue of Eq. 8-26 with resistance omitted and the $p\lambda_q$ term ignored.

The voltage E'_q, being directly proportional to λ_f, is constant whenever field flux linkages can be considered constant. It is shown on a phasor diagram in Fig. 8-5. If now the analysis in Art. 7-7 leading to Eq. 7-52 is repeated with E'_q and X'_{dT} replacing E_{af} and X_{dT}, the result is

$$P = \frac{E'_q E_e}{X'_{dT}}\sin\delta + E_e{}^2\frac{X'_{dT} - X_{qT}}{2X'_{dT}X_{qT}}\sin 2\delta \tag{8-79}$$

where

$$X'_{dT} = X'_d + X_e \tag{8-80}$$

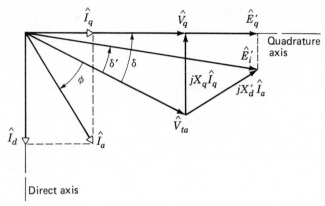

Fig. 8-5. Phasor diagram showing $\hat{E}_q'$.

and
$$X_{qT} = X_q + X_e \tag{8-81}$$

We now have in Eq. 8-79 a power-angle expression which is useful in the analysis of sudden disturbances or of sudden load application. It is applicable only over a period of time short compared with the adjusted time constant T_{d0}', but this covers the critical period in many dynamic analyses. On the other hand, the steady-state power-angle expression (Eq. 7-52) is applicable during changes which are sufficiently slow for the machine to be always in the steady state. Note that the coefficient of the second term on the right of Eq. 8-79 often is numerically negative because X_q for a normal machine is greater than X_d' (and hence $X_{qT} > X_{dT}'$).

The transient-power-angle curve will have an amplitude significantly greater than that of the steady-state curve. The relative amplitudes in a specific instance can be seen in Fig. 8-6. The suddenly induced component of field

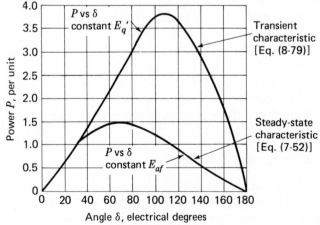

Fig. 8-6. Transient and steady-state power-angle curves.

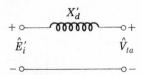

Fig. 8-7. Approximate transient equivalent circuit for a synchronous machine.

current following a disturbance is the physical reason for the greater transient amplitude. The practical consequence is that the machine is a stiffer system element in the transient state; e.g., it can withstand a large, suddenly applied power or torque overload if the duration of the overload is relatively short.

A further simplification of transient-power-angle representation is often made, one not unlike that in Art. 7-4, where cylindrical-rotor rather than salient-pole theory is used. Transient saliency is ignored by dropping the last term in Eq. 8-79. Not infrequently the simplification is carried one step further by using the equivalent circuit of Fig. 8-7 for the machine. The voltage $\hat{E}_i'$ in this circuit is called the *voltage behind transient reactance* and is shown on the phasor diagram of Fig. 8-5. The simplicity of this approximate representation is a valuable feature when transients and dynamics are to be investigated in complex multimachine networks, even when comprehensive analog or digital computers are used.

8–6 EFFECTS OF ADDITIONAL ROTOR CIRCUITS

Up to this point we have considered that there is only one effective circuit on the rotor of the synchronous machine, namely, the main-field winding in the direct axis. As a result, we have been able to concentrate on the most important aspects of machine performance without distracting algebraic complication. In many actual synchronous machines, however, there are additional circuits called *damper,* or *amortisseur, circuits.* In many salient-pole machines, they are formed by bar or cage windings embedded in the pole faces and connected at their ends by short-circuiting end rings. An additional equivalent rotor circuit may be formed by the bolts and iron of the rotor structure. Cylindrical-rotor machines may likewise have rotor circuits other than the main-field winding corresponding to conducting paths in the solid-iron rotor forgings, especially in the quadrature axis, where the rotor iron may form an equivalent circuit almost as effective as the main field for induced currents.

a. Effects on Power and Torque

Damper bars form the starting winding for synchronous motors. The principal reason for installing them in generators is, as their name implies, to produce torques which help to damp out oscillations of the machine rotor about its equilibrium position. When the speed of the rotor departs from synchronous speed, the damper circuits produce induction-motor torques, discussed in some detail in the next chapter. It is sufficient for present purposes to state that their magnitude is approximately proportional to the departure of rotor speed from synchronous speed as long as that departure is small. The torques produced are motor torques, i.e., in the direction of rotation, when the speed is below synchronous and generator torques when the speed is above.

A detailed analysis of synchronous-machine dynamics must include the effects of both synchronous and asynchronous (induction) torques. Thus, clearly the effects of these additional rotor circuits must be included. This can be done by representing them explicitly (see Art. 8-7) or approximately by including a term in the torque equation proportional to the deviation of the rotor speed from synchronous speed (Art. 8-8).

b. Effects on Current-Voltage Relations

The damper circuits, of course, have their own resistances and their own self-inductances and mutual inductances with respect to every other machine circuit. They therefore have an influence on the current-voltage performance of the machine following a disturbance. To reduce the complexity of the problem somewhat, analyses are usually confined to machines having but one effective rotor circuit other than the main field in the direct axis and one effective rotor circuit in the quadrature axis—in other words, one equivalent damper circuit in each axis.

If the analysis in Art. 8-4b were repeated with the damper circuits included, two additional equations would appear to express the damper-winding flux linkages and two further equations would be needed to indicate that the damper-circuit voltages are zero, i.e., the damper circuits closed on themselves. Additional terms would, of course, appear in the armature- and field-linkage relations to reflect the contributions from the damper circuits. The resulting phase current would be the same as in Eq. 8-55 except that X_d' and X_q would be replaced by the constants X_d'' and X_q'', respectively, representing new self- and mutual-inductance combinations including damper-circuit values. The reactance X_d'' is called the *direct-axis subtransient reactance,* and X_q'' is the *quadrature-axis subtransient reactance.* They are both lower than the respective reactances they replace. In effect, the initial short-circuit current is higher because induced damper currents appear as well as an induced field current.†

Further investigation shows that, because the damper circuits have relatively high resistance, the induced damper currents die out rapidly. After very few cycles, the phase current becomes that given by the methods of Art. 8-4. This is illustrated in Fig. 8-8, which shows a symmetrical trace of a short-

†For a detailed analysis, including damper circuits, see B. Adkins and R. G. Harley, "The General Theory of Alternating Current Machines: Application to Practical Problems," Chapman and Hall, London, 1975 (distributed in the U.S. by Halsted Press, a division of Wiley, New York); also C. Concordia, "Synchronous Machines Theory and Practice," General Electric Company, 1951.

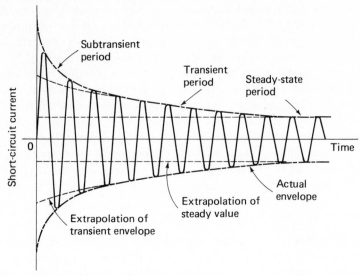

Fig. 8-8. Symmetrical trace of armature short-circuit current in a synchronous machine.

circuit current wave (dc component absent or taken out) such as might be obtained oscillographically. The wave, whose envelope is shown in Fig. 8-9, can be divided into three periods, or time regimes: the *subtransient period,* lasting only for the first few cycles, during which the current decrement is very rapid; the *transient period,* covering a relatively longer time, during which the current decrement is more moderate; and finally the *steady-state period.*

During the subtransient period, both the damper circuit and the field winding are trapping flux. As the induced damper currents decay away, only transient field current remains, corresponding to the transient period. The fundamental-frequency component of armature current can thus be expressed in a fashion similar to that of Eq. 8-70

$$i_a = \frac{\sqrt{2}E_{af0}}{X_d}\cos(\omega t + \theta_0) + \sqrt{2}E_{af0}\left(\frac{1}{X_d'} - \frac{1}{X_d}\right)\varepsilon^{-t/T_d'}\cos(\omega t + \theta_0)$$

$$+ \sqrt{2}E_{af0}\left(\frac{1}{X_d''} - \frac{1}{X_d'}\right)\varepsilon^{-t/T_d''}\cos(\omega t + \theta_0) \quad (8\text{-}82)$$

where T_d'' is the *direct-axis short-circuit subtransient time constant.*

As shown in Art. 8-4b, there will also be a dc and second harmonic component in the armature current following a sudden short circuit corresponding to the decay of trapped armature flux. These components do not appear in analyses which neglect the transformer-voltage terms $p\lambda_d$ and $p\lambda_q$.

It should be emphasized that representation of induced direct-axis rotor currents by a single circuit is an approximation to the actual situation, in

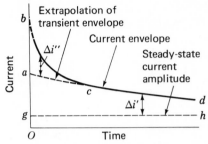

Fig. 8-9. Envelope of a synchronous-machine symmetrical short-circuit current.

which there are a number of conducting paths. However, this approximation has been found to be quite valid in many cases and is generally used to describe machine behavior. Thus, following a 3-phase short circuit from open-circuit conditions, the armature current can be shown to be of the form of Eq. 8-82 by appropriate semilogarithmic plots. The difference $\Delta i'$ (Fig. 8-9) between the transient envelope and the steady-state amplitude is plotted to a logarithmic scale as a function of time in Fig. 8-10. In similar fashion the difference $\Delta i''$ between the subtransient envelope and an extrapolation of the transient envelope is also plotted in Fig. 8-10. When the work is done carefully, both plots closely approximate straight lines, illustrating the essentially exponential nature of the decrement.

The subtransient reactance X_d'' determines the initial value Ob of the symmetrical subtransient envelope bc (Fig. 8-9); it is equal to the rms value of the prefault open-circuit phase voltage divided by $Ob/\sqrt{2}$, the factor $\sqrt{2}$ appearing because bc is the envelope of peak current values. The time constant T_d'' determines the decay of the subtransient envelope bc; it is equal to the time required for the envelope to decay to the point where the difference between it and the transient envelope acd is $1/\varepsilon$, or 0.368 times the initial difference ab.

Under the assumption of negligible transformer voltages, the short-cir-

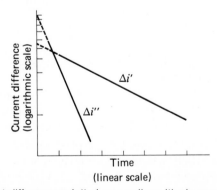

Fig. 8-10. Current differences plotted on semilogarithmic coordinates.

TABLE 8-1
TYPICAL VALUES OF MACHINE CONSTANTS†

Machine constant	Generators		Synchronous condensers	Salient-pole motors	
	Solid-rotor	Salient-pole		Low-speed	High-speed
X_d	1.9	1.7	2.2	1.5	1.4
X_d'	0.25	0.30	0.50	0.40	0.35
X_d''	0.20	0.18	0.30	0.25	0.20
X_q	1.85	1.0	1.3	1.0	1.0
X_q'	0.50	1.0	1.3	1.0	1.0
X_q''	0.20	0.25	0.35	0.30	0.25
T_d'	0.55	0.50	1.5	0.35	0.50
T_d''	0.02	0.02	0.03	0.01	0.01
T_a	0.17	0.05	0.25	0.04	0.04

†Reactances are per unit values based on the machine rating; time constants are in seconds.

cuit oscillogram (Fig. 8-8) of an unloaded machine involves only direct-axis quantities. In a more general case, such as that of a machine with an initial active-power loading, a comparable set of quadrature-axis constants must be available. In addition to X_q and X_q'' these include T_q'', the *quadrature-axis short-circuit subtransient time constant*, X_q', the *quadrature-axis transient reactance;* and T_q', the *quadrature-axis short-circuit transient time constant.* Typical values of synchronous machine reactances and time constants are given in Table 8-1.

These constants enter into the initial magnitude and decay of quadrature current i_q in the same manner as the direct-axis constants do for i_d. When X_q, X_q' and X_q'' all have different values, the implication is that there are two equivalent quadrature-axis rotor circuits. The induced currents in one are decremented at an appreciably slower rate than in the other. Greater insight into the effects of damper circuits will be gained during the study of induction machines in the next two chapters.

8–7 MODELS OF SYNCHRONOUS MACHINES FOR TRANSIENT ANALYSIS

Depending upon the type of analysis and the capability of the analysis program being used, a variety of representations can be used for synchronous-machine transient behavior. Among the simplest models is that given in Fig. 8-7, in which transient saliency is ignored and the machine is represented by a constant voltage E_i' behind transient reactance X_d'.

Based upon this model, the transient power-angle characteristic can be

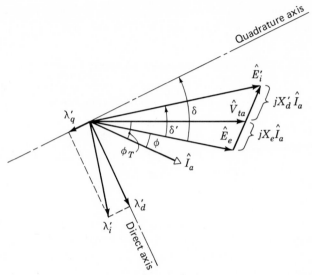

Fig. 8-11. Phasor diagram showing voltage E_i' and trapped flux in the direct axis $(\hat{\lambda}_d')$ and quadrature axis $\hat{\lambda}_q'$.

written by direct analogy from Eq. 8-79 with saliency neglected $(X_{qT} = X_{dT}')$

$$P = \frac{E_i' E_e}{X_{dT}} \sin \delta' \qquad (8\text{-}83)$$

Here the machine is considered to be connected to a constant voltage E_e through an impedance X_e. X_{dT}' is given by Eq. 8-80, and δ' is the angle between the voltages E_e and E_i'.

The validity of this model is based upon two assumptions: (1) During the time period of interest (the transient time period, generally on the order of 0.5 s), both the direct and quadrature axes of the rotor can be assumed to maintain their flux linkages. This is the constant-voltage assumption. (2) The apparent reactances of each axis under this condition are approximately equal. It is this assumption which permits us to neglect transient saliency and to model the machine by a single transient reactance.

Figure 8-11 shows a phasor diagram from which $\hat{E}_i'$ can be found for the general case of an initially loaded synchronous generator. In this figure ϕ_T is the power-factor angle at the generator terminals and ϕ is the power-factor angle at the external source. Note that the $\hat{E}_i'$ does not lie along the quadrature axis. The phasor diagram shows that the corresponding trapped flux $\hat{\lambda}_i'$ can be resolved into two components. The direct-axis component $\hat{\lambda}_d'$ represents flux trapped by the transient windings on the direct axis, and $\hat{\lambda}_q'$ represents flux trapped by the corresponding circuits on the quadrature axis. Note also that transient power angle δ' is smaller than the steady-state power angle δ and that the assumption of constant rotor flux linkages requires the angle between $\widehat{E}'_i$ and the quadrature axis to remain fixed during the transient time period.

EXAMPLE 8-2

The salient-pole generator of Example 7-5 has parameters (obtained from test data)

$$X_d = 1.00 \text{ per unit} \qquad X_q = 0.60 \text{ per unit} \qquad X'_d = 0.18 \text{ per unit}$$
$$X'_q = 0.15 \text{ per unit} \qquad T'_{d0} = 3.0 \text{ s} \qquad T'_{q0} = 1.6 \text{ s}$$
$$T'_d = 0.52 \text{ s} \qquad T'_q = 0.43 \text{ s}$$

For the loading of Example 7-5 (rated kilovoltamperes at 0.80 power factor, lagging current, and rated terminal voltage) calculate the steady-state and transient power-angle characteristics assuming the generator to be connected to an infinite bus directly at its armature terminals.

Solution

With reference to Example 7-5 and Fig. 8-12a,

$$\widehat{E}_{af} = 1.77\underline{/19.4°}$$

and thus the steady-state power-angle characteristic is (from Eq. 7-52)

$$P = \frac{V_{ta}E_{af}}{X_d}\sin\delta + V_{ta}^2\frac{X_d - X_q}{X_d X_q}\sin 2\delta = 1.77\sin\delta + 0.67\sin 2\delta$$

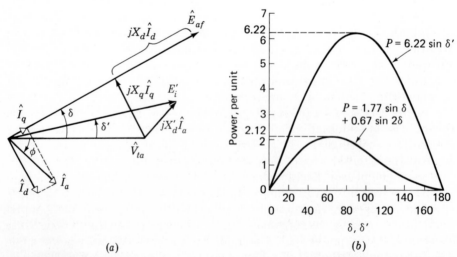

(a)   (b)

Fig. 8-12. Example 8-2: (a) generator phasor diagram and (b) steady-state and transient power-angle characteristics.

As shown in Fig. 8-12*b*, the peak of this curve occurs at $\delta = 63.4°$ and has magnitude $P_{max} = 2.12$ per unit.

The transient power-angle characteristic can be found from Eq. 8-83 under the assumption that transient saliency can be neglected and the generator represented simply as a voltage E_i' behind transient reactance X_d'. This assumption is reasonable in this case both because the direct- and quadrature-axis transient reactances are approximately equal and because their time constants are both reasonably long.

From Eq. 8-83

$$P = \frac{V_{ta} E_i'}{X_d'} \sin \delta'$$

and from Fig. 8-12*a* with $\widehat{I}_a = 0.80 - j0.60$

$$\widehat{E}_i' = \widehat{V}_{ta} + jX_d'\widehat{I}_a$$
$$= 1.00 + j(0.18)(0.80 - j0.60)$$
$$= 1.108 + j0.144 = 1.12\underline{/7.4°}$$
$$E_i' = 1.12 \text{ per unit}$$

and
$$P = 6.22 \sin \delta'$$

This transient characteristic is also plotted in Fig. 8-12*b*. Note that the effect of induced currents in the rotor windings is to increase the torque producing capability of the machine significantly. However, this effect lasts only through the transient time period and disappears as the induced transient rotor currents decay.

A logical extension of the model just presented would be to include the effects of transient saliency and to represent the machine by a separate voltage behind transient reactance on each rotor axis. Although this representation is perhaps somewhat more exact, it is still subject to the same limitations as the simpler model, specifically that its validity is limited by the extent to which the rotor fluxes can be assumed to remain constant. All these models suffer from the fact that they neglect damping effects, which can be of great significance in studies of synchronous-machine dynamics.

An alternative to these "trapped flux" models is to represent the effects of currents in each of the rotor windings specifically. For example, Fig. 8-13 shows the per unit equivalent circuits for a synchronous machine represented by a field winding and damper circuit on the rotor direct axis and a single damper circuit on the quadrature axis. The equations corresponding to this representation are similar to Eqs. 8-14 to 8-16 and 8-25 to 8-27 with the addition of equations representing the additional shorted damper windings. The torque expression of Eq. 8-31 continues to apply. Use of this type of represen-

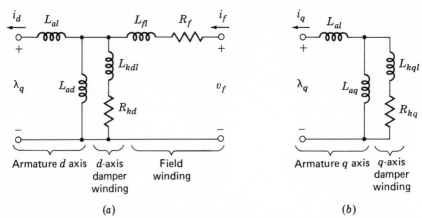

Fig. 8-13. Per unit synchronous-machine equivalent circuits including one damper circuit on each rotor axis: (a) direct-axis equivalent circuit, (b) quadrature-axis equivalent circuit.

tation (perhaps with additional damper windings) includes directly both synchronous and damping torques. When properly applied, it can lead to accurate representation of synchronous-machine dynamics in a wide variety of situations.

The choice of model for a given situation must in general depend upon the type of study being conducted as well as the data available and the type of results desired. This article has briefly indicated the types of models available. Many discussions of modeling are found in the growing literature on synchronous machines.

8–8 SYNCHRONOUS–MACHINE DYNAMICS

Important dynamic problems arise in synchronous-machine systems because successful operation of the machines demands equality of the mechanical speed of the rotor and the speed of the stator field, and because synchronizing forces tending to maintain this equality are brought into play whenever the relationship is disturbed. If the instantaneous speed of a synchronous machine in a system containing other synchronous equipment should decrease slightly, the decrease would be associated with a decrease in torque angle if the machine is a generator and an increase if it is a motor. For example, if a large load is suddenly applied to the shaft of a synchronous motor, the motor must slow down at least momentarily so that the torque angle can assume the increased value necessary to supply the added load. In fact, until the new angle is reached, an appreciable portion of the energy furnished to the load comes from stored energy in the rotating mass as it slows down. When the newly required value of angle is first reached, equilibrium is not yet attained, for the mechanical speed is then below synchronous speed. The angle must momentarily increase further in order to permit replacing the deficit of stored energy in the rotating mass. The ensuing processes involve a series of oscillations about the final position even when equilibrium is ultimately restored. Exact description of such events can be given only in terms of the associated electromechanical differential equation, and decisions on restoration of equilibrium can be based only on the solution of the equation.

Similar oscillations or hunting, with the accompanying power and current pulsations, may be particularly troublesome in synchronous motors driving loads whose torque requirements vary cyclically at a fairly rapid frequency, as in motors driving reciprocating air or ammonia compressors. If the natural frequency of mechanical oscillation of the synchronous motor approximates the frequency of an important torque harmonic in the compressor cycle, intolerable oscillations result. Electrodynamic transients of a very complicated form but of the same basic nature occur in electric-power systems. Unless they are carefully investigated during system planning, they may result in complete shutdowns over wide areas.

a. The Basic Electromechanical Equation

The electromechanical equation for a synchronous machine follows directly from equating the inertia torque (equal to the moment of inertia J times the angular acceleration) to the net mechanical and electric torque acting on the rotor. Thus

$$J\frac{d^2\theta}{dt^2} = T_{\text{mech}} - T_{\text{elec}} \tag{8-84}$$

where θ = angular position of rotor
T_{mech} = mechanical accelerating shaft torque applied to rotor
T_{elec} = electromagnetic torque acting to decelerate rotor

This equation, like those which follow, is written with a generator specifically in mind. The same equations can be applied to a motor although a change in sign of the torque terms is often convenient.

It is often convenient to express the rotor angular position θ in terms of a synchronously rotating reference frame as

$$\theta = \frac{2}{\text{poles}}(\omega t + \delta) \tag{8-85}$$

where ω is the synchronous electric frequency and δ is the electrical angle between a point on the rotor and the synchronous reference frame. δ is often taken equal to the synchronous-machine power angle. Equation 8-84 thus becomes

$$\frac{2}{\text{poles}}J\frac{d^2\delta}{dt^2} = T_{\text{mech}} - T_{\text{elec}} \tag{8-86}$$

Known as the *swing equation,* it can be used to solve for the electromechanical dynamics of the synchronous machine. The key to its use is the accurate representation of the torques acting on the rotor. The mechanical shaft torques must be determined from a representation of the dynamics of the prime mover (or load in the case of a motor).

As discussed in Art. 8-7, when a detailed representation of the synchronous machine is used, the electromagnetic torque T_{elec} can be obtained directly from Eq. 8-31. This type of representation, which finds torque directly from a knowledge of the machine fluxes and currents, includes both synchronous and damping torques. It can be expected to yield accurate results in many types of analyses, provided sufficient equivalent rotor circuits are included to represent the effects of rotor currents adequately and provided proper values for the model parameters can be obtained.

The model of constant voltage behind transient reactance, discussed in Arts. 8-6 and 8-7, is used in many situations where detailed analytical solutions are not possible or deemed unnecessary. Since this model results in an expression for electric-power output, Eq. 8-83, as a function of rotor angle, it is convenient to rewrite Eq. 8-86 in terms of power. This can be done by multiplying by the rotor mechanical velocity ω_m

$$\frac{2}{\text{poles}}J\omega_m\frac{d^2\delta}{dt^2} = P_{\text{mech}} - P_{\text{elec}} \tag{8-87}$$

Since for most situations of practical interest deviations of ω_m from synchronous speed are very small, ω_m can be replaced by the synchronous rotor velocity $\omega_s = (2/\text{poles})\omega$, giving

$$\frac{2}{\text{poles}}J\omega_s\frac{d^2\delta}{dt^2} = P_{\text{mech}} - P_{\text{elec}} \tag{8-88}$$

In this case the electromagnetic power P_{elec} can be represented by two components. One is the *damping power,* discussed in Art. 8-7 and often considered to vary linearly with the departure $d\delta/dt$ from synchronous speed. The second is the *synchronous power,* resulting from synchronous-machine action and characterized by Eqs. 7-42 or 7-52 in the steady state and Eqs. 8-79 or 8-83 in the transient state. The electromechanical equation 8-88 then becomes

$$\frac{2}{\text{poles}}J\omega_s\frac{d^2\delta}{dt^2} = P_{\text{mech}} - P_d\frac{d\delta}{dt} - P_s(\delta) \tag{8-89}$$

where P_d is the damping power per unit departure in speed from synchronous and $P_s(\delta)$ represents the synchronous power as a function of δ.

The specific nature of the external network must be known before the function $P_s(\delta)$ can be identified. When consideration is restricted to one machine connected directly to the terminals of a very large system (see Art. 7-4) and saliency is ignored, the function is $P_m \sin \delta$ (Eq. 7-42), where P_m is the amplitude of the sinusoidal power-angle curve. Equation 8-89 becomes

$$\frac{2}{\text{poles}} J\omega_s \frac{d^2\delta}{dt^2} = P_{\text{mech}} - P_d \frac{d\delta}{dt} - P_m \sin \delta \qquad (8\text{-}90)$$

Positive values of δ denote generator action and therefore energy conversion from mechanical to electric form, positive values of P_{mech} denote mechanical power input to the shaft, positive values of $d\delta/dt$ denote speeds above synchronous speed, and positive values of $d^2\delta/dt^2$ denote acceleration. Alternatively, the reverse convention may be used; i.e., positive values of δ denote motor action, positive values of P_{mech} mechanical power output from the shaft, positive values of $d\delta/dt$ speeds below synchronous speed, and positive values of $d^2\delta/dt^2$ deceleration.

It should be emphasized that for transients occurring in times on the order of the transient time constant or faster, the transient power-angle characteristic (Eqs. 8-79 or 8-83) should be used and Eq. 8-90 should be written in terms of the transient power angle δ'. However, for the purposes of this article the distinction need not be made.

Both Eqs. 8-89 and 8-90 are nonlinear. One method of analysis, applicable for small oscillations and illustrated below, is to linearize the power-angle expression. It is simply a special case of linearization about an operating point.

b. Linearized Analysis

When a single machine connected to a large system is under study, only a single differential equation rather than a group of such equations is involved. If, in addition, the variations of δ are small, the term $P_s(\delta)$ in Eq. 8-89 may be replaced by the equation for the slope of the power-angle curve at the operating point. For example, when δ varies between about $+\pi/6$ and $-\pi/6$ electrical radians, the sine of the angle is closely equal to the angle in radians. The term $P_m \sin \delta$ in Eq. 8-90 can then be replaced by the term $P_s\delta$, where P_s is the *synchronizing power* or slope of the power-angle characteristic evaluated at the origin; P_s has units of power per unit angle. Equation 8-90 then becomes

$$\frac{2}{\text{poles}} J\omega_s \frac{d^2\delta}{dt^2} = P_{\text{mech}} - P_d \frac{d\delta}{dt} - P_s\delta \qquad (8\text{-}91)$$

Since this equation is now linear, its solution for a particular case can readily be obtained, as illustrated in the following example.

EXAMPLE 8-3

A 200-hp 2300-V 3-phase 60-Hz 28-pole 257-r/min synchronous motor is directly connected to a large power system. The motor has the following characteristics:

$Wk^2 = 10,500 \text{ lb} \cdot \text{ft}^2$ (motor plus load)

Synchronizing power $P_s = 11.0 \text{ kW/elec deg}$

Damping torque $= 1770 \text{ lb} \cdot \text{ft/(mech rad/s)}$

(a) Investigate the mode of electromechanical oscillation of the machine.
(b) Rated mechanical load is suddenly thrown on the motor shaft at a time when it is operating in the steady state but unloaded. Study the electrodynamic transient which will ensue.

Solution

(a) Throughout this solution, the angle δ will be measured in electrical degrees rather than radians. This fact must be recognized in obtaining P_d and P_s from the given data.

The inertia is given as Wk^2 (weight times square of radius of gyration) in English units, a common practice for large machines. In SI units (see table of conversion factors in Appendix C)

$$J = \frac{Wk^2}{23.7} = \frac{10,500}{23.7} = 444 \text{ kg} \cdot \text{m}^2$$

When we use the factor $\pi/180$ to convert angular measurement to degrees, the coefficient of the angular acceleration term becomes

$$\frac{2}{\text{poles}} J\omega = \tfrac{2}{28}(444)\frac{2\pi(257)}{60}\frac{\pi}{180} = 14.9 \text{ W/(elec deg/s}^2)$$

The remaining motor constants in the appropriate units are

$$P_d = 2\pi(257)(1770)\frac{746}{33,000}\frac{\pi}{180}\frac{2}{28} = 80.6 \text{ W/(elec deg/s)}$$

$$P_s = 11.0(1000) = 11,000 \text{ W/elec deg}$$

The force-free equation which determines the mode of oscillation is then

$$14.9\frac{d^2\delta}{dt^2} + 80.6\frac{d\delta}{dt} + 11,000\delta = 0$$

The undamped angular frequency and damping ratio are, respectively,

$$\omega_n = \sqrt{\frac{11{,}000}{14.9}} = 27 \text{ rad/s}$$

$$\zeta = \frac{80.6}{2\sqrt{14.9(11{,}000)}} = 0.10$$

The magnitude of ζ places the transient response decidedly in the oscillatory region, as it does for all synchronous machines. Any operating disturbance will be followed by a relatively slowly damped oscillation, or swing, of the rotor before steady operation at synchronous speed is resumed. A large disturbance may, of course, be followed by complete loss of synchronism. The damped angular velocity of the motor is

$$\omega_d = 27\sqrt{1 - (0.10)^2} = 26.9 \text{ rad/s}$$

corresponding to a damped oscillation frequency of

$$f_d = \frac{26.9}{2\pi} = 4.3 \text{ Hz}$$

(b) The full load of 200 hp is equivalent to $200(746) = 149{,}200$ W. The steady-state operating angle is

$$\delta_\infty = \frac{149{,}000}{11{,}000} = 13.6 \text{ electrical degrees}$$

A detailed solution of the linearized swing equation would show that the angular excursions are characterized by the equation

$$\delta = 13.6°[1 - 1.004\varepsilon^{-2.7t} \sin (26.9t + 84.3°)]$$

c. Nonlinear Analysis: Equal-Area Methods

In most of the serious dynamic problems, the oscillations are of such magnitude that linearization is not permissible. The equations of motion must be retained in nonlinear form. Analog or digital computers are then often used to aid the analysis. For programming the study on a digital computer, use is made of numerical methods of solving sets of differential equations.† The

†See, for instance, G. W. Stagg and A. H. El-Abiad, "Computer Methods in Power System Analysis," McGraw-Hill, New York, 1968; also P. M. Anderson and A. A. Fouad "Power System Control and Stability," Iowa State Press, 1977.

object of the study is usually to find whether or not synchronism is maintained, i.e., whether or not the angle δ settles down to a steady operating value after the machine has been subjected to a sizable disturbance.

For simple synchronous-machine systems, damping may be neglected and use may be made of a graphical interpretation of the energy stored in the rotating mass as an aid to determining the maximum rotor angle following a disturbance and to settling the question of maintenance of synchronism. Because of the physical insight it gives to the dynamic process, application of the method to analysis of a single machine connected to a large system will be discussed.

Consider specifically a synchronous motor having the power-angle curve of Fig. 8-14. With the motor initially unloaded, the operating point is at the origin of the curve. When a mechanical load P_{mech} is suddenly applied, the operating point travels along the sinusoid ABC and, if synchronism is maintained, finally comes to rest at point B with a new torque angle δ_∞. To reach this new operating point, the motor must decelerate at least momentarily under the influence of the difference $P_{mech} - P_m \sin \delta$ between the power required by the load and that resulting from electromechanical energy conversion. Now recall from the thought process leading to Eq. 8-90 that both P_{mech} and $P_m \sin \delta$ are proportional to the corresponding torques. It can be shown that the integral $\int T \, d\delta$ of torque with respect to angle is energy. The area OAB in Fig. 8-14 is then seen to be proportional to the energy abstracted from the rotating mass during the initial period when electromagnetic energy conversion is insufficient to supply the shaft load. When point B is reached on the first excursion, therefore, the rotor has a momentum in the direction of deceleration. Acting under this momentum, the rotor must swing past point B until an equal amount of energy is recovered by the rotating mass. The result is that the rotor swings to point C and the angle δ_{max}, at which point

$$\text{Area } BCD = \text{Area } OAB \qquad (8\text{-}92)$$

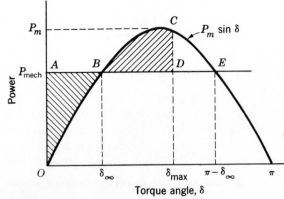

Fig. 8-14. Synchronous-motor power-angle curve and power required by load.

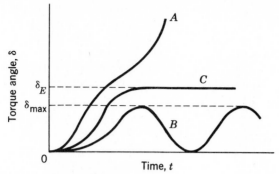

Fig. 8-15. Simple synchronous-machine swing curves showing instability (curve A), stability (curve B), and marginal or critical case (curve C).

Thereafter, in the absence of damping, the rotor would continue to oscillate between points O and C at its natural frequency. The damping present in any physical machine causes successive oscillations to be of decreasing amplitude and finally results in dynamic equilibrium at point B. The analogy to the oscillations of a pendulum may be noted.

This *equal-area method* provides a ready means of finding the maximum angle of swing. It also provides a simple indication of whether synchronism is maintained and a rough measure of the margin of stability. Thus, if area $BCED$ in Fig. 8-14 is less than area OAB, the decelerating momentum can never be overcome, the angle-time curve follows the course of curve A in Fig. 8-15, and synchronism is lost. On the other hand, if area $BCED$ is greater than area OAB, synchronism is maintained with a margin indicated by the difference in areas and the angle-time curve follows curve B of Fig. 8-15. Equality of areas $BCED$ and OAB yields a borderline solution of unstable equilibrium for which curve C is followed.

EXAMPLE 8-4

Determine the maximum shaft load which can be suddenly applied to the motor of Example 8-3 when it is initially operating unloaded. The synchronizing power of 11.0 kW/elec deg quoted in that example is the initial slope of the power-angle curve followed under these conditions. Damping is to be ignored.

Solution

The initial slope of the sinusoidal power-angle curve expressed in kilowatts per radian is equal to the amplitude of the curve in kilowatts. Hence

$$P_m = 11.0 \times \frac{180}{\pi} = 630 \text{ kW}$$

The load P_{mech} (Fig. 8-16) must be adjusted so that

$$\text{Area } OAB = \text{Area } BCD$$

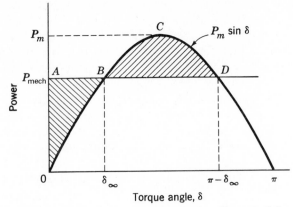

Fig. 8-16. Graphical application of equal-area criterion, Example 8-4.

or $\quad P_{\text{mech}}\delta_\infty - \int_0^{\delta_\infty} 630 \sin \delta \, d\delta = \int_{\delta_\infty}^{\pi - \delta_\infty} 630 \sin \delta \, d\delta - P_{\text{mech}}(\pi - 2\delta_\infty)$

Also, $\qquad\qquad\qquad\qquad P_{\text{mech}} = 630 \sin \delta_\infty$

Trial-and-error solution yields

$$P_{\text{mech}} = 455 \, \text{kW} = 610 \, \text{hp}$$

Notice that the result is independent of inertia when damping is neglected. Under these circumstances, inertia determines the period of oscillation but does not influence its amplitude.

8-9 SUMMARY

Electric-power systems are complex dynamic entities. Although they can often be considered to be operating under steady-state conditions, in actual fact, loads are continually changing and the system is continually reacting in response to these changes. Occasionally, disturbances occur, ranging from the loss of a transmission line due to a lightning strike or the flashover of a dirty insulator string to a multiple sequence of events resulting in a major blackout over a significant portion of the system. In each case, the synchronous generator plays a significant role in the subsequent transient response of the system; hence an accurate representation of its transient behavior is essential for both predicting and controlling the dynamic characteristics of electric-power systems.

Under transient conditions currents are induced in the rotor circuits of synchronous machines. In salient-pole machines these circuits are composed of damper bars arranged much like the rotor bars in a squirrel-cage induction motor. In solid-rotor machines these currents flow directly in the rotor iron. In both cases, currents are also induced in the field winding.

It is the nature of these currents which determines the transient behavior of synchronous machines. Simple transient models can be derived from the concept of trapped rotor flux. Based upon the concept that the circuits on the rotor keep their flux constant following a transient, simple mathematical models can be derived in which the machine is represented as a voltage proportional to this trapped flux in series with a transient reactance. The transient reactance is the apparent reactance seen from the armature terminals under the condition of constant rotor flux linkages.

Although such a model is useful for some simple studies of synchronous-machine dynamics, it is not adequate for many situations. It does not predict the damping of rotor oscillations. It cannot predict machine behavior as the various transient rotor currents decay. Thus, more sophisticated models of synchronous machines must explicitly include the effects of the various rotor circuits.

Analysis of synchronous machines is greatly facilitated by transforming armature quantities into a reference frame rotating at rotor speed via the dq0 transformation. In such a reference frame, the armature quantities can be resolved into two components, one along the field-winding axis, also known as the direct axis, and the second along an orthogonal axis known as the quadrature axis. The rotor can then be represented as a set of direct-axis windings consisting of the field winding and additional windings representing the damper or rotor-body circuits which produce flux along the direct-axis, and a set of quadrature-axis windings, which correspond to rotor circuits which produce flux along the quadrature-axis. In general such a model is cast in the form of a set of differential equations which can then be solved with a digital or analog computer.

The object of such modeling is to predict dynamic behavior. The basic electromechanical equation, often known as the swing equation, states that the product of the rotor moment of inertia times its angular acceleration is equal to the net torque applied to the rotor (mechanical shaft torque and electromagnetic torque). Under steady-state conditions, the net torque is zero and the rotor remains at constant (synchronous) speed. Transient analysis is performed to investigate the nature of the transients resulting from a disturbance and whether or not stable steady-state operation will be reestablished.

PROBLEMS

8-1 Consider a 2-winding transformer whose equivalent circuit is shown in Fig. 8-17. A dc voltage V_0 is applied to winding 1 at $t = 0$.

    (a) Assuming the secondary winding to be open-circuited ($i_2 = 0$), write an expression for the primary current as a function of time.

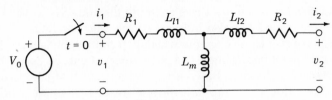

Fig. 8-17. Transformer equivalent circuit for Prob. 8-1.

(b) Assuming the secondary winding to be short-circuited ($v_2 = 0$) and the secondary resistance to be negligible ($R_2 = 0$), write an expression for (1) the primary current as a function of time and (2) the secondary current as a function of time. (3) What is the apparent inductance as seen from the primary terminals?

(c) Compare the primary open-circuit time constant found in part (a) with the primary short-circuit time constant found in part (b).

(d) Discuss in general terms the effects of a finite but small R_2.

8-2 A 60-Hz Y-connected salient-pole synchronous generator ($X_d = 1.70$ per unit, $X_q = 1.40$ per unit) is operating with a phase-a-to-neutral voltage of $8.0 \cos \omega t$ kV and a phase-a current of $650 \cos (\omega t - 20°)$ A; the currents and voltages are balanced 3-phase. The rotor angle is known to be $\theta = \omega t + 45°$.

(a) Draw a phasor diagram for this condition indicating the terminal voltage, the armature current, the location of the rotor direct and quadrature axes, and the direct- and quadrature-axis components of armature terminal voltage and current.

(b) Using the phasor diagram of part (a), calculate the magnitude of the direct- and quadrature-axis armature voltage and current.

(c) Repeat the calculation of part (b) using the dq0 transformation of Eq. 8-1.

8-3 This problem is concerned with analysis of a 2-phase synchronous machine instead of the 3-phase machine of the text. Consider that the machine is idealized as in Art. 8-3 except that there are two distributed windings a and b on the stator, one in each phase, with magnetic axes 90° apart. The salient-pole rotor has only the main-field winding f in the direct axis and no winding in the quadrature axis. The angle from the axis of the phase a to the direct axis is θ. That from the phase-b axis to the direct axis is $\theta + 270°$ or, what amounts to the same thing, $\theta - 90°$. Use the same general assumptions, conventions, and notation as in Arts. 7-2, 8-2 and 8-3.

(a) Write the flux-linkage equations in matrix form corresponding to Eq. 8-3.

(b) In the manner of Arts. 7-2 and 8-3, show that the inductance matrix

$[\mathcal{L}_{abf}]$ is given by

$$\begin{bmatrix} L_{aa0} + L_{g2}\cos 2\theta & L_{g2}\sin 2\theta & L_{af}\cos\theta \\ L_{g2}\sin 2\theta & L_{aa0} - L_{g2}\cos 2\theta & L_{af}\sin\theta \\ L_{af}\cos\theta & L_{af}\sin\theta & L_{ff} \end{bmatrix}$$

(c) Show that the appropriate dq transformation of variables is typified by

$$\begin{bmatrix} i_d \\ i_q \end{bmatrix} = \begin{bmatrix} \cos\theta & \sin\theta \\ -\sin\theta & \cos\theta \end{bmatrix}\begin{bmatrix} i_a \\ i_b \end{bmatrix}$$

Also write the relations for i_a and i_b in terms of i_d, i_q, and θ.
(d) Show that the flux linkages are given by

$$\lambda_f = L_{ff}i_f - L_{af}i_d \qquad \lambda_d = L_{af}i_f - L_d i_d \qquad \lambda_q = -L_q i_q$$

Also identify the direct- and quadrature-axis synchronous inductances L_d and L_q in terms of L_{aa0} and L_{g2}.
(e) Show that Eqs. 8-25 to 8-27 are correct for the voltages v_f, v_d, and v_q in this case.
(f) Show that the instantaneous power output from the 2-phase stator is

$$P_s = v_d i_d + v_q i_q$$

(g) Show that the motor torque is given by

$$T = \frac{\text{poles}}{2}(\lambda_d i_q - \lambda_q i_d)$$

8-4 (a) By carrying out the manipulations described in Art. 8-3 show that Eqs. 8-14 to 8-17 are correct.
(b) Similarly, show that Eqs. 8-25 to 8-28 are correct.

8-5 A 2-pole synchronous motor has a 2-phase winding with negligible resistance on the rotor. Its stator has salient poles with a field winding on the direct axis having the resistance R_f and excited by a direct voltage V_f. There is a short-circuited stator winding in the quadrature axis.

The motor is running in the steady state with balanced 2-phase voltages given by

$$v_a = -V\sin\omega t \qquad \text{and} \qquad v_b = -V\cos\omega t$$

applied to the rotor windings. The angle from the direct axis to the magnetic axis of rotor phase a is

$$\theta = \omega t + \delta$$

That between the phase-b axis and the direct axis is 90° greater than this. (The statement or solution of Prob. 8-3 can be used as a guide in the solution of this problem. Note very carefully the conditions stated here, however, in order to ensure the correct signs.)

(a) Show that the direct- and quadrature-axis rotor currents into the motor are given by

$$i_d = \frac{-\omega L_{af} I_f + V \cos \delta}{\omega L_d} \quad \text{and} \quad i_q = \frac{V \sin \delta}{\omega L_q}$$

(b) Show that the motor torque is given by

$$T = -\frac{L_{af} V_f V}{\omega L_d R_f} \sin \delta - \frac{V^2 (L_d - L_q)}{2\omega^2 L_d L_q} \sin 2\delta$$

(c) Show that the phase-a rotor current is

$$i_a = i_d \cos \theta + i_q \sin \theta$$

8-6 A 3-phase turbine generator is rated 13.8 kV (line to line), 110,000 kVA. Its constants, with reactances expressed in per unit on the machine rating as a base, are

$$X_d = 1.10 \qquad X_d' = 0.20 \qquad T_d' = 1.0 \text{ s}$$

It is operating unloaded at a terminal voltage of 1.00 per unit when a 3-phase short circuit occurs at its terminals. Except in part (g) ignore the dc component in the short-circuit current. Express numerical answers both in per unit and in amperes.

(a) What is the rms steady-state short-circuit current? Does it make sense physically for the steady-state short-circuit current to be lower than rated current, as it is here? Explain.
(b) Write the numerical equation for the instantaneous phase-a current as a function of time. Consider the fault to occur when the angle between the axis of phase a and the direct axis is 90°. Because of the neglect of the dc component, this is the *symmetrical* short-circuit current.
(c) Write the numerical equation for the envelope of the short-circuit current wave as a function of time.
(d) Using the result of part (c), write the numerical equation showing how the rms value of short-circuit current varies with time.

(e) What value is given by the expression in part (d) at $t = 0$? This is known as the *initial symmetrical rms short-circuit current*.

(f) Generalize the result in part (d) by writing the equation for rms symmetrical short-circuit current as a function of time, initial voltage behind transient reactance, and the machine constants.

(g) In part (b) suppose the fault occurs when the magnitude of the initial angle is other than 90°. The value of i_a at $t = 0$ would then be nonzero. But since the phase-a winding is a resistance-inductance circuit, the complete phase-a current cannot change instantaneously from zero. Hence a dc component must be present in i_a to reconcile the situation. This component dies away rapidly. Give the maximum possible initial magnitude of the dc component.

8-7 The machine in Prob. 8-6 has the constants

$$X_d'' = 0.10 \qquad T_d'' = 0.35 \text{ s}$$

in addition to those given there. Work Prob. 8-6, except for part (b), with the subtransient effects included as well as the transient effects. In part (f) the initial voltage behind subtransient reactance must also be reflected. In part (g), the principle still holds that the dc component must preserve continuity of instantaneous phase current just before and just after the short circuit.

8-8 While an alternator is undergoing a standard short-circuit test on a factor test floor, the short circuit is suddenly removed. Before this removal, rated steady-state short-circuit current was flowing. The machine constants are $X_d = 1.20$, $X_q = 0.80$, $X_d' = 0.40$, $X_q' = 0.80$, $T_{do}' = 5$ s, $T_d' = 1.67$ s. There are no effective damper circuits in either the direct or the quadrature axis, and the resistance and inductance of the exciter may be assumed negligible. Give the numerical expression for the per unit field current as a function of time after removal of the short circuit.

8-9 Particularly severe dips in generator terminal voltage are produced by the application of inductive loads at or close to zero power factor. The starting inrush to a large motor is one such type of load. Consider that a synchronous generator is operating initially unloaded at normal terminal voltage. A balanced inductive load X_L is suddenly applied to its terminals. The field voltage is not changed, and the generator continues to operate at synchronous speed. Ignore saturation.

(a) Show that the variation of terminal voltage with time after load application is given by

$$V_{ta} = E_{f0}\frac{X_L}{X_d + X_L} + E_{f0}\left(\frac{X_L}{X_d' + X_L} - \frac{X_L}{X_d + X_L}\right)\varepsilon^{-t/T_d'}$$

where E_{f0} is the preload excitation voltage and

$$T_d' = T_{do}' \frac{X_d' + X_L}{X_d + X_L}$$

(b) Give per unit voltage magnitudes when $X_d = 1.10$, $X_d' = 0.20$, $X_L = 1.25$, and $T_{do}' = 5.0\,\text{s}$. The reactances are per unit values. Those for the machine are typical of large turbine generators. That for X_L will, at normal voltage, load the generator to about its continuous reactive capability at zero power factor.

8-10 A non-salient-pole synchronous generator is characterized by the following parameters:

$X_d = X_q = 1.85$ per unit

$X_d' = 0.28$ per unit

$T_{do}' = 3.6\,\text{s}$

Field resistance $= 0.15\,\Omega$

Field voltage required to achieve rated open-circuit terminal voltage $= 450\,\text{V}$

For the purposes of this problem, neglect the effects of saturation and armature resistance.

(a) The machine is operating at rated speed, open-circuited, with no excitation applied to the field winding. A voltage of 450 V is suddenly applied to the field winding.

(1) What is the field current as a function of time?

(2) Write an expression for the per unit armature voltage as a function of time.

(b) The same machine with 450 V applied to the field winding has been operating into a 3-phase terminal short circuit long enough for all transients to have died out. What is the steady-state field current in amperes and the per unit armature current?

(c) The steady-state armature short circuit is suddenly removed. For the purposes of this analysis, assume that all three phase currents are simultaneously interrupted.

(1) What is the magnitude of the field current in amperes immediately after interrupting the armature current?

(2) Write expressions for the field current and terminal voltage as a function of time.

8-11 A 3-phase salient-pole generator has the following constants in per unit on the machine kVA rating as a base:

$$X_d = 1.0 \qquad X_q = 0.6 \qquad X_d' = 0.3$$

(a) The machine is connected to an infinite bus at its terminals with rated voltage and power output at 1.0 power factor. Compute and plot the transient power-angle curve of the machine. Consider E_q' constant and retain the true angle δ.

(b) Repeat part (a), using the approximate simplified representation of Fig. 8-7 for the machine.

(c) Instead of being at the terminals of the machine, the infinite bus is supplied from the generator over a line having negligible resistance and a reactance of 0.7 per unit. The generator continues to operate at rated terminal voltage, unity power factor at its terminals, and 1.00 power output. After determining the new value of the infinite-bus voltage, repeat parts (a) and (b) for the combination of machine and external reactance.

8-12 The ideal conditions for synchronizing an alternator with an electric-power system are that the alternator voltage be the same as that of the system bus in magnitude, phase, and frequency. Departure from these conditions results in undesirable current and power surges accompanying electromechanical oscillation of the alternator rotor. As long as the oscillations are not too violent, they can be investigated by a linearized analysis.

Consider that a 20-MVA 0.85-power-factor 60-Hz 2-pole alternator driven by a gas turbine is to be synchronized with a system large enough to be considered an infinite bus. The moment of inertia of the alternator and turbine is 3340 kg $\cdot$ m^2. The damping-power coefficient P_d is 16 kW/(elec deg/s), and the synchronizing-power coefficient P_s is 1.75 MW/elec deg. Both P_d and P_s may be assumed to remain constant. In all cases below, the terminal voltage is adjusted to its correct magnitude. The turbine governor is sufficiently insensitive not to act during the synchronizing period.

(a) Consider that the alternator is initially adjusted to the correct speed but that it is synchronized out of phase by 20 electrical degrees, with the alternator leading the bus. Obtain a numerical expression for the ensuing electromechanical oscillations. Also give the largest value of torque exerted on the rotor during the synchronizing period. Ignore losses, and express this torque as a percentage of that corresponding to the nameplate rating.

(b) Repeat part (a) with the alternator synchronized at the proper angle but with its speed initially adjusted 1.0 Hz fast.

(c) Repeat part (a) with the alternator initially leading the bus by 20 electrical degrees and its speed initially adjusted 1.0 Hz fast.

8-13 The following are the constants of a 35,000-kVA 13.8-kV 60-Hz water-wheel generator: $X_d = 1.00$, $X_d' = 0.35$, $X_d'' = 0.25$, $T_{do}' = 5.0$ s, $T_d'' = 0.04$ s. Reactances are in per unit on the generator rating as a base. This generator supplies a load over a line whose reactance is 0.50 per unit and whose resistance is negligible. Under normal conditions, the generator is fully loaded with a terminal power factor of 0.80 lagging and rated terminal voltage. A 3-phase short circuit is considered to occur at the receiving end of the line.

(a) Compute the prefault values of the voltages behind synchronous and transient reactances.

(b) Compute the largest possible initial value of the dc component of short-circuit current in the machine.

(c) Give the numerical equation for the generator rms short-circuit current as a function of time after fault occurrence. Ignore the dc component.

(d) The fault is cleared after 0.15 s. From the results of part (c) give the rms value of generator short-circuit current just before clearing.

8-14 Consider a synchronous machine with a direct-axis damper winding. The damper winding (labeled by subscript $d1$) is described by the following per unit parameters as referred to the stator:

$$\mathcal{L}_{d1d1} = L_{d1d1} = \text{damper self-inductance}$$
$$\mathcal{L}_{ad1} = L_{ad1} \cos \theta = \text{phase-a--to--damper mutual inductance}$$
$$\mathcal{L}_{fd1} = L_{fd1} = \text{field-to-damper mutual inductance}$$
$$R_{d1} = \text{damper resistance}$$

(a) Write the flux-current relationships for this machine corresponding to Eq. 8-3.

(b) Apply the dq0 transformation to the flux-current relationship of part (a). *Hint:* Note that the dq0 transformation treats both the damper and field windings in the same manner. As a result, it is possible to obtain the desired results directly from inspection of Eqs. 8-14 to 8-17.

(c) What are the transformed voltage equations for this machine? Note that the damper winding is short-circuited and thus $v_{d1} = 0$.

8-15 In Prob. 8-14 a synchronous machine with a direct-axis damper winding is analyzed. In this problem, repeat the analysis of Prob. 8-14 with the addition of a quadrature-axis damper winding (labeled by subscript $q1$). Note that the quadrature-axis leads the direct axis and thus the mutual inductance between phase a and the quadrature-axis damper winding is $\mathcal{L}_{aq1} = -L_{aq1} \sin \theta$.

8-16 Reciprocating air and ammonia compressors require a torque which fluctuates periodically about a steady average value. For a 2-cycle unit the torque harmonics have frequencies in hertz which are multiples of the speed in revolutions per second. When, as is commonly the case, the compressors are driven by synchronous motors, the torque harmonics cause periodic fluctuation of the torque angle δ and may result in undesirably high pulsations of power and current to the motor. It is therefore essential that, for the significant harmonics, the electrodynamic response of the motor be held to a minimum.

(a) To investigate the response of the motor to torque harmonics, use a linearized analysis. Let

$$P_{mech} = P_{sh,max} \cos \omega t$$

where $P_{sh,max}$ corresponds to the amplitude of the harmonic-torque pulsation whose angular frequency is ω. Then show that the differential equation can be written

$$\frac{d^2\delta}{dt^2} + 2\zeta\omega_n \frac{d\delta}{dt} - \omega_n^2 \delta = \frac{P_{sh,max}}{P_j} \sin \omega t$$

Identify these quantities in terms of P_s, P_d, and $P_j = (2/\text{poles})J\omega$.
(b) Show that the phasor expression for the steady-state solution is

$$\Delta = \frac{P_{sh,max}/P_j}{\omega_n^2 - \omega^2 + j2\zeta\omega_n\omega}$$

(c) The motor driving the compressor is that of Example 8-3. The compressor has a first-order torque harmonic with an amplitude of 580 lb · ft and a frequency of 4.3 Hz. Determine the maximum deviation in power angle δ and the corresponding pulsation of synchronous power.
(d) Consider that a flywheel is added to bring the total Wk^2 up to 16,500 lb · ft^2. Repeat the computation of part (c) and compare the results.

8-17 A 750-kVA, 4160-V 6-pole 60-Hz nonsalient synchronous generator is driven by a wind turbine and coupled to a 4160-V distribution system (considered here to be an infinite bus) through an impedance of 0.12 per unit (generator base). The generator parameters are

$$X_d = 1.57 \text{ per unit} \qquad X_d' = 0.28 \text{ per unit} \qquad T_{do}' = 2.7 \text{ s}$$

The total moment of inertia of the wind turbine and generator combination is $1.8 \times 10^3 \, \text{kg} \cdot \text{m}^2$.

The generator is initially operating with a terminal voltage of 4160 V at 500 kW output.

(a) Calculate the initial values (per unit magnitudes and angles with respect to the infinite bus) of the excitation voltage $\hat{E}_{af}$ and the voltage behind transient reactance $\hat{E}'_i$.

A sudden increase in wind speed causes a doubling of the wind torque. Assume that the voltage behind transient reactance remains constant throughout the ensuing transient.

(b) Using the equal-area criterion, calculate the maximum value of the transient rotor angle.

(c) Upon decay of the rotor oscillations, what is the final value of the transient rotor angle?

(d) Using a linearized analysis, estimate the frequency of the rotor oscillations. The linearization should be performed around the operating point found in part (c).

8-18 A synchronous motor whose input under rated operating conditions is 10,000 kVA is connected to an infinite bus over a short feeder whose impedance is purely reactive. The motor is rated at 60 Hz, 600 r/min, and has a total Wk^2 of 500,000 lb $\cdot$ ft^2 (including the shaft load). The power-angle curve under transient conditions is $2.00 \sin \delta$, where the amplitude is in per unit on a 10,000-kVA base.

(a) With the motor operating initially unloaded, a 10,000-kW shaft load is suddenly applied. Does the motor remain in synchronism?

(b) How large a shaft load can be applied suddenly without loss of synchronism?

(c) Consider now that the suddenly applied load is on for only 0.2 s, after which a comparatively long time elapses before any load is applied again. Determine the maximum value of such a load which will still allow synchronism to be maintained. Use the equal-area criterion as an aid in the process. For purposes of computing the angle δ at 0.2 s, ignore damping and use a linearized analysis in which the power-angle curve is approximated by a straight line through the origin and the 60° point.

Polyphase
Induction
Machines

The objects of this chapter are to develop equivalent circuits for the polyphase induction motor from which both the effects of the motor on its supply circuit and the characteristics of the motor itself can be determined and to study these effects and characteristics. The general form of equivalent circuit is suggested by the similarity of an induction machine to a transformer.

9-1 INTRODUCTION TO POLYPHASE INDUCTION MACHINES

As indicated in Art. 3-2b, an induction motor is one in which alternating current is supplied to the stator directly and to the rotor by induction or transformer action from the stator. As in the synchronous machine, the stator winding is of the type discussed in Art. 3-5. When excited from a balanced polyphase source, it will produce a magnetic field in the air gap rotating at

409

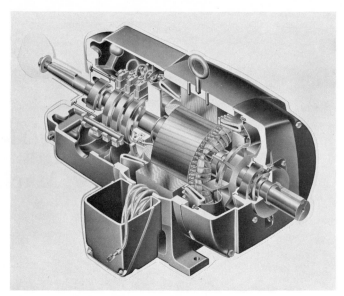

Fig. 9-1. Cutaway view of a 3-phase induction motor with a wound rotor and slip rings connected to the 3-phase rotor winding. (*General Electric Company.*)

synchronous speed as determined by the number of poles and the applied stator frequency f (Eq. 3-43). The rotor may be one of two types. A *wound rotor* carries a polyphase winding similar to, and wound for, the same number of poles as the stator. The terminals of the rotor winding are connected to insulated slip rings mounted on the shaft. Carbon brushes bearing on these rings make the rotor terminals available external to the motor, as shown in the cutaway view in Fig. 9-1. The motor in Fig. 3-14 has a *squirrel-cage rotor* with a winding consisting of conducting bars embedded in slots in the rotor iron and short-circuited at each end by conducting end rings. The extreme simplicity and ruggedness of the squirrel-cage construction are outstanding advantages of the induction motor.

Now assume that the rotor is turning at the steady speed n r/min in the same direction as the rotating stator field. Let the synchronous speed of the stator field be n_1 r/min as given by Eq. 3-43. The rotor is then traveling at a speed $n_1 - n$ r/min backward with respect to the stator field, or the *slip* of the rotor is $n_1 - n$ r/min. Slip is more usually expressed as a fraction of synchronous speed; i.e., the per unit slip s is

$$s = \frac{n_1 - n}{n_1} \qquad (9\text{-}1)$$

or

$$n = n_1(1 - s) \qquad (9\text{-}2)$$

This relative motion of flux and rotor conductors induces voltages of frequency sf, called *slip frequency*, in the rotor. Thus, the electrical behavior of an induction machine is similar to that of a transformer but with the additional feature of frequency transformation. A wound-rotor induction machine can be used as a frequency changer.

When used as an induction motor, the rotor terminals are short-circuited. The rotor currents are then determined by the magnitudes of the induced voltages and the rotor impedance at slip frequency. At starting, the rotor is stationary, the slip $s = 1$, and the rotor frequency equals the stator frequency f. The field produced by the rotor currents therefore revolves at the same speed as the stator field, and a starting torque results, tending to turn the rotor in the direction of rotation of the stator-inducing field. If this torque is sufficient to overcome the opposition to rotation created by the shaft load, the motor will come up to its operating speed. The operating speed can never equal the synchronous speed n_1, however, since the rotor conductors would then be stationary with respect to the stator field and no voltage would be induced in them.

With the rotor revolving in the same direction of rotation as the stator field, the frequency of the rotor currents is sf, and the component rotor field set up by them will travel at sn_1 r/min *with respect to the rotor* in the forward direction. But superimposed on this rotation is the mechanical rotation of the rotor at n r/min. The speed of the rotor field in space is the sum of these two speeds and equals

$$sn_1 + n = sn_1 + n_1(1 - s) = n_1 \qquad (9\text{-}3)$$

The stator and rotor fields are therefore stationary with respect to each other, a steady torque is produced, and rotation is maintained. Such a torque existing at any mechanical speed n other than synchronous speed is called an *asynchronous torque*.

Figure 9-2 shows a typical squirrel-cage induction-motor torque-speed characteristic. The factors influencing the shape of this characteristic can be appreciated in terms of the torque equation 3-86. In this equation recognize that the resultant air-gap flux Φ_{sr} is approximately constant when the stator-applied voltage and frequency are constant. Also recall that the rotor mmf F_r is proportional to the rotor current I_r. Equation 3-86 then reduces to

$$T = K I_r \sin \delta_r \qquad (9\text{-}4)$$

where K is a constant. The rotor current is determined by the voltage induced in the rotor and its leakage impedance, both at slip frequency. The rotor-induced voltage is proportional to slip. Under normal running conditions the slip is small—3 to 10 percent at full load in most squirrel-cage motors. The rotor frequency sf therefore is very low (of the order of 2 to 6 Hz in 60-Hz motors). Consequently, in this range the rotor impedance is largely resistive, and the rotor current is very nearly proportional to, and in phase with, the rotor voltage and is therefore very nearly proportional to slip. Furthermore,

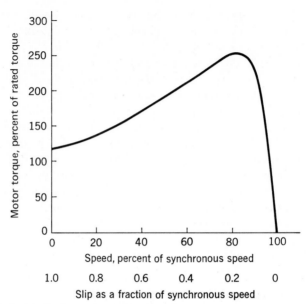

Fig. 9-2. Typical induction-motor torque-speed curve.

the rotor-mmf wave lags approximately 90 electrical degrees behind the resultant flux wave, and therefore $\sin \delta_r \approx 1$. (This point is discussed in Art. 9-2a.) Approximate linearity of torque as a function of slip is therefore to be expected in the range where the slip is small. As slip increases, the rotor impedance increases because of the increasing effect of rotor leakage inductance. Thus the rotor current is less than proportional to slip. Also the rotor current lags farther behind the induced voltage, the mmf wave lags farther behind the resultant flux wave, and $\sin \delta_r$ decreases. The result is that the torque increases with increasing slip up to a maximum value and then decreases, as shown in Fig. 9-2. The maximum torque, or *breakdown torque,* limits the short-time overload capability of the motor.

The squirrel-cage motor is substantially a constant-speed motor having a few percent drop in speed from no load to full load. Speed variation can be obtained by using a wound-rotor motor and inserting external resistance in the rotor circuit. In the normal operating range, external resistance simply increases the rotor impedance, necessitating a higher slip for a desired rotor mmf and torque.

In Art. 7-1 it was mentioned that a synchronous motor per se has no starting torque. To make a synchronous motor self-starting, a squirrel-cage winding, called an *amortisseur* or *damper winding,* is inserted in the rotor pole faces, as shown in Fig. 9-3. The rotor then comes up almost to synchronous speed by induction-motor action with the field winding unexcited. If the load and inertia are not too great, the motor will pull into synchronism when the field winding is energized from a dc source.

Fig. 9-3. Rotor of a 40-pole 180-r/min 4000-V 11,500-hp synchronous motor showing field coils, pole-face damper windings, and exciter. (*General Electric Company.*)

9–2 CURRENTS AND FLUXES IN INDUCTION MACHINES

For a coil-wound rotor, the flux-mmf situation can be seen with the aid of Fig. 9-4. This sketch shows a development of a simple 2-pole 3-phase rotor winding in a 2-pole field. It therefore conforms with the restriction that a wound rotor must have the same number of poles as the stator (although the number of phases need not be the same). The flux-density wave is moving to the right at slip speed with respect to the winding. It is shown in Fig. 9-4 in the position of maximum instantaneous voltage in phase a.

If rotor leakage reactance is very small compared with rotor resistance (which is very nearly the case at the small slips corresponding to normal operation), the phase-a current will also be a maximum. As shown in Art. 3-5, the rotor-mmf wave will then be centered on phase a. It is so shown in Fig. 9-4a. The displacement angle, or torque angle, δ under these conditions is at its optimum value of 90°.

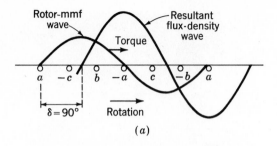

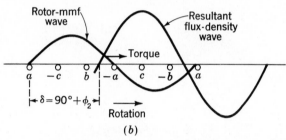

Fig. 9-4. Developed rotor winding of induction motor with flux-density and mmf waves in their relative positions for (a) zero and (b) nonzero rotor leakage reactance.

If the rotor leakage reactance is appreciable, however, the phase-a current lags the induced voltage by the power-factor angle ϕ_2 of the leakage impedance. The phase-a current will not be a maximum until a correspondingly later time. The rotor-mmf wave will then not be centered on phase a until the flux wave has traveled ϕ_2 degrees farther down the gap, as shown in Fig. 9-4b. The angle δ is now $90° + \phi_2$. In general, therefore, the torque angle of an induction motor is

$$\delta = 90° + \phi_2 \tag{9-5}$$

It departs from the optimum value by the power-factor angle of the rotor leakage impedance at slip frequency. The electromagnetic rotor torque is directed toward the right in Fig. 9-4, or in the direction of the rotating flux wave.

The comparable picture for a squirrel-cage rotor is given in Fig. 9-5. A 16-bar rotor placed in a 2-pole field is shown in developed form. For simplicity of drafting, only a relatively small number of rotor bars is chosen, and the number is an integral multiple of the number of poles, a choice normally avoided in order to prevent harmful harmonic effects. In Fig. 9-5a the sinusoidal flux-density wave induces a voltage in each bar which has an instantaneous value indicated by the solid vertical lines. At a somewhat later instant of time, the bar currents assume the instantaneous values indicated by the solid vertical lines in Fig. 9-5b, the time lag being the rotor power-factor angle ϕ_2. In this time interval, the flux-density wave has traveled in its direction of rotation with respect to the rotor through a space angle ϕ_2 and is then in the position shown in Fig. 9-5b. The corresponding rotor-mmf wave is shown by

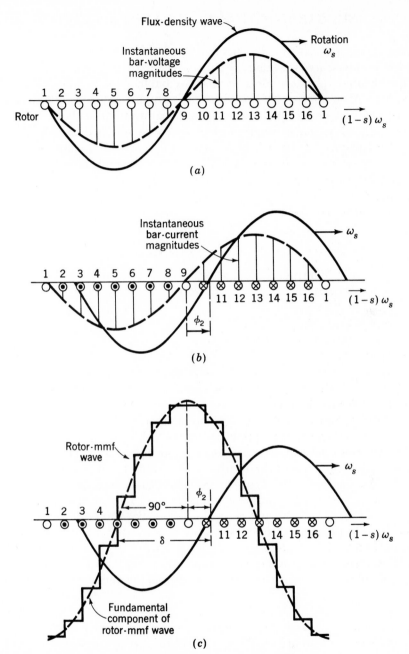

Fig. 9-5. Reactions of a squirrel-cage rotor in a 2-pole field.

the step wave of Fig. 9-5c. The fundamental component is shown by the dashed sinusoid and the flux-density wave by the solid sinusoid. Study of these figures confirms the general principle that the number of rotor poles in a squirrel-cage rotor is determined by the inducing flux wave.

9-3 THE INDUCTION–MOTOR EQUIVALENT CIRCUIT

The foregoing considerations of flux and mmf waves can readily be translated into the steady-state equivalent circuit for the machine. Only machines with symmetrical polyphase windings excited by balanced polyphase voltages are considered. As in many other discussions of polyphase devices, it may be helpful to think of 3-phase machines as being Y-connected, so that currents are always line values and voltages always line-to-neutral values.

First consider conditions in the stator. The synchronously rotating air-gap flux wave generates balanced polyphase counter emfs in the phases of the stator. The stator terminal voltage differs from the counter emf by the voltage drop in the stator leakage impedance, the phasor relation for the phase under consideration being

$$\widehat{V}_1 = \widehat{E}_1 + \widehat{I}_1 \, (R_1 + jX_1) \tag{9-6}$$

where $\widehat{V}_1$ = stator terminal voltage
$\widehat{E}_1$ = counter emf generated by resultant air-gap flux
$\widehat{I}_1$ = stator current
R_1 = stator effective resistance
X_1 = stator leakage reactance

The positive directions are shown in the equivalent circuit of Fig. 9-6.

The resultant air-gap flux is created by the combined mmf's of the stator and rotor currents. Just as in the transformer analog, the stator current can be resolved into two components, a load component and an exciting component. The load component $\widehat{I}_2$ produces an mmf which exactly counteracts the mmf of the rotor current. The exciting component $\widehat{I}_\varphi$ is the additional stator current required to create the resultant air-gap flux and is a function of the emf $\widehat{E}_1$. The exciting current can be resolved into a core-loss component $\widehat{I}_c$ in phase with $\widehat{E}_1$ and a magnetizing component $\widehat{I}_m$ lagging $\widehat{E}_1$ by 90°. In the equivalent circuit the exciting current can be accounted for by means of a shunt branch, formed by core-loss conductance G_c and magnetizing susceptance B_m in parallel, connected across $\widehat{E}_1$, as in Fig. 9-6. Both G_c and B_m are usually determined at rated stator frequency and for a value of E_1 close to the expected operating value; they are then assumed to remain constant for the

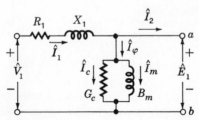

Fig. 9-6. Stator equivalent circuit for a polyphase induction motor.

small departures from that value associated with normal operation of the motor.

So far, the equivalent circuit representing stator phenomena is exactly like that for the primary of a transformer. To complete the circuit, the effects of the rotor must be incorporated. This is done by considering stator and rotor voltages and currents in terms of referred rotor quantities.

In Art. 9-2 we saw that, insofar as fundamental components are concerned, both squirrel-cage and wound rotors react by producing an mmf wave having the same number of poles as the inducing flux wave, traveling at the same speed as the flux wave, and with a torque angle 90° greater than the rotor power-factor angle. The reaction of the rotor-mmf wave on the stator calls for a compensating load component of stator current and thereby enables the stator to absorb from the line the power needed to sustain the torque created by the interaction of the flux and mmf waves. The only way the stator knows what is happening is through the medium of the air-gap flux and rotor-mmf waves. Consequently, if the rotor were replaced by one having the same mmf and power factor at the same speed, the stator would be unable to detect the change. Such replacement leads to the idea of referring rotor quantities to the stator, an idea which is of great value in translating flux-mmf considerations into an equivalent circuit for the motor.

Consider, for example, a coil-wound rotor, wound for the same number of poles and phases as the stator. The number of effective turns per phase in the stator winding is a times the number in the rotor winding. Compare the magnetic effect of this rotor with that of a magnetically equivalent rotor having the same number of turns as the stator. For the same flux and speed, the relation between the voltage $\widehat{E}_{\text{rotor}}$ induced in the actual rotor and the voltage $\widehat{E}_{2s}$ induced in the equivalent rotor is

$$\widehat{E}_{2s} = a\widehat{E}_{\text{rotor}} \tag{9-7}$$

If the rotors are to be magnetically equivalent, their ampere-turns must be equal, and the relation between the actual rotor current $\widehat{I}_{\text{rotor}}$ and the current $\widehat{I}_{2s}$ in the equivalent rotor must be

$$\widehat{I}_{2s} = \frac{\widehat{I}_{\text{rotor}}}{a} \tag{9-8}$$

Consequently the relation between the slip-frequency leakage impedance Z_{2s} of the equivalent rotor and the slip-frequency leakage impedance Z_{rotor} of the actual rotor must be

$$Z_{2s} = \frac{\widehat{E}_{2s}}{\widehat{I}_{2s}} = \frac{a^2\widehat{E}_{\text{rotor}}}{\widehat{I}_{\text{rotor}}} = a^2 Z_{\text{rotor}} \tag{9-9}$$

The voltages, currents, and impedances in the equivalent rotor are defined as their values *referred to the stator*. The thought process is essentially like that involved in referring secondary quantities to the primary in static-transformer theory (see Arts. 1-7 and 1-8). The referring factors are ratios of effective turns and are the same in essence as in transformer theory.

The referring factors must, of course, be known when one is concerned specifically with what is happening in the actual rotor circuits. From the viewpoint of the stator, however, the reflected effects of the rotor show up in terms of the referred quantities, and the theory of both coil-wound and cage rotors can be formulated in terms of the referred rotor. We shall assume, therefore, that the referred rotor constants are known.

Since the rotor is short-circuited, the phasor relation between the slip-frequency emf $\widehat{E}_{2s}$ generated in the reference phase of the referred rotor and the current $\widehat{I}_{2s}$ in this phase is

$$\frac{\widehat{E}_{2s}}{\widehat{I}_{2s}} = Z_{2s} = R_2 + jsX_2 \tag{9-10}$$

where Z_{2s} = slip-frequency rotor leakage impedance per phase referred to stator

R_2 = referred effective resistance

sX_2 = referred leakage reactance at slip frequency

The reactance is expressed in this way because it is proportional to rotor frequency and therefore to slip. Thus X_2 is defined as the value the referred rotor leakage reactance would have at *stator frequency*. The slip-frequency equivalent circuit of one phase of the referred rotor is shown in Fig. 9-7. This is the equivalent circuit of the rotor as seen in the rotor reference frame.

The stator sees a flux wave and an mmf wave rotating at synchronous speed. The flux wave induces the slip-frequency rotor voltage $\widehat{E}_{2s}$ and the stator counter emf $\widehat{E}_1$. If it were not for the effect of speed, the referred rotor voltage would equal the stator voltage, since the referred rotor winding is identical with the stator winding. Because the relative speed of the flux wave with respect to the rotor is s times its speed with respect to the stator, the

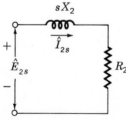

Fig. 9-7. Rotor equivalent circuit for a polyphase induction motor at slip frequency.

relation between the effective values of stator and rotor emf's is

$$\widehat{E}_{2s} = s\widehat{E}_1 \tag{9-11}$$

The rotor-mmf wave is opposed by the mmf of the load component $\widehat{I}_2$ of stator current, and therefore, for effective values,

$$\widehat{I}_{2s} = \widehat{I}_2 \tag{9-12}$$

Division of Eq. 9-11 by Eq. 9-12 then gives

$$\frac{\widehat{E}_{2s}}{\widehat{I}_{2s}} = \frac{s\widehat{E}_1}{\widehat{I}_2} \tag{9-13}$$

When we recognize that torque can be calculated in terms of mmf and the resultant air-gap flux, as in Eq. 3-86, and that equal and opposite torques act on the rotor and stator, we see that the mmf wave created by the stator load current $\widehat{I}_2$ must be space-displaced from the resultant flux wave by the same space angle as that between the rotor-mmf wave and the resultant air-gap flux, namely the torque angle δ. The time-phase angle between the stator voltage $\widehat{E}_1$ and the stator load current $\widehat{I}_2$ therefore must equal the corresponding time angle for the rotor, namely, the rotor power-factor angle ϕ_2. The fact that the rotor and stator torques are in opposition is accounted for, since the rotor current $\widehat{I}_{2s}$ is created by the rotor emf $\widehat{E}_{2s}$, whereas the stator current $\widehat{I}_2$ is flowing against the stator counter emf $\widehat{E}_1$. Therefore Eq. 9-13 is true, not only for effective values, but also in a phasor sense. Through substitution of Eq. 9-10 in the phasor equivalent of Eq. 9-13 we have

$$\frac{s\widehat{E}_1}{\widehat{I}_2} = \frac{\widehat{E}_{2s}}{\widehat{I}_{2s}} = R_2 + jsX_2 \tag{9-14}$$

Division by s then gives

$$\frac{\widehat{E}_1}{\widehat{I}_2} = \frac{R_2}{s} + jX_2 \tag{9-15}$$

That is, the stator sees magnetic conditions in the air gap which result in induced stator voltage $\widehat{E}_1$ and stator load current $\widehat{I}_2$, and by Eq. 9-15 these conditions are identical with the result of connecting an impedance $(R_2/s) + jX_2$ across $\widehat{E}_1$. Consequently, the effect of the rotor can be incorporated in the equivalent circuit of Fig. 9-6 by this impedance connected across the terminals ab. The final result is shown in Fig. 9-8. The combined effect of shaft load and rotor resistance appears as a reflected resistance R_2/s, a func-

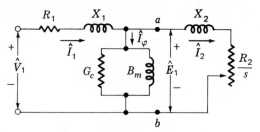

Fig. 9-8. Equivalent circuit for a polyphase induction motor.

tion of slip and therefore of the mechanical load. The current in the reflected rotor impedance equals the load component $\hat{I}_2$ of stator current; the voltage across this impedance equals the stator voltage $\hat{E}_1$. It should be noted that when rotor currents and voltages are reflected into the stator, their frequency is also changed to stator frequency. All rotor electrical phenomena, when viewed from the stator, become stator-frequency phenomena, because the stator winding simply sees mmf and flux waves traveling at synchronous speed.

9-4 ANALYSIS OF THE EQUIVALENT CIRCUIT

Among the important performance aspects in the steady state are the variation of current, speed, and losses as the load-torque requirements change, the starting torque, and the maximum torque. All these characteristics can be determined from the equivalent circuit.

The equivalent circuit shows that the total power P_{g1} transferred across the air gap from the stator is

$$P_{g1} = q_1 I_2^2 \frac{R_2}{s} \tag{9-16}$$

where q_1 is the number of stator phases. The total rotor I^2R loss is evidently

$$\text{Rotor } I^2R \text{ loss} = q_1 I_2^2 R_2 \tag{9-17}$$

The internal mechanical power P developed by the motor is therefore

$$P = P_{g1} - \text{rotor } I^2R \text{ loss} = q_1 I_2^2 \frac{R_2}{s} - q_1 I_2^2 R_2 \tag{9-18}$$

$$= q_1 I_2^2 R_2 \frac{1 - s}{s} \tag{9-19}$$

$$= (1 - s) P_{g1} \tag{9-20}$$

We see, then, that of the total power delivered to the rotor the fraction

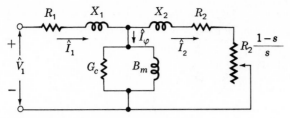

Fig. 9-9. Alternative form of equivalent circuit.

$1 - s$ is converted into mechanical power and the fraction s is dissipated as rotor-circuit I^2R loss. From this it is evident that an induction motor operating at high slip is an inefficient device. When power aspects are to be emphasized, the equivalent circuit is frequently redrawn in the manner of Fig. 9-9. The internal mechanical power per stator phase is equal to the power absorbed by the resistance $R_2(1 - s)/s$.

The internal electromagnetic torque T corresponding to the internal power P can be obtained by recalling that mechanical power equals torque times angular velocity. Thus, when ω_s is the synchronous angular velocity of the rotor in mechanical radians per second,

$$P = (1 - s)\omega_s T \qquad (9\text{-}21)$$

with T in newton-meters. Use of Eq. 9-19 leads to

$$T = \frac{1}{\omega_s} q_1 I_2^2 \frac{R_2}{s} \qquad (9\text{-}22)$$

the synchronous angular velocity ω_s being given by

$$\omega_s = \frac{4\pi f}{\text{poles}} \qquad (9\text{-}23)$$

The torque T and power P are not the output values available at the shaft because friction, windage, and stray load losses remain to be accounted for. It is obviously correct to subtract friction and windage effects from T or P, and it is generally assumed that stray load effects can be subtracted in the same manner. The final remainder is available in mechanical form at the shaft for useful work.

In static-transformer theory, analysis of the equivalent circuit is often simplified by either neglecting the exciting branch entirely or adopting the approximation of moving it out directly to the primary terminals. Such approximations are not permissible for the induction motor under normal running conditions because the presence of the air gap makes necessary a much higher exciting current—30 to 50 percent of full-load current—and because the leakage reactances are also necessarily higher. Some simplification of the

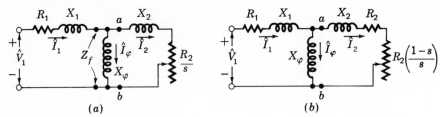

Fig. 9-10. Equivalent circuits.

induction-motor equivalent circuit results if the shunt conductance G_c is omitted and the associated core-loss effect deducted from T or P at the same time that friction, windage, and stray load effects are subtracted. The equivalent circuit then becomes that of Fig. 9-10a or b, and the error introduced is negligible. Such a procedure also has an advantage during motor testing, for no-load core loss need not then be separated from friction and windage. These last circuits will be used in subsequent discussions.

EXAMPLE 9-1

A 3-phase Y-connected 220-V (line-to-line) 10-hp 60-Hz 6-pole induction motor has the following constants in ohms per phase referred to the stator:

$$R_1 = 0.294 \quad R_2 = 0.144$$
$$X_1 = 0.503 \quad X_2 = 0.209 \quad X_\varphi = 13.25$$

The total friction, windage, and core losses may be assumed to be constant at 403 W, independent of load.

For a slip of 2.00 percent, compute the speed, output torque and power, stator current, power factor, and efficiency when the motor is operated at rated voltage and frequency. Neglect the impedance of the source.

Solution

The impedance Z_f (Fig. 9-10a) represents physically the per phase impedance presented to the stator by the air-gap field, both the reflected effect of the rotor and the effect of the exciting current being included therein. From Fig. 9-10a

$$Z_f = R_f + jX_f = \left(\frac{R_2}{s} + jX_2\right) \text{ in parallel with } jX_\varphi$$

Substitution of numerical values gives, for $s = 0.02$,

$$R_f + jX_f = 5.41 + j\,3.11$$
$$R_1 + jX_1 = 0.29 + j\,0.50$$
$$\text{Sum} = 5.70 + j\,3.61 = 6.75\underline{/32.4°}\ \Omega$$

$$\text{Applied voltage to neutral} = \frac{220}{\sqrt{3}} = 127 \text{ V}$$

$$\text{Stator current } I_1 = \frac{127}{6.75} = 18.8 \text{ A} \qquad \text{Power factor} = \cos 32.4° = 0.844$$

$$\text{Synchronous speed} = \frac{2f}{\text{poles}} = \frac{120}{6} = 20 \text{ r/s} = 1200 \text{ r/min}$$

$$\omega_s = 2\pi(20) = 125.6 \text{ rad/s}$$

$$\text{Rotor speed} = (1 - s)(\text{synchronous speed}) = 0.98(1200) = 1176 \text{ r/min}$$

From Eq. 9-16,

$$P_{g1} = q_1 I_2^2 \frac{R_2}{s} = q_1 I_1^2 R_f = 3(18.8)^2(5.41) = 5740 \text{ W}$$

From Eqs. 9-16 and 9-19 the internal mechanical power is

$$P = 0.98(5740) = 5630 \text{ W}$$

Deducting losses of 403 W gives

$$\text{Output power} = 5630 - 403 = 5230 \text{ W} = 7.00 \text{ hp}$$

$$\text{Output torque} = \frac{\text{output power}}{\omega_{\text{rotor}}} = \frac{5230}{0.98(125.6)} = 42.5 \text{ N} \cdot \text{m} = 31.4 \text{ lb} \cdot \text{ft}$$

The efficiency is calculated from the losses.

$$\text{Total stator } I^2R \text{ loss} = 3(18.8)^2(0.294) = \quad 312 \text{ W}$$
$$\text{Rotor } I^2R \text{ loss (from Eq. 9-17)} = 0.02(5740) = \quad 115$$
$$\text{Friction, windage, and core losses} = \quad 403$$
$$\text{Total losses} = \quad 830 \text{ W}$$
$$\text{Output} = 5230$$
$$\text{Input} = 6060 \text{ W}$$

$$\frac{\text{Losses}}{\text{Input}} = \frac{830}{6060} = 0.137$$

$$\text{Efficiency} = 1.000 - 0.137 = 0.863 = 86.3\%$$

The complete performance characteristics of the motor can be determined by repeating these calculations for other assumed values of slip.

9-5 TORQUE AND POWER BY USE OF THEVENIN'S THEOREM

When torque and power relations are to be emphasized, considerable simplification results from application of Thevenin's network theorem to the induction-motor equivalent circuit.

In its general form, Thevenin's theorem permits the replacement of any network of linear circuit elements and constant phasor voltage sources, as viewed from two terminals a and b in Fig. 9-11a, by a single phasor voltage source $\widehat{V}_s$ in series with a single impedance Z (Fig. 9-11b). The voltage $\widehat{V}_s$ is that appearing across terminals a and b of the original network when these terminals are open-circuited; the impedance Z is that viewed from the same terminals when all voltage sources within the network are short-circuited. For application to the induction-motor equivalent circuit, points a and b are taken as those so designated in Fig. 9-10a and b. The equivalent circuit then assumes the forms given in Fig. 9-12. So far as phenomena to the right of points a and b are concerned, the circuits of Figs. 9-10 and 9-12 are identical when the voltage $\widehat{V}_{1a}$ and the impedance $R_{e1} + jX_{e1}$ have the proper values. According to Thevenin's theorem, the equivalent source voltage $\widehat{V}_{1a}$ is the voltage that would appear across terminals a and b of Fig. 9-10 with the rotor circuits open and is

$$\widehat{V}_{1a} = \widehat{V}_1 - \widehat{I}_0(R_1 + jX_1) = \widehat{V}_1 \frac{jX_\varphi}{R_1 + jX_{11}} \tag{9-24}$$

where $\widehat{I}_0$ is the zero-load exciting current and

$$X_{11} = X_1 + X_\varphi \tag{9-25}$$

is the self-reactance of the stator per phase, which very nearly equals the reactive component of the zero-load motor impedance. For most induction motors, negligible error results from neglecting the stator resistance in Eq. 9-24. The Thevenin-equivalent stator impedance $R_{e1} + jX_{e1}$ is the impedance between terminals a and b of Fig. 9-10 viewed toward the source with the

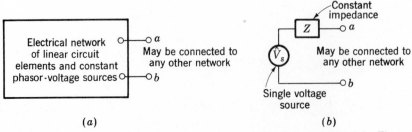

(a)   *(b)*

Fig. 9-11 (a) General linear network and (b) its equivalent at terminals ab by Thevenin's theorem.

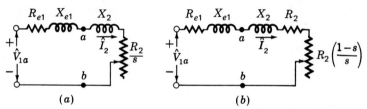

Fig. 9-12. Induction-motor equivalent circuits simplified by Thevenin's theorem.

source voltage short-circuited and therefore is

$$R_{e1} + jX_{e1} = (R_1 + jX_1) \text{ in parallel with } jX_{\varphi} \qquad (9\text{-}26)$$

From the Thevenin-equivalent circuit (Fig. 9-12) and the torque expression (Eq. 9-22) it can be seen that

$$T = \frac{1}{\omega_s} \frac{q_1 V_{1a}^2 (R_2/s)}{(R_{e1} + R_2/s)^2 + (X_{e1} + X_2)^2} \qquad (9\text{-}27)$$

The general shape of the torque-speed or torque-slip curve with the motor connected to a constant-voltage, constant-frequency source is shown in Figs. 9-13 and 9-14.

In normal motor operation the rotor revolves in the direction of rotation

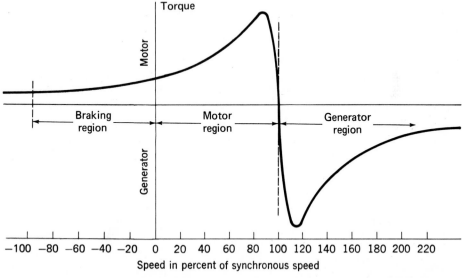

Fig. 9-13. Induction-machine torque-slip curve showing braking, motor, and generator regions.

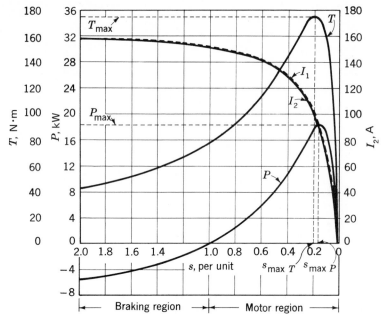

Fig. 9-14. Computed torque, power, and current curves for 10-hp induction motor in Examples 9-1 and 9-2.

of the magnetic field produced by the stator currents, the speed is between zero and synchronous speed, and the corresponding slip is between 1.0 and 0, as labeled "Motor region" in Fig. 9-13. Motor starting conditions are those of $s = 1.0$.

In order to obtain physical operation in the region of s greater than 1, the motor must be driven backward, against the direction of rotation of its magnetic field, by a source of mechanical power capable of counteracting the internal torque T. The chief practical usefulness of this region is in bringing motors to a quick stop by a method called *plugging*. By interchanging two stator leads in a 3-phase motor, the phase sequence, and hence the direction of rotation of the magnetic field, is reversed suddenly; the motor comes to a stop under the influence of the torque T and is disconnected from the line before it can start in the other direction. Accordingly, the region from $s = 1.0$ to $s = 2.0$ is labeled "Braking region" in Fig. 9-13.

The induction machine will operate as a generator if its stator terminals are connected to a constant-frequency voltage source and its rotor is driven above synchronous speed by a prime mover, as shown in Fig. 9-13. The source fixes the synchronous speed and supplies the reactive power input required to excite the air-gap magnetic field. The slip then is negative. One interesting application is that of an induction generator connected to a power system and driven by a wind turbine.

The *maximum internal*, or *breakdown, torque* T_{max}, indicated in Fig. 9-14, can be obtained readily from circuit considerations. Internal torque is a maximum when the power delivered to R_2/s in Fig. 9-12a is a maximum. Now, by the familiar impedance-matching principle in circuit theory, this power will be greatest when the impedance of R_2/s equals the magnitude of the imped-ance between it and the constant voltage V_{1a}, or at a value $s_{\max T}$ of slip for which

$$\frac{R_2}{s_{\max T}} = \sqrt{R_{e1}^2 + (X_{e1} + X_2)^2} \tag{9-28}$$

The slip $s_{\max T}$ at maximum torque is therefore

$$s_{\max T} = \frac{R_2}{\sqrt{R_{e1}^2 + (X_{e1} + X_2)^2}} \tag{9-29}$$

and the corresponding torque is, from Eq. 9-27,

$$T_{\max} = \frac{1}{\omega_s} \frac{0.5 q_1 V_{1a}^2}{R_{e1} + \sqrt{R_{e1}^2 + (X_{e1} + X_2)^2}} \tag{9-30}$$

EXAMPLE 9-2

For the motor of Example 9-1, determine: (a) the load component I_2 of the stator current, the internal torque T, and the internal power P for a slip $s = 0.03$; (b) the maximum internal torque and the corresponding speed; and (c) the internal starting torque and the corresponding stator-load current I_2.

Solution

First reduce the circuit to its Thevenin-theorem form. From Eq. 9-24, $V_{1a} = 122.3$; and from Eq. 9-26, $R_{e1} + jX_{e1} = 0.273 + j0.490$.

(a) At $s = 0.03$, $R_2/s = 4.80$. Then, from Fig. 9-12a,

$$I_2 = \frac{122.3}{\sqrt{(5.07)^2 + (0.699)^2}} = 23.9 \text{ A}$$

From Eq. 9-22

$$T = \frac{1}{125.6} (3)(23.9)^2 (4.80) = 65.5 \text{ N} \cdot \text{m}$$

From Eq. 9-19

$$P = 3(23.0)^2 (4.80)(0.97) = 7970 \text{ W}$$

Data for the curves of Fig. 9-14 were computed by repeating these calculations for a number of assumed values of s.

(b) At the maximum-torque point, from Eq. 9-29,

$$s_{\text{max } T} = \frac{0.144}{\sqrt{(0.273)^2 + (0.699)^2}} = \frac{0.144}{0.750} = 0.192$$

Speed at $T_{\text{max}} = (1 - 0.192)(1200) = 970 \text{ r/min}$

From Eq. 9-30,

$$T_{\text{max}} = \frac{1}{125.6} \frac{0.5(3)(122.3)^2}{0.273 + 0.750} = 175 \text{ N} \cdot \text{m}$$

(c) At starting, $s = 1$, and R_2 will be assumed constant. Therefore,

$$\frac{R_2}{s} = R_2 = 0.144 \qquad R_{e1} + \frac{R_2}{s} = 0.417$$

$$I_{2,\text{start}} = \frac{122.3}{\sqrt{(0.417)^2 + (0.699)^2}} = 150.5 \text{ A}$$

From Eq. 9-22

$$T_{\text{start}} = \frac{1}{125.6}(3)(150.5)^2(0.144) = 78.0 \text{ N} \cdot \text{m}$$

It is thus seen that the conventional induction motor with a squirrel-cage rotor is substantially a constant-speed motor having about 5 percent drop in speed from no load to full load. Speed variation can be obtained by using a wound-rotor motor and inserting external resistance in the rotor circuit. In the normal operating range, the external resistance simply increases the rotor impedance, necessitating a higher slip for a desired rotor mmf and torque. The influence of increased rotor resistance on the torque-speed characteristic is shown by the dashed curves in Fig. 9-15. Variation of starting torque with rotor resistance can be seen from these curves by noting the variation of the zero-speed ordinates.

Notice fom Eqs. 9-29 and 9-30 that the slip at maximum torque is directly proportional to rotor resistance R_2 but the value of the maximum torque is independent R_2. When R_2 is increased by inserting external resistance in the rotor of a wound-rotor motor, the maximum internal torque is therefore unaffected but the speed at which it occurs can be directly controlled.

In applying the induction-motor equivalent circuit, the idealizations on

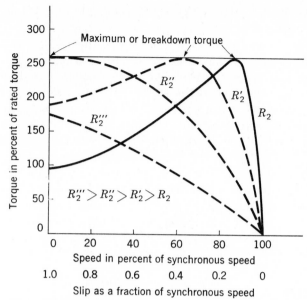

Fig. 9-15. Induction-motor torque-slip curves showing effect of changing rotor-circuit resistance.

which it is based should be kept in mind. This is particularly necessary when investigations are carried out over a wide speed range, as in motor-starting problems. Saturation under the heavy inrush currents associated with starting has a significant effect on the motor reactances. Moreover, the rotor currents are at slip frequency, which, of course, varies from stator frequency at zero speed to a low value at full-load speed. The current distribution in the rotor conductors and hence the rotor resistance may vary very significantly over this range. Errors from these causes can be kept to a minimum by using equivalent-circuit constants determined by simulating the proposed operating conditions as closely as possible.†

9–6 PERFORMANCE CALCULATIONS FROM NO–LOAD AND BLOCKED–ROTOR TESTS

The data needed for computing the performance of a polyphase induction motor under load can be obtained from the results of a no-load test, a blocked-rotor test, and measurements of the dc resistances of the stator windings. The stray load losses, which must be taken into account when accurate values of efficiency are to be calculated, can also be measured by tests which

---

†See, for instance, R. F. Horrell and W. E. Wood, A Method of Determining Induction Motor Speed-Torque-Current Curves from Reduced Voltage Tests, *Trans. AIEE,* **73**(III):670–674 (1954).

do not require loading the motor. The stray-load-loss tests will not be described here, however.†

Like the open-circuit test on a transformer, the no-load test on an induction motor gives information with respect to exciting current and no-load losses. The test is ordinarily taken at rated frequency and with balanced polyphase voltages applied to the stator terminals. Readings are taken at rated voltage, after the motor has been running long enough for the bearings to be properly lubricated. The total rotational loss at rated voltage and frequency under load usually is considered to be constant and equal to its no-load value.

At no load, the rotor current is only the very small value needed to produce sufficient torque to overcome friction and windage. The no-load rotor I^2R loss therefore is negligibly small. Unlike that of a transformer, whose no-load primary I^2R loss is negligible, the no-load stator I^2R loss of an induction motor may be appreciable because of its larger exciting current. The rotational loss P_R for normal running conditions is

$$P_R = P_{nl} - q_1 I_{nl}^2 R_1 \tag{9-31}$$

where P_{nl} = total polyphase power input
I_{nl} = current per phase
q_1 = number of stator phases
R_1 = stator resistance per phase

Because the slip at no load is very small, the reflected rotor resistance R_2/s_{nl} is very large. The parallel combination of rotor and magnetizing branches then becomes jX_φ shunted by a very high resistance, and the reactance of this parallel combination therefore very nearly equals X_φ. Consequently the apparent reactance X_{nl} measured at the stator terminals at no load very nearly equals $X_1 + X_\varphi$, which is the self-reactance X_{11} of the stator; i.e.,

$$X_{11} = X_1 + X_\varphi = X_{nl} \tag{9-32}$$

The self-reactance of the stator can therefore be determined from the instrument readings at no load. For a 3-phase machine considered to be Y-connected the magnitude of the no-load impedance Z_{nl} per phase is

$$Z_{nl} = \frac{V_{nl}}{\sqrt{3} I_{nl}} \tag{9-33}$$

where V_{nl} is the line-to-line terminal voltage in the no-load test. The no-load

---

†For information concerning test methods, see IEEE Test Procedures for Polyphase Induction Motors and Generators, no. 112, 1978, Institute of Electrical and Electronic Engineers.

resistance R_{nl} is

$$R_{nl} = \frac{P_{nl}}{3I_{nl}^2} \qquad (9\text{-}34)$$

where P_{nl} is the total 3-phase power input at no load; the no-load reactance X_{nl} then is

$$X_{nl} = \sqrt{Z_{nl}^2 - R_{nl}^2} \qquad (9\text{-}35)$$

Usually the no-load power factor is about 0.1, so that the no-load reactance very nearly equals the no-load impedance.

Like the short-circuit test on a transformer, the blocked-rotor test on an induction motor gives information with respect to the leakage impedances. The rotor is blocked so that it cannot rotate, and balanced polyphase voltages are applied to the stator terminals. Sometimes the blocked-rotor torque also is measured.

The equivalent circuit for blocked-rotor conditions is identical to that of a short-circuited transformer. An induction motor is more complicated than a transformer, however, because its leakage impedance may be affected by magnetic saturation of the leakage-flux paths and by rotor frequency. The blocked impedance may also be affected by rotor position, although this effect generally is small with cage rotors. The guiding principle is that the blocked-rotor test should be taken under conditions of current and rotor frequency approximately the same as those existing in the operating condition for which the performance is later to be calculated. For example, if one is interested in the characteristics at slips near unity, as in starting, the blocked-rotor test should be taken at normal frequency and with currents near the values encountered in starting. If, however, one is interested in the normal running characteristics, the blocked-rotor test should be taken at a reduced voltage which results in about rated current; the frequency also should be reduced, since the values of rotor effective resistance and leakage inductance at the low rotor frequencies corresponding to small slips may differ appreciably from their values at normal frequency, particularly with double-cage or deep-bar rotors, as discussed in Art. 10-1.

The IEEE Test Procedure suggests a frequency of 25 percent of rated frequency. The total leakage reactance at normal frequency can be obtained from this test value by considering the reactance to be proportional to frequency. The effects of frequency often are negligible for normal motors of less than 25 hp rating, and the blocked impedance can then be measured directly at normal frequency. The importance of maintaining test currents near their rated value stems from the fact that these leakage reactances are significantly affected by saturation.

If exciting current is neglected, the blocked-rotor reactance X_{bl}, corrected to normal frequency, equals the sum of the normal-frequency stator and rotor

TABLE 9-1

EMPIRICAL DISTRIBUTION OF LEAKAGE REACTANCES
IN INDUCTION MOTORS

Motor class	Description	Fraction of $X_1 + X_2$	
		X_1	X_2
A	Normal starting torque, normal starting current	0.5	0.5
B	Normal starting torque, low starting current	0.4	0.6
C	High starting torque, low starting current	0.3	0.7
D	High starting torque, high slip	0.5	0.5
Wound rotor		0.5	0.5

leakage reactances X_1 and X_2. The performance of the motor is relatively little affected by the way in which the total leakage reactance $X_1 + X_2$ is distributed between stator and rotor. The IEEE Test Procedure recommends the empirical distribution shown in Table 9-1.

The magnetizing reactance X_φ now can be determined from the no-load test and the value of X_1; thus

$$X_\varphi = X_{nl} - X_1 \qquad (9\text{-}36)$$

The stator resistance R_1 can be considered as its dc value. The rotor resistance then can be determined as follows. From the blocked-rotor test, the blocked resistance R_{bl} can be computed by means of a relation similar to Eq. 9-34. The difference between the blocked resistance and the stator resistance then can be determined from the test data. Denoting this resistance by R, we have

$$R = R_{bl} - R_1 \qquad (9\text{-}37)$$

From the equivalent circuit, with $s = 1$, the resistance R is the resistance of the combination of $R_2 + jX_2$ in parallel with jX_φ. For this parallel combination

$$R = R_2 \frac{X_\varphi^2}{R_2^2 + X_{22}^2} \approx R_2 \left(\frac{X_\varphi}{X_{22}}\right)^2 \qquad (9\text{-}38)$$

where $X_{22} = X_2 + X_\varphi$ is the self-reactance of the rotor. If X_{22} is greater than $10R_2$, as is usually the case, less than 1 percent error results from use of the approximate form of Eq. 9-38. Substitution of this approximate form in Eq. 9-37 and solution for R_2 then gives

$$R_2 = R\left(\frac{X_{22}}{X_\varphi}\right)^2 = (R_{bl} - R_1)\left(\frac{X_{22}}{X_\varphi}\right)^2 \qquad (9\text{-}39)$$

All the equivalent-circuit constants have now been determined, and the motor performance under load can be computed.

EXAMPLE 9-3

The following test data apply to a 7.5-hp 3-phase 220-V 19-A 60-Hz 4-pole induction motor with a double-squirrel-cage rotor of the high-starting-torque low-starting-current type (design class C):

Test 1: No-load test at 60 Hz

$$\text{Applied voltage } V = 219 \text{ V line to line}$$
$$\text{Average line current } I_{nl} = 5.70 \text{ A}$$
$$\text{Power (two wattmeters): } W_1 = 680 \text{ W} \qquad W_2 = -300 \text{ W}$$

Test 2: Blocked-rotor test at 15 Hz

$$V = 26.5 \text{ V} \qquad I = 18.57 \text{ A} \qquad W_1 = 215 \text{ W} \qquad W_2 = 460 \text{ W}$$

Test 3: Average dc resistance per stator phase (measured immediately after test 2)

$$R_1 = 0.262 \ \Omega/\text{phase (Y connection assumed)}$$

Test 4: Blocked-rotor test at 60 Hz

$$V = 212 \text{ V} \qquad I = 83.3 \text{ A} \qquad W_1 = 3300 \text{ W} \qquad W_2 = 16,800 \text{ W}$$
$$\text{Measured starting torque } T_{\text{start}} = 54.6 \text{ lb} \cdot \text{ft}$$

(a) Compute the no-load rotational loss and the equivalent-circuit constants applying to the normal running conditions. Assume the same temperature as in test 3.

(b) Compute the internal starting torque from the input measurements of test 4. Assume the same temperature as in test 3.

Solution

(a) From test 1, $P_{nl} = 380$ W, and by Eq. 9-31

$$P_R = 380 - 3(5.70)^2(0.262) = 354 \text{ W}$$

From test 1 and Eqs. 9-33 to 9-35

$$Z_{nl} = \frac{219}{\sqrt{3}(5.70)} = 22.2 \ \Omega/\text{phase Y}$$

$$R_{nl} = \frac{380}{3(5.70)^2} = 3.9 \ \Omega \qquad X_{nl} = 21.8 \ \Omega$$

The blocked-rotor test at reduced frequency and rated current reproduces approximately normal running conditions in the rotor. From test 2

$$Z'_{bl} = \frac{26.5}{\sqrt{3}(18.57)} = 0.825 \ \Omega/\text{phase at 15 Hz}$$

$$R_{bl} = \frac{675}{3(18.57)^2} = 0.654 \ \Omega \qquad X'_{bl} = 0.503 \ \Omega \text{ at 15 Hz}$$

where the primes indicate 15-Hz values. The blocked reactance referred to normal frequency then is

$$X_{bl} = \tfrac{60}{15}(0.503) = 2.01 \ \Omega/\text{phase at 60 Hz}$$

According to Table 9-1,

$$X_1 = 0.3(2.01) = 0.603 \qquad X_2 = 0.7(2.01) = 1.407 \ \Omega/\text{phase}$$

and by Eq. 9-36

$$X_\varphi = 21.8 - 0.6 = 21.2 \ \Omega/\text{phase}$$

From test 3 and Eqs. 9-37 and 9-39

$$R = 0.654 - 0.262 = 0.392 \qquad R_2 = 0.392\left(\frac{22.6}{21.2}\right)^2 = 0.445 \ \Omega/\text{phase}$$

The constants of the equivalent circuit for small values of slip have now been calculated.

(b) The internal starting torque can be computed from the input measurements in test 4. From the power input and stator I^2R losses, the air-gap power P_{g1} is

$$P_{g1} = 20{,}100 - 3(83.3)^2(0.262) = 14{,}650 \text{ W}$$

Synchronous speed $\omega_s = 188.5 \text{ rad/s}$, and

$$T_{\text{start}} = \frac{14{,}650}{188.5} = 77.6 \text{ N} \cdot \text{m} = 57.3 \text{ lb} \cdot \text{ft}$$

The test value, $T_{\text{start}} = 54.6 \text{ lb} \cdot \text{ft}$, is a few percent less than the calculated value because the calculations do not account for the power absorbed in stator core loss and in stray load losses.

9-7 SUMMARY

In a polyphase induction motor, slip-frequency currents are induced in the rotor windings as the rotor slips past the synchronously rotating stator flux wave. These rotor currents in turn produce a flux wave which rotates in synchronism with the stator flux wave; the torque stems from the interaction of these two flux waves. For increased load on the motor, the rotor speed decreases, resulting in larger slip and increased induced rotor currents, and greater torque.

Examination of the flux-mmf interactions in a polyphase induction motor shows that, electrically, the machine is a generalized transformer. The synchronously rotating air-gap flux wave in the induction machine is the counterpart of the mutual core flux in the transformer. The rotating field induces emf's of stator frequency in the stator windings and of slip frequency in the rotor windings for all rotor speeds other than synchronous speed. Thus, the induction machine transforms voltages and at the same time changes frequency. When viewed from the stator, all rotor electrical and magnetic phenomena are transformed to stator frequency. The rotor mmf reacts on the stator windings in the same manner that the mmf of the secondary current in a transformer reacts on the primary.

Pursuit of this line of reasoning leads to equivalent circuits for the machines. The effects of saturation on the equivalent circuit are less serious than in the corresponding steady-state circuit for synchronous machines. This is largely because, as in the transformer, the performance is determined to a considerably greater extent by the leakage impedances than by the magnetizing impedance. Care must be taken in both testing and analysis, however, to reflect the effects of saturation on leakage reactances as well as of nonuniformity of current distribution on rotor resistance.

One of the salient facts affecting induction-motor applications is that the slip at which maximum torque occurs can be controlled by varying the rotor resistance. A high rotor resistance gives optimum starting conditions but poor running performance. A low rotor resistance, on the other hand, may result in unsatisfactory starting conditions. The design of a squirrel-cage motor is therefore quite likely to be a compromise.

For applications requiring a substantially constant speed without excessively severe starting conditions, the squirrel-cage motor usually is unrivaled because of its ruggedness, simplicity, and relatively low cost. Its only disadvantage is its relatively low power factor (about 0.85 to 0.90 at full load for 4-pole 60-Hz motors and considerably lower at light loads and for lower-speed motors). The low power factor is a consequence of the fact that all the excitation must be supplied by lagging reactive power taken from the ac mains. At speeds below about 500 r/min and ratings above about 50 hp or at medium speeds (500 to 900 r/min) and ratings above about 500 hp, a synchronous motor may cost less than an induction motor.

In the next chapter issues of the dynamics and control of induction motors are discussed. The use of induction motors in variable-speed applications is considered. Marked improvement in the starting performance with relatively little sacrifice in running performance can be built into a squirrel-cage motor by using a deep-bar or double-cage rotor whose effective resistance increases with slip. A wound-rotor motor can be used for very severe starting conditions or when speed control by rotor resistance is required. A wound-rotor motor is more expensive than a squirrel-cage motor. Variable-frequency solid-state motor drives lend considerable flexibility to the application of induction motors in variable-speed applications.

PROBLEMS

9-1 A 4-pole 60-Hz induction motor is driving a load at 1740 r/min.

(a) What is the slip of the rotor?
(b) What is the frequency of the rotor currents?
(c) What is the angular velocity of the stator field with respect to the stator? With respect to the rotor?
(d) What is the angular velocity of the rotor field with respect to the rotor? With respect to the stator?

9-2 A 3-phase induction motor runs at almost 1200 r/min at no load and 1140 r/min at full load when supplied with power from a 60-Hz 3-phase line.

(a) How many poles has the motor?
(b) What is the percent slip at full load?
(c) What is the corresponding frequency of the rotor voltages?
(d) What is the corresponding speed of the rotor field with respect to the rotor? Of the rotor field with respect to the stator?
(e) What speed would the rotor have at a slip of 10 percent?
(f) What is the rotor frequency at this speed?
(g) Repeat part (d) for a slip of 10 percent.

9-3 Linear induction motors have been proposed for a variety of applications including high-speed ground transportation. A linear motor based on the induction-motor principle consists of a car riding on a track. The track is a developed squirrel-cage winding, and the car, which is 12 ft long, $3\frac{1}{2}$ ft wide, and only $5\frac{1}{2}$ in high, has a developed 3-phase 8-pole winding. The centerline distance between adjacent poles is $12/8 = 1\frac{1}{2}$ ft. Power at 60 Hz is fed to the car from arms extending through slots to rails below ground level.

(a) What is the synchronous speed in miles per hour?
(b) Will the car reach this speed? Explain your answer.
(c) To what slip frequency does a car speed of 75 mi/h correspond?

9-4 Describe the effect on the normal torque-speed characteristic of an induction motor produced by (*a*) halving the applied voltage with normal frequency and (*b*) halving both the applied voltage and frequency. Sketch the associated torque-speed characteristics in their approximate relative positions with respect to the normal one. Neglect the effects of stator resistance and leakage reactance.

9-5 Figure 9-16 shows a 3-phase wound-rotor induction machine whose shaft is rigidly coupled to the shaft of a 3-phase synchronous motor. The terminals of the 3-phase rotor winding of the induction machine are brought out to slip rings as shown. The induction machine is driven by the synchronous motor at the proper speed and in the proper direction of rotation so that 3-phase 120-Hz voltages are available at the slip rings. The induction machine has a 6-pole stator winding.

    (*a*) How many poles must the rotor winding of the induction machine have?

    (*b*) If the stator field in the induction machine rotates in a clockwise direction, what must the direction of rotation of its rotor be?

    (*c*) What must the speed in revolutions per minute be?

    (*d*) How many poles must the synchronous motor have?

9-6 The system shown in Fig. 9-16 is used to convert balanced 60-Hz voltages to other frequencies. The synchronous motor has two poles and drives the interconnecting shaft in the clockwise direction. The induction machine has 12 poles, and its stator windings are connected to the lines to produce a counterclockwise rotating field (in the opposite direction to the synchronous motor). The machine has a wound rotor whose terminals are brought out through slip rings.

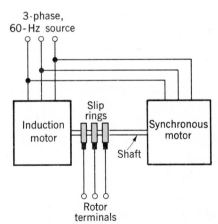

Fig. 9-16. Interconnected induction and synchronous machines, Probs. 9-5 and 9-6.

(a) At what speed does the motor run?

(b) What is the frequency of the rotor voltages in the induction machine?

9-7 A 3-phase 8-pole 60-Hz 4160-V 1000-hp squirrel-cage induction motor has the following equivalent-circuit constants in ohms per phase Y referred to the stator:

$$R_1 = 0.320 \qquad R_2 = 0.293 \qquad X_1 = 2.54 \qquad X_2 = 3.15 \qquad X_\varphi = 57.8$$

Determine the changes in these constants which will result from the following proposed design modifications. Consider each modification separately.

(a) Replace the stator winding with an otherwise identical winding with a wire size whose cross-sectional area is increased by 4 percent.

(b) Decrease the inner diameter of the stator laminations such that the air gap is decreased by 15 percent.

(c) Replace the aluminum rotor bars (conductivity 3.5×10^7 mhos/m) with copper bars (conductivity 5.8×10^7 mhos/m).

(d) The stator winding, originally connected in Y for 4160-V operation, is reconnected in Δ for 2.4-kV operation.

9-8 A 3-phase Y-connected 480-V (line-to-line) 30-hp 60-Hz 4-pole induction motor has the following equivalent-circuit constants in ohms per phase referred to the stator

$$R_1 = 0.24 \qquad R_2 = 0.20 \qquad X_1 = 1.3 \qquad X_2 = 1.2 \qquad X_\varphi = 37$$

The total friction, windage, and core losses may be assumed constant at 1340 W. The motor is connected directly to a 480-V source. Compute the speed, shaft output torque in newton-meters, power in horsepower, efficiency, and terminal power factor for slips of 1 and 2 percent.

9-9 A 10-hp 230-V 3-phase Y-connected 60-Hz 4-pole squirrel-cage induction motor develops full-load internal torque at a slip of 0.04 when operated at rated voltage and frequency. For the purposes of this problem rotational and core losses can be neglected. Impedance data on the motor in ohms per phase are as follows:

$$R_1 = 0.36 \qquad X_1 = X_2 = 0.47 \qquad X_\varphi = 15.5$$

Determine the maximum internal torque at rated voltage and frequency, the slip at maximum torque, and the internal starting torque at rated voltage and frequency. Express the torques in newton-meters.

9-10 Suppose the induction motor of Prob. 9-9 is supplied from a 240-V constant-voltage 60-Hz source through a feeder whose impedance is $0.50 + j\,0.30\,\Omega$/phase. Determine the maximum internal torque that the motor can deliver and the corresponding values of stator current and terminal voltage.

9-11 A 3-phase induction motor, at rated voltage and frequency, has a starting torque of 160 percent and a maximum torque of 200 percent of full-load torque. Neglect stator resistance and rotational losses and assume constant rotor resistance. Determine (a) the slip at full load, (b) the slip at maximum torque, and (c) the rotor current at starting, in per unit of full-load rotor current.

9-12 When operated at rated voltage and frequency, a 3-phase squirrel-cage induction motor (of the design classification known as a high-slip motor) delivers full load at a slip of 8.5 percent and develops a maximum torque of 250 percent of full-load torque at a slip of 50 percent. Neglect core and rotational losses, and assume that the resistances and inductances of the motor are constant.

Determine the torque and rotor current at starting with rated voltage and frequency. Express the torque and rotor current in per unit based on their full-load values.

9-13 A 250-kW 600-V (line-to-line) 6-pole, 60-Hz induction motor has the following impedance data in ohms per phase Y referred to the stator:

$$R_1 = 0.0245 \qquad R_2 = 0.0419 \qquad X_1 = 0.155 \qquad X_2 = 0.150 \qquad X_\varphi = 5.10$$

It achieves rated shaft output at a slip of 3.6 percent with an efficiency of 93 percent. The motor is to be used as a generator driven by a wind turbine. It will be connected to a distribution system which can be represented by a 600-V infinite bus.

(a) From the given data calculate the total rotational and core losses at rated load. Assume them to remain constant.

(b) With the wind turbine driving the induction machine at a slip of -3.6 percent calculate (1) the electric output power in kilowatts, (2) the efficiency (electric output power per shaft input power), and (3) the power factor measured at the machine terminals.

(c) The actual distribution system has an effective impedance of $0.02 + j\,0.053\,\Omega$/phase. For a slip of -3.6 percent, calculate (1) the induction-machine terminal voltage and power factor and (2) the electric output power in kilowatts measured at the machine terminals.

9-14 For a 25-hp 230-V 3-phase 60-Hz squirrel-cage motor operated at rated voltage and frequency, the rotor I^2R loss at maximum torque is 9.0 times that at full-load torque, and the slip at full-load torque is 0.030. Stator resistance and rotational losses may be neglected and the reactances and rotor resistance assumed to remain constant. Find (a) the slip at maximum torque, (b) the maximum torque, and (c) the starting torque. Express the torques in per unit of full-load torque.

9-15 A squirrel-cage induction motor runs at a slip of 5.0 percent at full load. The rotor current at starting is 5.0 times the rotor current at full load. The rotor resistance is independent of rotor frequency, and rotational losses, stray load losses, and stator resistance may be neglected. Compute (a) the starting torque and (b) the maximum torque and the slip at which maximum torque occurs. Express the torques in per unit of full-load torque.

9-16 A Δ-connected 50-hp 480-V (line-to-line) 3-phase 6-pole 60-Hz squirrel-cage induction motor has the following equivalent-circuit parameters in ohms per phase Y:

$$R_1 = 0.13 \qquad R_2 = 0.12 \qquad X_1 = 0.75 \qquad X_2 = 0.72 \qquad X_\varphi = 23$$

(a) Calculate the starting current and torque for this motor connected directly to a 480-V source.

(b) In order to limit the starting current, it is proposed to connect the stator winding in Y for starting and then to switch to the Δ connection for normal operation.

    (1) What are the equivalent-circuit parameters in ohms per phase for the Y connection?

    (2) With the motor connected directly to a 480-V source, calculate the starting current and torque.

9-17 The following test data apply to a 50-hp 2300-V 3-phase 4-pole 60-Hz squirrel-cage induction motor:

No-load test at rated voltage and frequency:

    Line current = 4.1 A

    3-phase power = 1550 W

Blocked-rotor test at 15 Hz:

    Line voltage = 268 V

    Line current = 25.0 A

    3-phase power = 9600 W

Stator resistance between line terminals = 5.80 Ω

Compute the stator current and power factor, horsepower output, and efficiency when this motor is operating at rated voltage and frequency with a slip of 3.00 percent.

9-18 Two 50-hp 440-V 59.8-A 3-phase 6-pole 60-Hz squirrel-cage induction motors have identical stators. The dc resistance measured between any pair of stator line terminals is 0.212 Ω. The blocked-rotor tests at 60 Hz are as follows:

Motor	Volts line to line	Amperes	3-phase power, kW
1	61.3	60.0	3.16
2	96.0	60.0	8.95

Determine the ratio of the internal starting torque developed by motor 2 to that developed by motor 1 (*a*) for the same current, (*b*) for the same applied voltage.

9-19 The results of a blocked-rotor test on a 25-hp 3-phase 220-V 60-Hz 6-pole squirrel-cage induction motor are

Line-to-line voltage = 110 V

Line current = 220 A

3-phase power = 21.0 kW

Torque = 65 lb · ft

Determine the starting torque at a line-to-line voltage of 220 V and 50 Hz.

9-20 A 220-V 3-phase 4-pole 60-Hz squirrel-cage induction motor develops a maximum internal torque of 250 percent at a slip of 16 percent when operating at rated voltage and frequency. If the effect of stator resistance is neglected, determine the maximum internal torque that this motor would develop if it were operated at 200 V and 50 Hz. Under these conditions, at what speed in revolutions per minute would maximum torque be developed?

Polyphase-Induction-Machine
Dynamics and Control

The polyphase induction motor is most often used to drive essentially constant loads at constant speed. As seen in Chap. 9, this operation is typically at low slips, where motor efficiency is high.

To complete our discussion of induction-motor performance, it is necessary to consider the behavior of such motors under non-steady-state conditions. This chapter investigates the start-up behavior of induction machines, as well as the dynamics associated with transient mechanical and electric disturbances. Various methods of speed control are also described.

Until recently, induction motors found little application in situations where widely varying or precision control of speed was required. The advent of power semiconductors has changed all that. Solid-state variable-frequency motor drives can now be used to permit induction motors to be used in situations previously considered the domain of dc motors. This flexibility, combined with the simplicity, low cost, and reliability of induction motors, guarantees their increased use in such applications as electric vehicles (trains and cars) and industrial processes. The use of variable-frequency solid-state drives is discussed in Art. 10-5.

10–1 EFFECTS OF ROTOR RESISTANCE; DOUBLE–SQUIRREL–CAGE ROTORS

A basic limitation of induction motors with constant rotor resistance is that the rotor design has to be a compromise. High efficiency under normal running conditions requires a low rotor resistance; but a low rotor resistance results in a low starting torque and high starting current at a low starting power factor.

a. Wound-Rotor Motors

The use of a wound rotor is one effective way of avoiding the need for compromise. The terminals of the rotor winding are connected to slip rings in contact with brushes. For starting, resistors may be connected in series with the rotor windings, the result being increased starting torque and reduced starting current at an improved power factor. The general nature of the effects on the torque-speed characteristics caused by varying rotor resistance is shown in Fig. 9-15. By use of the appropriate value of rotor resistance, the maximum torque can be made to occur at standstill if high starting torque is needed. As the rotor speeds up, the external resistances can be decreased, making maximum torque available throughout the accelerating range. Since most of the rotor I^2R loss is dissipated in the external resistors, the rotor temperature rise during starting is lower than it would be if the resistance were incorporated in the rotor winding. For normal running, the rotor winding can be short-circuited directly at the brushes. The rotor winding is designed to have low resistance so that running efficiency is high and full-load slip is low. Besides their use when starting requirements are severe, wound-rotor induction motors can be used for adjustable-speed drives. Their chief disadvantage is greater cost than squirrel-cage motors.

The principal effects of varying rotor resistance on the starting and running characteristics of induction motors can be shown quantitatively by means of the following example.

EXAMPLE 10-1

A 500-hp wound-rotor induction motor, with its slip rings short-circuited, has the following properties:

Full load slip = 1.5 percent

Rotor I^2R at full load torque = 5.69 kW

Slip at maximum torque = 6.0 percent

Rotor current at maximum torque = 2.82 I_{2fl}, where I_{2fl} is the full-load rotor current

Torque at 20 percent slip = 1.20 T_{fl}, where T_{fl} is the full-load torque.

Rotor current at 20 percent slip = 3.95 I_{2fl}

If the rotor-circuit resistance is increased to $5R_{\text{rotor}}$ by connecting nonin-ductive resistances in series with each rotor slip ring, determine (a) the slip at which the motor will develop the same full-load torque, (b) the total rotor-circuit I^2R loss at full-load torque, (c) the horsepower output at full-load torque, (d) the slip at maximum torque, (e) the rotor current at maximum torque, (f) the starting torque, and (g) the rotor current at starting. Express the torques and rotor currents in per unit based on the full-load torque values.

Solution

The solution involves recognition of the fact that the only way the stator becomes aware of the happenings in the rotor is through the effect of the resistance R_2/s. Examination of the equivalent circuit shows that for specified applied voltage and frequency everything concerning the stator performance is fixed by the value of R_2/s, the other impedance elements being constant. For example if R_2 is doubled and s is simultaneously doubled, the stator is unaware that any change has been made. The stator current and power factor, the power delivered to the air gap, and the torque are constant so long as the ratio R_2/s is the same.

Added physical significance can be given to the argument by examining the effects of simultaneously doubling R_2 and s from the viewpoint of the rotor. An observer on the rotor then sees the resultant air-gap flux wave traveling past him at twice the original slip speed, generating twice the original rotor voltage at twice the original slip frequency. The rotor reactance therefore is doubled, and since the original premise is that the rotor resistance also is doubled, the rotor impedance is doubled but the rotor power factor is unchanged. Since rotor voltage and impedance are both doubled, the effective value of the rotor current remains the same; only its frequency is changed. The air gap still has the same synchronously rotating flux and mmf waves with the same torque angle. The observer on the rotor therefore agrees with his counterpart on the stator that the torque is unchanged when both rotor resistance and slip are changed proportionally.

The observer on the rotor, however, is aware of two changes not apparent in the stator: (1) the rotor I^2R loss has doubled, and (2) the rotor is turning more slowly and therefore developing less mechanical power with the same torque. In other words, more of the power absorbed from the stator goes into I^2R heat in the rotor, and less is available for mechanical power.

The preceding thought processes now can readily be applied to the solution of this example.

(a) If the rotor resistance is increased 5 times, the slip must increase 5 times for the same value of R_2/s and therefore for the same torque. But the original slip at full load is 0.015. The new slip at full-load torque therefore is $5(0.015) = 0.075$.

(b) The effective value of the rotor current is the same as its full-load value before addition of the series resistance, and therefore the rotor I^2R loss is 5 times the full-load value of 5.69 kW, or

$$\text{Rotor } I^2R = 5(5.69) = 28.45 \text{ kW}$$

(c) The increased slip has caused the per unit speed at full-load torque to drop from $1 - s = 0.985$ down to $1 - s = 0.925$ with added rotor resistance. The torque is the same. The power output therefore has dropped proportionally, or

$$P = \frac{0.925}{0.985}(500) = 469.5 \text{ hp}$$

The decrease in output equals the increase in rotor I^2R loss.

(d) If rotor resistance is increased 5 times, the slip at maximum torque simply increases 5 times. But the original slip at maximum torque is 0.060. The new slip at maximum torque with the added rotor resistance therefore is

$$s_{\max T} = 5(0.060) = 0.30$$

(e) The effective value of the rotor current at maximum torque is independent of rotor resistance; only its frequency is changed when rotor resistance is varied. Therefore,

$$I_{2\,\max T} = 2.82 I_{2fl}$$

(f) With the rotor resistance increased 5 times, the starting torque will be the same as the original running torque at a slip of 0.20 and therefore equals the running torque without the series resistors, namely

$$T_{\text{start}} = 1.20 T_{fl}$$

(g) The rotor current at starting with the added rotor resistances will be the same as the rotor current when running at a slip of 0.20 with the slip rings short-circuited, namely,

$$I_{2,\text{start}} = 3.95 I_{2fl}$$

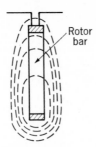

Fig. 10-1. Deep rotor bar and slot-leakage flux.

b. Deep-Bar and Double-Squirrel-Cage Rotors

An ingenious and simple way of obtaining a rotor resistance which will automatically vary with speed makes use of the fact that at standstill the rotor frequency equals the stator frequency; as the motor accelerates, the rotor frequency decreases to a very low value—perhaps 2 or 3 Hz at full load in a 60-Hz motor. With suitable shapes and arrangements for rotor bars, squirrel-cage rotors can be designed so that their effective resistance at 60 Hz is several times their resistance at 2 or 3 Hz. The various schemes all make use of the inductive effect of the slot-leakage flux on the current distribution in the rotor bars. The phenomena are basically the same as the skin and proximity effect in any system of conductors with alternating current in them.

Consider first a squirrel-cage rotor having deep, narrow bars like that shown in cross section in Fig. 10-1. The general character of the slot-leakage field produced by the current in the bar within this slot is shown in the figure. If the rotor iron had infinite permeability, all the leakage-flux lines would close in paths below the slot, as shown. Now imagine the bar to consist of an infinite number of layers of differential depth; one at the bottom and one at the top are indicated crosshatched in Fig. 10-1. The leakage inductance of the bottom layer is greater than that of the top layer because the bottom layer is linked by more leakage flux. But all the layers are electrically in parallel. Consequently, with alternating current, the current in the low-reactance upper layers will be greater than that in the high-reactance lower layers; the current will be forced toward the top of the slot, and the current in the upper layers will lead the current in the lower ones. The nonuniform current distribution results in an increase in the effective resistance and a smaller decrease in the effective leakage inductance of the bar. Since the distortion in current distribution depends on an inductive effect, the effective resistance is a function of the frequency. It is also a function of the depth of the bar and of the permeability and resistivity of the bar material. Figure 10-2 shows a curve of the ratio of ac effective resistance to dc resistance as a function of frequency computed for a copper bar 1.00 in deep. A squirrel-cage rotor with deep bars can readily be designed to have an effective resistance at stator frequency (standstill) several times greater than its dc resistance. As the motor acceler-

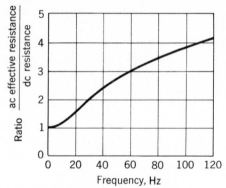

Fig. 10-2. Skin effect in a copper rotor bar 1.00 in deep.

ates, the rotor frequency decreases and therefore the rotor effective resistance decreases, approaching its dc value at small slips.

An alternative way of attaining similar results is the double-cage arrangement shown in Fig. 10-3. The squirrel-cage winding consists of two layers of bars short-circuited by end rings. The upper bars are of smaller cross-sectional area than the lower bars and consequently have higher resistance. The general nature of the slot-leakage field is shown in Fig. 10-3, from which it can be seen that the inductance of the lower bars is greater than that of the upper ones because of the flux crossing the slot between the two layers. The difference in inductance can be made quite large by properly proportioning the constriction in the slot between the two bars. At standstill, when rotor frequency equals stator frequency, there is relatively little current in the lower bars because of their high reactance; the effective resistance of the rotor at standstill then approximates that of the high-resistance upper layer. At the low rotor frequencies corresponding to small slips, however, reactance becomes negligible, and the rotor resistance then approaches that of the two layers in parallel.

Note that since the effective resistance and leakage inductance of double-cage and deep-bar rotors vary with frequency, the parameters R_2 and X_2 representing the referred effects of rotor resistance and leakage inductance as viewed from the stator are not constant. A more complicated form of equiva-

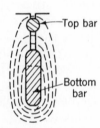

Fig. 10-3. Double-squirrel-cage rotor bars and slot-leakage flux.

lent circuit is required if the reactions of the rotor are to be represented by the effects of slip together with constant resistance and reactance elements.

The simple equivalent circuit derived in Art. 9-3 still correctly represents the motor, however, but now R_2 and X_2 are functions of slip. All the basic relations still apply to the motor if the values of R_2 and X_2 are properly adjusted with changes in slip. For example, in computing the starting performance, R_2 and X_2 should be taken as their effective values at stator frequency; in computing the running performance at small slips, however, R_2 should be taken as its effective value at a low frequency, and X_2 should be taken as the stator-frequency value of the reactance corresponding to a low-frequency effective value of the rotor leakage inductance. Over the normal running range of slips, the rotor resistance and leakage inductance usually can be considered constant at substantially their dc values.

c. Motor-Application Considerations

By use of double-cage and deep-bar rotors, squirrel-cage motors can be designed to have the good starting characteristics resulting from high rotor resistance and at the same time the good running characteristics resulting from low rotor resistance. The design is necessarily somewhat of a compromise, however, and the motor lacks the flexibility of the wound-rotor machine with external rotor resistance. The wound-rotor motor should be used when starting requirements are very severe. Alternatively, solid-state motor drives can be used, as discussed in Art. 10-5.

To meet the usual needs of industry, integral-horsepower 3-phase squirrel-cage motors are available from manufacturers' stock in a range of standard ratings up to 200 hp at various standard frequencies, voltages, and speeds. (Larger motors are generally regarded as special-purpose rather than general-purpose motors.) According to the terminology established by the NEMA, several standard designs are available to meet various starting and running requirements. Representative torque-speed characteristics of the four commonest designs are shown in Fig. 10-4. These curves are fairly typical of 1800 r/min (synchronous-speed) motors in ratings from 7.5 to 200 hp, although it should be understood that individual motors may differ appreciably from these average curves. Briefly, the characteristic features of these designs are as follows.

Design Class A: Normal Starting Torque, Normal Starting Current, Low Slip. This design usually has a low-resistance single-cage rotor. It emphasizes good running performance at the expense of starting. The full-load slip is low and the full-load efficiency high. The maximum torque usually is well over 200 percent of full-load torque and occurs at a small slip (less than 20 percent). The starting torque at full voltage varies from about 200 percent of full-load torque in small motors to about 100 percent in large motors. The high starting current (500 to 800 percent of full-load current when started at rated voltage) is the principal disadvantage of this design. In sizes below about

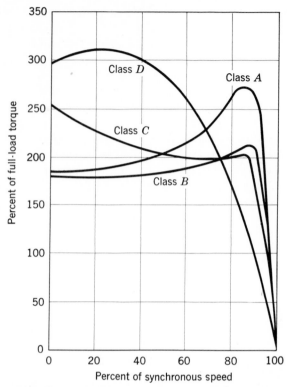

Fig. 10-4. Typical torque-speed curves for 1800-r/min general-purpose induction motors.

7.5 hp these starting currents usually are within the limits on inrush current which the distribution system supplying the motor can withstand, and across-the-line starting at full voltage then can be used; otherwise, reduced-voltage starting must be used. Reduced-voltage starting results in a decrease in starting torque because the starting torque is proportional to the voltampere input to the motor, which in turn is proportional to the square of the voltage applied to the motor terminals. The reduced voltage for starting is usually obtained from an autotransformer, called a *starting compensator,* which may be manually operated or automatically operated by relays which cause full voltage to be applied after the motor is up to speed. A circuit diagram of one type of compensator is shown in Fig. 10-5. If a smoother start is necessary, series resistance or reactance in the stator may be used.

The class A motor is the basic standard design in sizes below about 7.5 and above about 200 hp. It is also used in intermediate ratings wherein design considerations may make it difficult to meet the starting-current limitations of the class B design. Its field of application is about the same as that of the class B design described below.

Design Class B: Normal Starting Torque, Low Starting Current, Low Slip. This design has approximately the same starting torque as the class A

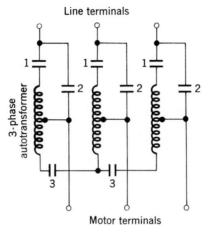

Starting sequence:
(*a*) Close 1 and 3
(*b*) Open 1 and 3
(*c*) Close 2

Fig. 10-5. Connections of a one-step starting autotransformer.

design with but 75 percent of the starting current. Full-voltage starting therefore may be used with larger sizes than with class A. The starting current is reduced by designing for relatively high leakage reactance, and the starting torque is maintained by use of a double-cage or deep-bar rotor. The full-load slip and efficiency are good—about the same as for the class A design. However, the use of high reactance slightly decreases the power factor and decidedly lowers the maximum torque (usually only slightly over 200 percent of full-load torque being obtainable).

This design is the commonest in the 7.5- to 200-hp range of sizes. It is used for substantially constant-speed drives where starting-torque requirements are not severe, such as in driving fans, blowers, pumps, and machine tools.

Design Class C: High Starting Torque, Low Starting Current. This design uses a double-cage rotor with higher rotor resistance than the class B design. The result is higher starting torque with low starting current but somewhat lower running efficiency and higher slip than the class A and class B designs. Typical applications are in driving compressors and conveyors.

Design Class D: High Starting Torque, High Slip. This design usually has a single-cage high-resistance rotor (frequently brass bars). It produces very high starting torque at low starting current, high maximum torque at 50 to 100 percent slip, but runs at a high slip at full load (7 to 11 percent) and consequently has low running efficiency. Its principal uses are for driving intermittent loads involving high accelerating duty and for driving high-impact loads such as punch presses and shears. When driving high-impact loads, the motor is generally aided by a flywheel which helps supply the impact and reduces the pulsations in power drawn from the supply system. A motor whose speed falls appreciably with increase in torque is required so that the flywheel can slow down and deliver some of its kinetic energy to the impact.

10–2 INDUCTION–MACHINE DYNAMICS

Among integral-horsepower induction motors, i.e., those used primarily for power purposes, the most common dynamic problems are associated with starting and stopping and with the ability of the motor to continue operation during serious disturbances of the supply system. For example, a typical problem in an industrial plant may concern the ability to start a large motor without causing other parallel motors to cease normal operation because of the voltage reduction caused by the heavy inrush current to the motor being started.

The methods of induction-motor representation in dynamic analyses depend to a considerable extent on the nature and complexity of the problem and the associated precision requirements. When the electric transients (Art. 10-4) in the motor are to be included, the equivalent circuit of Fig. 10-15 may be used. In many problems, however, the electric transients in the induction machine may be ignored. This simplification is possible, because as illustrated in Example 10-3, the electric transient subsides so rapidly—most commonly in a time short compared with the duration of the motional transient. (Among the principal exceptions to this statement are large 3600-r/min motors.) We shall confine ourselves here to this type of problem.

Representation of the machine under these conditions can be based on steady-state theory. The problem then becomes one of arriving at a sufficiently simple yet reasonably realistic representation which will not unduly complicate the dynamic analysis, particularly through the introduction of nonlinearities. One approach applicable to relatively simple problems is a graphical one. Both the torque produced by the motor and the torque required to turn the load are considered to be nonlinear functions of speed for which data are given in the form of curves. The procedure is illustrated in the following example.

EXAMPLE 10-2

A polyphase induction motor has the torque-speed curve for rated impressed voltage shown in Fig. 10-6. A curve of the torque required to maintain rotation of the load is also given in Fig. 10-6. The inertia of the load and rotor is J in SI units.

Consider that across-the-line starting at rated voltage is used and that the steady-state torque-speed curve represents the performance under tran-

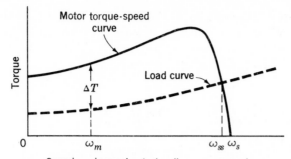

Fig. 10-6. Induction-motor torque-speed curve and curve of load torque.

sient conditions with sufficient accuracy. Show how a curve of speed as a function of time can be obtained.

Solution

At any motor speed ω_m mechanical radians per second, the torque differential ΔT between that produced by the motor and that required to turn the load is available to accelerate the rotating mass. Consequently,

$$J\frac{d\omega_m}{dt} = \Delta T$$

The time required to attain the speed ω_m is therefore

$$t = J\int_0^{\omega_m} \frac{1}{\Delta T}\, d\omega_m$$

This integral can be evaluated graphically by plotting a curve of $1/\Delta T$ as a function of ω_m and finding the area between the curve and the ω_m axis up to the value corresponding to the upper limit of the integral, a procedure illustrated in Fig. 10-7. The area can be found by planimeter, by counting small

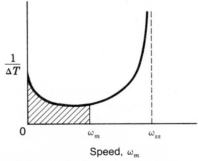

Fig. 10-7. Graphical analysis of induction-motor starting.

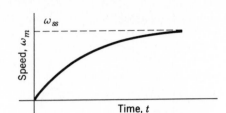

Fig. 10-8. Speed-time curve of induction motor during starting.

squares on the curve sheet, or by dividing it into uniform segments and using the average ordinate for each section. This area in units of radians per newton-meter-second times the inertia J in SI units gives the time t in seconds. The computations can be carried out conveniently in tabular form to yield a result of the type shown in Fig. 10-8.

Once the motor speed versus time is known, the corresponding current-versus-time characteristic can be obtained from either a current-versus-speed curve or from analytic expressions. The magnitude of armature current remains large until the speed increases to the point where the effective rotor impedance is on the order of the magnetizing reactance; at this point the current begins to drop toward its steady-state operating value. At no load, this current is essentially the magnetizing current of the motor. Under full-load conditions, the current includes both this magnetizing current and the stator-load-current component.

In some cases the rated-voltage motor starting current may exceed that considered desirable. In such cases, a number of options are available.

Series Impedance. A series reactive impedance can be used to limit starting current. After an appropriate time, this impedance can be automatically shorted out, connecting the motor directly to the source voltage.

Reduced Voltage. The motor can be connected to the source via an auto-transformer supplying reduced voltage, as shown in Fig. 10-5. The motor is then switched to full voltage at a predetermined operating condition.

Y-Δ Switching. An induction motor designed for a Δ armature connection can be started with the armature connect in Y. As shown in Eq. 1-82, this results in an armature impedance 3 times that of the Δ connection and hence one-third of the starting current (and torque).

Split-Winding Starting. In the split-winding scheme, multiple parallel circuits are provided for each phase and are connected in parallel for running; for starting, however, the circuits on one or more phases are connected in series.

These schemes provide reduced starting currents and correspondingly reduced starting torques. Their applicability to any particular situation must be carefully considered along with other issues related to their implementation. Alger† discusses these issues in some detail.

10–3 SPEED CONTROL OF INDUCTION MOTORS

The simple induction motor fulfills admirably the requirements of substantially constant-speed drives. Many motor applications, however, require several speeds, or even a continuously adjustable range of speeds. From the earliest days of ac power systems, engineers have been interested in the development of adjustable-speed ac motors.

The synchronous speed of an induction motor can be changed by (a) changing the number of poles or (b) varying the line frequency. The slip can be changed by (c) varying the line voltage, (d) varying the rotor resistance, or (e) inserting voltages of the appropriate frequency in the rotor circuits. The salient features of speed-control methods based on these five possibilities are discussed in the following five sections of this article. Control methods involving solid-state devices will be mentioned only briefly because they are more fully treated in Art. 10-5.

a. Pole-Changing Motors

The stator winding can be designed so that by simple changes in coil connections the number of poles can be changed in the ratio 2 to 1. Either of two synchronous speeds can be selected. The rotor is almost always of the squirrel-cage type. A cage winding always reacts by producing a rotor field having the same number of poles as the inducing stator field. If a wound rotor is used, additional complications are introduced because the rotor winding also must be rearranged for pole changing. With two independent sets of stator windings, each arranged for pole changing, as many as four synchronous speeds can be obtained in a squirrel-cage motor, for example, 600, 900, 1200, and 1800 r/min.

The basic principles of the pole-changing winding are shown in Fig. 10-9, in which aa and $a'a'$ are two coils comprising part of the phase-a stator winding. An actual winding would, of course, consist of several coils in each group. The windings for the other stator phases (not shown in the figure) would be similarly arranged. In Fig. 10-9a the coils are connected to produce a 4-pole field; in Fig. 10-9b the current in the $a'a'$ coil has been reversed by means of a controller, the result being a 2-pole field. At the same time that the controller reverses the $a'a'$ coils, the connections of the two groups of coils may be changed from series to parallel and the connections among the phases from Y to Δ, or vice versa. By these means the air-gap flux density can be adjusted to produce the desired torque-speed characteristics on the two con-

†P. L. Alger, "Induction Machines," 2d ed., chap. 8, Gordon and Breach, New York, 1970.

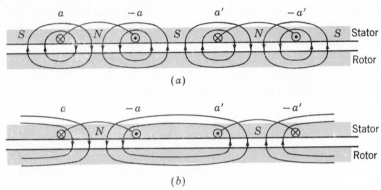

Fig. 10-9. Principles of the pole-changing winding.

nections. Figure 10-10 shows three possibilities and their corresponding torque-speed characteristics for three motors having identical characteristics on the high-speed connection. Figure 10-10a results in approximately the same maximum torque on both speeds and is applicable to drives requiring approximately the same torque on both speeds (loads in which friction predominates, for example). Figure 10-10b results in approximately twice the maximum torque on the low speed and is applicable to drives requiring approximately constant power (such as machine tools and winches). Figure 10-10c results in considerably less maximum torque on the low speed and is applicable to drives requiring less torque on the low speed (fans and centrifugal pumps, for example). The constant-horsepower type is the most expensive because it is the largest physically.

b. Line-Frequency Control

The synchronous speed of an induction motor can be controlled by varying the line frequency. In order to maintain approximately constant flux density, the line voltage should also be varied directly with the frequency. The maximum torque then remains very nearly constant. An induction motor used in this way has characteristics similar to those of a separately excited dc motor with constant flux and variable armature voltage.

The major problem is to determine the most effective and economical source of adjustable frequency. One method is to use a wound-rotor induction machine as a frequency changer. Another, considered in Art. 10-5, is to use solid-state frequency converters.

c. Line-Voltage Control

The internal torque developed by an induction motor is proportional to the square of the voltage applied to its primary terminals, as shown by the two torque-speed characteristics in Fig. 10-11. If the load has the torque-speed characteristic shown by the dashed line, the speed will be reduced from n_1 to

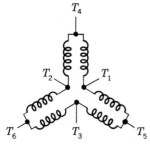

Speed	Lines			
	L_1	L_2	L_3	
Low	T_1	T_2	T_3	T_4, T_5, T_6 open
High	T_4	T_5	T_6	$T_1 - T_2 - T_3$ together

(a) Constant torque

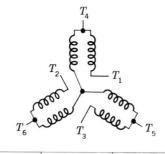

Speed	Lines			
	L_1	L_2	L_3	
Low	T_4	T_5	T_6	$T_1 - T_2 - T_3$ together
High	T_1	T_2	T_3	T_4, T_5, T_6 open

(b) Constant horsepower

Speed	Lines			
	L_1	L_2	L_3	
Low	T_1	T_2	T_3	T_4, T_5, T_6 open
High	T_4	T_5	T_6	$T_1 - T_2 - T_3$ together

(c) Variable torque

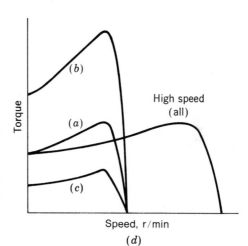

(d)

Fig. 10-10. Connections and torque-speed curves for three types of pole-changing induction motors.

n_2. This method of speed control is commonly used with small squirrel-cage motors driving fans.

d. Rotor-Resistance Control

The possibility of speed control of a wound-rotor motor by changing its rotor-circuit resistance has already been pointed out in Art. 10-1a. The

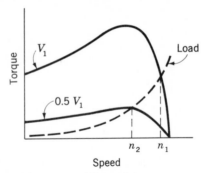

Fig. 10-11. Speed control by means of line voltage.

torque-speed characteristics for three different values of rotor resistance are shown in Fig. 10-12. If the load has the torque-speed characteristic shown by the dashed line, the speeds corresponding to each of the values of rotor resistance are n_1, n_2, and n_3. This method of speed control has characteristics similar to those of dc shunt-motor speed control by means of resistance in series with the armature.

The principal disadvantages of both line-voltage and rotor-resistance control are low efficiency at reduced speeds and poor speed regulation with respect to change in load.

e. Control of Slip by Auxiliary Devices

In considering schemes for speed control by varying the slip, the fundamental laws relating the flow of power in induction machines should be borne in mind. The fraction s of the power absorbed from the stator is transformed by electromagnetic induction to electric power in the rotor circuits. If the rotor circuits are short-circuited, this power is wasted as rotor I^2R loss and operation at reduced speeds is inherently inefficient.

Numerous schemes have been invented for recovering this slip-frequency electric power. Although some of them are rather complicated in detail, they all comprise a means for introducing adjustable voltages of slip frequency into

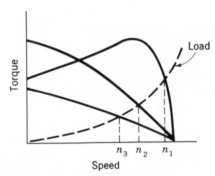

Fig. 10-12. Speed control by means of rotor resistance.

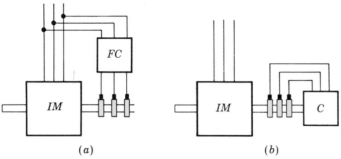

Fig. 10-13. Two basic schemes for induction-motor speed control by auxiliary machines.

the rotor circuits of a wound-rotor induction motor. They can be broadly classified in two types, as shown in Fig. 10-13, where *IM* represents a 3-phase wound-rotor induction motor whose speed is to be regulated. In Fig. 10-13a the rotor circuits of *IM* are connected to auxiliary frequency-changing apparatus, represented by the box *FC,* in which the slip-frequency electric power generated in the rotor of the main motor is converted into electric power at line frequency and returned to the line. In Fig. 10-13b the rotor circuits of *IM* are connected to auxiliary apparatus, represented by the box *C,* in which the slip-frequency electric power is converted into mechanical power and added to the shaft power developed by the main motor. In both these schemes the speed and power factor of the main motor can be adjusted by controlling the magnitude and phase of the slip-frequency voltages of the auxiliary machines. The auxiliary apparatus may be a fairly complicated system of rotating machines and adjustable-ratio transformers or, in the case of Fig. 10-13a, may consist of solid-state frequency conversion.

10–4 ELECTRIC TRANSIENTS IN INDUCTION MACHINES

The transient behavior of induction machines can be examined by following substantially the same approach followed in Art. 8-4 for synchronous machines. The results will be very similar to those in the synchronous case.

Consider that the induction machine is acting as either a motor or a generator when a 3-phase short circuit takes place at its terminals. In either case, the machine will feed current into the fault because of the "trapped" flux linkages with the rotor circuits. This current will, in time, decay to zero. In addition to an ac component, it will in general have a decaying dc component in order to keep the flux linkages with the associated phase initially constant. The ac component of the short-circuit current as a function of time appears as in Fig. 10-14.

The analysis of a symmetrical 3-phase short circuit on an initially unloaded synchronous machine in Art. 8-4 is greatly simplified by the use of the dq0 transformation, which refers all stator quantities to a synchronously ro-

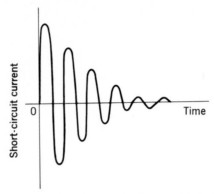

Fig. 10-14. Waveform of symmetrical short-circuit current in induction machine.

tating reference frame. A similar transformation is possible for an induction machine; however, the choice of reference frame is not so obvious. One choice is a reference frame rotating with the rotor. In this case the transformed steady-state stator quantities vary at slip frequency, as do the rotor quantities.

A second choice of reference frame is one which rotates at synchronous speed. In this case there would be speed-voltage terms in both the stator and rotor equations; mechanical speed would multiply the stator fluxes, and slip speed would multiply the rotor fluxes. In either case, analysis of a 3-phase short on an induction machine would yield results similar to those found in Art. 8-4 for the synchronous machine. Thus, instead of rederiving this solution for the induction machine, we can use the synchronous-machine solution to obtain our solution directly.

The initial magnitude of the ac component of stator current can be determined in terms of a *transient reactance* X' and a voltage E'_1 behind that reactance, assumed equal to its prefault value. The decay of the ac component can be characterized in terms of a *short-circuit transient time constant* T'. Since the ac armature currents are driven by dc decaying rotor currents, the frequency of the armature currents is determined by the rotor angular velocity. For low slips this frequency is essentially the synchronous electric frequency; for higher slips the frequency will be determined accordingly. It is usually assumed that the machine speed remains constant during the short-circuit transient since the transient is of short duration.

Much of the analytical development of Art. 8-4 can be applied by making the appropriate changes in notation. The currents i_d and i_f, for example, become the stator and rotor currents i_1 and i_2. The reactances X_{al}, X_f, and $X_{\varphi d}$ become the stator and rotor leakage reactances X_1 and X_2 and the magnetizing reactance X_φ, respectively. Because of the cylindrical structure of the induction-machine rotor, the distinction between direct and quadrature axis need not be maintained. The results corresponding to Eqs. 8-45, 8-56, 8-64,

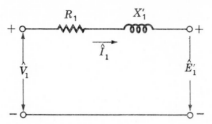

Fig. 10-15. Simplified transient equivalent circuit of an induction machine.

and 8-65 are, respectively,

$$X' = X_1 + X_\varphi - \frac{X_\varphi^2}{X_\varphi + X_2} \tag{10-1}$$

$$= X_1 + \frac{X_\varphi X_2}{X_\varphi + X_2} \tag{10-2}$$

$$T_0' = \frac{X_2 + X_\varphi}{2\pi f R_2} \tag{10-3}$$

and

$$T' = T_0' \frac{X'}{X_1 + X_\varphi} \tag{10-4}$$

where T_0' is the *open-circuit transient time constant* of the induction machine and R_2 is the rotor-circuit resistance. The initial transient-current magnitude is given by

$$I_1 = \frac{E_1'}{X'} \tag{10-5}$$

The induction machine can then be represented by the simple transient equivalent circuit of Fig. 10-15. The reactance is the transient reactance X', defined by Eqs. 10-1 and 10-2. Although it has been ignored so far, the stator resistance R_1 can also be added if somewhat greater precision is desired. The voltage E_1' behind transient reactance is a voltage proportional to rotor link-ages. It changes with these linkages and, for a 3-phase short circuit, decreases to zero at a rate determined by the time constant T' (Eq. 10-4). Adjustment of the time constant for external reactance between the machine terminals and the fault can readily be carried out by adding the reactance to the numerator and denominator of the fraction in Eq. 10-4.

EXAMPLE 10-3

A 400-hp 440-V (line-to-line) 60-Hz Y-connected 6-pole squirrel-cage induc-tion motor has a full-load efficiency of 93 percent and power factor of 90 percent. The motor constants in ohms per phase referred to the stator are

$$X_1 = 0.060 \qquad X_2 = 0.060 \qquad X_\varphi = 2.50$$
$$R_1 = 0.0073 \qquad R_2 = 0.0064$$

While the motor is operating in the steady state under rated conditions, a 3-phase short circuit occurs on its supply line near the motor terminals. Determine the motor rms short-circuit current.

Solution

From Eq. 10-1 the motor transient reactance is seen to be

$$X' = 0.060 + 2.50 - \frac{(2.50)^2}{0.06 + 2.50} = 0.12\ \Omega/\text{phase}$$

The prefault stator current is

$$I_1 = \frac{400(746)}{0.90(0.93)(440\sqrt{3})} = 467\ \text{A}$$

With terminal voltage as the reference phasor, the prefault voltage behind transient reactance is then

$$\widehat{E}_1' = \frac{440}{\sqrt{3}} - (0.0073 + j0.12)(467\underline{/-\cos^{-1}0.90}) = 232\underline{/-12.2°}$$

From Eq. 10-5, the initial rms short-circuit current is

$$\frac{232}{0.12} = 1940\ \text{A}$$

The open-circuit time constant (Eq. 10-3) is

$$T_0' = \frac{2.50 + 0.060}{2\pi(60)(0.0064)} = 1.06\ \text{s}$$

and the short-circuit time constant (Eq. 10-4) is

$$T' = 1.06\frac{0.12}{2.56} = 0.050\ \text{s}$$

The rms short-circuit current is therefore

$$I_1 = 1940\varepsilon^{-t/0.050} \qquad \text{A}$$

The short-circuit time constant is three cycles on a 60-Hz base. Accordingly, in three cycles the short-circuit current decreases to 36.8 percent of its initial value. It has substantially disappeared in about 10 cycles. It is thus seen that, while the initial short-circuit current of an induction machine is relatively high compared with its normal current, the transient usually disappears rapidly. Because of this fact, the electric transients in induction machines are often neglected.

10–5 APPLICATION OF ADJUSTABLE–SPEED SOLID–STATE AC MOTOR DRIVES

As discussed briefly in Art. 10-3, induction-motor speed control can be obtained by such means as frequency and voltage control as well as recovering the slip-frequency rotor power via auxiliary devices. The use of solid-state electronics in implementing these techniques in ac motor drive systems has led to the increased use of induction machines in situations where speed control is required. Solid-state ac drive systems are considerably more complex than their dc counterparts (Art. 6-6); however, although dc motors have dominated the adjustable-speed drive field, ac motors are used in drive systems where their special features, such as the absence of commutators and brushes, must be utilized.

Adjustable-frequency power is generated by a thyristor circuit called an *inverter*. Inverters are used to transfer energy from a dc source to an ac load of arbitrary frequency and phase. They are used in ac drive systems to provide power of adjustable frequency to ac motors and to regenerate rotor-circuit power back to the ac line in wound-rotor induction motors. A typical motor drive consists of a rectifier to convert power from the ac supply to dc form, an inverter to form the adjustable-frequency alternating current from the dc bus, and a control system to set the frequency, adjust the motor voltage, and ensure that the maximum torque of the motor will not be exceeded.

Inverters for providing power at adjustable frequency for ac motor drive systems are usually 3-phase inverters. A battery may be floated on the dc bus to ensure an uninterrupted supply of energy for critical applications if power from the ac supply to the rectifier is interrupted. The 3-phase inverter has a minimum of six thyristors arranged in a bridge configuration, as shown in schematic form in Fig. 10-16; these thyristors (labeled *TR*) are switched sequentially to synthesize at the ac terminals of the inverter bridge a set of 3-phase voltages which are applied to the motor. Practical inverter circuits must include provisions for commutating current from thyristor to thyristor as the thyristors are switched.†

---

†For a description of practical inverter circuits, see B. D. Bedford and R. G. Hoft, "Principles of Inverter Circuits," Wiley, New York, 1964; also S. B. Dewan and A. Straughen, "Power Semiconductor Circuits," Wiley, New York, 1975.

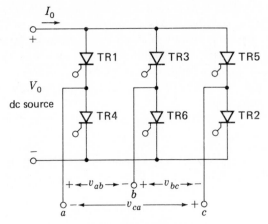

Three phase inverter output

Fig. 10-16. Three-phase inverter-bridge configuration.

Typical waveforms for the line-to-neutral voltages v_{a0} and v_{b0} and the line-to-line voltage v_{ab} are shown in Fig. 10-17, which is drawn for the case where each thyristor conducts for exactly one-half cycle (180° conduction). Table 10-1 shows the status of each thyristor in the bridge of Fig. 10-16 at the various steps of the waveform of Fig. 10-17. Each step corresponds to 60° on the output waveform. The line-to-neutral voltages are three-step-per-half-cycle approximations to sine waves while the line-to-line voltages are positive and negative pulses 120° wide. When the inverter voltages are applied to a motor, the motor responds to the fundamental and the harmonics of the waveforms to produce positive and braking torques and both normal and additional losses. Inverters can be built that produce output waveforms with less harmonic content by various techniques. For example, the outputs of two inverters whose waveforms are phase-displaced can be added, resulting in cancellation of some of the harmonics.

The output of a 3-phase inverter is applied to synchronous and induction motors to obtain adjustable-speed operation by means of frequency control.† Where closely regulated speed is required from one or more motors, as in textile applications, synchronous motors are used; they will operate in synchronism with the oscillator that controls the inverter independently of their loading. On the other hand, where nominally regulated speed is required, as in high-speed grinders or special vehicle drives, induction motors are used. Special types of motors are frequently employed because the operation is different from the usual fixed-frequency application. The synchronous motors are usually hybrid induction-synchronous motors which synchronize by reluctance torque or by using permanent-magnet fields. The induction motors are either squirrel-cage or solid-iron rotor types. Examples are shown in Fig.

†See A. Kusko, "Solid-State AC Motor Drives," M.I.T. Press, Cambridge, Mass., 1971.

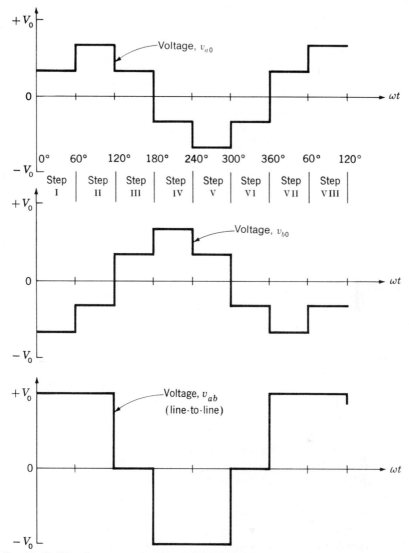

Fig. 10-17. Waveforms generated by a 3-phase inverter, 180° conduction.

10-18. Operation is frequently carried both above and below the nominal frequency of the motor, for example, 10 to 120 Hz for a 60-Hz motor.

The synchronous speed is directly proportional to frequency. Of primary interest is the shape of the torque-speed curve of the induction motor at each frequency in the operating range, including the starting and breakdown torques. The torques are determined by how the terminal voltage of the motor is managed as the frequency is adjusted. It is clear that to secure operation of a given motor at the maximum air-gap flux density over its speed range, the air-gap voltage must be adjusted proportional to frequency f. By so

TABLE 10-1

Step	Thyristor					
	TR1	TR2	TR3	TR4	TR5	TR6
I	On	Off	Off	Off	On	On
II	On	On	Off	Off	Off	On
III	On	On	On	Off	Off	Off
IV	Off	On	On	On	Off	Off
V	Off	Off	On	On	On	Off
VI	Off	Off	Off	On	On	On

doing we obtain the maximum available torque from the induction motor at each frequency setting. In a practical system, the terminal voltage V_1 is controlled so that V_1/f is constant; this type of control is termed *constant volts per hertz*. For a voltage waveform that changes shape as a function of frequency, the half-wave average voltage rather than the rms voltage is controlled in the above-described manner. Present practice is to maintain V_1 constant below a specific frequency to overcome the $I_1 R_1$ drop; in some drives, the voltage V_1 is held constant for operation above the nominal motor frequency and for obtaining constant-horsepower operation.

The torque-speed curve of an induction motor for a given frequency can be calculated using the methods of Chap. 9 within the accuracy of the motor parameters at that frequency. The torque-speed curves for the motor operated over a range of frequency, but at constant air-gap flux density are practi-

Fig. 10-18. Cutaway views of a rotor of a Synchrospeed (trademark) motor. (*Left*) The magnetic laminations are shown dark, the conductors light. (*Right*) Only the rotor conductors are shown; the iron has been etched away. (*The Louis Allis Company.*)

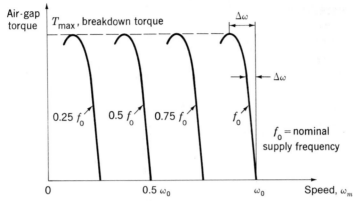

Fig. 10-19. Idealized torque-speed curves of induction motor under adjustable-frequency control.

cally the same on a slip-speed scale as shown in Fig. 10-19. This can be confirmed by viewing the air gap from the rotor; the rotor behavior is determined only by the slip speed, which sets the rotor induced voltage, frequency, current, and the torque produced. Note that a speed regulation of 3 percent at nominal frequency ω_0 becomes a speed regulation of 30 percent at $\omega_s = 0.1\omega_0$.

An adjustable-frequency drive can be started in several ways. It can be started at the lowest frequency and be brought up to speed by raising the frequency. It can be started at any frequency in the range as a constant-frequency induction motor is started. Or, it can be started from the ac supply line and transferred to the inverter after it is up to speed. The problems of starting are the dual ones of obtaining sufficient starting torque and of not exceeding the current rating of the inverter, which is usually based on the ability of the inverter to commutate the current, rather than the thyristor ratings.

The measured torque-speed curves on an actual adjustable-frequency drive system correspond closely with the curves of Fig. 10-19 for the ideal motor. The terminal voltage of the motor is usually held constant for operation above nominal frequency. Hence, the inverter-rectifier system requires no modification other than to provide thyristor gating pulses for such operation. The breakdown torque, instead of remaining constant as in Fig. 10-19, declines. The drive operates in a constant-horsepower mode. The operation is directly analogous to field-weakening control of a dc drive with fixed armature voltage over the nominal speed.

The waveform of each half cycle of the motor voltage must be controlled below nominal speed to maintain constant air-gap flux density. The method of such control should minimize the harmonic content to avoid undue losses and braking torques. It should also be done without complicating or penalizing the inverter. The waveforms for two typical methods for such control are shown in Fig. 10-20.

Figure 10-20a shows control of a line-to-line voltage by amplitude varia-

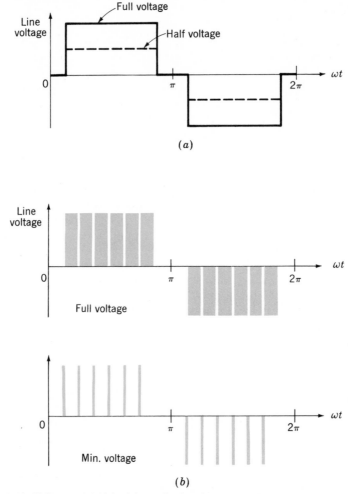

Fig. 10-20. Voltage control by (a) amplitude; (b) pulse-width control.

tion of a 120° wide pulse. Such control is obtained by varying the dc voltage from the rectifier portion of the drive. If self-commutation is employed in the inverter, it may not be possible to charge the commutation capacitors from the dc bus to commutate the full motor current properly at low frequencies. The inverter may have to be derated or a separate fixed-voltage dc bus set up just to supply the commutating capacitors.

Figure 10-20b shows waveform control by pulse-width control of each half cycle of the ac voltage. The 120° width of the waveform is divided typically into six 20° wide pulses where maximum voltage is obtained at 20° and minimum voltage at, say, 2° for a 10-to-1 speed range. The inverter dc voltage level remains unchanged, but auxiliary commutation must be used to turn off the thyristors at prescribed points in the half cycle.

10–6 SUMMARY

The dynamic performance of induction machines can often be analyzed by use of a steady-state speed-torque characteristic, as developed in Art. 9-5. Such an analysis neglects the effects of electric transients which generally occur much faster than the electromechanical dynamics of interest.

The design of an induction machine must often be a compromise between running requirements (speed regulation, efficiency, etc.) and those of start-up. High torque at start-up generally requires relatively high rotor resistance, while acceptable running characteristics dictate lower resistances.

Engineers have a range of options available to achieve these objectives. They may use a double squirrel-cage design, in which the effective rotor resistance changes with slip. Alternatively, they may choose a wound-rotor design, in which the rotor resistance can be varied via external resistors or a regeneration scheme which absorbs slip-frequency rotor power (as a simple resistor does) and feeds this power back to the ac supply at the armature terminals.

Solid-state variable-frequency motor drives permit the application of induction motors in situations which previously were the almost exclusive domain of dc motors. This allows the engineer to benefit from the simplicity, reliability, and lower cost of induction motors. On the other hand, because of the necessity to generate variable-frequency and/or variable-voltage alternating current, these solid-state drive systems are more complex and expensive than their dc counterparts.

PROBLEMS

10-1 A 50-hp 440-V 3-phase 4-pole 60-Hz wound-rotor induction motor develops a maximum internal torque of 250 percent at a slip of 16 percent when operating at rated voltage and frequency with its rotor short-circuited directly at the slip rings. Stator resistance and rotational losses may be neglected, and the rotor resistance may be assumed to be constant, independent of rotor frequency. Determine (a) the slip at full load in percent, (b) the rotor I^2R loss at full load in watts, and (c) the starting torque at rated voltage and frequency in newton-meters.

If the rotor resistance is now doubled (by inserting external series resistances), determine (d) the torque in newton-meters when the stator current has its full-load value and (e) the corresponding slip.

10-2 A 50-hp 3-phase 440-V 4-pole wound-rotor induction motor develops its rated full-load output at a speed of 1746 r/min when operated at rated voltage and frequency with its slip rings short-circuited. The maximum torque it can develop at rated voltage and frequency is 200 percent of full-load torque. The resistance of the rotor winding is 0.10 Ω/phase Y. Neglect rotational and stray load losses and stator resistance.

(a) Compute the rotor I^2R loss at full load.

(b) Compute the speed at maximum torque.

(c) How much resistance must be inserted in series with the rotor to produce maximum starting torque?

The motor is now run from a 50-Hz supply with the applied voltage adjusted so that the air-gap flux wave has the same amplitude at the same torque as on 60 Hz.

(d) Compute the 50-Hz applied voltage.

(e) Compute the speed at which the motor will develop a torque equal to its 60-Hz full-load value with its slip rings short-circuited.

10-3 A 220-V 3-phase 4-pole 60-Hz wound-rotor induction motor develops an internal torque of 150 percent with a line current of 155 percent at a slip of 5.0 percent when running at rated voltage and frequency with its rotor terminals short-circuited. (Torque and current are expressed as percentages of their full-load values.) The rotor resistance is 0.100 Ω between each pair of rotor terminals and may be assumed to be constant. What should be the resistance of each of three balanced Y-connected resistors inserted in series with each rotor terminal if the starting current at rated voltage and frequency is to be limited to 155 percent? What internal starting torque will be developed?

10-4 The resistance measured between each pair of slip rings of a 3-phase 60-Hz 300-hp 16-pole induction motor is 0.035 Ω. With the slip rings short-circuited, the full-load slip is 0.025, and it may be assumed that the slip-torque curve is a straight line from no load to full load. This motor drives a fan which requires 300 hp at the full-load speed of the motor. The torque required to drive the fan varies as the square of the speed. What resistances should be connected in series with each slip ring so that the fan will run at 300 r/min?

10-5 A polyphase induction motor has negligible rotor rotational losses and is driving a pure-inertia load. The moment of inertia of the rotor plus load is J SI units.

(a) Obtain an expression for the rotor energy loss during starting. Express the result in terms of J and the synchronous angular velocity ω_s.

(b) Obtain an expression for the rotor energy loss associated with reversal from full speed forward by reversing the phase sequence of the voltage supply (a process known as *plugging*). Express the result in terms of J and ω_s.

(c) State and discuss the degree of dependence of the results in parts (a) and (b) on the current-limiting scheme which can be used during starting and reversal.

(d) A 5-hp 3-phase 4-pole 60-Hz squirrel-cage induction motor has a full-load efficiency of 85.0 percent. The total inertia of rotor plus load is 6.33×10^{-2} kg·m². The total motor losses for a reversal may be assumed to be 2.25 times the rotor losses. The impairment of ventilation arising from the lower average speed during reversing is to be ignored.

Using the result in part (b), compute the number of times per minute that the motor can be reversed without exceeding its allowable temperature rise.

(e) Discuss the optimism or pessimism of the result in part (d).

10-6 A 230-V 3-phase Y-connected 6-pole 60-Hz wound-rotor induction motor has a stator-plus-rotor leakage reactance of $0.50\ \Omega$/phase referred to the stator, a rotor-plus-load moment of inertia of 1.0 kg·m², negligible losses (except for rotor I^2R loss), and negligible exciting current. It is connected to a balanced 230-V source and drives a pure-inertia load. Across-the-line starting is used, and the rotor-circuit resistance is to be adjusted so that the motor brings its load from rest to one-half synchronous speed in the shortest possible time.

Determine the value of the rotor resistance referred to the stator and the minimum time to reach one-half of synchronous speed.

10-7 Consider an induction machine with equivalent-circuit parameters which can be considered to be constant independent of slip.

(a) Show that in the limit of negligible armature resistance R_1, the torque expression of Eq. 9-27 can be written as

$$\frac{T}{T_{\max}} = \frac{2}{(s/s_{\max T}) + (s_{\max T}/s)}$$

where $T_{\max}$ is the maximum torque and $s_{\max T}$ is the slip at maximum torque, defined by Eqs. 9-29 and 9-30.

(b) The period of heavy current inrush to an induction motor during starting usually lasts for about the time required to reach the slip $s_{\max T}$ at maximum torque, after which it decreases to the normal running value. Develop an expression for the time t required to reach the speed corresponding to $s_{\max T}$ for an induction motor being started at full voltage. The combined inertia of the rotor and the connected mechanical equipment is J. The motor is unloaded at starting, and rotational losses are to be neglected.

10-8 A 3-phase induction motor is operating in the steady state at the slip s_1. By suddenly interchanging two stator leads, the motor is to be plugged to a quick stop. Use the notation and assumptions of Prob. 10-7 and consider the motor to be unloaded.

> (a) Develop an expression for the braking time of the motor.
> (b) Suppose that the motor is to be reversed instead of simply brought to a stop. Develop an expression for the reversing time.

10-9 A 4160-V 2500-hp 2-pole 3-phase 60-Hz squirrel-cage induction motor is driving a boiler-feed pump in an electric generating plant. The table below lists points on the motor torque-speed curve at rated voltage, speed being in percent of synchronous speed and torque in percent of rated torque. Also listed are the torque requirements of the boiler-feed pump expressed in percent of motor rated torque. The drive is started at rated voltage with the discharge valve open but working against a check valve until the pump head equals the system head. The straight-line portion of the pump characteristic between 0 and 10 percent speed represents the breakaway-torque requirements. At 92 percent speed the check valve opens, and there is a discontinuity in the slope of the pump curve. At 98 percent speed the motor produces its maximum torque.

Percent speed	0	10	30	50	70	90	92	98	99.5
Percent motor torque	75	75	75	75	80	125	155	240	100
Percent pump torque	15	0+	4	12	26	42	44	87	100

The inertia of the drive is such that 5400 kW · s of energy is stored in the rotating mass at synchronous speed. Determine (a) the time required for the check valve to open after rated voltage is applied to the motor and (b) the time required to reach the motor maximum-torque point.

10-10 For the boiler-feed-pump drive in Prob. 10-9, consider that the motor must occasionally be started with the motor voltage at 80 percent of its rated value. The motor torque may be assumed to be proportional to the square of the voltage.

> (a) Determine the time required for the check valve to open.
> (b) To what value may the motor starting voltage be reduced without making it impossible to reach a speed at which water is delivered to the system?

10-11 A frequency-changer set is to be designed for supplying variable-frequency power to induction motors driving the propellers on scale-model airplanes for wind-tunnel testing, as described in Art. 10-3b. The frequency changer is a wound-rotor induction machine driven by a dc motor whose speed can be controlled. The 3-phase stator winding of the induction machine is excited from a 60-Hz source, and variable-frequency 3-phase power is taken from its rotor winding. The set must meet the following specifications:

Output frequency range = 120 to 450 Hz

Maximum speed 3000 r/min

Maximum power output = 80 kW at 0.80 power factor and 450 Hz

The power required by the induction-motor load drops off rapidly with decreasing frequency, so that the maximum-speed condition determines the sizes of the machines.

On the basis of negligible exciting current, losses, and voltage drops in the induction machine, find (a) the minimum number of poles for the induction machine, (b) the corresponding maximum and minimum speeds, (c) the kVA rating of the stator winding of the induction machine, and (d) the horsepower rating of the dc machine.

10-12 An adjustable-speed drive is to be furnished for a large fan in an industrial plant. The fan is to be driven by two wound-rotor induction motors coupled mechanically to the fan shaft and arranged so that lower speeds will be carried on the smaller motor and higher speeds on the larger motor. The speed control is to be arranged in steps so that the control will be uninterrupted from minimum to maximum speed and so that there will not be a sudden change in speed during the transfer from one motor to the other.

The larger motor is to be a 2300-V 3-phase 500-hp 60-Hz 6-pole motor; the smaller motor is to be a 200-hp 60-Hz 8-pole motor. The following motor data are furnished by the manufacturer:

Constant	200-hp motor	500-hp motor
Stator resistance, Ω	0.57	0.14
Rotor resistance, Ω	0.93	0.24
Stator plus rotor leakage reactance, Ω	2.6	0.98

These values are per phase values (Y connection) referred to the stator. Motor rotational losses and exciting requirements may be ignored.

The minimum operating speed of the fan is to be 450 r/min; the maximum speed is to be approximately 1170 r/min. The fan requires 450 hp at 1170 r/min, and the power required at other speeds varies nearly as the cube of the speed.

The proposed control scheme is based on the stators of both motors being connected to the line at all times when the drive is in operation. In the lower speed range, the rotor of the 500-hp motor is open-circuited. This lower range is obtained by adjustment of external resistance in the rotor of the 200-hp motor. Above this lower speed range, the rotor of the 200-hp motor is open-circuited; speed adjustment in the upper range is obtained by adjustment of external resistance in the rotor of the 500-hp motor.

The transition between motors from the lower to the upper speed range is to be handled as follows:

1 All external rotor resistance is cut out of the 200-hp motor.
2 By means of a closed-before-open type of contactor, the rotor circuit of the 500-hp motor is closed, and then the rotor circuit of the 200-hp motor is opened. The external rotor resistance in the 500-hp motor for this step has such a value that the speed will be the same as in the first step.
3 Higher speeds are obtained by cutting out rotor resistance in the 500-hp motor.

This procedure is essentially reversed in going from the higher to the lower speed range.

As plant engineer, you are asked to give consideration to some features of this proposal. In particular, you are asked to (a) determine the range of external rotor resistance (referred to the stator) which must be available for insertion in the 200-hp motor, (b) determine the range of external rotor resistance which must be available for insertion in the 500-hp motor, and (c) discuss any features of the scheme which you may not like and suggest alternatives.

10-13 A 4-pole 440-V 400-hp 3-phase 60-Hz Y-connected wound-rotor induction motor has the following constants in ohms per phase referred to the stator: $X_1 = X_2 = 0.055$, $X_\varphi = 2.23$, $R_1 = 0.0054$, $R_2 = 0.0071$. The motor is supplied at normal terminal voltage through a series reactance of $0.03\ \Omega$/phase representing a step-down transformer bank. It is fully loaded, the slip rings are short-circuited, and the efficiency and power factor are 90.5 and 90.0 percent, respectively.

A 3-phase short circuit occurs at the high-voltage terminals of the transformer bank. Determine the initial symmetrical short-circuit current in the motor, and show how it is decremented.

10-14 Following are points on the torque-speed curve of a 3-phase squirrel-cage induction motor with balanced rated voltage impressed:

Torque, per unit	0	1.00	2.00	3.00	3.50	3.25	3.00
Speed, per unit	1.00	0.97	0.93	0.80	0.47	0.20	0

Unit speed is synchronous speed; unit torque is rated torque. The motor is coupled to a machine tool which requires rated torque regardless of speed. The inertia of the motor plus load is such that it requires 1.2 s to bring them to rated speed with a constant accelerating torque equal to rated torque.

With the motor driving the load under normal steady conditions, the voltage at its terminals suddenly drops to 50 percent of rated value because of a short circuit in the neighborhood. It remains at this reduced value for 0.6 s and is then restored to its full value by clearing of the short circuit. The undervoltage release on the motor does not operate. Will the motor stop? If not, what is its lowest speed? Neglect any effects of secondary importance.

10-15 Induction-motor transient and dynamic considerations are important factors in the industrial application of these motors, as in refineries, chemical process plants, auxiliary drives in steam stations, etc. It is common practice to provide two sources of electric power to critical drives, a normal source and an emergency source to which the motor bus is automatically transferred in the event of failure of the normal source. When the normal source has been tripped, flux linkages are "trapped" with the closed rotor circuits of the operating motors. A motor terminal voltage (often called the *residual voltage*) is therefore produced which may require an appreciable time to decay to zero. If the motor is transferred almost immediately to the emergency source, the residual voltage and the emergency-source voltage may be so phase-displaced that the resulting heavy inrush current may damage the motor. One solution is to install a voltage-sensitive relay which delays transfer until the residual voltage has decayed to about 25 percent of normal motor voltage. In the meantime the motor is slowing down and may have real difficulty in re-accelerating when the emergency source is connected.

To investigate these problems, consider a 4160-V 2500-hp 2-pole 3-phase 3580-r/min 60-Hz squirrel-cage induction motor driving a boiler-feed pump. The motor has an open-circuit transient time constant of 3.6 s, short-circuit time constant of 0.12 s, and transient reactance of 1.3 Ω/phase Y. The motor is operating fully loaded at rated voltage and draws 2160 kVA at 0.90 power factor from the mains. (This problem is based on a single motor. In a practical case, the effects of other parallel motors must be considered.†)

(*a*) Determine the time required after interruption of the normal supply for the residual voltage to decay to 25 percent of the normal bus voltage.

(*b*) Determine the speed to which the motor decelerates in this interval. The inertia of the motor rotor and pump impeller is 76.0 kg · m². Consider that the load torque varies as the square of the speed, and neglect losses.

(*c*) Assume that the emergency source consists of a constant 100 percent voltage behind a reactive impedance of 0.10 Ω (representing a step-down transformer bank). Assume also that torque-slip and current-slip curves at rated voltage are available for the motor. Outline a method by which a speed-time curve can be obtained for the re-

†See D. G. Lewis and W. D. Marsh, Transfer of Steam-Electric Generating Station Auxiliary Buses, *Trans. AIEE*, **74**:322–330, 1955.

acceleration period. A step-by-step method arranged for either hand computation or digital-computer programming is suggested.

(d) Let the torque-slip curve be that given for the motor in Prob. 10-9. Corresponding data for the current-slip curve are given below, the current being percent of current under rated conditions.

Percent speed	0	30	50	70	90	92	98	99.5
Percent current	620	620	620	620	580	540	390	100

By the method you have outlined in part (c), plot motor speed as a function of time during reacceleration.

(e) The practical problem may be to determine the maximum allowable impedance and hence the size of the emergency step-down transformer bank which will permit both motor starting and successful reacceleration of all the motor drives after transfer. Describe how you would carry out such an investigation.

10-16 An induction motor is being selected to drive a pump. The speed of the motor and pump will be controlled by adjustment of the primary stator voltage. The motor slip is 0.05 per unit for rated torque.

Sketch the profile of the allowable motor torque as a function of slip from $s = 0$ to 1.0 per unit for the following levels of rotor power dissipation: (a) nominal, corresponding to dissipation at $s = 0.05$ per unit; (b) twice the value for part (a); (c) 4 times the value for part (a).

10-17 A variable-frequency variable-voltage 3-phase ac motor drive is used to drive a squirrel-cage induction motor. The output frequency can be varied from 30 to 60 Hz and, in order to maintain essentially constant flux in the motor, the output voltage varies linearly with frequency (constant volts per hertz).

(a) For this motor-drive–motor combination, rewrite Eq. 9-27 in terms of ω_s, the synchronous speed, ω_m, the shaft mechanical speed, V_{t0}, the applied voltage at 60 Hz, and $\omega_0 = 2\pi \times 60$, the 60-Hz angular frequency.

(b) Show that in the limit of negligible stator resistance the magnitude of the peak torque remains constant, independent of the applied frequency.

(c) Show that in the limit of negligible stator resistance the shape of the torque-speed characteristic is independent of the applied frequency and is simply shifted up or down in speed as determined by the applied frequency.

(d) Express quantitatively the criterion for the validity of negligible-stator-resistance approximation used in parts (b) and (c).

10-18 The motor drive of Prob. 10-17 is used to drive the 3-phase 220-V (line-to-line) 10 hp 60-Hz 6-pole induction motor of Examples 9-1 and 9-2. This problem repeats the calculations of Examples 9-1 and 9-2 with the motor-drive output at 30 Hz and 110 V (line to line). Assume that the total friction, windage, and core loss under these conditions is 100 W, independent of load.

(*a*) Based upon the results of part (*c*) of Prob. 10-17, the shaft torque at a slip of 4 percent should correspond to the 2 percent 60-Hz 220-V case of Example 9-1. Calculate the motor speed, output torque and power, stator current, power factor, and efficiency for a slip of 4 percent.

(*b*) Calculate the maximum torque and the corresponding speed and slip.

(*c*) Calculate the starting torque and current.

Fractional–Horsepower
AC Motors

Fractional-horsepower motors supply the motive power for all types of equipment in the home, office, and factory. They differ from integral-horsepower motors in the wide variety of designs and characteristics available, prompted by the economics and special requirements of their application. In addition, fractional-horsepower ac motors are nearly all designed for, and used on, single-phase lines.

Although fractional-horsepower ac motors are relatively simple in construction, they are considerably more difficult to analyze than larger 3-phase motors. Much design is carried out by building and testing prototype motors until the required design is achieved. Computer design programs are being used to realize more accurate paper designs and to reduce the extent of cut-and-try to obtain required performance. In this chapter, we shall describe the various types of motors qualitatively in terms of the rotating-field theory and develop a method of analysis appropriate for 2-phase motors from which the starting and running characteristics of single-phase induction motors can be calculated.†

The series ac motor, also termed the universal motor, is widely used in portable tools, vacuum cleaners, and kitchen equipment, where its high speed results in high horsepower per unit motor size. These motors are frequently used with solid-state devices for speed control and will be treated in this chapter.

11–1 SINGLE–PHASE INDUCTION MOTORS: QUALITATIVE EXAMINATION

Structurally, the commonest types of single-phase induction motors resemble polyphase squirrel-cage motors except for the arrangement of the stator windings. An induction motor with a cage rotor and a single-phase stator winding is represented schematically in Fig. 11-1. Instead of being a concentrated coil, the actual stator winding is distributed in slots to produce an approximately sinusoidal space distribution of mmf. Such a motor inherently has no starting torque, but if started by auxiliary means it will continue to run. Before considering auxiliary starting methods, the basic properties of the elementary motor of Fig. 11-1 will be described.

For the single-phase stator winding of Fig. 11-1, if the stator current is a cosinusoidal function of time, the mmf wave can be written as a function of space and time as

$$\mathcal{F}_1 = F_{1,\text{max}} \cos \omega t \cos \theta \tag{11-1}$$

which, as shown in Eq. 3-22, can be written as the sum of a positive- and negative-traveling mmf wave of equal magnitude. The positive-traveling wave

†For an extensive treatment of fractional-horsepower motors, see C. G. Veinott, "Fractional- and Subfractional Horsepower Electric Motors," McGraw-Hill, New York, 1970.

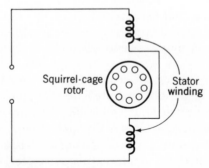

Fig. 11-1. Elementary single-phase induction motor.

is given by

$$\mathcal{F}^+ = \tfrac{1}{2}F_{1,\max} \cos (\theta - \omega t) \tag{11-2}$$

and the negative-traveling wave is given by

$$\mathcal{F}^- = \tfrac{1}{2}F_{1,\max} \cos (\theta + \omega t) \tag{11-3}$$

Each of these component mmf waves produces induction-motor action, but the corresponding torques are in opposite directions. With the rotor at rest, the forward and backward air-gap flux waves created by the combined mmf's of stator and rotor currents are equal, the component torques are equal, and no starting torque is produced. If the forward and backward air-gap flux waves remained equal when the rotor is revolving, each of the component fields would produce a torque-speed characteristic similar to that of a polyphase motor with negligible stator leakage impedance, as illustrated by the dashed curves f and b in Fig. 11-2a. The resultant torque-speed characteristic, which is the algebraic sum of the two component curves, shows that if the motor were started by auxiliary means, it would produce torque in whatever direction it was started.

The assumption that the air-gap flux waves remain equal when the rotor is in motion is a rather drastic simplification of the actual state of affairs. In the first place, the effects of stator leakage impedance are ignored. Furthermore the effects of induced rotor currents are not properly accounted for. Both these effects will ultimately be included in the detailed quantitative theory of Art. 11-3. The following qualitative explanation shows that the performance of a single-phase induction motor is considerably better than would be predicted on the basis of equal forward and backward flux waves.

When the rotor is in motion, the component rotor currents induced by the backward field are greater than at standstill and their power factor is lower. Their mmf, which opposes that of the stator current, results in a reduction of the backward flux wave. Conversely, the magnetic effect of the component currents induced by the forward field is less than at standstill because the rotor currents are less and their power factor is higher. As speed increases, therefore, the forward flux wave increases while the backward flux wave decreases, their sum remaining roughly constant since it must induce the stator counter emf, which is approximately constant if the stator leakage-impedance voltage drop is small. Hence, with the rotor in motion, the torque of the forward field is greater and that of the backward field is less than in Fig. 11-2a, the true situation being about as shown in Fig. 11-2b. In the normal running region at a few percent slip, the forward field is several times greater than the backward field, and the flux wave does not differ greatly from the constant-amplitude revolving field in the air gap of a balanced polyphase motor. In the normal running region, therefore, the torque-speed characteristics of a single-

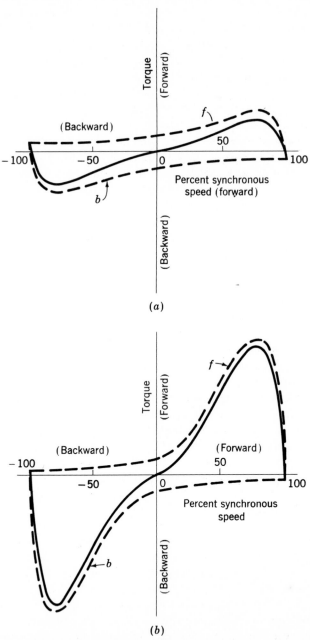

Fig. 11-2. Torque-speed characteristic of a single-phase induction motor: (a) On the basis of constant forward and backward flux waves; (b) taking into account changes in the flux waves.

phase motor is not too greatly inferior to that of a polyphase motor having the same rotor and operating with the same maximum air-gap flux density.

In addition to the torques shown in Fig. 11-2, double-stator-frequency torque pulsations are produced by the interactions of the oppositely rotating flux and mmf waves which glide past each other at twice synchronous speed. These interactions produce no average torque, but they tend to make the motor noisier than a polyphase motor. Such torque pulsations are unavoidable in a single-phase motor because of the pulsations in instantaneous power input inherent in a single-phase circuit. The effects of the pulsating torque can be minimized by using an elastic mounting for the motor. The torque referred to on the torque-speed curves is the time average of the instantaneous torque.

11–2 STARTING AND RUNNING PERFORMANCE OF SINGLE–PHASE INDUCTION AND SYNCHRONOUS MOTORS

Single-phase induction motors are classified in accordance with the methods of starting and are usually referred to by names descriptive of these methods. Selection of the appropriate motor is based upon the starting- and running-torque requirements of the load, the duty cycle, and the limitations on starting and running current from the supply line for the motor. The cost of single-phase motors increases with the horsepower and with the performance, such as starting-torque-to-current ratio; hence, the application engineer selects the lowest horsepower and performance motor that meets his requirement to minimize the cost. Where a large volume of motors is to be used for a specific purpose, a special motor is designed for the application to ensure least cost. In the fractional-horsepower motor business, differences of cost in pennies are important. Starting methods and resulting torque-speed characteristics will be considered qualitatively in this article.

a. Split-Phase Motors

Split-phase motors have two stator windings, a main winding m and an auxiliary winding a, with their axes displaced 90 electrical degrees in space. They are connected as shown in Fig. 11-3a. The auxiliary winding has a higher resistance-to-reactance ratio than the main winding, so that the two currents are out of phase, as indicated in the phasor diagram of Fig. 11-3b, which is representative of conditions at starting. Since the auxiliary-winding current $\widehat{I}_a$ leads the main-winding current $\widehat{I}_m$, the stator field first reaches a maximum along the axis of the auxiliary winding and then somewhat later in time reaches a maximum along the axis of the main winding. The winding currents are equivalent to unbalanced 2-phase currents, and the motor is equivalent to an unbalanced 2-phase motor. The result is a rotating stator field which causes the motor to start. After the motor starts, the auxiliary winding is

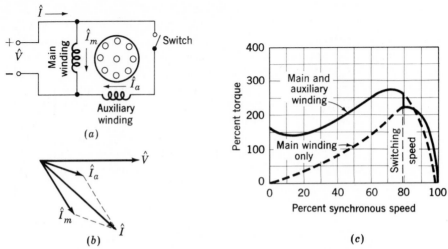

Fig. 11-3. Split-phase motor: (a) connections, (b) phasor diagram at starting, and (c) typical torque-speed characteristic.

disconnected, usually by means of a centrifugal switch that operates at about 75 percent of synchronous speed. The simple way to obtain the high resistance-to-reactance ratio for the auxiliary winding is to wind it with smaller wire than the main winding, a permissible procedure because this winding is in circuit only during starting. Its reactance can be reduced somewhat by placing it in the tops of the slots. A typical torque-speed characteristics is shown in Fig. 11-3c.

Split-phase motors have moderate starting torque with low starting current. Typical applications include fans, blowers, centrifugal pumps, and office equipment. Typical ratings are $\frac{1}{20}$ to $\frac{1}{2}$ hp; in this range they are the lowest-cost motors available.

b. Capacitor-Type Motors

Capacitors can be used to improve the starting performance, running performance, or both, of the motor, depending upon the size and connection of the capacitor.

The capacitor-start motor is also a split-phase motor, but the time-phase displacement between the two currents is obtained by means of a capacitor in series with the auxiliary winding, as shown in Fig. 11-4a. Again the auxiliary winding is disconnected after the motor has started, and consequently the auxiliary winding and capacitor can be designed at minimum cost for intermittent service. By using a starting capacitor of appropriate value the auxiliary-winding current $\hat{I}_a$ at standstill could be made to lead the main-winding current $\hat{I}_m$ by 90 electrical degrees, as it would in a balanced 2-phase motor (see Fig. 11-4b). Actually, the best compromise between the factors of starting torque, starting current, and costs results with a phase angle somewhat less

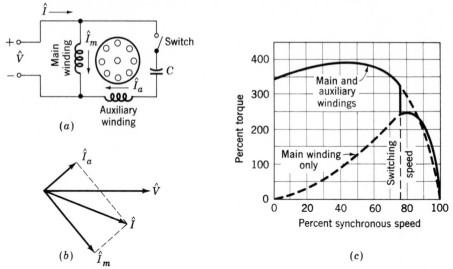

Fig. 11-4. Capacitor-start motor: (a) connections, (b) phasor diagram at starting, and (c) typical torque-speed characteristic.

than 90°. A typical torque-speed characteristic is shown in Fig. 11-4c, high starting torque being an outstanding feature. These motors are used for compressors, pumps, refrigeration and air-conditioning equipment, and other hard-to-start loads. A cutaway view of a capacitor-start motor is shown in Fig. 11-5.

In the permanent-split-capacitor motor, the capacitor and auxiliary winding are not cut out after starting; the construction can be simplified by

Fig. 11-5. Cutaway view of a capacitor-start induction motor. The starting switch is at the right of the rotor. The motor is of dripproof construction. (*General Electric Company.*)

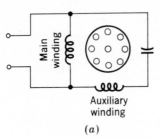 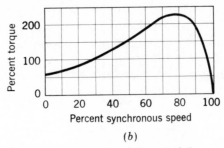

Auxiliary
winding

(a) *(b)*

Fig. 11-6. Permanent-split-capacitor motor and typical torque-speed characteristic.

omission of the switch, and the power factor, efficiency, and torque pulsations improved. For example, the capacitor and auxiliary winding could be designed for perfect 2-phase operation at any one desired load. The backward field would then be eliminated, with resulting improvement in efficiency. The double stator-frequency torque pulsations also would be eliminated, the capacitor serving as an energy-storage reservoir for smoothing out the pulsations in power input from the single-phase line. The result is a quiet motor. Starting torque must be sacrificed because the capacitance is necessarily a compromise between the best starting and running values. The resulting torque-speed characteristic and a schematic diagram are given in Fig. 11-6.

If two capacitors are used, one for starting and one for running, theoretically optimum starting and running performance can both be obtained. One way of accomplishing this result is shown in Fig. 11-7*a*. The small value of capacitance required for optimum running conditions is permanently connected in series with the auxiliary winding, and the much larger value required for starting is obtained by a capacitor connected in parallel with the running capacitor. The starting capacitor is disconnected after the motor starts.

The capacitor for a capacitor-start motor has a typical value of 300 μF for a $\frac{1}{2}$-hp motor. Since it must carry current for just the starting time, the capac-

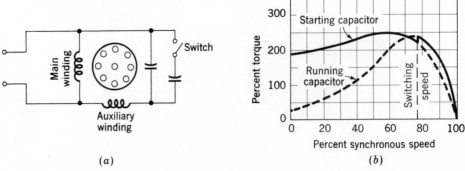

(a) *(b)*

Fig. 11-7. Two-value-capacitor motor and typical torque-speed characteristic.

itor is a special compact ac electrolytic type made for motor-starting duty. The capacitor for the same motor permanently connected has a typical rating of 40 μF; since it operates continuously, the capacitor is an ac paper, foil, and oil type. The cost of the motors is related to the performance: the permanent-capacitor motor is the lowest cost, the capacitor-start motor next, and the two-value-capacitor motor the highest cost.

EXAMPLE 11-1

A $\frac{1}{3}$-hp 120-V 60-Hz capacitor-start motor has the following constants for the main and auxiliary windings (at starting):

$$\text{Main winding, } Z_m = 4.5 + j\,3.7\,\Omega$$
$$\text{Auxiliary winding, } Z_a = 9.5 + j\,3.5\,\Omega$$

Find the value of starting capacitance that will place the main and auxiliary winding currents in quadrature at starting.

Solution

The currents $\widehat{I}_m$ and $\widehat{I}_a$ are as shown in Fig. 11-4a and b. The impedance angle of the main winding is

$$\phi_m = \tan^{-1}\frac{3.7}{4.5} = 39.6°$$

The impedance angle of the auxiliary winding must be

$$\phi_a = 39.6° - 90.0° = -50.4°$$

The reactance X_c of the capacitor must satisfy the relationship

$$\tan^{-1}\frac{3.5 - X_c}{9.5} = -50.4°$$

$$\frac{3.5 - X_c}{9.5} = -1.21$$

$$X_c = 1.21(9.5) + 3.5 = 15.0\,\Omega$$

The capacitance C is

$$C = \frac{10^6}{15.0(377)} = 177\,\mu\text{F}$$

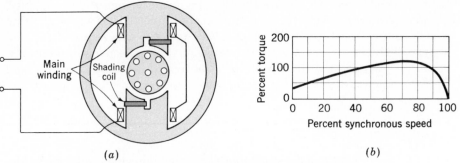

Fig. 11-8. Shaded-pole motor and typical torque-speed characteristic.

c. Shaded-Pole Motors

As illustrated schematically in Fig. 11-8a the shaded-pole motor usually has salient poles with one portion of each pole surrounded by a short-circuited turn of copper called a *shading coil*. Induced currents in the shading coil cause the flux in the shaded portion of the pole to lag the flux in the other portion. The result is like a rotating field moving in the direction from the unshaded to the shaded portion of the pole, and a low starting torque is produced. A typical torque-speed characteristic is shown in Fig. 11-8b. The efficiency is low. Shaded-pole motors are the least expensive type of fractional-horsepower motor and are built up to about $\frac{1}{20}$ hp.

d. Self-Starting Reluctance Motors

Any one of the induction-motor types described above can be made into a self-starting synchronous motor of the reluctance type. Anything which makes the reluctance of the air gap a function of the angular position of the rotor with respect to the stator coil axis will produce reluctance torque when the rotor is revolving at synchronous speed. For example, suppose some of the teeth are removed from a squirrel-cage rotor, leaving the bars and end rings intact, as in an ordinary squirrel-cage induction motor. Figure 11-9a shows a lamination for such a rotor designed for use with a 4-pole stator. The stator may be polyphase or any one of the single-phase types described above. The motor will start as an induction motor and at light loads will speed up to a small value of slip. The reluctance torque arises from the tendency of the rotor to try to align itself in the minimum-reluctance position with respect to the synchronously revolving forward air-gap flux wave, in accordance with the principles explained in Chap. 2. At a small slip, this torque alternates slowly in direction; the rotor is accelerated during a positive half cycle of the torque variation and decelerated during the succeeding negative half cycle. If, however, the moment of inertia of the rotor and its mechanical load is sufficiently small, the rotor will be accelerated from slip speed up to synchronous speed during an accelerating half cycle of the reluctance torque. The rotor

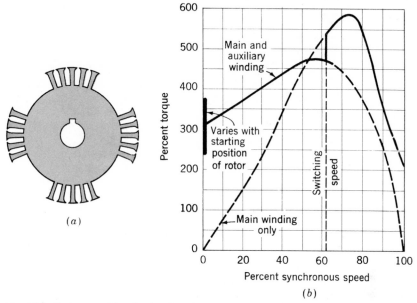

Fig. 11-9. Rotor punching for 4-pole reluctance-type synchronous motor and typical starting characteristics.

will then pull into step and continue to run at synchronous speed. The torque of the backward-revolving field affects the synchronous-motor performance in the same way as an additional shaft load.

A typical torque-speed characteristic for a split-phase-start synchronous motor of the single-phase reluctance type is shown in Fig. 11-9b. Notice the high values of induction-motor torque. The reason for this is that in order to obtain satisfactory synchronous-motor characteristics it has been found necessary to build reluctance-type synchronous motors on frames which would be suitable for induction motors of 2 or 3 times the synchronous-motor rating. Also notice that the principal effect of the salient-pole rotor on the induction-motor characteristic is at standstill, where considerable "cogging" is evident; i.e., the torque varies considerably with rotor position.

e. Hysteresis Motors

The phenomenon of hysteresis can be used to produce mechanical torque. In its simplest form, the rotor of a hysteresis motor is a smooth cylinder of magnetically hard steel, without windings or teeth. It is placed inside a slotted stator carrying distributed windings designed to produce as nearly as possible a sinusoidal space distribution of flux, since undulations in the flux wave greatly increase the losses. In single-phase motors, the stator windings usually are the permanent-split-capacitor type, as in Fig. 11-6. The capacitor is chosen so as to result in approximately balanced 2-phase conditions within

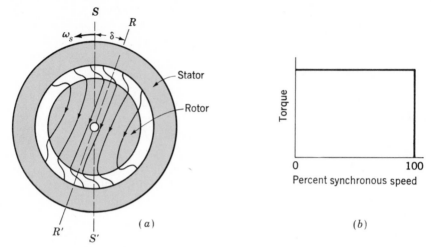

Fig. 11-10. (*a*) General nature of the magnetic field in the air gap and rotor of a hysteresis motor; (*b*) idealized torque-speed characteristic.

the motor windings. The stator then produces a rotating field, approximately constant in space waveform and revolving at synchronous speed.

Instantaneous magnetic conditions in the air gap and rotor are indicated in Fig. 11-10*a* for a 2-pole stator. The axis SS' of the stator-mmf wave revolves at synchronous speed. Because of hysteresis, the magnetization of the rotor lags behind the inducing mmf wave, and therefore the axis RR' of the rotor flux wave lags behind the axis of the stator-mmf wave by the hysteretic lag angle δ (Fig. 11-10*a*). If the rotor is stationary, starting torque is produced proportional to the product of the fundamental components of the stator mmf and rotor flux and the sine of the torque angle δ. The rotor then accelerates if the counter torque of the load is less than the developed torque of the motor. So long as the rotor is turning at less than synchronous speed, each particle of the rotor is subjected to a repetitive hysteresis cycle at slip frequency. While the rotor accelerates, the lag angle δ remains constant if the flux is constant, since the angle δ depends merely on the hysteresis loop of the rotor and is independent of the rate at which the loop is traversed. The motor therefore develops constant torque right up to synchronous speed, as shown in the idealized torque-speed characteristic of Fig. 11-10*b*. This feature is one of the advantages of the hysteresis motor. In contrast with a reluctance motor, which must "snap" its load into synchronism from an induction-motor torque-speed characteristic, a hysteresis motor can synchronize any load which it can accelerate, no matter how great the inertia. After reaching synchronism, the motor continues to run at synchronous speed and adjusts its torque angle so as to develop the torque required by the load.

The hysteresis motor is inherently quiet and produces smooth rotation of its load. Furthermore, the rotor takes on the same number of poles as the stator field. The motor lends itself to multispeed synchronous-speed operation where the stator is wound with several sets of windings and utilizes pole-changing connections. The hysteresis motor can accelerate and synchronize high-inertia loads because its torque is uniform from standstill to synchronous speed.

11-3 REVOLVING-FIELD THEORY OF SINGLE-PHASE INDUCTION MOTORS

In Art. 11-1 the stator-mmf wave of a single-phase induction motor is shown to be equivalent to two constant-amplitude mmf waves revolving in opposite directions at synchronous speed. Each of these component stator-mmf waves induces its own component rotor currents and produces induction-motor action just as in a balanced polyphase motor. This double-revolving-field concept not only is useful for qualitative visualization but also can be developed into a quantitative theory applicable to a wide variety of induction-motor types. A simple and important case is that of the single-phase induction motor running on only its main winding.

First consider conditions with the rotor stationary and only the main stator winding m excited. The motor then is equivalent to a transformer with its secondary short-circuited. The equivalent circuit is shown in Fig. 11-11a, where R_{1m} and X_{1m} are, respectively, the resistance and leakage reactance of the main winding, X_φ is the magnetizing reactance, and R_2 and X_2 are the standstill values of the rotor resistance and leakage reactance referred to the

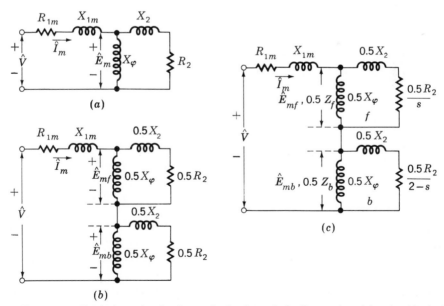

Fig. 11-11. Equivalent circuits for a single-phase induction motor: (a) rotor blocked; (b) rotor blocked, showing effects of forward and backward fields; (c) running conditions.

main stator winding by use of the appropriate turns ratio. Core loss, which is omitted here, will be accounted for later as if it were a rotational loss. The applied voltage is $\widehat{V}$, and the main-winding current is $\widehat{I}_m$. The voltage $\widehat{E}_m$ is the counter emf generated in the main winding by the stationary pulsating air-gap flux wave produced by the combined action of the stator and rotor currents.

In accordance with the double-revolving-field concept of Art. 11-1, the stator mmf can be resolved into half-amplitude forward and backward rotating fields. At standstill the amplitudes of the forward and backward resultant air-gap flux waves both equal half the amplitude of the pulsating field. In Fig. 11-11b the portion of the equivalent circuit representing the effects of the air-gap flux is split into two equal portions, representing the effects of the forward and backward fields, respectively.

Now consider conditions after the motor has been brought up to speed by some auxiliary means and is running on only its main winding in the direction of the forward field at a per unit slip s. The rotor currents induced by the forward field are of slip frequency sf, where f is the stator frequency. Just as in any polyphase motor with a symmetrical polyphase or cage rotor, these rotor currents produce an mmf wave traveling forward at slip speed with respect to the rotor and therefore at synchronous speed with respect to the stator. The resultant of the forward waves of stator and rotor mmf creates a resultant forward wave of air-gap flux which generates a counter emf $\widehat{F}_{mf}$ in the main winding m of the stator. The reflected effect of the rotor as viewed from the stator is like that in a polyphase motor and can be represented by an impedance $0.5R_2/s + j\,0.5X_2$ in parallel with $j\,0.5X_\varphi$, as in the portion of the equivalent circuit (Fig. 11-11c) labeled f. The factors of 0.5 come from the resolution of the pulsating stator mmf into forward and backward components.

Now consider conditions with respect to the backward field. The rotor is still turning at a slip s with respect to the forward field, and its per unit speed n in the direction of the forward field is

$$n = 1 - s \qquad (11\text{-}4)$$

The relative speed of the rotor with respect to the backward field is $1 + n$, or its slip with respect to the backward field is

$$1 + n = 2 - s \qquad (11\text{-}5)$$

The backward field then induces rotor currents whose frequency is $(2 - s)f$. For small slips, these rotor currents are of almost twice stator frequency. At a small slip, an oscillogram of rotor current therefore will show a high-frequency component from the backward field superposed on a low-frequency component from the forward field. As viewed from the stator, the rotor-mmf wave of the backward-field rotor currents travels at synchronous speed but in the backward direction. The equivalent circuit representing these internal reactions from the viewpoint of the stator is like that of a polyphase motor whose slip is $2 - s$ and is shown in the portion of the equivalent circuit (Fig. 11-11c) labeled b. As with the forward field, the factors of 0.5 come from the resolution of the pulsating stator mmf into forward and backward components. The voltage $\widehat{E}_{mb}$ across the parallel combination representing the backward field is the counter emf generated in the main winding m of the stator by the resultant backward field.

By use of the equivalent circuit of Fig. 11-11c, the stator current, power input, and power factor can be computed for any assumed value of slip when the applied voltage and the motor impedances are known. To simplify the notation, let

$$Z_f \equiv R_f + jX_f \equiv \left(\frac{R_2}{s} + jX_2\right) \text{ in parallel with } jX_\varphi \qquad (11\text{-}6)$$

$$Z_b \equiv R_b + jX_b \equiv \left(\frac{R_2}{2 - s} + jX_2\right) \text{ in parallel with } jX_\varphi \qquad (11\text{-}7)$$

The impedances representing the reactions of the forward and backward fields from the viewpoint of the single-phase stator winding m are $0.5Z_f$ and $0.5Z_b$, respectively, in Fig. 11-11c.

Examination of the equivalent circuit (Fig. 11-11c) confirms the conclusion, reached by qualitative reasoning in Art. 11-1 (Fig. 11-2b), that the forward air-gap flux wave increases and the backward wave decreases when the rotor is set in motion. When the motor is running at a small slip, the reflected effect of the rotor resistance in the forward field, $0.5R_2/s$, is much larger than its standstill value, while the corresponding effect in the backward field, $0.5R_2/(2 - s)$, is smaller. The forward-field impedance therefore is larger than its standstill value, while that of the backward field is smaller. The forward-field counter emf $\widehat{E}_{mf}$ therefore is larger than its standstill value, while the backward-field counter emf $\widehat{E}_{mb}$ is smaller; i.e., the forward air-gap flux wave increases, while the backward flux wave decreases.

Moreover, mechanical-output conditions can be computed by application of the torque and power relations developed for polyphase motors in Chap. 9. The torques produced by the forward and backward fields can each be treated in this manner. The interactions of the oppositely rotating flux and mmf waves cause torque pulsations at twice stator frequency but produce no average torque.

As in Eq. 9-22, the internal torque T_f of the forward field in newton-meters equals $1/\omega_s$ times the power P_{gf} in watts delivered by the stator winding to the forward field, where ω_s is the synchronous angular velocity in mechanical radians per second; thus

$$T_f = \frac{1}{\omega_s} P_{gf} \qquad (11\text{-}8)$$

When the magnetizing impedance is treated as purely inductive, P_{gf} is the power absorbed by the impedance $0.5Z_f$; that is,

$$P_{gf} = I_m^2 0.5R_f \qquad (11\text{-}9)$$

where R_f is the resistive component of the forward-field impedance defined in Eq. 11-6. Similarly, the internal torque T_b of the backward field is

$$T_b = \frac{1}{\omega_s}P_{gb} \qquad (11\text{-}10)$$

where P_{gb} is the power delivered by the stator winding to the backward field, or

$$P_{gb} = I_m^2 0.5R_b \qquad (11\text{-}11)$$

where R_b is the resistive component of the backward-field impedance Z_b defined in Eq. 11-7. The torque of the backward field is in the opposite direction to that of the forward field, and therefore the net internal torque T is

$$T = T_f - T_b = \frac{1}{\omega_s}(P_{gf} - P_{gb}) \qquad (11\text{-}12)$$

Since the rotor currents produced by the two component air-gap fields are of different frequencies, the total rotor I^2R loss is the numerical sum of the losses caused by each field. In general, as shown by comparison of Eqs. 9-16 and 9-17, the rotor I^2R loss caused by a rotating field equals the slip of the field times the power absorbed from the stator, whence

$$\text{Forward-field rotor } I^2R = sP_{gf} \qquad (11\text{-}13)$$
$$\text{Backward-field rotor } I^2R = (2 - s)P_{gb} \qquad (11\text{-}14)$$
$$\text{Total rotor } I^2R = sP_{gf} + (2 - s)P_{gb} \qquad (11\text{-}15)$$

Since power is torque times angular velocity and the angular velocity of the rotor is $(1 - s)\omega_s$, the internal power P converted to mechanical form, in watts, is

$$P = (1 - s)\omega_s T = (1 - s)(P_{gf} - P_{gb}) \qquad (11\text{-}16)$$

As in the polyphase motor, the internal torque T and internal power P are not the output values because rotational losses remain to be accounted for. It is obviously correct to subtract friction and windage effects from T or P, and it is usually assumed that core losses can be treated in the same manner.

For the small changes in speed encountered in normal operation, the rotational losses are often assumed to be constant.†

EXAMPLE 11-2

A $\frac{1}{4}$-hp 110-V 60-Hz 4-pole capacitor-start motor has the following constants in ohms and losses in watts.

$$R_{1m} = 2.02 \qquad X_{1m} = 2.79 \qquad R_2 = 4.12 \qquad X_2 = 2.12 \qquad X_\varphi = 66.8$$
$$\text{Core loss} = 24 \text{ W} \qquad \text{Friction and windage} = 13 \text{ W}$$

For a slip of 0.05, determine the stator current, power factor, power output, speed, torque, and efficiency when this motor is running as a single-phase motor at rated voltage and frequency with its starting winding open.

Solution

The first step is to determine the values of the forward- and backward-field impedances at the assigned value of slip. The following relations, derived from Eq. 11-6, simplify the computations:.

$$R_f = \frac{X_\varphi^2}{X_{22}} \frac{1}{sQ_2 + 1/sQ_2} \qquad X_f = \frac{X_2 X_\varphi}{X_{22}} + \frac{R_f}{sQ_2}$$

where
$$X_{22} = X_2 + X_\varphi \qquad \text{and} \qquad Q_2 = \frac{X_{22}}{R_2}$$

Substitution of numerical values gives, for $s = 0.05$,

$$R_f + jX_f = 31.9 + j40.3 \ \Omega$$

Corresponding relations for the backward-field impedance Z_b are obtained by substituting $2 - s$ for s in these equations. When $(2 - s)Q_2$ is greater than 10, as is usually the case, less than 1 percent error results from use of the following approximate forms:

$$R_b = \frac{R_2}{2 - s} \left(\frac{X_\varphi}{X_{22}} \right)^2 \qquad X_b = \frac{X_2 X_\varphi}{X_{22}} + \frac{R_b}{(2 - s)Q_2}$$

Substitution of numerical values gives, for $s = 0.05$,

$$R_b + jX_b = 1.98 + j2.12 \ \Omega$$

---

†For a treatment of the experimental determination of motor constants and losses, see Veinott, op. cit., chap. 18.

Addition of the series elements in the equivalent circuit of Fig. 11-11c gives

$$
\begin{aligned}
R_{1m} + jX_{1m} &= 2.02 + j2.79 \\
0.5(R_f + jX_f) &= 15.95 + j20.15 \\
\underline{0.5(R_b + jX_b)} &= \underline{0.99 + j1.06} \\
\text{Input } Z = \text{sum} &= 18.96 + j24.00 = 30.6\underline{/51.7^\circ}
\end{aligned}
$$

$$
\text{Stator current } I_m = \frac{110}{30.6} = 3.59 \text{ A}
$$

$$
\text{Power factor} = \cos 51.7^\circ = 0.620
$$

$$
\text{Power input} = 110(3.59)(0.620) = 244 \text{ W}
$$

The power absorbed by forward field (Eq. 11-19) is

$$
P_{gf} = (3.59)^2(15.95) = 206 \text{ W}
$$

The power absorbed by backward field (Eq. 11-11) is

$$
P_{gb} = (3.59)^2(0.99) = 12.8 \text{ W}
$$

Internal mechanical power (Eq. 11-16) is

$$
P = 0.95(206 - 13) = 184
$$

$$
\text{Rotational loss} = 24 + 13 = 37
$$

$$
\text{Power output} = \text{difference} = 147 \text{ W} = 0.197 \text{ hp}
$$

$$
\text{Synchronous speed} = 1800 \text{ r/min} = 30 \text{ r/s}
$$

$$
\omega_s = 2\pi(30) = 188.5 \text{ rad/s}
$$

$$
\text{Rotor speed} = (1 - s)(\text{synchronous speed})
$$

$$
= 0.95(1800) = 1710 \text{ r/min}
$$

$$
= 0.95(188.5) = 179 \text{ rad/s}
$$

$$
\text{Torque} = \frac{\text{power}}{\text{angular velocity}} = \frac{147}{179} = 0.821 \text{ N} \cdot \text{m} = 0.605 \text{ lb} \cdot \text{ft}
$$

$$
\text{Efficiency} = \frac{\text{output}}{\text{input}} = \frac{147}{244} = 0.602
$$

As a check on the power bookkeeping, compute the losses:

$$
\text{Stator } I_m^2 R_{1m} = (3.59)^2(2.02) = 26.0
$$

$$
\text{Forward-field rotor } I^2R, \text{ Eq. 11-13} = 0.05(206) = 10.3
$$

$$
\text{Backward-field rotor } I^2R, \text{ Eq. 11-14} = 1.95(12.8) = 25.0
$$

$$
\text{Rotational losses} = \underline{37.0}
$$

$$
98.3 \text{ W}
$$

From input − output, total losses = 97 W, which checks within accuracy of computations.

Examination of the order of magnitude of the numerical values in Example 11-2 suggests approximations which usually can be made. These approximations pertain particularly to the backward-field impedance. Note that the impedance $0.5(R_b + jX_b)$ is only about 5 percent of the total motor impedance for a slip near full load. Consequently, an approximation as large as 20 percent of this impedance would cause only about 1 percent error in the motor current. Although, strictly speaking, the backward-field impedance is a function of slip, very little error usually results from computing its value at any convenient slip in the normal running region—say, 5 percent—and then assuming R_b and X_b to be constants. With a slightly greater approximation, the shunting effect of jX_φ on the backward-field impedance can often be neglected, whence

$$Z_b \approx \frac{R_2}{2-s} + jX_2 \tag{11-17}$$

This equation gives values of the backward-field resistance that are a few percent high, as can be seen by comparison with the exact expression given in Example 11-2. Neglecting s in Eq. 11-17 would tend to give values of the backward-field resistance that would be too low, and therefore such an approximation would tend to counteract the error in Eq. 11-17. Consequently, for small slips

$$Z_b \approx \frac{R_2}{2} + jX_2 \tag{11-18}$$

In the polyphase motor (Art. 9-5), maximum internal torque and the slip at which it occurs can easily be expressed in terms of the motor constants; the maximum internal torque is independent of rotor resistance. No such simple relations exist for the single-phase motor. The single-phase problem is much more involved because of the presence of the backward field, the effect of which is twofold: (1) it absorbs some of the applied voltage, thus reducing the voltage available for the forward field and decreasing the forward torque developed; and (2) the backward field then absorbs some of the forward-field torque. Both these effects depend on rotor resistance as well as leakage reactance. Consequently, unlike the polyphase motor, the maximum internal torque of a single-phase motor is influenced by rotor resistance; increasing the rotor resistance decreases the maximum torque and increases the slip at which maximum torque occurs.

Principally because of the effects of the backward field, a single-phase induction motor is somewhat inferior to a polyphase motor using the same rotor and the same stator core. The single-phase motor has a lower maximum torque which occurs at a lower slip. For the same torque, the single-phase motor has a higher slip and greater losses, largely because of the backward-field rotor I^2R loss. The voltampere input to the single-phase motor is greater, principally because of the power and reactive voltamperes consumed by the backward field. The stator I^2R loss also is somewhat higher in the single-phase motor, because one phase, rather than several, must carry all the current. Because of the greater losses, the efficiency is lower, and the temperature rise for the same torque is higher. A larger frame size must be used for a single-phase motor than for a polyphase motor of the same power and speed rating. Because of the larger frame size, the maximum torque can be made comparable with that of a physically smaller but equally rated polyphase motor. In spite of the larger frame size and the necessity for auxiliary starting arrangements, general-purpose single-phase motors in the standard fractional-horsepower ratings cost approximately the same as correspondingly rated polyphase motors because of the much greater volume of production of the former.

11-4 UNBALANCED OPERATION OF SYMMETRICAL TWO-PHASE MACHINES; THE SYMMETRICAL-COMPONENT CONCEPT

Unbalanced operation of an induction motor occurs either when the voltages applied to the stator do not constitute a balanced polyphase set or when the stator or rotor windings are not symmetrical with respect to the phases. The single-phase motor is the extreme case of a motor operating under unbalanced stator-voltage conditions. In some cases, unbalanced voltages are produced in the supply network to a motor, e.g., when a line fuse is blown. In other cases, unbalanced voltages are produced by the starting impedances of single-phase motors, as described in Art. 11-2. An important use of unbalanced voltages is made in the control of 2-phase servomotors, as described in Art. 11-5. The purpose of this article is to develop the symmetrical-component theory of 2-phase induction motors from the double-revolving-field concept and to show how the theory can be applied to a variety of problems involving induction motors having two stator windings in space quadrature.

First consider in review what happens when balanced 2-phase voltages are applied to the stator terminals of a 2-phase machine having a uniform air gap, a symmetrical polyphase or cage rotor, and two identical stator windings a and m in space quadrature. The stator currents are equal in magnitude and in time quadrature. When the current in winding a has its instantaneous maximum, the current in winding m is zero and the stator-mmf wave is centered on the axis of winding a. Similarly, the stator-mmf wave is centered on the axis of winding m at the instant when the current in winding m has its instantaneous maximum. The stator-mmf wave therefore travels 90 electrical degrees in space in an interval of 90° in time, the direction of its travel depend-

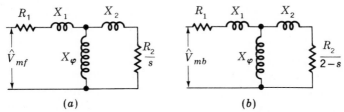

Fig. 11-12. Equivalent circuits for a 2-phase motor under unbalanced conditions: (a) forward field; (b) backward field.

ing on the phase sequence of the currents. A more complete analysis in the manner of Art. 3-5 proves that the traveling wave has constant amplitude and constant angular velocity. This fact is, of course, the basis of the whole theory of balanced operation of induction machines.

The behavior of the motor for balanced 2-phase applied voltages of either phase sequence can readily be determined. Thus, if the rotor is turning at a per unit speed n in the direction from winding a toward winding m, the terminal impedance per phase is given by the equivalent circuit of Fig. 11-12a when the applied voltage v_a leads the applied voltage v_m by 90°. Throughout the rest of this treatment, this phase sequence will be called *positive sequence* and will be designated by subscript f since positive-sequence currents result in a forward field. With the rotor still forced to run at the same speed and in the same direction, the terminal impedance per phase is given by the equivalent circuit of Fig. 11-12b when v_a lags v_m by 90°. This phase sequence will be called *negative sequence* and will be designated by subscript b, since negative-sequence currents produce a backward field.

Suppose now that *two* balanced 2-phase voltage sources *of opposite phase sequence* are connected in series and applied simultaneously to the motor, as indicated in Fig. 11-13a, where phasor voltages $\widehat{V}_{mf}$ and $j\widehat{V}_{mf}$ applied, respectively, to windings m and a, form a balanced system of positive sequence and phasor voltages $\widehat{V}_{mb}$ and $-j\widehat{V}_{mb}$ form another balanced system but of negative sequence. The resultant voltage $\widehat{V}_m$ applied to winding m is, as a phasor,

$$\widehat{V}_m = \widehat{V}_{mf} + \widehat{V}_{mb} \tag{11-19}$$

and that applied to winding a is

$$\widehat{V}_a = j\widehat{V}_{mf} - j\widehat{V}_{mb} \tag{11-20}$$

If, for example, the forward, or positive-sequence, system is given by the phasors $\widehat{V}_{mf}$ and $j\widehat{V}_{mf}$ in Fig. 11-13b and the backward, or negative-sequence, system is given by the phasors $\widehat{V}_{mb}$ and $-j\widehat{V}_{mb}$ then the resultant voltages are given by the phasors $\widehat{V}_m$ and $\widehat{V}_a$. An unbalanced 2-phase system of applied voltages $\widehat{V}_m$ and $\widehat{V}_a$ has thus been synthesized by combining two symmetrical systems of opposite phase sequence.

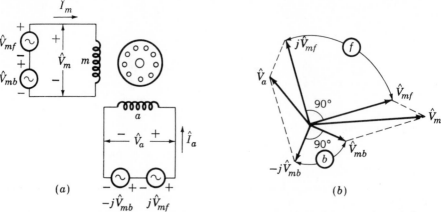

Fig. 11-13. Synthesis of an unbalanced 2-phase system from the sum of two balanced systems of opposite phase sequence.

The symmetrical component systems are, however, much easier to work with than their unbalanced resultant system. Thus, it is easy to compute the component currents produced by each symmetrical component system of applied voltages because the induction motor operates as a balanced 2-phase motor for each component system. By superposition, the actual current in a winding then is the sum of its components. Thus, if $\widehat{I}_{mf}$ and $\widehat{I}_{mb}$ are, respectively, the positive- and negative-sequence component phasor currents in winding m, then the corresponding positive- and negative-sequence component phasor currents in winding a are, respectively, $j\widehat{I}_{mf}$ and $-j\widehat{I}_{mb}$ and the actual winding currents $\widehat{I}_m$ and $\widehat{I}_a$ are

$$\widehat{I}_m = \widehat{I}_{mf} + \widehat{I}_{mb} \tag{11-21}$$

$$\widehat{I}_a = j\widehat{I}_{mf} - j\widehat{I}_{mb} \tag{11-22}$$

The inverse operation of finding the symmetrical components of specified voltages or currents must often be performed. Solution of Eqs. 11-19 and 11-20 for the phasor components $\widehat{V}_{mf}$ and $\widehat{V}_{mb}$ in terms of known phasor voltages $\widehat{V}_m$ and $\widehat{V}_a$ gives

$$\widehat{V}_{mf} = \tfrac{1}{2}(\widehat{V}_m - j\widehat{V}_a) \tag{11-23}$$

$$\widehat{V}_{mb} = \tfrac{1}{2}(\widehat{V}_m + j\widehat{V}_a) \tag{11-24}$$

These operations are illustrated in the phasor diagram of Fig. 11-14. Obviously, similar relations give the phasor symmetrical components $\widehat{I}_{mf}$ and $\widehat{I}_{mb}$ of the current in winding m in terms of specified phasor currents $\widehat{I}_m$ and $\widehat{I}_a$ in the 2 phases; thus

$$\widehat{I}_{mf} = \tfrac{1}{2}(\widehat{I}_m - j\widehat{I}_a) \tag{11-25}$$

$$\widehat{I}_{mb} = \tfrac{1}{2}(\widehat{I}_m + j\widehat{I}_a) \tag{11-26}$$

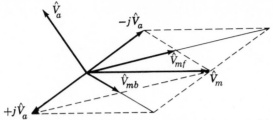

Fig. 11-14. Resolution of unbalanced 2-phase voltages into symmetrical components.

EXAMPLE 11-3

The equivalent-circuit constants of a 5-hp 220-V 60-Hz 2-phase squirrel-cage induction motor in ohms per phase are

$$R_1 = 0.534 \quad X_1 = 2.45 \quad X_\varphi = 70.1 \quad R_2 = 0.956 \quad X_2 = 2.96$$

This motor is operated from an unbalanced 2-phase source whose phase voltages are, respectively, 230 and 210 V, the smaller voltage leading the larger by 80°. For a slip of 0.05, find (a) the positive- and negative-sequence components of the applied voltages, (b) the positive- and negative-sequence components of the stator phase currents, (c) the effective values of the phase currents, and (d) the internal mechanical power.

Solution

(a) Let $\widehat{V}_m$ and $\widehat{V}_a$ denote the voltages applied to the two phases, respectively. Then

$$\widehat{V}_m = 230\underline{/0°} = 230 + j0 \text{ V}$$
$$\widehat{V}_a = 210\underline{/80°} = 36.4 + j207 \text{ V}$$

From Eqs. 11-23 and 11-24 the forward and backward components of voltages are, respectively,

$$\widehat{V}_{mf} = \tfrac{1}{2}(230 + j0 + 207 - j36.4) = 218.5 - j18.2 = 219.5\underline{/-4.8°} \text{ V}$$
$$\widehat{V}_{mb} = \tfrac{1}{2}(230 + j0 - 207 + j36.4) = 11.5 + j18.2 = 21.5\underline{/57.7°} \text{ V}$$

(b) As in Example 11-2, the forward-field impedance is, for a slip of 0.05,

$$Z_f = 16.46 + j7.15 \ \Omega$$
$$R_1 + jX_1 = \underline{0.53 + j2.45 \ \Omega}$$
$$16.99 + j9.60 = 19.50\underline{/29.4°} \ \Omega$$

Hence, the forward component of stator current is

$$\widehat{I}_{mf} = \frac{219\,/-4.8°}{19.50\,/29.4°} = 11.26\,/-34.2°\ \text{A}$$

For the same slip, again as in Example 11-2, the backward-field impedance is

$$Z_b = 0.451 + j2.84\ \Omega$$
$$R_1 + jX_1 = 0.534 + j2.45\ \Omega$$
$$\overline{\hphantom{R_1 + jX_1 =}}\ 0.985 + j5.29 = 5.38\,/79.5°\ \Omega$$

Hence, the backward component of stator current is

$$\widehat{I}_{mb} = \frac{21.5\,/57.7°}{5.38\,/79.5°} = 4.00\,/-21.8°\ \text{A}$$

(c) By Eqs. 11-21 and 11-22 the currents in the two phases are, respectively,

$$\widehat{I}_m = 13.06 - j7.79 = 15.2\,/-31°\ \text{A}$$
$$\widehat{I}_a = 4.81 + j5.64 = 7.40\,/49.2°\ \text{A}$$

Note that the currents are much more unbalanced than the applied voltages. Even though the motor is not overloaded insofar as shaft load is concerned, the losses are appreciably increased by the current unbalance and the stator winding with the greatest current may overheat.

(d) The power delivered to the forward field by the two stator phases is

$$P_{gf} = 2I_{mf}^2 R_f = 2(126.8)(16.46) = 4175\ \text{W}$$

and the power delivered to the backward field is

$$P_{gb} = 2I_{mb}^2 R_b = 2(16.0)(0.451) = 15\ \text{W}$$

Thus, according to Eq. 11-16, the internal mechanical power developed is

$$P = 0.95(4175 - 15) = 3950\ \text{W}$$

If the core losses, friction and windage, and stray load losses are known, the shaft output can be found by subtracting them from the internal power. The friction and windage losses depend solely on the speed and are the same as they would be for balanced operation at the same speed. The core and stray load losses, however, are somewhat greater than they would be for balanced operation with the same positive-sequence voltage and current. The increase is caused principally by the $(2 - s)$-frequency core and stray losses in the rotor caused by the backward field.

11–5 TWO–PHASE CONTROL MOTORS

Control systems utilize fractional-horsepower electromagnetic components as motors to drive the loads and as sensors to measure speed and position of the controlled elements. In this article we describe the operation of a control motor, which operates as an unbalanced 2-phase induction motor and is suitable for control systems up to a few hundred watts. In Art. 11-7 we shall describe the operation of stepper motors, which are suitable for digital control systems. In Arts. 11-8 and 11-9 we shall describe the operation of speed and position sensors. More information on these and other control-system components can be found in the technical literature.

A schematic diagram of a 2-phase control motor is given in Fig. 11-15. Phase m of the motor is the *fixed,* or *reference, phase.* The voltage $\hat{V}_m$ is a fixed voltage applied from a constant-voltage constant-frequency source. Phase a is the *control phase.* The voltage $\hat{V}_a$ is supplied from an amplifier, usually of magnetic-amplifier or solid-state construction, and has an amplitude proportional to the input control signal. The voltages $\hat{V}_m$ and $\hat{V}_a$ must be in synchronism, which means that they must be derived from the same ultimate ac source. They must also be made to be approximately in time quadrature either by introducing a 90° phase shift in the amplifier or by connecting a suitable capacitor in series with the reference phase m. The amplifier derives its power from the same ac source that supplies the reference phase, so that the amplifier output voltage $\hat{V}_a$ is a modulated ac wave having its fundamental component at the same frequency as $\hat{V}_m$. When $\hat{V}_a$ has a nonzero value and

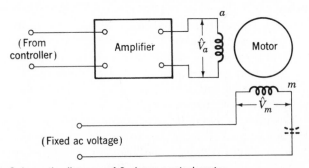

Fig. 11-15. Schematic diagram of 2-phase control motor.

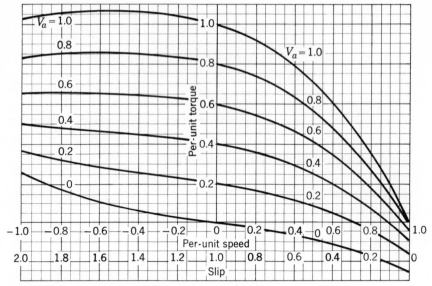

Fig. 11-16. Typical torque-speed curves of 2-phase control motor.

its phase leads $\widehat{V}_m$ by approximately 90°, rotation in one direction is obtained; when $\widehat{V}_a$ lags $\widehat{V}_m$, rotation in the other direction results. Since the torque is a function of both $\widehat{V}_m$ and $\widehat{V}_a$, changing the magnitude of $\widehat{V}_a$ changes the developed torque of the motor.

Typical torque-speed curves for a 2-phase control motor are given in Fig. 11-16 for a series of values of control voltage and unity reference voltage. They are based on a motor having identical 2-phase stator windings and on negligible source impedances.

The requirements of a control system very nearly specify the shape of the torque-speed characteristic of a 2-phase induction motor suitable for such use. As with any motor for such service, the torque should be high at speeds near zero, and the slope of the torque-speed characteristic should be negative in the normal operating range around zero speed in order to provide a stabilizing feature for the control system. Both these requirements can be met by a high-resistance rotor designed so that maximum torque is developed at a reverse speed of approximately one-half synchronous speed, as shown by the torque-speed characteristic labeled $V_a = 1.0$ in Fig. 11-16. Normal operation near zero speed is then in the stable region to the right of the maximum-torque points. A further requirement is that the motor must not tend to run as a single-phase motor when the error signal is zero. It can be shown that this requirement also is met by use of a high-resistance rotor.

Although the squirrel-cage induction motor prefers to run at a small slip and therefore is most readily adaptable to constant-speed drives, this type of motor has other features which are sufficiently attractive for it to be used extensively in low-power control systems. The ruggedness and simplicity of

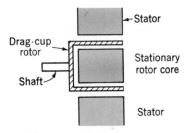

Fig. 11-17. Cross section of drag-cup rotor.

the cage rotor are a great advantage, for both economic and technical reasons. There are no brushes riding on sliding contacts and requiring inspection and maintenance, as in other types of motors. Because the rotor windings do not require insulation, the rotor temperature is limited only by mechanical considerations and indirectly by its effect on the stator-winding temperature. If suitable means are provided for cooling the stator windings, higher rotor losses can be tolerated than in other types of motors. Because there is relatively little inactive material, the inertia of a squirrel-cage motor can be made less than that of a correspondingly rated dc motor. When the maximum power output is below a few watts, inertia can be minimized by using a thin metallic cup as the rotor. As shown in the simplified sketch of Fig. 11-17, the rotating member is then like a can with one end removed. A stationary iron core, like a plug inside the cup, completes the magnetic circuit. The construction is known as a *drag-cup rotor*.

The principal disadvantage of the 2-phase control motor is the inherent inefficiency of a squirrel-cage induction motor running at a large slip. As pointed out in Art. 9-4, the efficiency of a polyphase induction motor with short-circuited rotor windings is like that of a slipping mechanical clutch; the slip is a direct measure of the rotor losses. The inherent limit on efficiency will be even lower in a speed-control scheme which involves unbalancing the applied voltages, because of the decrease in output and increase in rotor losses.

EXAMPLE 11-4

It may be noted from Fig. 11-16 that in the low-speed range the torque of a 2-phase control motor is linearly proportional to the rms control-winding voltage. In this range, the torque-speed curves have a negative slope which is approximately constant and substantially the same for the lower values of control voltage. Under these conditions, the torque-speed curves can be represented empirically by the relation

$$T = k_1 V_a - k_2 \omega_0$$

where ω_0 is the shaft mechanical angular velocity and k_1 and k_2 are constants determined from the curves and the units involved.

The motor is driving a load requiring a viscous-friction torque $f_v \omega_0$ proportional to speed. The combined inertia of rotor and load is J. Determine the transfer functions relating (a) motor-shaft position and control voltage and (b) motor-shaft velocity and control voltage.

Solution

(a) The signal to the motor consists of variation of the rms value V_a of the control-winding voltage. Let the value of this voltage at any time t be v_s. The electrodynamic differential equation in terms of the shaft position angle θ_0 is then

$$J \frac{d^2\theta_0}{dt^2} + f_v \frac{d\theta_0}{dt} = k_1 v_s - k_2 \frac{d\theta_0}{dt}$$

In terms of the phasor amplitudes $\widehat{\Theta}_0$ and $\widehat{V}_s$ of sinusoidal variations at the angular frequency ω, this equation becomes

$$(j\omega)^2 J \widehat{\Theta}_0 + j\omega(f_v + k_2)\widehat{\Theta}_0 = k_1 \widehat{V}_s$$

which yields the transfer function

$$\frac{\widehat{\Theta}_0}{\widehat{V}_s} = \frac{k_1}{(j\omega)^2 J + j\omega(f_v + k_2)}$$

(b) The differential equation in terms of shaft angular velocity ω_0 is

$$J \frac{d\omega_0}{dt} + f_v \omega_0 = k_1 v_s - k_2 \omega_0$$

Then

$$j\omega J \widehat{\Omega}_0 + (f_v + k_2)\widehat{\Omega}_0 = k_1 \widehat{V}_s$$

or the transfer function is

$$\frac{\widehat{\Omega}_0}{\widehat{V}_s} = \frac{k_1}{f_v + k_2 + (j\omega)J}$$

The characteristics of the 2-phase control motor can be determined by test on a dynamometer capable of driving it in the slip range from 1.0 to 2.0 and loading it from slip of 1.0 to zero. The characteristics can also be calculated using the symmetrical-component method of Art. 11-4. The analysis is greatly simplified if the currents and voltages are considered to be sinusoidal, if the effects of the source impedances are neglected, and if the motor is assumed to have identical 2-phase stator windings. The motor then is simply a symmetrical 2-phase motor operating from an unbalanced 2-phase source.

11–6 SERIES UNIVERSAL MOTORS

A series motor has the convenient ability to run on either alternating or direct current and with similar characteristics, provided both stator and rotor cores are laminated. Such a single-phase series motor therefore is commonly called a *universal motor.* The torque angle is fixed by the brush position and is normally at its optimum value of 90°. If alternating current is supplied to a series motor, the stator and rotor field strengths will vary in exact time phase. Both will reverse at the same instant, and consequently the torque will always be in the same direction, though pulsating in magnitude at twice line frequency. Average torque will be produced, and the performance of the motor will be generally similar to that with direct current.

Small universal motors are used where light weight is important, as in vacuum cleaners, kitchen appliances, and portable tools, and usually operate at high speeds (1500 to 15,000 r/min). Typical characteristics are shown in Fig. 11-18. The ac and dc characteristics differ somewhat for two reasons: (1) with alternating current, reactance-voltage drops in the field and armature absorb part of the applied voltage, and therefore for a specified current and torque the rotational counter emf generated in the armature is less than with direct current and the speed tends to be lower; (2) with alternating current, the magnetic circuit may be appreciably saturated at the peaks of the current wave, and the rms value of the flux may thus be appreciably less with alternating current than with the same rms value of direct current; the torque therefore tends to be less and the speed higher with alternating than with direct current. The universal motor provides the highest horsepower per dollar in the fractional-horsepower range, at the expense of noise, relatively short life, and high speed.

To obtain control of the speed and torque of the series motor, the ac voltage may be applied in series with a solid-state device which controls the

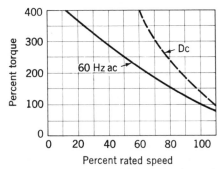

Fig. 11-18. Typical torque-speed characteristics of a series universal motor.

voltage applied to the motor. A thyristor (Art. 6-5) can control alternate half cycles of the voltage so that the motor carries unidirectional current. A bidirectional switch (Triac) controls both half cycles of the applied voltage. These control circuits are similar to those described in Art. 6-6. The firing angle can be manually adjusted, as in a trigger-controlled electric drill, or can be controlled by a speed-control circuit, as in some portable tools and appliances. The combination of solid-state device and series motor provides an economical, controllable motor package.

EXAMPLE 11-5

Show that the time-average torque of an ac series motor is proportional to the square of the rms current when the magnetic structure is not saturated.

Solution

The instantaneous magnetic flux φ produced by the motor current i_a in the field poles is

$$\varphi = K_f i_a$$

The instantaneous torque is $K_t \varphi i_a$; hence the time-average torque T_{av} is

$$T_{av} = \frac{\omega}{\pi} \int_0^{\pi/\omega} K_t K_f i_a^2 \, dt = K_t K_f I_a^2$$

where I_a is the rms value of the current i_a. This relationship is independent of the waveform of the current i_a.

11–7 STEPPER MOTORS

The stepper motor is a form of synchronous motor which is designed to rotate a specific number of degrees for each electrical pulse received by its control unit. Typical steps are 2, 2.5, 5, 7.5, and 15° per pulse. The stepper motor is used in digital control systems, where the motor receives open-loop commands as a train of pulses to turn a shaft or move a plate by a specific distance. A typical application for the motor is positioning a worktable in two dimensions for automatic drilling in accordance with hole-location instructions on tape. With a stepper motor a position sensor and feedback system is not normally required to make the output member follow input instructions. Stepper motors are built to follow signals as rapid as 1200 pulses per second and with equivalent power ratings up to several horsepower.

Stepper motors are usually designed with a multipole, multiphase stator

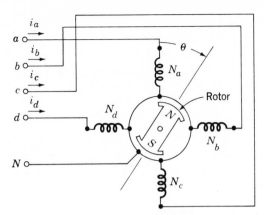

Fig. 11-19. Elementary diagram of 4-phase stepper motor.

winding that is not unlike the windings of conventional machines. They typically use 3- and 4-phase windings, the number of poles being determined by the required angular change per input pulse. The rotors are either of the variable-reluctance or permanent-magnet type. Stepper motors operate with an external drive logic circuit; as a train of pulses is applied to the input of the drive circuit, the circuit delivers appropriate currents to the stator windings of the motor to make the axis of the air-gap field step around in coincidence with the input pulses. Depending upon the pulse rate and the load torque, including inertia effects, the rotor follows the axis of the air-gap magnetic field by virtue of the reluctance torque and/or the permanent-magnet torque.

The elementary operation of a 4-phase stepper motor with a 2-pole rotor is shown in the sequence of Fig. 11-19. The rotor can be either a ferromagnetic element or a permanent magnet. The rotor assumes the angles $\theta = 0, 45, 90°, \ldots$ as the windings are excited in the sequence N_a, $N_a + N_b$, $N_b, \ldots$. The stepper motor of Fig. 11-19 can also be used for 90° steps by exciting the coils singly. In the latter case, only the permanent-magnet rotor can be used. The torque-angle curves for the two types of rotors are shown in Fig. 11-20; whereas the permanent-magnet rotor has its peak torque when the excitation is shifted 90°, the ferromagnetic rotor has zero torque and can move in either direction. The permanent-magnet rotor has the additional feature that the rotor position θ is defined by the winding currents with no ambiguity, whereas the ferromagnetic rotor has two possible positions for each winding-current pattern. Winding patterns can be visualized for steps of 22.5°, 11.25°, and smaller, per pulse to the input circuit.

To obtain relatively small steps of angle, a differential construction, as shown in Fig. 11-21 can be used. The stator has a 2-phase winding, while the rotor has five projecting poles. The position shown is for current in winding N_a. If the current is now transferred to winding N_b, the rotor will turn by $\theta = 90° - 72° = 18°$ to align pole 2 with the N_b axis. The rotor can be ferromagnetic or can utilize permanent-magnet construction. The permanent

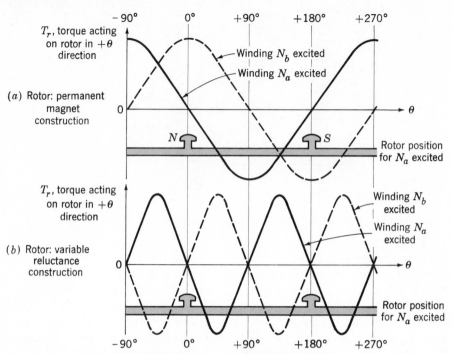

Fig. 11-20. Torque-displacement curves for winding pattern of stepper motor of Fig. 11-19: (a) permanent-magnet rotor; (b) variable-reluctance rotor.

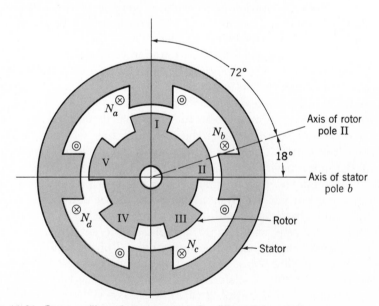

Fig. 11-21. Cross section of stepper motor for differential operation.

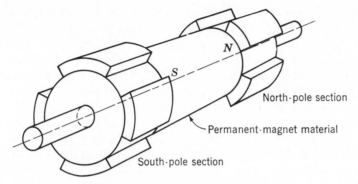

Fig. 11-22. Rotor of stepper motor with axial permanent magnet.

magnet is usually placed in the axial position as shown in Fig. 11-22 with ferromagnetic end plates to form the poles. Such construction tends to reduce the reluctance torque and makes the motor smoothly responsive to the winding currents. It also provides a uniform magnetic loading on the permanent-magnet material.†

The characteristics of a stepper motor are frequently presented as the torque versus stepping rate of the pulses applied to the drive unit, as shown in Fig. 11-23. As the stepping rate is increased, the motor can provide less torque because the rotor has less time to drive the load from one position to the next as the stator-winding-current pattern is shifted. The start range of Fig. 11-23 is that in which the load position follows the pulses without losing steps. The slew range is that in which the load velocity follows the pulse rate without losing steps but cannot start, stop, or reverse on command. The maximum

†See A. E. Snowden and E. W. Madsen, Characteristics of a Synchronous Inductor Motor, *IEEE Trans. Appl. Ind.*, **81**:1–5 (1962).

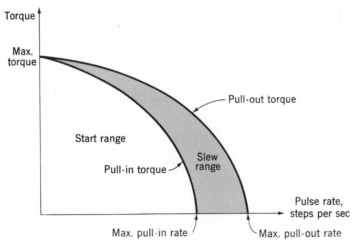

Fig. 11-23. Stepper-motor characteristics of torque versus pulse rate.

torque point is the maximum holding torque of the excited motor to a steady load. At light loads, the maximum slew rate can be as much as 10 times the position response rate.

The advantage of the stepper motor is the smaller size and lower cost of the motor-drive unit package compared with the corresponding parts of a proportional position or velocity servo system. Typical variable-reluctance stepper motors operate at small steps, 15° or less, or at maximum position response rates up to 1200 pulses per second. Typical permanent-magnet types operate at larger steps, up to 90°, and at maximum response rates of 300 pulses per second. Applications include table positioning for machine tools, line printers, tape drives, recorder pen drives, and X-Y plotters; very small stepper motors are used to feed paper out of hand-held printing calculators.†

11–8 AC TACHOMETERS

For feedback control systems, it is frequently necessary to measure the angular velocity of a shaft, and it is often desirable that this measure be in the form of an alternating voltage of constant frequency. A small 2-phase induction motor can be used for this purpose. The connections are shown in Fig. 11-24. Winding m, often referred to as the *fixed field* or *reference field,* is energized from a suitable alternating voltage of constant magnitude and frequency. A voltage of the same frequency is then generated in the auxiliary winding, or *control field a.* This voltage is applied to the high-impedance input circuit of an amplifier, and therefore winding a can be considered as open-circuited. The electrical requirements are, ideally, that the magnitude of the signal voltage generated in winding a be linearly proportional to the speed and that the phase of this voltage be fixed with respect to the applied voltage V_m.

The operation of the ac tachometer can be visualized in terms of the double-revolving-field theory of Art. 11-3.‡ As viewed from the reference

†For a description of various types of stepper motors see Stepper Motor and Controls, *Electromech. Des.*, 107–119 (1969). An extensive bibliography is given by D. J. Robinson and C. K. Taft, A Dynamic Analysis of Magnetic Stepping Motors, *IEEE Trans. Ind. Electron. Control Instrum.*, **16**(2):111–125 (1969).

‡For a quantitative analysis, see R. H. Frazier, Analysis of the Drag-Cup AC Tachometer, *Trans. AIEE*, **70**(1951); also H. H. Woodson and J. R. Melcher, "Electromechanical Dynamics, part II, Fields, Forces and Motion," chap. 7, Wiley, New York, 1968.

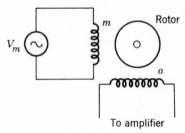

Fig. 11-24. Schematic diagram of a 2-phase tachometer.

winding m, the tachometer is equivalent to a small single-phase induction motor, and the equivalent circuit of Fig. 11-11c therefore applies to conditions as viewed from this winding. The voltages across the impedances $0.5Z_f$ and $0.5Z_b$ in Fig. 11-11c are the voltages generated in winding m by the forward and backward flux waves, respectively. These flux waves also generate voltages in the auxiliary winding a. If the ratio of effective turns in winding a to effective turns in winding m is a, the voltages generated in winding a are a times the corresponding voltages generated in winding m. By effective turns is meant the number of turns corrected for the effects of winding distribution insofar as fundamental space distributions of flux and mmf are concerned. If the direction of rotation is such that the forward field revolves past winding a a quarter cycle in time before it passes winding m, then the voltage $\widehat{E}_{af}$ generated by the forward field in winding a leads the corresponding voltage $\widehat{E}_{mf}$ generated in winding m by 90°, or, as phasors,

$$\widehat{E}_{af} = ja\widehat{E}_{mf} = ja\widehat{I}_m 0.5Z_f \tag{11-27}$$

where $\widehat{I}_m$ is the phasor current in winding m and is determined by the equivalent circuit of Fig. 11-11c. The backward field revolves in the opposite direction, and therefore the voltage $\widehat{E}_{ab}$ generated by it in winding a lags the corresponding voltage $\widehat{E}_{mb}$ generated in winding m by 90°, or

$$\widehat{E}_{ab} = -ja\widehat{E}_{mb} = -ja\widehat{I}_m 0.5Z_b \tag{11-28}$$

The total voltage $\widehat{E}_a$ generated in winding a is the sum of the components generated by each field, or

$$\widehat{E}_a = ja\widehat{I}_m 0.5(Z_f - Z_b) \tag{11-29}$$

At standstill, the forward and backward fields are equal, and no voltage is generated in winding a. When the rotor is revolving, however, the impedance of the forward field increases while that of the backward field decreases, the difference between them being a function of the speed. The voltage generated in winding a is therefore a function of speed. Reversal of the direction of rotation reverses the phase of the auxiliary-winding voltage.

The shapes of the curves of voltage magnitude and phase angle as functions of speed depend on the speed range and tachometer constants, primarily on the rotor self-reactance-to-resistance ratio Q_2. It can be shown that either a low-Q_2 rotor (X_{22}/R_2 less than about 0.1) or a high-Q_2 rotor (X_{22}/R_2 greater than about 10) will provide nearly a constant phase angle and nearly a linear relation between the auxiliary-winding voltage and speed. The sensitivity in volts per revolution per minute is sacrificed if a low-Q_2 rotor is used, but the linear speed range is wide. On the other hand, if a high-Q_2 rotor is used, the speed range around zero speed is limited to a fairly small fraction of synchronous speed when the requirements for linearity of voltage and constancy of phase angle are strict. These restrictions on rotor Q_2 should not be taken too literally, however, since satisfactory performance can be obtained with inter-mediate values of Q_2 if the requirements for linearity of voltage and constancy of phase angle are not too severe.

In common with other measuring instruments, the ac tachometer should have as little effect as possible on the system into which it is inserted. In other words, its torque should be small compared with other torques acting in the system, and its inertia should be small when rapid speed variations are en-countered, as in automatic control systems. To minimize the inertia, ac tach-ometers are often built with a thin, metallic drag-cup rotor like that shown in the simplified sketch of Fig. 11-17. Because of the relatively long air gap, this construction inherently gives a fairly low Q_2, which can be made still lower, if desired, by making the drag cup of high-resistivity material.

Ac tachometers require precise workmanship and care in design and as-sembly in order to maintain concentricity and to eliminate direct coupling through the leakage fluxes between the excited winding and the output wind-ing. Such coupling would result in signal voltage at zero speed. Sometimes soft-iron shields are provided to minimize pickup from stray fields. Frequently ac tachometers are used in 400-Hz systems.

11–9 SYNCHROS AND CONTROL TRANSFORMERS

Synchros are used in control systems for transmitting shaft-position informa-tion, for maintaining synchronism between two or more shafts, and for per-forming arithmetic operations with angular information. The designation *sel-syn* is also used from the combination "self-synchronous." A control transformer is a particular type of synchro that provides an electric error signal proportional to the deviation of its shaft from the synchronous posi-tion; it is widely used in position-feedback control systems. Synchros are gen-erally constructed like miniature versions of ac synchronous machines; they have three distributed stator windings and a 2-pole single-phase rotor winding.

A basic arrangement of two synchros in which the two shafts maintain synchronism is shown in Fig. 11-25. In most respects, the construction of the *synchro generator,* or *transmitter,* is similar to that of the *synchro motor,* or *receiver.* Both have a single-phase winding (usually on the rotor) connected to a common ac voltage source. On the other member (usually the stator), both have three windings with axes 120° apart and connected in Y; these windings on the transmitter and motor have their corresponding terminals connected together. When the single-phase rotor windings are excited, voltages are in-duced by transformer action in the Y-connected stator windings. If the two rotors are in the same space position relative to their stator windings, the transmitter and motor stator-winding voltages are equal, no current circu-lates in these windings, and no torque is transmitted. If, however, the two rotor space positions do not correspond, the stator-winding voltages are un-

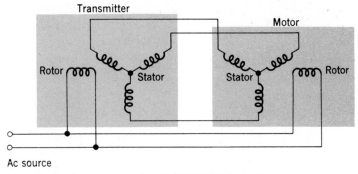

Fig. 11-25. Single-phase synchro-transmitter-motor system.

equal and currents circulate in the stator windings. These currents, in conjunction with the air-gap magnetic fields, produce torques tending to align the two rotors.

Mechanically, synchros have the same general construction features as small motors. The rotor structure of a synchro motor can be seen in Fig. 11-26. The rotor and stator are laminated, and ball bearings are used to minimize friction. Dampers are used to force the motors to settle quickly at their angular positions.

The motor torque at standstill or for slow rotation can be shown to depend closely upon the sine of the relative angular difference in position of the transmitter and motor shafts. Torque gradients developed by electrically identical transmitters and motors interconnected as in Fig. 11-25 range from $0.07 \text{ in} \cdot \text{oz/deg}$ for the smaller units to $1.75 \text{ in} \cdot \text{oz/deg}$ for the larger units.

A modification of the synchro system of Fig. 11-25 can be introduced by including a *differential synchro,* permitting the rotation of a shaft to be a function of the sum or difference of the rotation of two other shafts. In Fig. 11-27 the differential synchro acts as a differential transmitter. The voltages impressed on its stator windings induce corresponding voltages in the rotor windings. The relative magnitudes of the three rotor voltages are the same as would exist if the differential were removed and the transmitter turned through an angle equal to the sum or difference of the transmitter and differential angles. Such differential transmitters usually have a bank of three ca-

Fig. 11-26. Two dampers (front and rear) and wound rotor, with damper, for synchro motor. (*General Electric Company.*)

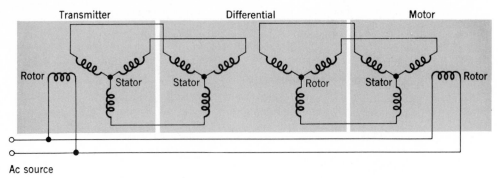

Fig. 11-27. Synchro generator-motor system with differential.

pacitors connected across the primary terminals to improve power factor and hence minimize the possibility of overheating in the system. Alternatively, the differential may be used as a motor supplied from two separate synchro transmitters and producing a rotation dependent upon the sum or difference of the two transmitter rotations.

The maximum static error for a system consisting of a transmitter and a single motor of the same size is of the order of 1° and is caused largely by friction in the motor bearings. The error increases as additional synchros are added or as the line impedance between tansmitter and motor becomes appreciable. Dynamic errors, created by mechanical oscillation of the motor shaft about the correct position, may be 2 or 3 times the static errors. To minimize dynamic errors, mechanical dampers are built into the rotors of motor units, as shown in Fig. 11-26. When the motor is called upon to supply significant torque, the error increases because of the need of a definite angular displacement between transmitter and motor shafts for torque transmission. This fact, together with heating of the synchro equipment, definitely limits the torque magnitudes and makes the use of a feedback control system operated from a control transformer more desirable.

The basic circuit by which a control transformer produces an error signal proportional to rotor-angle deviation is shown in Fig. 11-28. Two interconnected units are again involved, one a transmitter and the other a very similar unit called a *control transformer*. The rotor of the synchro transmitter is excited from a single-phase source, producing a magnetic field in the transmitter and voltages in the stator windings of both the transmitter and the control transformer. If the voltage drops caused by exciting current are neglected, the induced voltages in the two stator windings must be equal. Therefore, the distribution of flux about the control-transformer stator must be similar to that about the transmitter stator. The effect is consequently the same as if the two rotor windings were on the same magnetic circuit and arranged so that their axes could be given any arbitrary displacement angle in space. The arrangement is thus the equivalent of an adjustable mutual inductance between the two rotor windings, but with the added feature that geographical

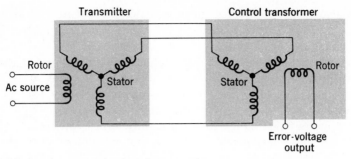

Fig. 11-28. Synchro generator-transformer system.

separation of the two windings is possible. When the angle is 90 electrical degrees, corresponding to a 90-electrical-degree displacement of the two shafts, no voltage is induced in the transformer rotor; this displacement is the equilibrium position of the two shafts. When the angle has any value except 90 and 270°, a voltage is induced in the transformer rotor. The magnitude of the voltage is a function of the angular discrepancy between the two shafts, and the instantaneous polarity depends on the direction of the displacement. In an actual control system, the control transformer operates with an error angle of a fraction of a degree. In this range the error voltage is proportional to error angle. Differential synchros can also be incorporated between the transmitters and control transformers of these systems.

Figure 11-29, for example, illustrates an application to control of the angular position of an output shaft in accordance with an input shaft. The rotor of the synchro transmitter is mechanically connected to the input shaft. The rotor of the control transformer is mechanically connected to the output

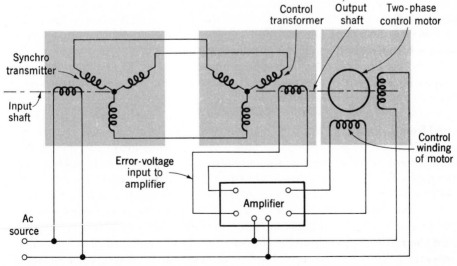

Fig. 11-29. Application of synchro transmitter and transformer to angular position control.

shaft and electrically connected to the input of a servo amplifier. Mechanical power to turn the output shaft and its associated load is furnished by a 2-phase control motor. The input to the control winding of the motor is supplied by the amplifier, which includes phase-splitting capacitors in its circuitry. When the output shaft is in the correct position, 90° from the input shaft position, the voltage input to the amplifier, and hence the power input to the control winding or the motor, is zero, and the motor does not turn. When an angular discrepancy exists, a definite error voltage appears at the amplifier input. Its relative polarity is such that the motor is caused to turn in the direction to correct the angular discrepancy.

11–10 SUMMARY

One theme of this chapter is a continuation of the induction-machine theory of Chap. 9 and its application to the single-phase induction motor. This theory is expanded by a step-by-step reasoning process from the simple revolving-field theory of the symmetrical polyphase induction motor. The basic concept is the resolution of the stator-mmf wave into two constant-amplitude traveling waves revolving around the air gap at synchronous speed in opposite directions. If the slip for the forward field is s, then that for the backward field is $2 - s$. Each of these component fields produces induction-motor action, just as in a symmetrical polyphase motor. From the viewpoint of the stator, the reflected effects of the rotor can be visualized and expressed quantitatively in terms of simple equivalent circuits. The ease with which the internal reactions can be accounted for in this manner is the essential reason for the usefulness of the double-revolving-field theory.

For a single-phase winding, the forward- and backward-component mmf waves are equal, and their amplitude is half the maximum value of the peak of the stationary pulsating mmf produced by the winding. The resolution of the stator mmf into its forward and backward components then leads to the physical concept of the single-phase motor described in Art. 11-1 and finally to the quantitative theory developed in Art. 11-3 and to the equivalent circuits of Fig. 11-11.

The next step is investigation of the possibilities of applying the double-revolving-field resolution to a symmetrical 2-phase motor with unbalanced applied voltages, as in Art. 11-4. This investigation leads to the symmetrical-component concept, whereby an unbalanced 2-phase system of currents or voltages can be resolved into the sum of two balanced 2-phase component systems of opposite phase sequence. Resolution of the currents into symmetrical-component systems is equivalent to resolving the stator-mmf wave into its forward and backward components, and therefore the internal reactions of the rotor for each symmetrical-component system are the same as those which we have already investigated. A very similar reasoning process, not considered here, leads to the well-known 3-phase symmetrical-component method for treating problems involving unbalanced operation of 3-phase rotating machines. The ease with which the rotating machine can be analyzed in terms of revolving-field theory is the chief reason for the usefulness of the symmetrical-component method.

This chapter also introduces a number of other commonly used small-motor types. Although they work on the same principles as their larger counterparts, their analysis is complex and their design is often largely empirical. In each motor, however, the electromechanical transduction is achieved by the spatial displacement of the flux distributions associated with the stationary and rotating members.

Two of the motor types discussed in this chapter are found in a wide variety of common applications. The universal motor is simply a series motor with a laminated rotor and stator so that it can be run from both a dc and an ac source. It is found in most small ac appliances such as blenders, mixers, vacuum cleaners, and small electric handtools. Besides being inexpensive to manufacture, it is readily adaptable to variable-speed applications, either by the insertion of series resistance or by the use of solid-state voltage controllers.

Stepper motors are electromechanical companions to digital electronics. By proper application of voltage pulses to the stator windings these motors can be made to rotate in well-defined steps of a few degrees or less per pulse. They are thus essential components of digitally controlled electromechanical systems where a high degree of precision is required. They are found in numerically controlled machine tools, in line printers and disk drives in computer systems, and in X-Y plotters. Very small stepper motors are used to advance the paper in hand-held printing calculators.

Finally, this chapter introduces a number of applications of small rotating machines as sensors and transducers in control systems. They are used to sense speed and position as well as to maintain the relative position between two remote rotating systems. Clearly, in the space of one chapter one cannot begin to include all the various types of electric machines and their applications—pancake motors, linear motors, printed-circuit motors, etc., but each of these many devices can be understood on the basis of principles presented in this volume.

PROBLEMS

11-1 What type of motor would you use in the following applications? Give reasons. Vacuum cleaner, refrigerator, washing machine, domestic oil burner, desk fan, sewing machine, bench grinder, clock, food mixer, record player, portable electric drill, water pump.

11-2 At standstill the currents in the main and auxiliary windings of a capacitor-start induction motor are $I_m = 14.14$ A and $I_a = 7.07$ A. The auxiliary-winding current leads the main-winding current by 60°. The effective turns per pole, i.e., the number of turns corrected for the effects of winding distribution, are $N_m = 80$ and $N_a = 100$. The windings are in space quadrature.

Determine the amplitudes of the forward and backward stator-mmf waves. Suppose it were possible to adjust the magnitude and phase of the auxiliary-winding current. What should its magnitude and phase be to produce a pure forward mmf wave?

11-3 Find the mechanical power output of the $\frac{1}{2}$-hp 4-pole 110-V 60-Hz single-phase induction motor, whose constants in ohms are given below, at a slip of 0.05:

$$R_{1m} = 1.93 \qquad X_{1m} = 2.58 \qquad X_\varphi = 57.8 \qquad R_2 = 3.42 \qquad X_2 = 2.58$$
$$\text{Core loss} = 37 \text{ W} \qquad \text{Friction and windage} = 12.4 \text{ W}$$

11-4 For the single-phase induction motor of Example 11-2 running at a slip of 0.05, determine the ratio of the backward flux wave to the forward flux wave. Plot a half wave of the resultant flux distribution for instants of time corresponding to $\omega t = 0$, 45, 90, 135, and 180°, zero time being chosen as the instant when the forward and backward flux waves are in space phase. If the forward and backward flux waves are represented by space phasors, draw a diagram showing the components and the resultant for the same five instants of time. Sketch the locus of the tip of the phasor representing the resultant air-gap flux wave. What kind of curve do you think this locus is?

11-5 Derive an expression in terms of Q_2 for the nonzero speed of a single-phase induction motor at which the internal torque is zero (see Example 11-2).

11-6 A small 2-phase 2-pole induction motor has the following constants in ohms at 60 Hz:

$$R_{1m} = 375 \qquad R_2 = 255 \qquad X_{1m} = X_2 = 50 \qquad X_\varphi = 920$$

The main and auxiliary windings have the same number of turns. This motor is used as a tachometer with a 60-Hz reference voltage applied to its main winding, as in Fig. 11-24. Compute the speed-voltage sensitivity in volts output per volt input per radian per second near zero speed. Also compute the phase angle of the output voltage relative to the input voltage.

11-7 (a) Find the starting torque of the motor given in Example 11-3 for the conditions specified.

(b) Compare the result of part (a) with the torque which the motor would develop at starting when balanced 2-phase voltages of 220 V are applied.

(c) Show, in general, that, if the stator voltages V_m and V_a of a 2-phase induction motor are in quadrature but unequal, the starting torque is the same as that developed when balanced 2-phase voltages of $\sqrt{V_m V_a}$ are applied.

11-8 The induction motor of Example 11-3 is supplied from an unbalanced 2-phase source by a 4-wire feeder having an impedance of $1.0 + j3.0 \, \Omega/\text{phase}$. The source voltages can be expressed as

$$V_m = 240\underline{/0°} \text{ V} \qquad V_a = 200\underline{/75°} \text{ V}$$

For a slip of 0.05, show that the induction-motor performance is such that the motor's terminal voltages correspond more nearly to those of a balanced 2-phase system than those at the source.

11-9 The equivalent-circuit constants in ohms per phase referred to the stator for a 2-phase 1.5-hp 220-V 4-pole 60-Hz squirrel-cage induction motor are given below. The no-load rotational loss is 200 W.

$$R_1 = 3.2 \qquad R_2 = 2.4 \qquad X_1 = X_2 = 3.2 \qquad X_\varphi = 100$$

(a) The voltage applied to phase m is $220\underline{/0°}$ V, and the voltage applied to phase a is $220\underline{/60°}$ V. At a slip $s = 0.04$, $Z_f = 41.9 + j27.2 \, \Omega$, and $Z_b = 1.20 + j3.2 \, \Omega$. What is the net air-gap torque?

(b) What is the starting torque with the applied voltages of part (a)?

(c) The applied voltages are readjusted so that $\widehat{V}_m = 220\underline{/0°}$ and $\widehat{V}_a = 220\underline{/90°}$. Full load on the machine occurs at $s = 0.04$. At what value of slip does maximum torque occur? What is the value of maximum air-gap torque in newton-meters?

(d) While the motor is running as in part (c), phase a is open-circuited. What is the horsepower developed by the machine at slip $s = 0.04$?

(e) What voltage appears across the open phase-a terminals under the conditions of part (d) at $s = 0.04$?

11-10 This problem demonstrates the derivation of a simplified dynamic analysis for 2-phase control motors. The control motor has identical 2-phase stator windings and a high-resistance rotor and is supplied from a low-impedance source. When balanced sinusoidal voltages are applied, the torque-slip curve is linear over the range of interest

$$T = kV_1^2 s$$

where k is a constant and V_1 is the balanced stator voltage.

(a) For control-motor usage as in Art. 11-5, show that the forward and backward torques are given by

$$T_f = k\left(\frac{V_a + V_m}{2}\right)^2 \left(1 - \frac{\omega_0}{\omega_s}\right)$$

$$T_b = k\left(\frac{V_a - V_m}{2}\right)^2 \left(1 + \frac{\omega_0}{\omega_s}\right)$$

where ω_0 is the shaft angular velocity and ω_s is the synchronous angular velocity.

(b) The motor drives a load having moment of inertia J and a torque requirement $f_v\omega_0$ proportional to speed. Neglect motor losses. With the motor at rest and the voltage V_m on the reference field, the rms voltage V_a is suddenly applied to the control field. Determine the velocity ω_0 as a function of time. Ignore electric transients.

(c) For small values of control-field voltage, V_a^2 can be neglected in comparison with V_m^2. What is the time constant in part (b) under this assumption?

(d) Using the assumption in part (c), determine the transfer functions relating shaft velocity and shaft position angle to the control-field signal.

11-11 The motor of Prob. 11-6 is used as a 2-phase control motor. When the reference-field voltage is 100 V and the control-field voltage is 70 V and leads the reference voltage by 90° (both voltages at 60 Hz), compute (a) the ratio of the backward flux wave to the forward flux wave at standstill, (b) the ratio of the backward flux wave to the forward flux wave at a slip $s = 0.80$, and (c) the internally developed mechanical power in watts at $s = 0.80$.

11-12 For the 2-phase induction motor of Example 11-6, plot a family of curves of rotor power loss and internal power developed for per unit values of control-phase voltage of 1.00, 0.50, and 0, covering a speed range from -1 to $+1$ per unit. The applied voltages are in quadrature. Express the power in per unit based on the power delivered to the air gap by the stator windings at standstill when balanced 2-phase voltages of 1.00 per unit are applied to the two stator phases.

11-13 A small 2-pole squirrel-cage induction motor for use in servo systems has symmetrical 2-phase stator windings. At standstill, the input impedance measured at the terminals of each stator winding at 60 Hz is $305 + j51$ Ω. For the purposes of this problem rotational and core losses may be neglected. Three points on the torque-slip characteristic of this motor with balanced 2-phase voltages of 100 V at 60 Hz applied to its stator terminals are given below:

Torque, N · m	0.064	0.082	0.088
Slip, per unit	0.50	1.00	1.50

If the reference-phase voltage is held constant at 100 V, 60 Hz, and the control-phase voltage is reduced to 50 V (the two voltages being in time quadrature), compute (a) the standstill torque in newton-meters, (b) the power input to the reference phase at standstill, (c) the power input to the control phase at standstill, (d) the total rotor I^2R loss at standstill, and (e) the torque at $s = 0.50$.

11-14 For the 2-phase control motor of Prob. 11-13 at standstill with 100 V applied to the reference phase and variable voltage applied to the control phase, plot curves of the following variables as functions of the standstill torque: (a) total rotor I^2R loss, (b) control-phase stator I^2R loss, (c) reference-phase stator I^2R loss, (d) power input to control phase, and (e) power input to reference phase.

11-15 A symmetrical 2-phase control motor produces a torque of 1.25 lb · ft at standstill with balanced voltages of 100 V applied.

If the motor is required to produce an acceleration at zero speed of 64.4 rad/s^2 in a load having no friction but having an inertia of 0.5 lb · ft^2, what voltage must be supplied to the auxiliary winding when the main winding is supplied with 100 V in time quadrature?

11-16 Consider a 4-phase stepper motor with a permanent-magnet rotor, as shown in Fig. 11-19. The motor is to be controlled using a 4-bit digital signal from a minicomputer. The 4 bits represent phases a to d respectively and a 1 indicates that the corresponding phase is to be excited. Thus the output 0100 will cause phase b to be excited, 1001 will cause phases a and d to be excited, etc. The digital signal is the input to an electronic circuit which provides the actual excitation.

   (a) Make a table of the 4-bit signals which correspond to rotor angular positions of 0, 45, . . . , 315°.
   (b) By sequencing through the digital signals found in part (a) the motor can be made to rotate at a constant speed. For a speed of 1000 r/min at what time interval (in milliseconds) should the digital signal be changed?

appendix **A**

Three-Phase Circuits

Generation, transmission, and heavy-power utilization of ac electric energy almost invariably involve a type of system or circuit called a *polyphase system* or *polyphase circuit*. In such a system, each voltage source consists of a group of voltages having related magnitudes and phase angles. Thus, an n-phase system will employ voltage sources which, conventionally, consist of n voltages substantially equal in magnitude and successively displaced by a phase angle of $360°/n$. A *3-phase system* will employ voltage sources which, conventionally, consist of three voltages substantially equal in magnitude and displaced by phase angles of $120°$. Because it possesses definite economic and operating advantages, the 3-phase system is by far the most common, and consequently emphasis is placed on 3-phase circuits in this appendix.

The three individual voltages of a 3-phase source may each be connected to its own independent circuit. We would then have three separate *single-phase systems*. Alternatively, as will be shown in Art. A-1, symmetrical electric connections can be made between the three voltages and the associated circuitry to form a 3-phase system. It is the latter alternative that we are concerned with in this appendix. Note that the word *phase* now has two distinct meanings. It may refer to a portion of a polyphase system or circuit, or, as in the familiar steady-state circuit theory, it may be used in reference to the angular displacement between voltage or current phasors. There is very little possibility of confusing the two.

A–1 GENERATION OF THREE–PHASE VOLTAGES

Consider the elementary 3-phase 2-pole generator of Fig. A-1. On the armature are three coils, aa', bb', and cc', whose axes are displaced 120° in space from each other. This winding can be represented schematically as shown in Fig. A-2. When the field is excited and rotated, voltages will be generated in the three phases in accordance with Faraday's law. If the field structure is so designed that the flux is distributed sinusoidally over the poles, the flux linking any phase will vary sinusoidally with time and sinusoidal voltages will be induced in the three phases. As shown in Fig. A-3, these three waves will be displaced 120 electrical degrees in time as a result of the phases being displaced 120° in space. The corresponding phasor diagram is shown in Fig. A-4. In general, the time origin and the reference axis in diagrams such as Figs. A-3 and A-4 are chosen on the basis of analytical convenience.

There are two possibilities for the utilization of voltages generated in this manner. The six terminals a, a', b, b', c, and c' of the winding may be connected to three independent single-phase systems, or the three phases of the winding may be interconnected and used to supply a 3-phase system. The latter procedure is adopted almost universally. The three phases of the winding may be interconnected in two possible ways, as shown in Fig. A-5. Terminals a', b',

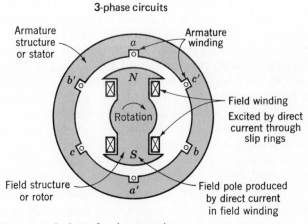

Fig. A-1. Elementary 3-phase 2-pole generator.

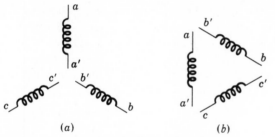

(a)   (b)

Fig. A-2. Schematic representation of windings of Fig. A-1.

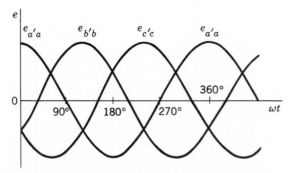

Fig. A-3. Voltage waves generated in windings of Figs. A-1 and A-2.

and c' may be joined to form the neutral o, yielding a *Y connection,* or terminals a and b', b and c', and c and a' may be joined individually, yielding a Δ *connection.* In the Y connection, a *neutral conductor,* shown dashed in Fig. A-5a, may or may not be brought out. If a neutral conductor exists, the system is a *4-wire 3-phase system;* if not, it is a *3-wire 3-phase system.* In the Δ connection (Fig. A-5b), no neutral exists and only a 3-wire 3-phase system can be formed.

The three phase voltages, Figs. A-3 and A-4, are equal and phase-displaced by 120 electrical degrees, a general characteristic of a *balanced 3-phase system.* Furthermore, the impedance in any one phase is equal to that in either of the other two phases, so that the resulting phase currents are equal and phase displaced from each other by 120 electrical degrees. Likewise, equal

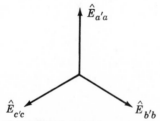

Fig. A-4. Phasor diagram of generated voltages.

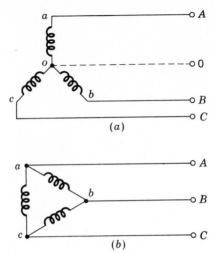

Fig. A-5. Three-phase connections: (a) Y connection; (b) Δ connection.

power and equal reactive power flow in each phase. An *unbalanced 3-phase system,* on the other hand, may lack any or all of the equalities and 120° displacements. It is important to note that *only balanced systems are treated in this appendix and that none of the methods developed or conclusions reached apply to unbalanced systems.* Most practical problems are concerned with balanced systems. Many industrial loads are 3-phase loads and therefore inherently balanced, and in supplying single-phase loads from a 3-phase source definite efforts are made to keep the 3-phase system balanced by assigning approximately equal single-phase loads to each of the three phases.

A–2 THREE–PHASE VOLTAGES, CURRENTS, AND POWER

When the three phases of the winding in Fig. A-1 are Y-connected, as in Fig. A-5a, the phasor diagram of voltages is that of Fig. A-6. The *phase order* or *phase sequence* in Fig. A-6 is *abc;* that is, the voltage of phase *a* reaches its maximum 120° before that of phase *b*. The use of *double-subscript notation* in Fig. A-6 greatly simplifies the task of drawing the complete diagram. The subscripts indicate the points between which the voltage exists, and the order of subscripts indicates the direction in which the voltage rise is taken. Thus, $\widehat{E}_{ao} = -\widehat{E}_{oa}$.

The 3-phase voltages, $\widehat{E}_{oa}$, $\widehat{E}_{ob}$, and $\widehat{E}_{oc}$, are also called *line-to-neutral voltages.* The three voltages $\widehat{E}_{ab}$, $\widehat{E}_{bc}$, and $\widehat{E}_{ca}$, called *line voltages* or, more specifically, *line-to-line voltages,* are also important. By Kirchhoff's voltage law, the line voltage $\widehat{E}_{ab}$ is

$$\widehat{E}_{ab} = \widehat{E}_{ao} + \widehat{E}_{ob} = -\widehat{E}_{oa} + \widehat{E}_{ob} = \sqrt{3}\widehat{E}_{ob}\underline{/-30°} \qquad \text{(A-1)}$$

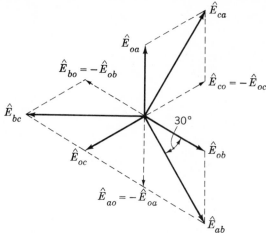

Fig. A-6. Voltage phasor diagram for Y connection.

as shown in Fig. A-6. Similarly,

$$\hat{E}_{bc} = \sqrt{3}\,\hat{E}_{oc}\underline{/-30^\circ} \tag{A-2}$$

and

$$\hat{E}_{ca} = \sqrt{3}\,\hat{E}_{oa}\underline{/-30^\circ} \tag{A-3}$$

Stated in words, these equations show that, *for a Y connection, the line voltage is $\sqrt{3}$ times the phase voltage, or the line-to-line voltage is $\sqrt{3}$ times the line-to-neutral voltage.*

The corresponding current phasors for the Y connection of Fig. A-5a are given in Fig. A-7. Obviously, *for a Y connection, the line currents and phase currents are equal.*

When the three phases are Δ-connected, as in Fig. A-5b, the phasor diagram of voltages is that of Fig. A-8. Obviously, *for a Δ connection, the line voltages and phase voltages are equal.*

The corresponding phasor diagram of currents is given in Fig. A-9. The 3-phase currents are $\hat{I}_{ab}$, $\hat{I}_{bc}$, and $\hat{I}_{ca}$, the order of the subscripts indicating the

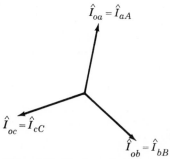

Fig. A-7. Current phasor diagram for Y connection.

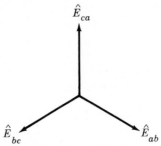

Fig. A-8. Voltage phasor diagram for Δ connection.

current directions. By Kirchhoff's current law, the line current $\widehat{I}_{aA}$ is

$$\widehat{I}_{aA} = \widehat{I}_{ba} + \widehat{I}_{ca} = -\widehat{I}_{ab} + \widehat{I}_{ca} = \sqrt{3}\widehat{I}_{ca}\underline{/30°} \qquad \text{(A-4)}$$

as shown in Fig. A-9. Similarly,

$$\widehat{I}_{bB} = \sqrt{3}\widehat{I}_{ab}\underline{/30°} \qquad \text{(A-5)}$$

and

$$\widehat{I}_{cC} = \sqrt{3}\widehat{I}_{bc}\underline{/30°} \qquad \text{(A-6)}$$

Stated in words, Eqs. A-4 to A-6 show that *for a Δ connection, the line current is $\sqrt{3}$ times the phase current.* Evidently, the relations between phase and line currents of a Δ connection are similar to those between phase and line voltages of a Y connection.

For both Y- and Δ-connected systems it can be shown that the total of the instantaneous power for all three phases of a balanced 3-phase circuit does not pulsate with time. Thus, with the time origin taken at the maximum positive point of the phase-*a* voltage wave, the instantaneous voltages of the three phases are

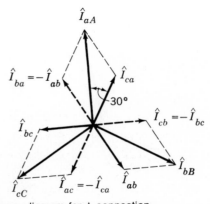

Fig. A-9. Current phasor diagram for Δ connection.

$$e_a = \sqrt{2}\,E_p \cos \omega t \qquad\qquad\qquad (A\text{-}7)$$
$$e_b = \sqrt{2}\,E_p \cos(\omega t - 120°) \qquad\quad (A\text{-}8)$$
$$e_c = \sqrt{2}\,E_p \cos(\omega t - 240°) \qquad\quad (A\text{-}9)$$

where E_p is the rms value of the phase voltage. When the phase currents are displaced from the corresponding phase voltages by the angle θ, the instantaneous phase currents are

$$i_a = \sqrt{2}\,I_p \cos(\omega t + \theta) \qquad\qquad (A\text{-}10)$$
$$i_b = \sqrt{2}\,I_p \cos(\omega t + \theta - 120°) \qquad (A\text{-}11)$$
$$i_c = \sqrt{2}\,I_p \cos(\omega t + \theta - 240°) \qquad (A\text{-}12)$$

where I_p is the rms value of the phase current.

The instantaneous power in each phase then becomes

$$p_a = e_a i_a = E_p I_p[\cos(2\omega t + \theta) + \cos\theta] \qquad\qquad (A\text{-}13)$$
$$p_b = e_b i_b = E_p I_p[\cos(2\omega t + \theta - 240°) + \cos\theta] \qquad (A\text{-}14)$$
$$p_c = e_c i_c = E_p I_p[\cos(2\omega t + \theta - 480°) + \cos\theta] \qquad (A\text{-}15)$$

The total instantaneous power for all three phases is

$$p = p_a + p_b + p_c = 3E_p I_p \cos\theta \qquad\qquad (A\text{-}16)$$

Notice that the sum of the cosine terms which involve time in Eqs. A-13 to A-15 (the first terms in the brackets) is zero. The total instantaneous power is accordingly independent of time. This situation is depicted graphically in Fig. A-10. Instantaneous powers for the three phases are plotted, together with the total instantaneous power, which is the sum of the three individual waves. *The total instantaneous power for a balanced 3-phase system is constant and is equal to 3 times the average power per phase.*

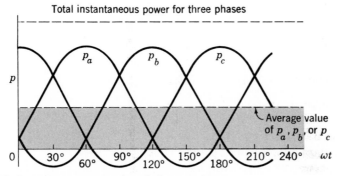

Fig. A-10. Instantaneous power in a 3-phase system.

In general, it can be shown that the total instantaneous power for any balanced polyphase system is constant. This is one of the outstanding advantages of polyphase systems. It is of particular advantage in the operation of polyphase motors, for example, for it means that the shaft-power output is constant and that torque pulsations, with the consequent tendency toward vibration, do not result from pulsations inherent in the supply system.

On the basis of single-phase considerations, the average power per phase P_p for either a Y- or Δ-connected system is

$$P_p = E_p I_p \cos \theta = I_p^2 R_p \tag{A-17}$$

where E_p, I_p, and R_p are the voltage, current, and resistance, respectively, per phase. The total 3-phase power P is

$$P = 3P_p \tag{A-18}$$

Similarly, for reactive power per phase Q_p and total 3-phase reactive power Q,

$$Q_p = E_p I_p \sin \theta = I_p^2 X_p \tag{A-19}$$

and
$$Q = 3Q_p \tag{A-20}$$

The voltamperes per phase $(VA)_p$ and total 3-phase voltamperes VA are

$$(VA)_p = E_p I_p = I_p^2 Z_p \tag{A-21}$$

and
$$VA = 3(VA)_p \tag{A-22}$$

In Eqs. (A-17) and (A-19) θ is the angle between phase voltage and phase current. As in the single-phase case, it is given by

$$\theta = \tan^{-1}\frac{X_p}{R_p} = \cos^{-1}\frac{R_p}{Z_p} = \sin^{-1}\frac{X_p}{Z_p} \tag{A-23}$$

The power factor of a balanced 3-phase system is therefore equal to that of any one phase.

A–3 Y– AND Δ–CONNECTED CIRCUITS

Three specific examples will be given to illustrate the computational details of Y- and Δ-connected circuits. Explanatory remarks which are generally applicable are incorporated in the solutions.

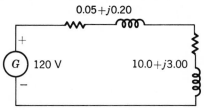

0.05+j0.20

G 120 V

10.0+j3.00

Fig. A-11. Circuit for part (a) of Example A-1.

EXAMPLE A-1

In Fig. A-11 is shown a 60-Hz transmission system consisting of a line having the impedance $Z_l = 0.05 + j0.20\ \Omega$, at the receiving end of which is a load of equivalent impedance $Z_L = 10.0 + j3.00\ \Omega$. The impedance of the return conductor should be considered zero.

(a) Compute (1) the line current I; (2) the load voltage E_L; (3) the power, reactive power, and voltamperes taken by the load; and (4) the power and reactive-power loss in the line.

Suppose now that three such identical systems are to be constructed to supply three such identical loads. Instead of drawing the diagrams one below the other, let them be drawn in the fashion shown in Fig. A-12, which is, of course, the same electrically.

(b) For Fig. A-12 give the current in each line; the voltage at each load; the power, reactive power, and voltamperes taken by each load; the power and reactive-power loss in each of the three transmission systems; the total power, reactive power, and voltamperes taken by the loads; and the total power and reactive-power loss in the three transmission systems.

Next consider that the three return conductors are combined into one

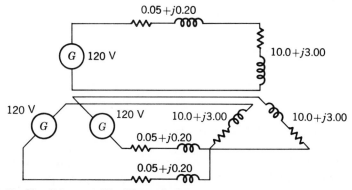

Fig. A-12. Circuit for part (b) of Example A-1.

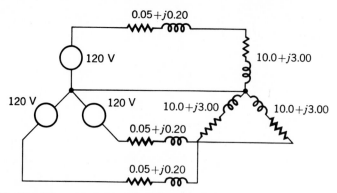

Fig. A-13. Circuit for parts (c) to (e) of Example A-1.

and that the phase relationship of the voltage sources is such that a balanced 4-wire 3-phase system results, as in Fig. A-13.

(c) For Fig. A-13 give the line current; the load voltage, both line-to-line and line-to-neutral; the power, reactive power, and voltamperes taken by each phase of the load; the power and reactive-power loss in each line; the total 3-phase power, reactive power, and voltamperes taken by the load; and the total power and reactive-power loss in the lines.

(d) In Fig. A-13 what is the current in the combined return or neutral conductor?

(e) Can this conductor be dispensed with in Fig. A-13 if desired?

Assume now that this neutral conductor is omitted. This results in the 3-wire 3-phase system of Fig. A-14.

(f) Repeat part (c) for Fig. A-14.

(g) On the basis of the results of this example, outline briefly the method of reducing a balanced 3-phase Y-connected circuit problem to its

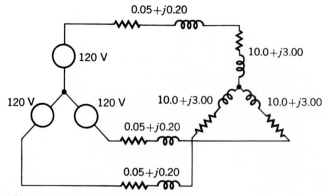

Fig. A-14. Circuit for part (f) of Example A-1.

equivalent single-phase problem. Be careful to distinguish between the use of line-to-line and line-to-neutral voltages.

Solution

(*a*)

$$I = \frac{120}{\sqrt{(0.05 + 10.0)^2 + (0.20 + 3.00)^2}} = 11.4 \text{ A}$$

$$E_L = I|Z_L| = 11.4 \sqrt{(10.0)^2 + (3.00)^2} = 119 \text{ V}$$
$$P_L = I^2R_L = (11.4)^2(10.0) = 1300 \text{ W}$$
$$Q_L = I^2X_L = (11.4)^2(3.00) = 390 \text{ VA reactive}$$
$$(\text{VA})_L = I^2|Z_L| = (11.4)^2 \sqrt{(10.0)^2 + (3.00)^2} = 1360 \text{ VA}$$
$$P_l = I^2R_l = (11.4)^2(0.05) = 6.5 \text{ W}$$
$$Q_l = I^2X_l = (11.4)^2(0.20) = 26 \text{ VA reactive}$$

(*b*) The first four obviously have the same values as in part (*a*)

$$\text{Total power} = 3P_L = 3(1300) = 3900 \text{ W}$$
$$\text{Total reactive power} = 3Q_L = 3(390) = 1170 \text{ VA reactive}$$
$$\text{Total VA} = 3(\text{VA})_L = 3(1360) = 4080 \text{ VA}$$
$$\text{Total power loss} = 3P_l = 3(6.5) = 19.5 \text{ W}$$
$$\text{Total reactive-power loss} = 3Q_l = 3(26) = 78 \text{ VA reactive}$$

(*c*) The results obtained in part (*b*) are unaffected by this change. The voltage in parts (*b*) and (*a*) is now the line-to-neutral voltage. The line-to-line voltage is

$$\sqrt{3}(119) = 206 \text{ V}$$

(*d*) By Kirchhoff's current law, the neutral current is the phasor sum of the three line currents. These line currents are equal and phase-displaced 120°. Since the phasor sum of three equal phasors 120° apart is zero, the neutral current is zero.

(*e*) The neutral current being zero, the neutral conductor can be dispensed with if desired.

(*f*) Since the presence or absence of the neutral conductor does not affect conditions, the values are the same as in part (*c*).

(*g*) A neutral conductor can be assumed, regardless of whether one is physically present. Since the neutral conductor in a balanced 3-phase circuit carries no current and hence has no voltage drop across it, the neutral conductor should be considered to have zero impedance. Then one phase of the Y, together with the neutral conductor, can be removed for study. Since this phase is uprooted at the neutral, *line-to-neutral voltages must be used.* This procedure yields the equivalent single-phase circuit, in which all quantities correspond to those in one phase of the 3-phase circuit. Conditions in the other two phases being the same (except for the 120° phase displacements in the currents and voltages), there is no need for investigating them individually. Line currents in the 3-phase system are the same as in the single-phase circuit, and total 3-phase power, reactive power, and voltamperes are 3 times the corresponding quantities in the single-phase circuit. If line-to-line voltages are desired, they must be obtained by multiplying voltages in the single-phase circuit by $\sqrt{3}$.

EXAMPLE A-2

Three impedances of value $Z_p = 4.00 + j3.00 = 5.00\underline{/36.9°}\ \Omega$ are connected in Y, as shown in Fig. A-15. For balanced line-to-line voltages of 208 V, find the line current, the power factor, and the total power, reactive power, and voltamperes.

Solution

The line-to-neutral voltage across any one phase, such as *ao*, is

$$E_p = \frac{208}{\sqrt{3}} = 120 \text{ V}$$

Hence

$$I_l = I_p = \frac{E_p}{|Z_p|} = \frac{120}{5.00} = 24.0 \text{ A}$$

$$\text{Power factor} = \cos\theta = \cos 36.9° = 0.80 \text{ lagging}$$

$$P = 3P_p = 3I_p^2 R_p = 3(24.0)^2(4.00) = 6910 \text{ W}$$

$$Q = 3Q_p = 3I_p^2 X_p = 3(24.0)^2(3.00) = 5180 \text{ VA reactive}$$

$$\text{VA} = 3(VA)_p = 3E_p I_p = 3(120)(24.0) = 8640 \text{ VA}$$

It should be noted that phases *a* and *c* (Fig. A-15) do not form a simple series circuit. Consequently, the current cannot be found by dividing 208 V by the sum of the phase-*a* and -*c* impedances. To be sure, an equation can be written for voltage between points *a* and *c* by Kirchhoff's voltage law, but this must be a phasor equation taking account of the 120° phase displacement between the phase-*a* and phase-*c* currents. As a result, the method of thought outlined in Example A-1 leads to the simplest solution.

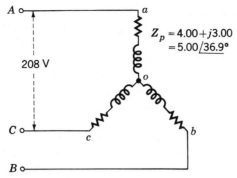

Fig. A-15. Circuit for Example A-2.

EXAMPLE A-3

Three impedances of value $Z_p = 12.00 + j9.00 = 15.00\underline{/36.9°}$ Ω are connected in Δ, as shown in Fig. A-16. For balanced line-to-line voltages of 208 V, find the line current, the power factor, and the total power, reactive power, and volt-amperes.

Solution

The voltage across any one phase, such as ca, is evidently equal to the line-to-line voltage. Consequently,

$$E_p = 208 \text{ V}$$

and
$$I_p = \frac{E_p}{|Z_p|} = \frac{208}{15.00} = 13.87 \text{ A}$$

$$\text{Power factor} = \cos \theta = \cos 36.9° = 0.80 \text{ lagging}$$

From Eq. A-4

$$I_l = \sqrt{3} I_p = \sqrt{3}(13.87) = 24.0 \text{ A}$$

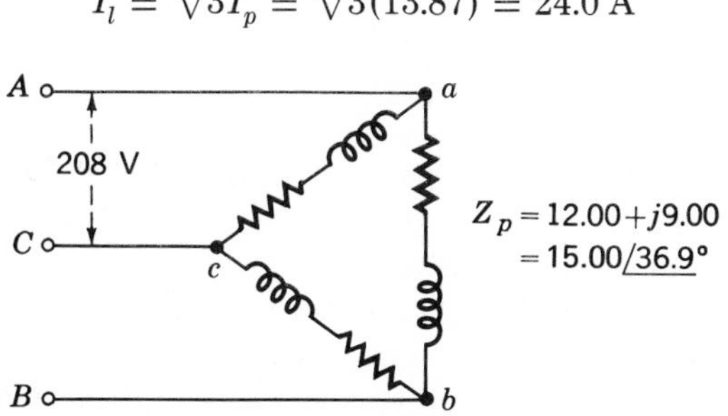

A o—
208 V
C o—

$Z_p = 12.00 + j9.00$
$= 15.00\underline{/36.9°}$

B o—

Fig. A-16. Circuit for Example A-3.

Also
$$P = 3P_p = 3I_p^2 R_p = 3(13.87)^2(12.00) = 6910 \text{ W}$$
$$Q = 3Q_p = 3I_p^2 X_p = 3(13.87)^2(9.00) = 5180 \text{ VA reactive}$$

and
$$\text{VA} = 3(\text{VA})_p = 3E_p I_p = 3(208)(13.87) = 8640 \text{ VA}$$

It should be noted that phases ab and bc do not form a simple series circuit, nor does the path cba form a simple parallel combination with the direct path through the phase ca. Consequently, the line current cannot be found by dividing 208 V by the equivalent impedance of Z_{ca} in parallel with $Z_{ab} + Z_{bc}$. Kirchhoff's-law equations involving quantities in more than one phase can be written, but they must be phasor quantities taking account of the 120° phase displacement between phase currents and between phase voltages. As a result, the method outlined above leads to the simplest solution.

Comparison of the results of Examples A-2 and A-3 leads to a valuable and interesting conclusion. It will be noted that the line-to-line voltage, line current, power factor, total power, reactive power, and voltamperes are precisely equal in the two cases; in other words, conditions viewed from the terminals A, B, and C are identical, and one cannot distinguish between the two circuits from their terminal quantities. It will also be seen that the impedance, resistance, and reactance per phase of the Y connection (Fig. A-15) are exactly one-third of the corresponding values per phase of the Δ connection (Fig. A-16). Consequently, a balanced Δ connection can be replaced by a balanced Y connection providing that the circuit constants per phase obey the relation

$$Z_Y = \tfrac{1}{3} Z_\Delta \qquad\qquad\qquad (\text{A-24})$$

Conversely, a Y connection can be replaced by a Δ connection provided Eq. A-24 is satisfied. The concept of this Y-Δ equivalence stems from the general Y-Δ transformation and is not the accidental result of a specific numerical case.

Two important corollaries follow from this equivalence: (1) A general computational scheme for balanced circuits can be based entirely upon Y-connected circuits or entirely on Δ-connected circuits, whichever one prefers. Since it is frequently more convenient to handle a Y connection, the former scheme is the one usually adopted. (2) In the frequently occurring problems in which the connection is not specified and is not pertinent to the solution, either a Y or a Δ connection may be assumed. Again the Y connection is more commonly selected. In analyzing 3-phase motor performance, for example, the actual winding connections need not be known unless the investigation is to include detailed conditions within the coils themselves. The entire analysis can be based on an assumed Y connection.

A–4 ANALYSIS OF BALANCED THREE–PHASE CIRCUITS; SINGLE–LINE DIAGRAMS

By combining the principle of Δ-Y equivalence with the technique revealed by Example A-1 a simple method of reducing a balanced 3-phase-circuit problem to its corresponding single-phase problem can be developed. All the methods of single-phase-circuit analysis thus become available for its solution. The end results of the single-phase analysis are then translated back into 3-phase terms to give the final results.

In carrying out this procedure, phasor diagrams need be drawn for only one phase of the Y connection, the diagrams for the other two phases being unnecessary repetition. Furthermore, circuit diagrams can be simplified by drawing only one phase. Examples of such *single-line diagrams* are given in Fig. A-17, showing two 3-phase generators with their associated lines or cables supplying a common substation load. Specific connections of apparatus can be indicated if desired. Thus, Fig. A-17b shows that G_1 is Y-connected and G_2 is Δ-connected. Impedances are given in ohms per phase.

When dealing with power, reactive power, and voltamperes, it is sometimes more convenient to deal with the entire 3-phase circuit at once instead of concentrating on one phase. This possibility arises because simple expressions for 3-phase power, reactive power, and voltamperes can be written in terms of line-to-line voltage and line current regardless of whether the circuit is Y- or Δ-connected. Thus, from Eqs. A-17 and A-18, 3-phase power is

$$P = 3P_p = 3E_pI_p \cos \theta \qquad (A\text{-}25)$$

For a Y connection, $I_p = I_{\text{line}}$ and $E_p = E_{\text{line}}/\sqrt{3}$. For a Δ connection, $I_p = I_{\text{line}}/\sqrt{3}$ and $E_p = E_{\text{line}}$. In either case, Eq. A-25 becomes

$$P = \sqrt{3}E_{\text{line}}I_{\text{line}} \cos \theta \qquad (A\text{-}26)$$

Similarly,

$$Q = \sqrt{3}E_{\text{line}}I_{\text{line}} \sin \theta \qquad (A\text{-}27)$$

and

$$\text{VA} = \sqrt{3}E_{\text{line}}I_{\text{line}} \qquad (A\text{-}28)$$

It should be borne in mind, however, that the power-factor angle θ, given by Eq. A-23, is the angle between $\widehat{E}_p$ and $\widehat{I}_p$ and not that between $\widehat{E}_{\text{line}}$ and $\widehat{I}_{\text{line}}$.

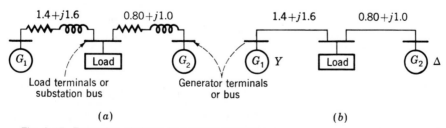

(a)   (b)

Fig. A-17. Examples of single-line circuit diagrams.

EXAMPLE A-4

Figure A-17 is the equivalent circuit of a load supplied from two 3-phase generating stations over lines having the impedances per phase given on the diagram. The load requires 30 kW at 0.80 power factor lagging. Generator G_1 operates at a terminal voltage of 797 V line to line and supplies 15 kW at 0.80 power factor lagging. Find the load voltage and the terminal voltage and power and reactive power output of G_2.

Solution

Let I, P, and Q, respectively, denote line current and 3-phase active and reactive power. The subscripts 1 and 2 denote the respective branches of the system; the subscript r denotes a quantity measured at the receiving end of the line. We then have

$$I_1 = \frac{P_1}{\sqrt{3}E_1 \cos \theta_1} = \frac{15,000}{\sqrt{3}(797)(0.80)} = 13.6 \text{ A}$$

$$P_{r1} = P_1 - 3I_1^2 R_1 = 15,000 - 3(13.6)^2(1.4) = 14,220 \text{ W}$$

$$Q_{r1} = Q_1 - 3I_1^2 X_1 = 15,000 \tan (\cos^{-1} 0.80) - 3(13.6)^2(1.6)$$
$$= 10,350 \text{ VA reactive}$$

The factor 3 appears before $I_1^2 R_1$ and $I_1^2 X_1$ in the last two equations because the current I_1 exists in all three lines. The load voltage is

$$E_L = \frac{\text{VA}}{\sqrt{3}(\text{current})} = \frac{\sqrt{(14,220)^2 + (10,350)^2}}{\sqrt{3}(13.6)}$$

$$= 748 \text{ V line to line}$$

Since the load requires 30,000 W and 30,000 tan $(\cos^{-1} 0.80)$ or 22,500 VA reactive,

$$P_{r2} = 30,000 - 14,220 = 15,780 \text{ W}$$

and
$$Q_{r2} = 22,500 - 10,350 = 12,150 \text{ VA reactive}$$

$$I_2 = \frac{\text{VA}}{\sqrt{3}(\text{voltage})} = \frac{\sqrt{(15,780)^2 + (12,150)^2}}{\sqrt{3}(748)} = 15.4 \text{ A}$$

$$P_2 = P_{r2} + 3I_2^2 R_2 = 15,780 + 3(15.4)^2(0.80) = 16,350 \text{ W}$$

$$Q_2 = Q_{r2} + 3I_2^2 X_2 = 12,150 + 3(15.4)^2(1.0) = 12,870 \text{ VA reactive}$$

$$E_2 = \frac{\text{VA}}{\sqrt{3}(\text{current})} = \frac{\sqrt{(16,350)^2 + (12,870)^2}}{\sqrt{3}(15.4)}$$

$$= 780 \text{ V line to line}$$

A–5 OTHER POLYPHASE SYSTEMS

Although 3-phase systems are by far the most common of all polyphase systems, other numbers of phases are used for specialized purposes. The 5-wire 4-phase system (Fig. A-18) is sometimes used for low-voltage distribution. It has the advantage that for a phase voltage of 115 V, single-phase voltages of 115 (between a, b, c, or d and o, Fig. A-18) and 230 V (between a and c or b and d) are available, as well as a system of polyphase voltages. Essentially the same advantages are possessed by 4-wire 3-phase systems having a line-to-neutral voltage of 120 V and a line-to-line voltage of 208 V, however.

Four-phase systems are obtained from 3-phase systems by means of special transformer connections. Half of the 4-phase system—the part aob (Fig. A-18), for example—constitutes a 2-phase system. In mercury-arc rectifiers, 6-, 12-, 18-, and 36-phase connections are used for the conversion of alternating to direct current. These systems are also obtained by transformation from 3-phase systems.

When the loads and voltages are balanced, the methods of analysis for 3-phase systems can be adapted to any of the other polyphase systems by considering one phase of that polyphase system. Of course, the basic voltage, current, and power relations must be modified to suit the particular polyphase system.

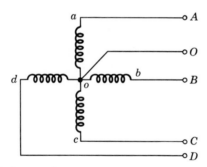

Fig. A-18. A 5-wire 4-phase system.

Voltages, Magnetic Fields, and Inductances of Distributed AC Windings

Both amplitude and waveform of the generated voltage and armature mmf's in machines are determined by the winding arrangements and general machine geometry. These configurations in turn are dictated by economic use of space and materials in the machine and by suitability for the intended service. In this appendix we shall supplement the introductory discussion of these considerations in Chap. 3 by analytical treatment of ac voltages and mmf's in the balanced steady state. Attention will be confined to the time-fundamental component of voltages and the space-fundamental component of mmf's.

B-1 GENERATED VOLTAGES

In accordance with Eq. 3-55, the rms generated voltage per phase for a concentrated winding having N_{ph} turns per phase is

$$E = 4.44 f N_{\mathrm{ph}} \Phi \tag{B-1}$$

f being the frequency and Φ the fundamental flux per pole.

A more complex and practical winding will have coil sides for each phase distributed in several slots per pole. Equation B-1 can then be used to compute the voltage distribution of individual coils. To determine the voltage of an entire phase group, the voltages of the component coils must be added as phasors. Such addition of fundamental-frequency voltages is the subject of this article.

a. Distributed Fractional-Pitch Windings

A simple example of a distributed winding is illustrated in Fig. B-1 for a 3-phase 2-pole machine. This case retains all the features of a more general one with any integral number of phases, poles, and slots per pole per phase. At the same time, a *double-layer winding* is shown. Double-layer windings usually lead to simpler end connections and to a machine which is more economical to manufacture and are found in all machines except some small motors below 10 hp in size. Generally, one side of a coil, such as a_1, is placed in the bottom of a slot, and the other side, $-a_1$, is placed in the top of another slot. Coil sides such as a_1 and a_3 or a_2 and a_4 which are in adjacent slots and associated with the same phase constitute a *phase belt*. All phase belts are alike when an integral number of slots per pole per phase are used, and for the normal machine the peripheral angle subtended by a phase belt is 60 electrical degrees for a 3-phase machine and 90 electrical degrees for a 2-phase machine.

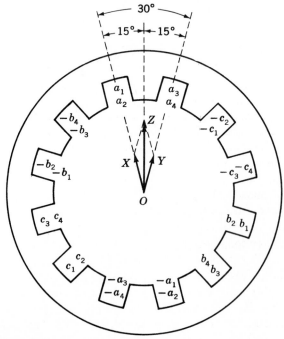

Fig. B-1. Distributed 3-phase 2-pole full-pitch armature winding with voltage phasor diagram.

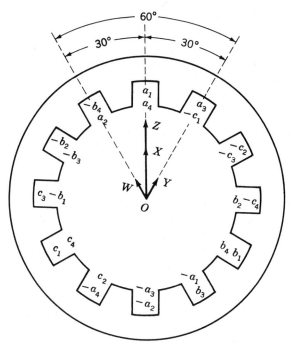

Fig. B-2. Distributed 3-phase 2-pole fractional-pitch armature winding with voltage phasor diagram.

Individual coils in Fig. B-1 all span a full pole pitch, or 180 electrical degrees; accordingly, the winding is a *full-pitch winding*. Suppose now that all coil sides in the tops of the slots are shifted one slot counterclockwise, as in Fig. B-2. Any coil, such as a_1, $-a_1$, then spans only five-sixths a pole pitch or $\frac{5}{6}(180) = 150$ electrical degrees, and the winding is a *fractional-pitch,* or *chorded, winding.* Similar shifting by two slots yields a $\frac{2}{3}$-pitch winding, and so forth. Phase groupings are now intermingled, for some slots contain coil sides in phases a and b, a and c, and b and c. Individual phase groups, such as that formed by a_1, a_2, a_3, a_4 on one side and $-a_1$, $-a_2$, $-a_3$, $-a_4$ on the other, are still displaced by 120 electrical degrees from the groups in other phases so that 3-phase voltages are produced. Besides the minor feature of shortening the end connections, fractional-pitch windings will be found to decrease the harmonic content of both the voltage and mmf waves.

The end connections between the coil sides are normally in a region of negligible flux density, and hence altering them does not significantly affect the mutual flux linkages of the winding. Allocation of coil sides in slots is then the factor determining the generated voltages, and only that allocation need be specified in Figs. B-1 and B-2. The only requisite is that all coil sides in a phase be included in the interconnection in such a manner that individual voltages make a positive contribution to the total. The practical consequence is that end connections can be made according to the dictates of manufacturing simplicity; the theoretical consequence is that when computational advantages result, the coil sides in a phase can be combined in an arbitrary fashion to form equivalent coils.

One sacrifice is made in using the distributed and fractional-pitch windings of Figs. B-1 and B-2 compared with a concentrated full-pitch winding: for the same number of turns per phase, the fundamental-frequency generated voltage is lower. The harmonics are, in general, lowered by an appreciably greater factor, however, and the total number of turns which can be accommodated on a fixed iron geometry is increased. The effect of distributing the winding in Fig. B-1 is that the voltages of coils a_1 and a_2 are not in phase with those of coils a_3 and a_4. Thus, the voltage of coils a_1 and a_2 can be represented by phasor OX in Fig. B-1, and that of coils a_3 and a_4 by the phasor OY. The time-phase displacement between these two voltages is the same as the electrical angle between adjacent slots, so that OX and OY coincide with the center lines of adjacent slots. The resultant phasor OZ for phase a is obviously smaller than the arithmetic sum of OX and OY.

In addition, the effect of fractional pitch in Fig. B-2 is that a coil links a smaller portion of the total pole flux than if it were a full-pitch coil. The effect can be superimposed on that of distributing the winding by regarding coil sides a_2 and $-a_1$ as an equivalent coil with the phasor voltage OW (Fig. B-2), coil sides a_1, a_4, $-a_2$, and $-a_3$ as two equivalent coils with the phasor voltage OX (twice the length of OW), and coil sides a_3 and $-a_4$ as an equivalent coil with phasor voltage OY. The resultant phasor OZ for phase a is obviously smaller than the arithmetic sum of OW, OX, and OY and is also smaller than OZ in Fig. B-1.

The combination of these two effects can be included in a *winding factor* k_w to be used as a reduction factor in Eq. B-1. Thus, the generated voltage per phase is

$$E = 4.44 k_w f N_{\text{ph}} \Phi \tag{B-2}$$

where N_{ph} is the total turns in series per phase and k_w inserts the departure from the concentrated full-pitch case. For a 3-phase machine, Eq. B-2 yields the line-to-line voltage for a Δ-connected winding and the line-to-neutral voltage for a Y-connected winding. As in any balanced Y connection, the line-to-line voltage of the latter winding is $\sqrt{3}$ times the line-to-neutral voltage.

b. Breadth and Pitch Factors

By separately considering the effects of distributing and of chording the winding, reduction factors can be obtained in generalized form convenient for quantitative analysis. The effect of distributing the winding in n slots per phase belt is to yield n voltage phasors phase-displaced by the electrical angle γ between slots, γ being equal to 180 electrical degrees divided by the number of slots per pole. Such a group of phasors is shown in Fig. B-3a and, in a more

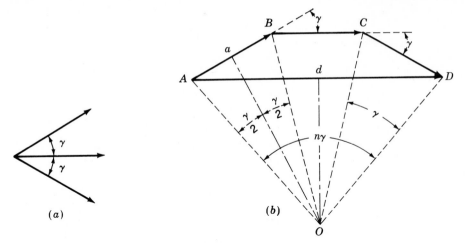

Fig. B-3. (a) Coil voltage phasors and (b) phasor sum.

convenient form for addition, again in Fig. B-3b. Each phasor AB, BC, and CD is the chord of a circle with center at O and subtends the angle γ at the center. The phasor sum AD subtends the angle $n\gamma$, which, as noted previously, is 60 electrical degrees for the normal, uniformly distributed 3-phase machine and 90 electrical degrees for the corresponding 2-phase machine. From triangles OAa and OAd, respectively,

$$OA = \frac{Aa}{\sin(\gamma/2)} = \frac{AB}{2\sin(\gamma/2)} \tag{B-3}$$

$$OA = \frac{Ad}{\sin(n\gamma/2)} = \frac{AD}{2\sin(n\gamma/2)} \tag{B-4}$$

Equating these two values of OA yields

$$AD = AB\frac{\sin(n\gamma/2)}{\sin(\gamma/2)} \tag{B-5}$$

But the arithmetic sum of the phasors is $n(AB)$. Consequently, the reduction factor arising from distributing the winding is

$$k_b = \frac{AD}{n(AB)} = \frac{\sin(n\gamma/2)}{n\sin(\gamma/2)} \tag{B-6}$$

The factor k_b is called the *breadth factor* of the winding.

The effect of chording on the coil voltage can be obtained by first determining the flux linkages with the fractional-pitch coil. Thus, in Fig. B-4 coil side $-a$ is only ρ electrical degrees from side a instead of the full 180°. The

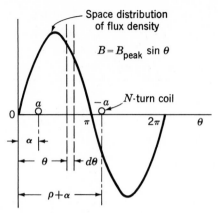

Fig. B-4. Fractional-pitch coil in sinusoidal field.

flux linkages with the coil are

$$\lambda = NB_{\text{peak}}lr\frac{2}{P}\int_{\alpha}^{\rho+\alpha}\sin\theta\ d\theta \tag{B-7}$$

$$\lambda = NB_{\text{peak}}lr\frac{2}{P}[\cos\alpha - \cos(\alpha + \rho)] \tag{B-8}$$

where l = axial length of coil side
  r = coil radius
  P = number of poles

With α replaced by ωt to indicate rotation at ω electrical radians per second, Eq. B-8 becomes

$$\lambda = NB_{\text{peak}}lr\frac{2}{P}[\cos\omega t - \cos(\omega t + \rho)] \tag{B-9}$$

The addition of cosine waves required in the brackets of Eq. B-9 may be performed by a phasor diagram as indicated in Fig. B-5, from which it follows that

$$\cos\omega t - \cos(\omega t + \rho) = 2\cos\frac{\pi - \rho}{2}\cos\left(\omega t - \frac{\pi - \rho}{2}\right) \tag{B-10}$$

a result which can also be obtained directly from the terms in Eq. B-9 by the appropriate trigonometric transformations. The flux linkages are then

$$\lambda = NB_{\text{peak}}lr\frac{4}{P}\cos\frac{\pi - \rho}{2}\cos\left(\omega t - \frac{\pi - \rho}{2}\right) \tag{B-11}$$

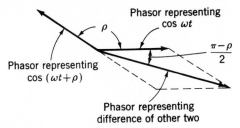

Fig. B-5. Phasor addition for fractional-pitch coil.

and the instantaneous voltage is

$$e = \omega N B_{\text{peak}} lr \frac{4}{P} \cos \frac{\pi - \rho}{2} \sin \left(\omega t - \frac{\pi - \rho}{2} \right) \qquad (B\text{-}12)$$

The phase angle $(\pi - \rho)/2$ in Eq. B-12 merely indicates that the instantaneous voltage is no longer zero when α in Fig. B-4 is zero. The factor $\cos [(\pi - \rho)/2]$ is an amplitude-reduction factor, however, so that the rms voltage of Eq. B-1 is modified to

$$E = 4.44 k_p f N_{\text{ph}} \Phi \qquad (B\text{-}13)$$

where the *pitch factor* k_p is

$$k_p = \cos \frac{\pi - \rho}{2} \qquad (B\text{-}14)$$

When both the breadth and pitch factors apply, the rms voltage is

$$E = 4.44\, k_b k_p f N_{\text{ph}} \Phi \qquad (B\text{-}15)$$

which is an alternate form of Eq. B-2; the winding factor k_w is seen to be the product of the pitch and breadth factors.

EXAMPLE B-1

Calculate the breadth, pitch, and winding factors for the distributed fractional-pitch winding of Fig. B-2.

Solution

The winding of Fig. B-2 has two coils per phase belt, separated by an electrical angle of 30°. From Eq. B-6 the breadth factor is

$$k_b = \frac{\sin (n\gamma/2)}{n \sin (\gamma/2)} = \frac{\sin [2(30°)/2]}{2 \sin (30°/2)} = 0.966$$

The fractional-pitch coils span $150° = 5\pi/6$ rad, and from Eq. B-14 the pitch factor is

$$k_p = \cos\frac{\pi - \rho}{2} = \cos\frac{\pi - 5\pi/6}{2} = 0.966$$

The winding factor is

$$k_w = k_b k_p = 0.933$$

B-2 ARMATURE–MMF WAVES

Distribution of a winding in several slots per pole per phase and the use of fractional-pitch coils influence not only the emf generated in the winding but also the magnetic field produced by it. Space-fundamental components of the mmf distributions will be examined in this article.

a. Concentrated Full-Pitch Windings

We have seen in Art. 3-3 that a concentrated winding of N turns in a P-pole machine produces a rectangular mmf wave around the air-gap circumference. With excitation by a sinusoidal rms current I, the time-maximum height of the space-fundamental component of the wave is, in accordance with Eq. 3-6,

$$\frac{4}{\pi}\frac{N}{P}(\sqrt{2}I) \qquad \text{A} \cdot \text{turns/pole} \tag{B-16}$$

For a polyphase concentrated winding, the amplitude for one phase becomes

$$\frac{4}{\pi}\frac{N_{ph}}{P}(\sqrt{2}I) \qquad \text{A} \cdot \text{turns/pole} \tag{B-17}$$

where N_{ph} is the number of series turns per phase.

Each phase of a polyphase concentrated winding creates such a pulsating standing mmf wave in space. This situation forms the basis of the analysis leading to Eq. 3-41. For concentrated windings, Eq. 3-41 can be rewritten

$$\mathcal{F}(\theta,t) = \frac{3}{2}\frac{4}{\pi}\frac{N_{ph}}{P}(\sqrt{2}I)\cos(\theta - \omega t) \tag{B-18}$$

The amplitude of the resultant mmf wave in a 3-phase machine in ampere-turns per pole is then

$$F_A = \frac{3}{2}\frac{4}{\pi}\frac{N_{ph}}{P}(\sqrt{2}I) = 0.90\frac{3N_{ph}}{P}I \tag{B-19}$$

Similarly, it can be shown that for a q-phase machine, the amplitude is

$$F_A = \frac{q}{2}\frac{4}{\pi}\frac{N_{ph}}{P}(\sqrt{2}I) = 0.90\frac{qN_{ph}}{P}I \qquad (B\text{-}20)$$

In Eqs. B-19 and B-20, I is the rms current per phase. The equations include only the fundamental component of the actual distribution and apply to concentrated full-pitch windings with balanced excitation.

b. Distributed Fractional-Pitch Winding

When the coils in each phase of a winding are distributed among several slots per pole, the resultant space-fundamental mmf can be obtained by superposition from the preceding simpler considerations for a concentrated winding. The effect of distribution can be seen from Fig. B-6, which is a reproduction of the 3-phase 2-pole full-pitch winding with 2 slots per pole per phase given in Fig. B-1. Coils a_1 and a_2, b_1 and b_2, and c_1 and c_2 by themselves constitute the equivalent of a 3-phase 2-pole concentrated winding because they form three sets of coils excited by polyphase currents and mechanically displaced 120° from each other. They therefore produce a rotating space-fundamental mmf; the amplitude of this contribution is given by Eq. B-19 when N_{ph} is taken as the sum of the series turns in coils a_1 and a_2 only. Similarly, coils a_3 and a_4, b_3 and b_4, and c_3 and c_4 produce another identical mmf wave, but one which is phase-displaced in space by the slot angle γ from the former wave. The resultant fundamental-mmf wave for the winding can be obtained by adding these two sinusoidal contributions.

The contribution from the $a_1a_2b_1b_2c_1c_2$ coils can be represented by the phasor OX in Fig. B-6. Such phasor representation is appropriate because the waveforms concerned are sinusoidal, and phasor diagrams are simply conven-

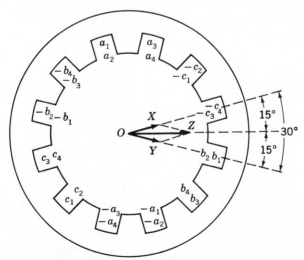

Fig. B-6. Distributed 3-phase 2-pole full-pitch armature winding with mmf phasor diagram.

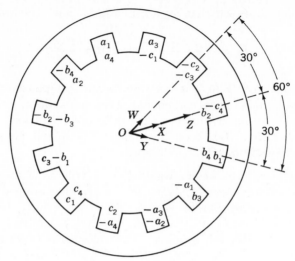

Fig. B-7. Distributed 3-phase 2-pole fractional-pitch armature winding with mmf phasor diagram.

ient means for adding sine waves. These are space sinusoids, however, not time sinusoids. Phasor OX is drawn in the space position of the mmf peak for an instant of time when the current in phase a is a maximum. The length of OX is proportional to the number of turns in the associated coils. Similarly, the contribution from the $a_3 a_4 b_3 b_4 c_3 c_4$ coils may be represented by the phasor OY. Accordingly, the phasor OZ represents the resultant mmf wave. Just as in the corresponding voltage diagram, the resultant mmf is seen to be smaller than if the same number of turns per phase were concentrated in one slot per pole.

In like manner, mmf phasors can be drawn for fractional-pitch windings as illustrated in Fig. B-7, which is a reproduction of the 3-phase 2-pole $\frac{5}{6}$-pitch winding with two slots per pole per phase given in Fig. B-2. Phasor OW represents the contribution for the equivalent coils formed by conductors a_2 and $-a_1$, b_2 and $-b_1$, and c_2 and $-c_1$; OX for $a_1 a_4$ and $-a_3 -a_2$, $b_1 b_4$ and $-b_3 -b_2$, and $c_1 c_4$ and $-c_3 -c_2$; and OY for a_3 and $-a_4$, b_3 and $-b_4$, and c_3 and $-c_4$. The resultant phasor OZ is, of course, smaller than the algebraic sum of the individual contributions and is also smaller than OZ in Fig. B-6.

By comparison with Figs. B-1 and B-2, these phasor diagrams can be seen to be identical with those for generated voltages. It therefore follows that pitch and breadth factors previously developed can be applied directly to the determination of resultant mmf. Thus, for a distributed fractional-pitch polyphase winding, the amplitude of the space-fundamental component of mmf can be obtained by using $k_b k_p N_{\text{ph}}$ instead of simply N_{ph} in Eqs. B-19 and B-20. These equations then become

$$F_A = \frac{3}{2}\frac{4}{\pi}\frac{k_b k_p N_{\text{ph}}}{P}(\sqrt{2}I) = 0.90\frac{3 k_b k_p N_{\text{ph}}}{P}I \qquad \text{(B-21)}$$

for a 3-phase machine and

$$F_A = \frac{q}{2}\frac{4}{\pi}\frac{k_b k_p N_{ph}}{P}(\sqrt{2}I) = 0.90\frac{qk_b k_p N_{ph}}{P}I \qquad \text{(B-22)}$$

for a q-phase machine, where F_A is in ampere-turns per pole.

B–3 AIR–GAP INDUCTANCES OF DISTRIBUTED WINDINGS

Figure B-8a shows an N-turn full-pitch concentrated armature winding in a magnetic structure with a concentric cylindrical rotor. The mmf of this configuration is shown in Fig. B-8b. Since the air-gap length g is much smaller than the average air-gap radius r, the air-gap radial magnetic field can be considered uniform and equal to the mmf divided by g.

From Eq. 3-5 the space-fundamental mmf is given by

$$\mathcal{F}_{a1} = \frac{4}{\pi}\frac{Ni}{2}\cos\theta \qquad \text{(B-23)}$$

and the corresponding air-gap flux density is

$$\mathcal{B} = \mu_0\frac{\mathcal{F}_{a1}}{g} = \frac{2\mu_0 Ni}{\pi g}\cos\theta \qquad \text{(B-24)}$$

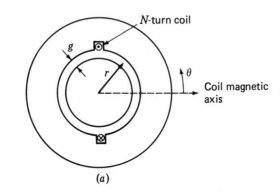

(a)

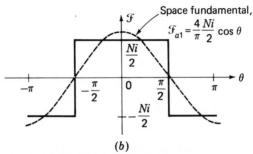

(b)

Fig. B-8. (a) N-turn concentrated coil; (b) resultant mmf.

Equation B-24 can be integrated to find the fundamental air-gap flux per pole (Eq. 1-3)

$$\Phi = l \int_{-\pi/2}^{\pi/2} \mathcal{B} \, r \, d\theta = \frac{4\mu_0 N l r i}{\pi g} \tag{B-25}$$

where l is the axial length of the air gap. The air-gap inductance of the coil can be found from Eq. 1-18

$$L = \frac{\lambda}{i} = \frac{N\Phi}{i} = \frac{4\mu_0 N^2 l r}{\pi g} \tag{B-26}$$

For a distributed P-pole coil with N_{ph} series turns and a winding factor $k_w = k_b k_p$, the air gap inductance can be found from Eq. B-26 by substituting for N the effective turns per pole pair $(2k_w N_{\text{ph}}/P)$

$$L = \frac{4}{\pi}\mu_0 \left(\frac{2k_w N_{\text{ph}}}{P}\right)^2 \frac{lr}{g} = \frac{16\mu_0 (k_w N_{\text{ph}})^2 lr}{\pi g P^2} \tag{B-27}$$

Finally, Fig. B-9 shows schematically two coils (labeled 1 and 2) with winding factors k_{w1} and k_{w2} and with $2N_1/P$ and $2N_2/P$ turns per pole pair, respectively; their magnetic axes are separated by an electrical angle α (equal to $P/2$ times their spatial angular displacement). The mutual inductance between these two windings is given by

$$L_{12} = \frac{4}{\pi}\mu_0 \frac{2k_{w1}N_1}{P} \frac{2k_{w2}N_2}{P} \frac{lr}{g} \cos \alpha$$

$$= \frac{16\mu_0 (k_{w1}N_1)(k_{w2}N_2)lr}{\pi g P^2} \cos \alpha \tag{B-28}$$

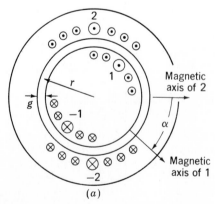

(a)

Fig. B-9. Two distributed windings separated by electrical angle α.

Although the figure shows one winding on the rotor and the second on the stator, Eq. B-28 is equally valid for the case where both windings are on the same member.

EXAMPLE B-2

The 2-pole stator-winding distribution of Fig. B-2 is found on an induction motor with an air-gap length of 0.015 in (3.81×10^{-4} m), an average rotor radius of 2.5 in (6.35×10^{-2} m), and an axial length of 8 in (0.203 m). Each stator coil has 15 turns, and the coil phase connections are as shown in Fig. B-10). Calculate the phase-a air-gap inductance L_{aao} and a to b mutual inductance L_{ab}.

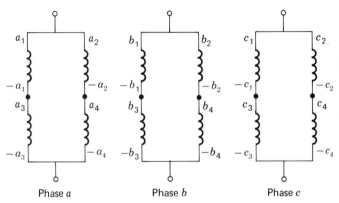

Fig. B-10. Coil phase connections of Fig. B-2 for Example B-2.

Solution

Note that the placement of the coils around the stator is such that the flux linkages of each of the two parallel paths are equal. In addition, the air-gap flux distribution is unchanged if, rather than dividing equally between the two legs, as actually occurs, one path were disconnected and all the current were to flow in the remaining path. Thus, the phase inductances can be calculated by calculating the inductances associated with only one of the parallel paths.

This result may appear to be somewhat puzzling because the two paths are connected in parallel, and thus it would appear that the parallel inductance should be one-half that of the single-path inductance. However, the inductances share a common magnetic circuit, and their combined inductance must reflect this fact. It should be pointed out that the phase resistance is one-half that of each of the paths.

The winding factor has been calculated in Example B-1. Thus, from Eq. B-27,

$$L_{aao} = \frac{16\mu_0 k_w^2 N_{ph}^2 lr}{\pi g P^2}$$

$$= \frac{16(4\pi \times 10^{-7})(0.933)^2(30)^2(0.203)(6.35 \times 10^{-2})}{\pi(3.81 \times 10^{-4})(2^2)}$$

$$= 42.4\,\text{mH}$$

The winding axes are separated by $\alpha = 120°$, and thus from Eq. B-28

$$L_{ab} = \frac{16\mu_0 k_w^2 N_{ph}^2 lr}{\pi g P^2} \cos \alpha = -21.2\,\text{mH}$$

appendix

Table of Constants and Conversion Factors for SI Units

CONSTANTS

Permeability of free space $\qquad$ $\mu_0 = 4\pi \times 10^{-7}$ H/m
Permittivity (capacitivity) of free space $\quad$ $\epsilon_0 = 8.854 \times 10^{-12}$ F/m
Acceleration of gravity $\qquad$ $g = 9.807$ m/s^2

CONVERSION FACTORS

Length $\qquad$ 1 m = 3.281 ft = 39.37 in
Mass $\qquad$ 1 kg = 0.0685 slug = 2.205 lb (mass)
Force $\qquad$ 1 N = 0.225 lb = 7.23 poundals
Torque $\qquad$ 1 N · m = 0.738 lb · ft
Energy $\qquad$ 1 J (W · s) = 0.738 ft · lb
Power $\qquad$ 1 W = 1.341 $\times 10^{-3}$ hp
Moment of inertia $\qquad$ 1 kg · m^2 = 0.738 slug · ft^2 = 23.7 lb · ft^2
Magnetic flux $\qquad$ 1 Wb = 10^8 maxwells (lines)
Magnetic flux density $\qquad$ 1 Wb/m^2 = 1 T = 10,000 gauss = 64.5 kilolines/in^2
Magnetizing force $\qquad$ 1 A · turn/m = 0.0254 A · turn/in = 0.0126 oersted

INDEX

INDEX